The Palaeobiogeography of China

OXFORD BIOGEOGRAPHY SERIES

1. T. C. Whitmore (ed.): *Wallace's line and plate tectonics*
2. Christopher J. Humphries and Lynne R. Parenti: *Cladistic biogeography*
3. T. C. Whitmore and G. T. Prance (ed.): *Biogeography and Quaternary history in tropical America*
4. T. C. Whitmore (ed.): *Biogeographical evolution of the Malay archipelago*
5. S. R. Aiken and C. H. Leigh: *Vanishing rain forests: the ecological transition in Malaysia*
6. Paul Adam: *Australian rainforests*
7. Wilma George and René Lavocat (ed.): *The Africa–South America connection*
8. Yin Hongfu (ed.): *The Palaeobiogeography of China*

The Palaeobiogeography of China

Edited by

Yin Hongfu
Professor of Geology and Palaeontology
China University of Geosciences,
Wuhan, China

Clarendon Press · Oxford · 1994

Oxford University Press, Walton Street, Oxford OX2 6DP
Oxford New York Toronto
Delhi Bombay Calcutta Madras Karachi
Petaling Jaya Singapore Hong Kong Tokyo
Nairobi Dar es Salaam Cape Town
Melbourne Auckland
and associated companies in
Berlin Ibadan

Oxford is a trade mark of Oxford University Press

Published in the United States
by Oxford University Press, New York

Chinese edition published by China Geosciences University Press, 1988

A catalogue record for this book is available from the British Library

Library of Congress Cataloging-in-Publication Data
The Palaeobiography of China /edited by Yin Hongfu.
(Oxford biogeography series; 8)
Includes bibliographical references and indexes.
1. Palaeobiogeography—China. I. Yin, Hongfu. II. Series.
QE721.2.P24P35 1993 560.9′51—dc20 93-2109
ISBN 0-19-854671-8

Typeset by EXPO Holdings, Malaysia
Printed in Great Britain by
Martin's The Printers Ltd, Berwick upon Tweed

FOREWORD TO THE ENGLISH EDITION
PROFESSOR A. HALLAM

The territory of China occupies a substantial portion of our planet's continental surface, and contains a succession of richly fossiliferous strata extending back through the whole of the Phanerozoic. Comprehension of Chinese geology has undergone a profound transformation in recent years as a result of interpretation based on plate tectonic models. Some of the most crucial information bearing on such models comes from within the realm of palaeobiogeography. Yet most of us in the West remain largely ignorant of these developments, because most of the primary source material is in the Chinese language and hardly any relevant syntheses in English have been published. For the most part, one has had to depend on research papers on particular topics by collaborative groups of Chinese, American, or European scientists. Unless Westerners are themselves actively involved in Chinese geology and palaeontology such work is likely to be overlooked.

This English edition of *The Palaeobiogeography of China* by Professor Yin and his colleagues is therefore greatly to be welcomed. Its outlook, in being thoroughly receptive to plate tectonics and displaced terrains, is not something that one could take for granted in the Chinese geological community until very recently. Thus the splendidly produced *Atlas of the Palaeogeography of China*, by members of the Wuhan College of Geology (Beijing), published in 1985 by the Cartographic Publishing House, Beijing, shows a series of beautiful multicolour palaeogeographic reconstructions back to the late Pre-Cambrian preserving the outlines of modern China. Valuable though the succession of plates may be for indicating something of the depositional conditions and climate of different times in the geological past, in a province such as Sichuan, it does not help us to locate where Sichuan might have been with respect to other parts of China.

It has become clear, however, that present-day China is an amalgam of pieces of continent that have been progressively accreted over an immense period of time, dating back well into the Palaeozoic, as recorded by indicators of continental suturing such as ophiolite belts, together with data from palaeomagnetism and palaeobiogeography. To take a familiar example, the Angara, Cathaysia, and Gondwana floras of the late Palaeozoic, all well represented in China, are now thought to signify development on widely separated lands with substantially different climates. India, furthermore, is merely the last of a succession of continental blocks that have collided with eastern Eurasia in the course of the Phanerozoic.

By accepting plate tectonics as being, together with climate, the major factor controlling palaeobiogeographic distributions, an extra degree of freedom is, of course, introduced into the problem of reconstructing the former positions of continental masses, a problem that becomes progressively more acute with increasing age, such that Palaeozoic global reconstructions still have a decidedly speculative air about them. Professor Yin and his colleagues face up to this problem and attempt to point out the areas where most controversy persists. They also indicate a full awareness of the other major problem, which is to disentangle differences due to facies variations reflecting differing depositional conditions in a range of marine and continental regimes, from differences due to varying geographical location.

Such is the complexity of the subject-matter, and the wealth of data presented, that, even in a book of quite substantial length, it is not always possible for the reader to obtain an adequate understanding of the justification for the many interpretations put forward in successive chapters, although Professor Yin does his best, in his general chapters, to present in a lucid fashion the principles that guided the authors. There can, however, be little doubt that the book will prove to be an invaluable guide to a subject of global significance.

PREFACE

Since the propagation of plate tectonics into China during the 1970s there has been an increasing demand for the demonstration of plate amalgamation, separation, and drift of China in terms of palaeobiogeography. Meanwhile, rigorous developments in sedimentology called for corresponding palaeoecological, palaeoclimatic, and palaeogeographical interpretations of environmental reconstructions in Chinese geological history. Palaeobiogeographical research projects in China up to the early 1980s were few, and were generally restricted to a certain group of a certain period. The conclusions were not uncommonly controversial, and the more regrettable aspect is that most of them were derived from the standpoint of a fixed, unmovable Earth's crust, thus failing to satisfy the demands mentioned above. This situation inspired our plan to make a comprehensive investigation of the palaeobiogeography of China. Few others in China were in a better position to do so than us: a group of nearly 20 colleagues who had collaborated closely in teaching and research in palaeontology and historical geology for more than two decades in the same university, each of us specializing in a particular geological period.

This project aimed for a comprehensive investigation of the major groups for all the Phanerozoic periods in China; to provide biotic data and interpretations for the palaeogeography, palaeoclimate, and plate tectonics of China in the Phanerozoic that would be as uncontroversial as possible. The work was done in four stages, beginning in 1982. First, through a series of personal communications, panel discussions, seminars, and symposia, the research group reached a basic concensus on the general pattern of China's Phanerozoic plate tectonics and on the important theoretical problems, as stated in Chapter 1. In the second stage we worked out the general plan and format of each chapter, trying to achieve the result in the form of a monograph instead of a symposium. The third stage was the essential and practical one, during which each author started from the collection of all available data up to 1984, sorted, analysed, and summarized, and ended up with submission of their drafts for group reviews and revisions; during which a rough sum of 12 468 species and genera, registered on 10 170 cards and 557 sheets of distribution charts were compiled. In the final stage, these statistics were sent for computerized cluster analysis, with an input set of about 217 000 items, which resulted in 1–5 cluster diagrams for each period, of which 19 were adopted in the text to complement the statements made in traditional form.

The Chinese version was completed and approved by a high-level appraisal committee consisting of academicians Wang Hongzhen, Lu Yanhao, Mu Enzi, Guo Linzhi, and other senior specialists. Its publication by the China University of Geosciences Press in 1988 attracted wide attention in Chinese geological circles and the senior author was urged to write an English version. In 1990, Oxford University Press came to an agreement with the chief author for the publication of this book in English. It took him 2 years not only to translate but also to delete introductory sections of the text that would be redundant for western readers; to add explanatory contents, tables, and a geographical index for items that are common knowledge for Chinese geologists but will be unfamiliar to others; and, most importantly, to update the text in the light of recent developments in biogeography and plate tectonics. The manuscript was written during his 6 month stay in the United Kingdom, which was supported by the Royal Society of London and the National Committee of Education, China.

Special acknowledgement should be made to the China University of Geosciences Press for its far-sightedness and generosity. This press not only published the Chinese version but also gave the initial impetus for its English translation. In order to promote the introduction of Chinese geosciences to the outer world it encouraged the publication of this book in the United Kingdom without any claim on copyright.

I feel honoured to have Professor A. Hallam of Birmingham University as the English editor of this book. His authoritative knowledge in this domain has helped greatly both in evaluating and promoting the publication of this book and especially in raising its academic level.

Thanks are also due to the Oxford University Press for patient and careful editorial and other help, without which it would be impossible for a Chinese to cope with the University Press standard.

Professor R. C. Selley of the Department of Geology, Imperial College, University of London, kindly provided computer facilities for my compiling work. Facilities were also provided by the British Museum (Natural History). Dr Tong Jingnan assisted in compilation, especially of the figures and tables. Mrs Ge Xiafei and Zhou Caofan made primary translations of a few chapters. Dr Chen Meng'e, Dr Fu Lipu, and Dr Zhu Hong kindly provided their unpublished material. Some text-figures or tables are reproduced by courtesy of Professors A. N. Strahler, M. F. Glaessner, S. Kojima, C. K. Seyfert, Huang Jiqing, and Xu Ren. Finally, I wish to acknowledge the encouragement and help of Professor A. Ziegler, Dr Chu Songhe, the staff of the Drawing office of the China University of Geosciences, and the authors of all the separate chapters of this book.

Beijing Y.H.
1993

CONTENTS

CONTRIBUTORS

Professor Cao Bexun
Faculty of Physical Geology, China University of Geosciences, Yujiashan, Wuhan, Hubei, 430074, China

Professor Chen Feng
Palaeontology Laboratory, China University of Geosciences, 29 Xueyuan Road, Beijing, 100083, China

Professor Chen Yiyuan
Faculty of Physical Geology, China University of Geosciences, Yujiashan, Wuhan, Hubei, 430074, China

Professor He Xinyi
Palaeontology Laboratory, China University of Geosciences, 29 Xueyuan Road, Beijing, 100083, China

Professor Li Zhiming
Palaeontology Laboratory, China University of Geosciences, Yujiashan, Wuhan, Hubei, 430074, China

Professor Liang Yuji
Faculty of Mathematics, South China Construction University, Lu Jing Road, Guangzhou, Guangdong, 510405, China

Dr Lin Qiuxian
Guangdong Institute of Geological Researches, 739 Dongfeng Donglu (eastern road), Guangzhou, Guangdong, 510080, China

Professor Liu Benpei
Faculty of Historical Geology, China University of Geosciences, 29 Xueyuan Road, Beijing, 100083, China

Dr Tong Jinnan
Palaeontology Laboratory, China University of Geosciences, Yujiashan, Wuhan, Hubei, 430074, China

Professor Tong Linfeng
Palaeontology Laboratory, China University of Geosciences, Yujiashan, Wuhan, Hubei, 430074, China

Professor Wu Shunbao
Palaeontology Laboratory, China University of Geosciences, Yujiashan, Wuhan, Hubei, 430074, China

Professor Xu Guirong
Palaeontology Laboratory, China University of Geosciences, Yujiashan, Wuhan, Hubei, 430074, China

Professor Xu Yuling
Palaeontology Laboratory, China University of Geosciences, 29 Xueyuan Road, Beijing, 100083, China

Dr Yang Weiping
Faculty of Historical Geology, China University of Geosciences, 29 Xueyuan Road, Beijing, 100083, China

Professor Yang Fengqing
Palaeontology Laboratory, China University of Geosciences, Yujiashan, Wuhan, Hubei, 430074, China

Professor Yang Jialu
Palaeontology Laboratory, China University of Geosciences, Yujiashan, Wuhan, Hubei, 430074, China

Professor Yin Hongfu
Palaeontology Laboratory, China University of Geosciences, Yujiashan, Wuhan, Hubei, 430074, China

Professor Zhao Xiwen
Faculty of Historical Geology, China University of Geosciences, Yujiashan, Wuhan, Hubei, 430074, China

1 INTRODUCTION

Yin Hongfu, Liang Yuji, and Xu Guirong

Palaeobiogeographical studies began late in China (Grabau 1932; Sun 1963). During the 1970s, after the acceptance of plate tectonics, research in this area acquired a great impetus. Among the pioneers were Lu (Lu *et al.* 1974) who, basing his conclusions on the Cambrian biota of China and the world, argued that palaeobiogeographical distribution and provincialization were controlled by biotic–environmental factors such as source regions, parallel evolution, and physical environments. Later Mu (1983) claimed that palaeobiogeographical provincialization resulted from ecological differentiation. He summarized and subdivided the Ordovician marine biotas of China into several modes of life adapted to different marine environments and occupying different tectonic units of China. Application of palaeobiogeographical results to the interpretation of plate tectonics resulted in a large number of publications during the past decade, including a symposium (Editorial Committee of Treatise on Basic Theories of Palaeontology, ECTBTP 1983) dealing with various fossil groups from the Phanerozoic of China. Notwithstanding these advances, there are few summaries on the whole biota of any particular period and even fewer discussions on the general concepts and methods of palaeobiogeography, although discrepancies of viewpoints are obvious in the publications.

On the basis of general agreement on the plate tectonic framework, and a consensus on the classification and interpretation of the palaeobiogeographical provinces, the authors of this book co-operated to work out the palaeobiogeography of China for all the Phanerozoic periods. Data for 12 468 species, covering all major groups, were recorded on 10 170 cards and 557 distribution charts. Some 217 000 items of data were dealt with by computer. In this chapter we shall not reiterate the generally accepted principles in this area, but will simply demonstrate the criteria that we use when working on the palaeobiogeography of China, notably: the recognition of climatically demarcated palaeobiogeographical provincialization, the classification of the palaeobiogeographical provinces of China, and the application of cluster analysis to the palaeobiogeography of China.

1.1 CRITERIA FOR THE RECOGNITION OF CLIMATICALLY DEMARCATED PALAEOBIOGEOGRAPHICAL PROVINCIALIZATION

1.1.1 TROPICAL AND SUBTROPICAL BIOTAS

High diversity

In many extant groups the ratios of diversity of temperate to tropical–subtropical biotas are around 1 : 1.5–3.0, with a few reaching 1 : 10 (Table 1.1). Similar proportions are shown in the fossil groups of China (Table 1.2). Biologists generally agree that there is a sharp change in diversity at the tropical–temperate boundary. The diversity gradient of gastropods along the Atlantic coast of North America (see Fig. 9.20 in Raup and Stanley 1979) serves as an example. This can be used as a criterion in delimiting palaeobiogeographic realms in geological history.

High percentage of endemism

High endemism of recent tropical organisms has long been known (Table 1.3). A fossil example is shown in Table 1.4. According to the classification of Kauffman (1973), the tropical Silurian Yangtze region reaches the province level (endemism 25–50 per cent). This character of the Yangtze region persisted into other periods, e.g. the endemism percentage of Emsian brachiopods reaches 30 per cent (Wang and Rong 1983), and Devonian fish show similar endemism.

Table 1.1 Number of modern organism groups in different temperature zones

<table>
<tr><th></th><th>Tropical</th><th>Subtropical</th><th>Temperate</th><th>Microthermal</th><th>Reference</th></tr>
<tr><td>Planktonic foraminifer species (northern hemisphere)</td><td>24</td><td>20</td><td>18</td><td>8 (Subarctic) 5 (Arctic)</td><td>Bé (1977)</td></tr>
<tr><td>Shallow benthic foraminifers (North America, average)</td><td colspan="2">46.1 (28°N)</td><td>24.2 (43°N)</td><td>17.2 (75°N)</td><td>Bé (1977)</td></tr>
<tr><td>Coral genera (Great Barrier Reef)</td><td colspan="2">60 (9°S)</td><td>1 (35°S)</td><td></td><td>Wells (1957)</td></tr>
<tr><td>Prosobranch snail species (average)</td><td colspan="2">550</td><td>200</td><td>100</td><td>Thorson (1957)</td></tr>
<tr><td>Bivalve species (North America)</td><td colspan="2">388 (Central America)</td><td>176 (California)</td><td>38 (Alaska)</td><td>Stehli et al. (1967)</td></tr>
<tr><td>Bivalve species (Asia)</td><td colspan="2">1037 (Philippines)</td><td>415 (Japan)</td><td>46 (Arctic coast)</td><td>Stehli et al. (1967)</td></tr>
<tr><td>Shrimp species (China)</td><td>290 (South China Sea)</td><td>96 (East China Sea)</td><td>54 (Yellow Sea) 22 (Gulf of Chihli)</td><td></td><td>Liu Ruiyu (1963)</td></tr>
<tr><td>Shrimp species (North America)</td><td colspan="2">293 (Mexico)</td><td>126 (USA)</td><td>22 (Canada)</td><td>Fischer (1960)</td></tr>
<tr><td>Reptile species (CIS)</td><td></td><td></td><td>47.5–27.3 (35–50°N)</td><td>7.9–2.4 (50–70°N) 0 (70–90°N)</td><td>Terengiev (1963)</td></tr>
<tr><td>Nesting bird species (North America)</td><td colspan="2">1100 (Panama)</td><td>195 (New York)</td><td>56 (Greenland)</td><td>Fischer (1960)</td></tr>
<tr><td>Mammalian species (North American forest area)</td><td colspan="2">70 (Panama)</td><td>35 (Michigan)</td><td>15 (Alaska)</td><td>Cox et al. (1976)</td></tr>
<tr><td>Seaweed species (China)</td><td>999 (South China Sea)</td><td>372 (East China Sea)</td><td>221 (Yellow Sea and Gulf of Chihli)</td><td></td><td>Zeng Chenkui and Zhang Junfu (1959)</td></tr>
<tr><td>Flowering plant species (average)</td><td colspan="2">5000 (Southern Asia)</td><td>2900 (Southern CIS)</td><td>117 (Central CIS) 222 (Arctic CIS)</td><td>Boblinski (1956)</td></tr>
<tr><td>Marine species</td><td>40 000 (around Malaya)</td><td>7000 (Mediterranean)</td><td>3000 (around British Isles)</td><td>2000 (Barents Sea) 1200 (Kara Sea); 400 (Laptev Sea)</td><td>Wei Mochao (1988)</td></tr>
</table>

Table 1.2 Numbers of fossil organism groups in different temperature zones of China

	Boreal	Tethys	Gondwana	Reference
Late Devonian ammonoid species (genera)	13(9) (Unul)	34(18) (Guizhou and Guangxi)		Zhao Xiwen*
Early Carboniferous invertebrate genera	261 temperate (Tianshan–Xingan)	369 tropical (Yangtze)	125 cool-cold (Gangdise–Himalaya)	Yang Fengqing*
Middle and Upper Carboniferous invertebrate genera	305 temperate (Tianshan–Xingan)	443 tropical (Yangtze–Tarim–W. Yunnan)	90 cool–cold (Gangdise–Nianqingtanggula–Himalaya)	Yang Fengqing*
Common genera of Carboniferous corals brachiopods, and fusulinids	20	60	25	Wang Naiwen (1984)
Common genera of Permian corals brachiopods, and fusulinids	55	110	65	Wang Naiwen (1984)
Triassic ammonoid genera (world-wide)	88	368	232 mainly temperate (Pacific)	Tozer (1981)
Late Triassic plant species	16(10) temperate (Tianshan–Junggar)	116(46) warm–temperate (Yanchang)	191(80) tropical (Xujiahe)	Yin and Lin*
Important marine invertebrate species of the Late Triassic		279 tropical (Changdo)	207 subtropical (Qomolangma)	Yin and Lin*

*Data source cited is in this book.

Table 1.3 Endemism difference between tropical and microthermal organisms (Vinogradov 1977)

	Tropical Pacific		Arctic–Boreal
Endemic families	11		0
Endemic genera	51		0
Endemic percentage of species	c. 50%		≪ 50%
Endemism ratio at species level	17	:	1

Table 1.4 Percentage of endemism of major Silurian groups (genera) of Yangtze province

	Early Silurian			Middle Silurian			Upper Silurian		
	Total	Endemic	Endemic percentage	Total	Endemic	Endemic percentage	Total	Endemic	Endemic percentage
Rugose corals	51	21	41	35	13	38	17	3	18
Tabulate corals	41	22	53.3	46	15	32.6	19	12	10.5
Trilobites	31	13	42	15	10	67	5	1	20
Cephalopods	11	5	45	49	22	49	19	8	47

Totals: genera, 339; endemic genera, 145. Endemic percentage calculated from totals, 42.7%.

Appearance of typical tropical organisms

Analyses of fossil groups, lithofacies, and inferred palaeolatitudes and palaeotemperatures of water bodies have established the presence of typical tropical and subtropical marine organisms during most Phanerozoic periods in China. These are shown as the Cathaysian Tethys faunas of Table 14.2, and the North China and Tarim faunas of Table 14.3 and part of Table 15.1. Generally speaking, besides reef-building organisms, the following groups are typically tropical and subtropical: a large proportion of conodonts; the majority of Early Palaeozoic trilobites, nautiloids, and rostroconchs; the South China type of brachiopods, corals, and goniatitines (clymenids exclusively) of the Devonian and Carboniferous; specialized and cemented brachiopods, Iranophyllidae (corals), and Verbeekinidae (forams) of the Permian; heteromorphic ammonoids, megalodontids, rudists, nummulitids, and many echinoderms of the Mesozoic and Cenozoic, etc. Tropical conditions are usually implied by bivalves when there is a high proportion of epifauna versus infauna, for example when pectinaceans exceed one-third of the total genera in a diversified fauna.

The recognized tropical and subtropical floras of China include: the *Drepanophycus–Zosterophyllum* flora (Early Devonian), the *Protolepidodendron scharyanum–Barrandeina dusliana* flora (Middle Devonian), the *Lepidophloeum rhombicum* flora (Lower Devonian), the Euramerican flora (Early–Middle Carboniferous), the Cathaysian flora (Lower Carboniferous–Permian; for details see Chapters 7 and 8 and Tables 14.2 and 14.3), the *Voltzia–Pleuromeia* flora (Early–Middle Trias.), the *Dictyophyllum–Clathropteris* flora (Early Trias.), the *Ptilophyllum–Coniopteris* flora (Early–Middle Jurassic), and other later Jurassic to Tertiary tropical and subtropical floras (Table 15.1). Lycopsids, Palaeozoic arthrophytes, Mesozoic cycadophytes, and many Cenozoic evergreen plants are generally indicators of tropical and subtropical conditions.

Preliminary recognition of Mesozoic and Cenozoic tropical–subtropical biotas

In China during the Mesozoic and Cenozoic, there existed a roughly longitudinal climatic zonation from wet and wet–dry tropical, arid subtropical and humid subtropical, to humid temperate (see Chapter 15). The arid tropical–subtropical belt is marked by widespread terrestrial red beds, usually interbanded by gypsum and salt deposits. This belt lacks abundant vegetation, but usually yields plants and palynomorphs indicative of arid conditions, such as high percentages (>50 per cent) of Jurassic to Cretaceous *Classopollis* and Tertiary *Palibinia*. Animals of this belt include many red-bed reptiles (Early to Middle Jurassic *Lufengosaurus*, *Mamenchisaurus*, etc.; Late Cretaceous *Nanshiungosaurus*, *Shantungosaurus*, etc.) and thick-shelled unionaceans (Middle Jurassic *Eolamprotula–Psilunio–Cuneopsis* fauna; Early

Cretaceous *Trigonioides–Plicatunio–Nippononaia* fauna, etc.). The Mesozoic southern-type floras (Table 15.1) are taken as humid subtropical.

1.1.2 TEMPERATE (MESOTHERMAL) BIOTAS

Intermediate diversity between those of the tropical and cold (microthermal) biotas

The boundary between temperate and tropical–subtropical biotas coincides approximately with the 15 °C winter isotherm. Diversity statistics of permanent zooplankton species in modern oceans (Sverdrup *et al.* 1946) show a great diversity difference between tropical–subtropical biotas (1378 species) and non-tropical biotas (180 species). The intermediate diversities of modern and fossil temperate biotas are shown in Tables 1.1 and 1.2.

The tropical–temperate change can be observed not only in marine invertebrates, but also in terrestrial vertebrates and plants. On these grounds, we hold that the tropical–temperate boundary should always be a biogeographical boundary at realm level in geological history. Wherever an abrupt decrease of biotic diversity or change of taxonomic composition occurs latitudinally, the tropical–temperate boundary should be taken into consideration (see, for example, Stehli 1973).

Recognizing temperate and tropical biotas according to bipolar and tropical distributions

Bipolarity exists in many temperate and psychrophilic organisms (growing best at low temperatures). There are bipolarities of taxonomy, of relationships (bipolar similarities in form and structure, e.g. white colour and hairiness in polar animals), and of phenomena (e.g. the sudden bipolar flourishing of diatoms during the spring and autumn breeding seasons). Causes of bipolarity have been discussed by Hedgpeth (1957). It serves as an index to recognize temperate and psychrophilic biotas. Bipolar assemblages distributed on both sides of tropically distributed biotas are generally recognized as temperate biotas.

Bipolarity appeared at least as early as the Silurian. From Fu's synthesis on Silurian brachiopods (Table 5.4), five genera (nos. 1–5) show bipolarity. This coincides with the palaeomagnetic reconstruction whereby, during the Silurian, these two realms had already drifted into southern and northern middle latitudes.

Bipolarity became most conspicuous in the Late Palaeozoic. The dominance of small, solitary, rugose corals in the Boreal and Gondwanan realms, in contrast to the flourishing of large, colonial, rugose corals in the Tethyan realm is a typical example. During the Devonian, such solitary corals flourished in the Boreal and Gondwanan parts of China, whereas the region of the Tethyan realm in Central China (including Xainza of Central Tibet) yields abundant colonial corals, such as Disphyllidae, Hexagonariidae, etc. Devonian bipolarity is also demonstrated by the lack of brachiopods typical of the central region of China *(Orientospirifer*, *Stringocephalus*, *Yunnanella*, etc.) in Himalaya and Junggar–Xingan. During the Carboniferous, bipolar distribution existed not only in small, solitary corals (*Siphonophyllia*, *Humboldtia*, *Hapsiphyllum*, *Zaphrentis*, *Cyathaxonia*, etc.), but also in brachiopods (*Syringothyris*, *Pseudosyrinx*, *Marginatia*) (for details see Chapter 7). The Permian is the Palaeozoic period with the strongest bipolarity, reflecting most pronounced latitudinal differentiation of plates. Bipolar distributions at generic level include the small, solitary corals *Lytvolasma* and *Wannerophyllum*; the brachiopods *Strophalosia*, *Wyndhamia*, *Aulosteges*, *Neospirifer*, and *Spiriferella*; the foram *Monodiexodina*; the ammonoid *Uraloceras*; and the bivalve *Atomodesma*. These existed in the Gondwanan and Boreal parts of China and adjacent areas. Each has its sychronous Tethyan counterpart: the corals *Ipciphyllum–Iranophyllum*; the brachiopods *Tyloplecta yangtzeensis–Cathaysia*; the forams *Verbeekina–Neoschwagerina*; the ammonoids *Shouchangoceras–Pseudohalorites*; and the bivalves *Guizhoupecten–Hunnanopecten*. These pairs provide reasonable indices for demarcating tropical and temperate realms. A similar situation exists for plants. A few elements of the Gondwanan

Glossopteris flora have been found in the Boreal realm (for example in Primorye, Russia and Kaishantun, Jilin province), whereas the *Gigantopteris* flora is distributed in the intervening areas.

The Triassic displays considerable bipolar distribution. Examples include the ammonoids *Otoceras*, *Keyserlingites*, *Nordophiceras*, Popanoceratidae, typical of the Gondwanan Tethys and Boreal realms; the bivalve *Entomonotis*, a genus of the Gondwanan Tethys, Maorian, and Boreal regions, has been discovered from the Qomolangma Mountain and Nadanhada Mountains, Heilongjiang province, but not in the Cathaysian Tethys region of China. Lystrosaur and kannemeyerian vertebrates also developed in north and south middle palaeolatitudes, avoiding tropical and polar regions (Cox 1973). A few elements of the Gondwanan *Dicroidium* flora occur in the northern *Danaeopsis–Bernoullia* (*Symopteris*) flora. After the Jurassic, the main part of China became land and the Himalayan region no longer remained in southern middle palaeolatitudes. Bipolarity then became scarce in China with a few exceptions, e.g. Late Jurassic–Early Cretaceous *Buchia* (Qomolangma Mountain and the Nadanhada Mountains of Heilongjiang province). This phenomenon still occurred frequently in other parts of the world, however.

Lower proportion of endemic and stenothermal organisms in temperate than in tropical biotas

This will be discussed further on p. 7.

Tentatively recognized temperate biotas in the geological history of China

Large parts of the Boreal and Gondwana areas were located within temperate zones during most Phanerozoic periods. Analyses of fossil faunas and floras according to the biogeographical principles given above, together with lithofacies and palaeoclimatic data, have led to preliminary recognition of the following biotas as temperate (some of them may even be psychrophilic): the Silurian *Tuvaella* assemblage; the Late Palaeozoic small, solitary coral assemblages, the syringothyrid brachiopods, and the monodiexodinid forams; the bivalves of the Atomodesmiidae (Permian) and Buchiidae (Jurassic-Cretaceous); a few, probably nektonic, ammonoids, e.g. the Devonian *Sporadoceras*, the Permian *Uraloceras*, the Triassic *Otoceras*, *Keyserlingites*, and *Amphipopanoceras*; the Late Jurassic Jehol fauna (*Lycoptera*, *Eosestheria middendorfii*, *Eupheme-ropsis trisetalis*); the Late Palaeozoic Angara flora; and the Triassic *Dicroidium* flora. During the Mesozoic and Cenozoic, as mentioned above, there existed in China a humid temperate belt north of north-east of the tropical–subtropical belts. The humid temperate biota can be recognized through their zonal distribution associated with the corresponding lithofacies, characterized by coal-bearing strata. They consist of 'northern' floras of Mesozoic and Cenozoic age. The Late Triassic *Danaeopsis Bernoullia* (*Symopteris*) flora, the Early–Middle Jurassic *Coniopteris–Phoenicopsis* flora, and the Middle–Late Jurassic flora in north-eastern China are categorized as part of the Siberian phytogeographical realm, representing lower humidity and temperature than their more southerly counterparts, and can thus be regarded as warm temperate. The 'northern' *Ruffordia–Onychiopsis* flora (Cao *et al.* 1982, Chapter 15) occupied roughly the same area as the synchronous Jehol fauna; the former belongs to the Siberian phytogeographical realm and yields few or no *Classopollis* (Li Wenben 1983), while the latter coexisted with buchiids in the adjacent Wandashan palaeo-gulf; thus the biota should also reflect a humid temperate climate. The tropical–temperate boundary during these periods should be drawn along the demarcation line between southern and northern floras (Fig. 15.1). These temperate floras have the following characters: trees with annual rings; absence of typical tropical pteridophytes such as dipteridaceans; dominance of Coniferales and Ginkgoales over Cycadales and Cycadeoidales in gymnosperms; temperate floras also include mixed forests of conifers and broad-leaved trees in geologically recent times, the broad-leaved trees usually being deciduous.

Palaeoclimatic implications of mixed biotas

During some periods in geological history there existed intermediates between tropical and temperate biotas, either dominated by tropical organisms with subordinate temperate ones or vice versa. In this book the former is regarded as subtropical (see, for example, the second and third columns of Table 15.1); the latter is taken as warm temperate, for example the Late Triassic flora of the North–Northwestern China province dominated by *Danaeopsis–Bernoullia* (*Symopteris*) but mixed with a few *Dictyophyllum–Clathropteris* elements (Fig. 9.5). We do not know the temperature at which these mixed biotas grew, and therefore these designations are still speculative.

1.1.3 COLD CLIMATE (MICROTHERMAL) BIOTAS

Very low diversity accompanied by high abundance of dominant species

The low diversity of microthermal biotas is shown in Tables 1.1 and 1.2, from which an average tropical : microthermal ratio as high as 10 : 1 can be calculated. Statistics indicate that the tropical : cold-water species diversity ratio of epifaunal communities occupying same-sized areas is 5–10 : 1 (Parker 1964). The abundance of dominant species can be exemplified by *Euphausia superba*, reaching $6 \times 10^4/m^3$ in some water bodies and about 5–6 billion tons *en masse* in Antarctic seas. In the south-eastern areas of the Japanese northern seas, influenced by the cold Oya Current, the density of cold-water microplankton cells reaches $10^9/m^3$ during the season of maximum production (spring), while in the tropical eastern sea areas of Japan, influenced by the warm Kuro Current, the density is only 103–105 cells/m^3 during the season of maximum production (winter). Similar conditions probably occurred in the geological past for some cold-water faunas.

Low endemism and high cosmopolitanism

Fewer stenothermal organisms and more eurythermal ones are common features of both temperate and microthermal biotas, albeit to different degrees. Tropical biotas are characterized by a high percentage of endemic organisms, which are generally stenothermal. In temperate biotas the endemic percentage is reduced, and is even less in microthermal biotas (Table 1.3). It has been reported that in the Late Palaeozoic glacio-marine biotas of the Tibet–Qinghai plateau a majority of the genera are eurythermal and are commonly seen in South China, whereas stenothermal psychrophilic organisms (*Stepanoviella*, *Costiferina*, *Eurydesma*, etc.) are much fewer. This had led some palaeontologists to doubt their Gondwanan attribution, although this relationship is in fact normal in extra-tropical biotas (Wang Naiwen 1984, Chapter 14). The Silurian *Tuvaella* fauna of the Junggar–Xingan region is also composed mostly of eurythermal cosmopolitans, whereas the endemics (*Tuvaella*, *Tannuspirifer*, etc.) comprise a minority. In the sense of Kauffman's (1973) classification of biogeographical provincialization based on the percentage of endemism, some authors have claimed that this fauna could constitute an endemic centre or subprovince, thus having no palaeoclimatic or paleobiogeographical significance (Su 1981; Zhang *et al.* 1983; Wang *et al.* 1984; see Chapter 5 for further references). We suggest that the applicability of Kauffman's scheme is limited mainly to tropical regions, because it did not consider seriously the great difference of endemic proportions in tropical, temperate, and microthermal biotas. Analyses of lithological, palaeomagnetic, and palaeontological data have convinced us that the *Tuvaella* fauna reflects a temperate–cool climate, showing bipolarity of relationship with the Malvinokaffric *Clarkeia* fauna (Table 5.4). Another example is that of the early Triassic Boreal bivalve fauna which represents a temperate–cool climate. It is composed mainly of cosmopolitans such as *Claraia stachei*, *Eumorphotis multiformis*, *Promyalina*, plus a few endemics like *Posidonia mimer* and *Posidonia aranea*. In treating palaeobiotas consisting of a eurythermal majority together with a few cold-water organisms, it is advisable to be cautious in interpreting them as mixed biotas; one should consider seriously the possibility of their being extra-tropical. Indeed, to reach such a conclusion needs further evidence, because the same situation may be due not only to latitudinal factors, but also to cold ocean currents and tropical deep water.

Other features of microthermal biotas

The dominance of siliceous microplankton is a conspicuous feature of microthermal marine biotas. Diatoms are most abundant in cold water. The circum-Antarctic microplankton south of the Antarctic Convergent Zone (*c.* 50°S) are mainly siliceous, whereas those of the subarctic temperate water bodies north of this zone are mostly calcareous. This has been used as a criterion to demarcate Recent and Cenozoic microthermal marine biotas.

Microthermal floras

Modern Arctic and Antarctic floras are generally lacking in trees and ferns, and are dominated by tundras consisting of grasses, sedges, and lichens, together with shrubs of willow. It is not surprising that microthermal floras are seldom identified in the geological record. This is largely because the conditions for their fossilization were rarely favourable. In subarctic regions, however, the vegetation changes into birch–lichen woodland and, traced further southward, coniferous forests cover a large part of the cold climate zone. Fossilized coniferous forests have also been discovered in polar regions in, for example, the early Tertiary of Greenland, Spitsbergen, and the Novosibirsk Islands.

Microthermal biotas were few in the geological past and are difficult to recognize

The palaeoclimates of Cambrian, Early Ordovician, Middle–Late Silurian, Devonian, Mesozoic, and Early Tertiary times were warmer than at present. Whether microthermal climates existed during these periods remains to be demonstrated. Periods with palaeoclimates cooler than, or similar to, the present are the Late Ordovician–Early Silurian, Late Carboniferous–Early Permian, Neogene, and Quaternary. Ice-caps formed and ice ages existed in these periods, and the existence of microthermal palaeoclimates cannot therefore be doubted. However, what the microthermal biotas of these periods were remains uncertain, except for the Neogene and Quaternary, for which microthermal biotas have been extrapolated on the evidence of the abundance of cold-water radiolarians, forams, sporopollen, and dinoflagellates and the migration of the diatom-facies boundary. The literature on earlier periods generally refers to cold-water faunas and cool- or cold-climate floras. The palaeotemperatures represented have not been determined accurately; it is commonly postulated that they correspond to those now existing between 40°N and the polar circle. The following biotas have been suggested to be cold-water faunas or cool- to cold-climate floras: the Late Ordovician *Dalmanitina–Hirnantia* fauna;[1] the Silurian and Devonian Malvinokaffric fauna (including *Clarkeia* and *Australospirifer*); the Late Carboniferous–Early Permian *Eurydesma–Stanoviella* fauna; the *Glossopteris* flora; the Pleistocene *Mammuthus–Coelodonta antiquitatis* fauna; the Quaternary dextral *Globigerina pachyderma* fauna.

1.1.4 SOME PROBLEMS IN RECOGNIZING THE PALAEOCLIMATIC IMPLICATIONS OF FOSSIL BIOTAS

Influences of sea depth and continental altitude

The ocean is layered with respect to temperature. At low latitudes throughout the year and in middle latitudes in the summer there develops a three-layered structure, the depths of which are variable: warm surface layer, main thermocline, and cool deep water. Hence, the statements made in earlier sections are relevant mainly to the shallow benthos and surface plankton and nekton, but not to those organisms living in deeper layers. Tropical submergence exists in many taxa. The chaetognath *Eukrochnia hamata* is distributed near the ocean

[1] Recently Rong Jiayu and Chen Xue (personal communication) questioned the cold-water attribution of this fauna because the habitat of this group of organisms, the Yangtze Epicontinental Sea, appears to have been quite warm.

surface at 50–60°N but it submerges to 800 m depth near the equator (Ekman 1967). There are 17 permanent plankton species that change from epiplankton in polar regions to mesoplankton in tropical areas (Sverdrup *et al.* 1946). In a geological column where deposits of deep and shallow water alternate, psychrophilic and thermophilic assemblages may also replace each other accordingly. In a Jurassic section at Nyalam, southern Tibet, two *Buchia*-bearing black shale sequences alternate with three classic sequences yielding bivalve assemblages of Ethiopian type. *Buchia* is known to be psychrophilic, while the bivalves represent warmer water. Judging from the accompanying lithofacies, Gou and Li (1982) concluded that these assemblages denoted shifts of water depth rather than palaeoclimate changes.

Alpine floras bear many similarities to boreal and Arctic floras because of the differentiation of vegetation with altitude, consequently much of the literature treats them in one category. In the Late Triassic, the *Danaeopsis–Bernoullia* (*Symopteris*) flora generally occupied North China while the *Dictyophyllum–Clathropteris* flora occupied South China, but in southern Jilin province (Sun Ge 1981) and Pamir (Kiparisova *et al.* 1973), the two occurred very close to each other and sometimes the former appeared in a more southerly location than the latter. This distribution is thought to be due to ecological (habitat) factors rather than to biogeographical factors. Altitude and distance from the sea are the major habitat factors considered.

Interference of anomalous conditions, especially abnormal salinity

In hyperhaline or brackish-water bodies, the faunas commonly show lower diversity accompanied by high abundance of a few dominant species, similar to those of the microthermal faunas. One example is the *Leptochondria illyrica–Costatoria goldfussi mansuyi* fauna of the Anisian hyperhaline sea of South China. Such abnormal situations can be recognized through synthetic study of environmental factors.

Errors in uniformitarianism

These should also be taken into consideration using traditional methods, Valentine and Meade (1981) distinguished a subtropical and a warm temperate fauna in the Pleistocene bivalves of North America. A reinvestigation using oxygen isotope data showed that the water temperature of the subtropical fauna was 17–20 °C, approximately that of the modern subtropics, whereas the water temperature in which the warm temperature fauna lived was 5–12 °C, remarkably lower than modern warm temperate water. The climates of the geological past do not necessarily correspond accurately to those of today.

1.2 SCHEME OF PALAEOBIOGEOGRAPHICAL PROVINCIALIZATION IN THE GEOLOGICAL HISTORY OF CHINA

It is widely accepted that there are three schools of biogeography (Ross 1974*b*): the ecological, the taxonomic, and the phylogenetic or historic (including vicariance). In compiling this book, we have tried in vain to use the taxonomic scheme as a unified system, and we have hence been forced to use both taxonomic and ecological biogeographical units. The first, second, and third orders of taxonomic biogeographical units used in these studies, namely realm, region, and province of China, are quite useful, but this is not the case for further subdivisions. Although the use of subprovince and endemic centre has been encouraged wherever possible, in most cases further subdivisions have to be ecological rather than taxonomic biogeographical units. The common cases are:

1. Marine biogeographical provinces usually cannot be further subdivided taxonomically by barriers. Instead, they often display a more-or-less circumferencial distribution of benthos-dominated, transitional, and plankton-dominated magnafacies from the province proper to its periphery. The subdivisions thus distinguished are magnafacies (or biome (*sensu stricto*), see p. 12), not subprovinces.

2. Biogeographical provinces consisting of both marine and terrestrial environments are usually further subdivided into terrestrial (often occupying the central parts), deltic-lagoonal, and epeiric bio- and lithofacies. Subdivisions thus demarcated are also magnafacies (or biome (*sensu stricto*)), not subprovinces.

3. A terrestrial biogeographical province is usually too small to be subdivided by climatic or geographical barriers. Its subdivision is therefore mainly based on different vegetation types, reflecting different relief and precipitation. These types are again biomes.

Consequently, according to the practical situation in China, we have adopted a dual (taxonomic and ecological) palaeobiogeographical system.

1.2.1 TAXONOMIC PALAEOBIOGEOGRAPHICAL UNITS

Realm, region, and province have been used for the first-, second-, and third-order taxonomic palaeobiogeographical units (Ekman 1967; Kauffman 1973; Boucot 1974, 1981; Ross 1974*a*, *b*).

Realm

A realm is the first-order unit, covering a large part of the world. It should have a characteristic biota. This biota has a long independent history of development, formed in accordance with plate tectonics history and palaeogeographical change; undergoing, for example, ecological replacements through long-term parallel evolution, such as the Australian marsupials compared with the Eurasian placentals. Characteristic taxa on which a realm is based are generally benthic families and planktonic genera.

Realms are demarcated primarily by temperature-latitude boundaries (such as the tropical–temperate boundary) and, secondly, by great geographical barriers (such as continents and oceans). The total number of realms in the whole world may be two, three, or up to six or seven, depending on the extents to which the palaeoclimate was differentiated and the lands and seas were separated. The number of realms and regions have tended to increase since the Phanerozoic. During some periods of the Early Palaeozoic, because of the aggregation of most palaeocontinents and their marginal seas along the palaeoequator, realms were few, sometimes only two in the whole world. After the Silurian, because palaeocontinents and their marginal seas gradually scattered to high, middle, and low latitudes, there were commonly three realms—Boreal (or Siberian), Gondwana (or Malvinokaffric), and Tethys, representing south and north temperate–cold realms and tropical–subtropical realm, respectively—plus one or two more realms in a few periods. The Pacific Plate and its predecessor have formed a longitudinal barrier ever since their emergence in the Permian and thus have established a new realm. The modern continents are perhaps more scattered in latitude and more separated by oceans than during any earlier period, with six or seven realms being established.

In China, using cluster analysis, the similarity coefficients between realms are commonly below 0.1 (Otsuka coefficient). In a few periods and fossil groups it can reach 0.1–0.2 (Lower Devonian brachiopods, Upper Devonian corals, Lower Carboniferous plants, Lower and Middle Triassic biota, Palaeogene biota). Areas with similarity coefficients exceeding 0.2 should not be established as separate realms.

Region

The term 'region' is used in this book as a subdivision of a realm, reflecting major geographical barriers or climatic differentiation which is generally latitudinal but may be deviated by ocean currents. Characteristic taxa are benthic families and genera as well as planktonic genera or species groups. The establishment of regions usually meets the following criteria:

1. Two or more plates, or groups of microplates, may each form a biogeographical region. Examples are the Ordovician North America and Siberia regions within the North America–Siberia realm, as well as the China and Baltic regions within the Southern (Gondwana) realm (Fig. 4.1); also the

Upper Palaeozoic–Triassic Cathaysia region within the Tethys realm (Fig. 9.1). They reflect major geographical barriers.

2. Two or more latitudinal–palaeoclimatic belts within a realm each form a region. Examples are, in the Triassic, the regions of Southern Gondwana and Gondwanan Tethys within the Gondwanan realm, and the Middle and North regions of the realm of the Laurasian Continent (Fig. 9.1). They reflect major palaeoclimatic and palaeolatitudinal belts.

Regions are not significant in the geological history of China due to the following factors:

1. During most of the Phanerozoic periods the main part of China only constituted one region (China or Cathaysia) within one realm. The boundary of this region usually outreached the national boundary. In such a case, provinces can be attributed directly to realm, and their regional associations are self-evident.

2. The south-western and northern borderlands of China have been attributed to two other realms during many periods. The Chinese parts of the two realms are relatively small, and only constitute a portion of one region in each realm. The category of region is thus somewhat redundant and can be disregarded.

3. In the cluster analysis of all geological periods of China, there is no distinctive range of similarity coefficients for regions. The rank of region in China is difficult to distinguish by mathematic methods.

In short, during many geological periods, the main part, south-western and northern borders of China were each assigned to one region of one realm, thus making the rank of region self-evident in China.

Province

A province, the third rank of provincialization, or subdivision, of a region, is the essential unit in the palaeobiogeography of China. It is distinguished by characteristic species or, in fewer cases, genera. The area of a province would correspond to a small plate (Yangtze, North China), a group of neighbouring microplates (e.g. Gangdise, Qiangtang, West Yunnan), or part of a larger plate. It usually consists of only one palaeoclimatic belt (tropical, temperate) and an orderly set of biomes from centre to margin (benthic–transitional–planktonic).

The following suggestions have been reported for the number of modern shallow-water biotic provinces: 21 (Gurianova 1964), 25 (Ross 1974*a*, *b*), and 31 (Valentine 1973); and for modern bathyal biotic provinces and subregions: 27 (Menzies *et al.* 1973), 10 (Kusakin 1971), and 11 (Vinogradova 1956), plus 10 hadal provinces (Beliaev 1966). For the Cretaceous, 17 provinces and subprovinces have been suggested (Kauffman 1973), and for the Devonian, fewer than 10 (Boucot 1975). Modern China consists of 7–8 biotic provinces (see Chapter 12; ECPGAS 1979). In consideration of these data, the numbers of provinces in the geological history of China are usually set at 4–5; in the Cenozoic the number may have increased to 6–8.

Results of cluster analysis for various geological periods of China have revealed that interprovincial similarity ratios are usually less than, or nearly, 0.2, a few reaching 0.2–0.25, scarcely up to 0.25–0.3. Two adjacent areas with similarity ratio exceeding 0.3 would be better incorporated into one province, although in the Early Ordovician and Late Devonian there were exceptional adjacent provinces with similarity ratios reaching 0.31.

Subprovince and endemic centre

As fourth and fifth ranks of taxonomic biogeographical provincialization, these are less commonly used because, as mentioned above, further subdivision of provinces is mainly based on ecological factors rather than on palaeoclimatic belts or palaeogeographical barriers. Provinces are not large enough to span various climatic belts or barriers such as oceans and continents.

Kauffman's criteria (Kauffman 1973) are not applied here because the percentage of endemism varies greatly from tropical to polar regions, as discussed on p. 7 of this chapter.

1.2.2 ECOLOGICAL BIOGEOGRAPHICAL UNITS

The most widely used ecological biogeographical unit is that of biofacies. This is a vaguely defined term meaning a definite type of environment plus adapted biota. This term is only used here for general reference and has no strict status in this book.

Biome

The term 'biome' was first applied on land as an ecological biogeographical unit. Adapted plant communities, and the corresponding adaptive animal communities, are associated with a range of environmental, mainly climatic, factors (such as temperature, precipitation, and daily and seasonal changes). Such associations of biotic communities adapted to certain environments covering large areas of land are called biomes. In much of the literature biome is used *sensu lato*, that is, type of biome. Traditionally there are seven major types of biome on modern continents: tundra, taiga, deciduous forest, grassland, desert, high plateau, and tropical forest. Walter (1985) established nine types and Ziegler (1990) added a tenth for the Permian. A biome (*sensu stricto*) means an individual ecological biogeographical unit belonging to a certain type of biome but restricted to an area separated from other units of the same type of biome not only by distance, but also by taxonomic composition. The North American oak–pine–juniper–squirrel biome is different from that of the same woodland biome type in Russia. The dimensions of biomes vary greatly, from those much smaller than a province in tropical areas to those equal or even larger than a province in polar regions.

'Magnafacies' is a term in geology corresponding to biome in biology. Some of the literature takes them as synonymous (Fairbridge 1979, p. 516).

Following the precedents of applying realms and provinces in geological history, we suggest that this ecological biogeographical unit, biome, be endowed with a sense of geological history and be applied in palaeobiogeography. In modern biogeography, the concept of biome infers the simultaneous existence of different communities (simultaneity), while in palaeobiogeography it is a time concept, spanning, for example, an epoch, and thus actually means a succession of biomes of the same type. Because terms such as 'realms' and 'provinces' have undergone some conceptual alteration when applied in palaeobiology, we deem it appropriate also to use the term 'biome' in a manner specific to this setting.

The term biome as used in this book represents a type of magnafacies or community association occurring over continuous areas of identical tectonic and palaeogeographical setting (e.g. epeiric sea or continental slope), and reflecting a series of interrelated lithofacies (e.g. from shallow-water carbonate platform to restricted evaporitic sea), commonly reaching an area of tens and hundreds of thousands, or even more than one million, square kilometres. Because biomes are based on biofacies, in cluster analysis within one province, biomes of the same type may have quite high similarity ratios, for example 0.54 between Lower Devonian brachiopods of Xingan and Junggar and 0.38 between Lower Jurassic bivalves of Upper and Lower Yangtze, while biomes of different types may be quite low in similarity ratio, for example 0.05–0.1 between Lower Ordovician cephalopods of central and marginal Yangtze biomes. Besides, there are cases where biomes of the same type but assigned to different provinces may be clustered together (Fig. 9.7a, b). Biofacies are thus first-order factors in demarcating biomes using computer data.

The term 'biome' has long been applied to marine biotas, such as ocean, rocky shore, and sandy or muddy shore biomes. However, such modern usage does not suit the purpose of geological analysis. For the palaeobiogeography of China we suggest the following types of marine biomes.

Platformal benthic biome

Known as shelly facies in the Chinese literature, this commonly covers the central part of platforms, occupying a sizeable field in massive or irregular form. It is always associated with shallow-water carbonates, clastics, and other relevant lithofacies.

Main groups are Early Palaeozoic trilobites, nautiloids, corals, and brachiopods; Late Palaeozoic brachiopods, corals, and forams; Mesozoic and Cenozoic bivalves, gastropods, and forams, as well as algae.

Plankton biome

This type of biome commonly covers basin and trough areas, massive or belt-shaped, and is related to black shale, flysch, and other deep-water or still-water lithofacies. Main groups are Early Palaeozoic graptolites, agnostids, and hyolithids; Late Palaeozoic nektonic and planktonic ammonoids and bivalves, and tentaculitids; Mesozoic and Cenozoic nektonic ammonoids, halobiids, other planktonic molluscs, radiolarians, tintinids, deep-water forams, diatoms, and other phytoplankton.

Transitional biome

This usually occupies platform margins, and is related to platform-margin lithofacies. Transitional biotas are composed of groups from both of the two above-mentioned biomes and are usually the most highly diversified in geological periods (Fig. 3.13).

The *reef biome* is commonly known as a kind of transitional biome. However, this only applies to reefs reaching a sizeable dimension. Patch-like or narrowly band-like reef biotas are here regarded as communities. Reef biomes are made of Pre-Cambrian stromatolites, Early Palaeozoic archaeocyathids, porifers, bryozoans and algae; Late Palaeozoic corals, stromatoporoids, sphinctozoans, bryozoans, and phylloid and other algae; Mesozoic and Cenozoic corals, sphinctozoans and other calcareous sponges, rudists, and algae.

Community

Community, the assemblage of organisms inhabiting a specified area, is used here as a subdivision of biome. A biome may consists of several geographically connected and ecologically associated communities (Figs 3.10, 3.11). In the geological history of China, a community is deemed to occupy a much smaller area than a biome, usually tens to more than a hundred thousand square kilometres. Relatively small-scale biofacies, such as the Devonian Nandan shelly biofacies and the Xiangzhou planktonic biofacies of Guangxi province, as well as the Hirnantian Guanyinqiao shelly biofacies and Ashgill Wufeng planktonic biofacies, are here regarded as communities rather than biomes. We have so far failed to demarcate communities by cluster analysis.

The ecological subdivisions of communities, such as association and biotope, are beyond the scope of palaeobiogeographical provincialization. Also, for the sake of avoiding complications, such biogeographical terms as biotic province (Kendeigh 1961), vegetation type, biochore are not used in this book. As in the case of biome, the term 'community' in geological history is not a concept of simultaneity, but represents an evolutionary succession of biotic assemblages with similar composition spanning considerable time intervals, i.e. occurring in one or more successive horizons. For example, in Fig. 3.11 the *Prosaukia* community (in the appearance of an assemblage) spans the whole Fengshanian Stage (Cm), which actually consists of several fossil horizons yielding analogous assemblages.

Distribution pattern of ecological biogeographical units

Because these units are demarcated according to biofacies, they display the following characteristics:

The first is their repeatability. The same type of biome or community can recur in different continents or oceans wherever there are analogous environments. Thus, the same type of biome may be assigned to different provinces, while the same region or province may consist of various types of biome.

The second is their zonal or symmetrical distribution along continental or oceanic plates. There are cases of circumcratonic or circumplatformal distributions of benthic, transitional, to planktonic biomes or communities from platform shelf, slope, to basin. Examples are the Middle–Late Cambrian biofacies distribution of North America (Palmer 1969). The circum-Yangtze distribution of Cambrian shallow benthic, transitional, and planktonic biomes is another example (Figs 3.5, 3.6). Often the zones are not perfectly circumferential, and thus display a parallel or nearly concentrically arcuate distribution. Widely known parallel communities of five benthic biomes plus one planktonic assemblage is exemplified in the Early Silurian of Wales (Ziegler 1965). In China, there are, for example, the Silurian–Permian marginal belt of North China (I_2^2 of Figs 6.6, 7.7, I_3 of Fig. 8.3), the Triassic Qingyan–Cehang reef belt (Fig 9.2), and the Yarlung Zangbo belt (Fig. 9.5). If there are similar tectonic and lithofacies environments on both sides of a continent or an ocean, amphicratonic or amphioceanic distributions may occur. It must be noted that due to tectonic and depositional complications, many of the ecological biogeographical units do not show zonation but are simply massive or irregularly shaped.

1.2.3 THE RELATIONSHIP BETWEEN TAXONOMIC AND ECOLOGICAL BIOGEOGRAPHICAL UNITS

Units of the two different systems are simultaneously applied in the practical provincialization of Chinese palaeobiogeography. Often a plate would constitute a region or a province (taxonomic biogeography). Then it would be further subdivided, not by climatic belts or geographical barriers since they were very rare within such a small area, but into biomes and communities, according to litho- and biofacies zonation from the central part toward the margins, as shown in Table 1.5 (ecological biogeography). This situation can be compared with the simultaneous application of both chronostratigraphic and lithostratigraphic units in the practice of stratigraphy. In all regional stratigraphic schemes of China, formations and members follow after systems and series. Everyone knows that they belong to different categories, but the real situation often does not allow a further subdivision into stages and biozones; application of formations and members are thus natural and practical.

Notwithstanding these practical applications, care must be taken not to confuse them conceptually.

Table 1.5 Relationship between province and biome type

	Province I_2 includes	Province II_1 includes
Biome type 1 (shelly)	Biome I_2^1	Biome II_1^1
Biome type 2 (transitional)	Biome I_2^2	Biome II_1^2
Biome type 3 (planktonic)	Biome I_2^3	Biome II_1^3

Valentine's suggestion (1973) of five orders of units (marine biota, province, community, population, phenotype) is not recommended. A biome may be either equal to a province, or part of it, or spanning two adjacent provinces; with its boundary coinciding or not coinciding with a provincial boundary, just like the relationship between formation and series. In cluster analysis there are cases where isofacial biomes or communities pertaining to different provinces share higher similarity coefficients than that between anisofacial units of one province, so if we consider only ratio quantities but leave out the influence of facies, we may confuse isofacial biomes of different provinces into one province, or mistake anisofacial biomes of the same province as two provinces. Therefore, in treating the results of cluster analysis, it is necessary to compare only isofacial units.

1.3 THE PROBLEM OF MIXED BIOTAS

This is an important problem because mixed biotas can be interpreted in disparate ways, resulting in greatly varied geological reconstructions. For example, a mixed fauna of both Tethyan and Boreal elements may be interpreted as due to tectonic melange, palaeo-oceanic currents, or intermediate location, resulting in different reconstructions of terranes, currents, or palaeolatitudes. In the compilation of this book, much attention has been paid to the following aspects.

1.3.1 DISTINCTION BETWEEN MIXED BIOTA AND REWORKED OR TECTONICALLY MINGLED FOSSILS

Mixed biota result from the blending of biogeographically separated organisms during their lifetime and taphonomy. At least four patterns of mixture can be cited here.

Mixture by filter or sweepstake route

This results in the occurrence of a small percentage of anisofacial elements in certain biota. The occurrence of a few shell fragments of psychrophilic *Otoceras* in the tropical Early Triassic fauna of southern Jiangsu province is interpreted as due to transportation by the Boreal Paleo-Pacific current. Other media, such as land bridges, seaways, floating wood, may also provide means of transportation (Simpson 1953).

Intermingling

In this pattern, elements of two different biotas are mixed so that each constitutes a significant proportion of the whole biota. From the Late Triassic mixed flora of northern *Danaeopsis–Bernoullia (Symopteris)* type and southern *Dictyophyllum–Clathropteris* type, to modern broadleaf–coniferous mixed forest, there are several Mesozoic and Cenozoic intermingling floras in the middle palaeolatitudes. The same situation occurs in marine biotas. Suggested causes include geographical corridors, intermediate temperatures, and ocean currents.

Alternating in sequence

Ecologically or biogeographically inharmonious biotas may occur repeatedly in sequence. The overlying of the Tethyan-type Maokouan (Guadalupian) *Verbeekina* fauna on cold-water Early Permian *Lytvolasma* fauna in many sections of southern Tibet is interpreted as due to the spread of warm-water conditions following the retreat of cold climatic conditions. Another example is the alternate occurrence of cool-water *Buchia* and warm-water Ethiopian-type molluscs in Jurassic sections of the Qomolangma region. This is interpreted as due to eustatic fluctuations of sea-level, during which the thermophilic molluscs were associated with shallow clastics, while buchiids appeared in deep-water black shales resulting from tropical submergence.

Noah's ark

This concept (McKenna 1973) denotes biota carried by a drifting plate which later accreted onto another continent already inhabited by a disparate biota, thus forming a mixed biota. An example is given in Section 1.3.4.

Reworked or tectonically mingled biotas can form mixtures of biogeographically and even chronologically incompatible fossils. These should not be confused with mixed biotas. Depositional reworking has been reported frequently in the Chinese literature. In the western Qinling Mountains, Permo-Triassic olistostromes consist of thick deposits of carbonate debris flow, the limestone gravels of which yield Carboniferous and Permian fusulinids and corals, and sometimes even Devonian fossils (Yin *et al.* 1988).

Tectonics may also introduce intermingling. The melange belt south of the Yarlung Zangbo Ophiolites consists of Triassic matrix and Cretaceous blocks containing radiolarians and forams (Yin Jixiang and Wen Chuanfen 1988). This situation once led to the erroneous identification of the radiolarians as a Late Triassic assemblage.

Another proposed source of mixing is known as the Viking funeral ship — the rafting of already fossilized assemblages to a new continent by passage aboard a 'floating' plate. In this case intermingling occurs after fossilization, and is therefore heterochronic, whereas in the Noah's ark model organisms are mixed during their lifetime or shortly thereafter. For example, the interfingering distribution of Tethyan and indigenous Permian fusulinids in western North America (Bostwick and Nestell 1967) has been interpreted as due to Mesozoic terrain accretion.

One of the characteristics of reworking is the mirror-image effect. Because rock lumps and fossils are originally eroded and slide down from a sequence of older to younger rocks, the appearance of exotic fossils or fossil-bearing blocks in olistostromes should be from younger to older in ascending order, forming a mirror-image of the source sequence. This has been observed in favourable circumstances in western Qingling.

1.3.2 MIXED BIOTAS DUE TO PALAEOGEOGRAPHICAL CAUSES

Simpson's summary (1953) on this aspect is still useful. Paths of biotic interchange can be divided into three categories: the corridor, with a high similarity coefficient; the filter, with quite a low similarity coefficient, e.g. 0.2 in the case of the Panama isthmus filter; and the sweepstake route, with a very low similarity coefficient, e.g. the odds per year for the interchange of insects via the Hawaiian sweepstake route is 1 in 20 000. This criterion can be used in interpreting palaeobiogeographical situations. The discovery of a few Permian Cathaysian floral elements in the Gondwanan flora of New Guinea can thus be explained by passage via a certain filter route. Stauriid corals originated in the middle Llandovery of Yangtze where they reached a diversity of six genera. Later they migrated to the Kunlun Mountains (Lower–Middle Silurian), Kazakhstan (Lower Silurian), and Gotland (Middle Silurian) with only one genus each (He Xinyi, personal communication). This seaway migration is another type of filter.

1.3.3 MIXED BIOTAS DUE TO PALAEOCLIMATIC CAUSES

Palaeobiogeographical provincialization is controlled to a large extent by palaeoclimatic conditions: temperature and precipitation on land, water temperature related to latitude, currents, and depth in the sea. Because climatic changes are generally gradual, the provincialization they control should not be sharply delineated. In climate-controlled palaeobiogeographical units, mixed biotas are normal and inevitable. The mixed belts correspond to transitional palaeoclimatic zones. During the Triassic there were mixed biotas occupying intermediate areas between the temperate *Danaeopsis–Bernoullia (Symopteris)* flora and the tropical *Dictyophyllum–Clathropteris* flora, as well as between the Tethyan fauna and the Boreal fauna. These are given intermediate interpretations, such as warm temperate or subtropical. On the other hand, direct contact of climatically disparate biotas is abnormal; there must have been something, such as plate collision, causing the subduction of transitional belts.

The alternately overlying type of biotic mixture (see p. 15) can be brought about either by climatic or by tectonic causes. Psychrophilic organisms may be introduced by cool temperature or chilly currents to overlie strata with thermophilic biota, or vice versa. Such repetition of climatic changes will introduce stratigraphically succeeding or geographically interfingering patterns of incompatible biotas. Waterhouse (1976, Fig. 3) showed a latitudinal shift of warm- and cool-water biomes due to fluctuating temperature throughout the Permian; the main alternation belts were located in transitional climate zones (20–45° palaeolatitude) and the scope of change could reach 25° of latitude. In the Palaeozoic of China, based on the alternating nature of thermophilic and psychrophilic biotas, Wang Naiwen (1984) distinguished a southern and a northern transitional province located between Tethys and Gondwana or Xingan–Mongolia, respectively, and proposed that it reflected a temperate zone. To the south of the Yarlung Zangbo River, along the Gondwanan margin of Tethys, the warm-water *Verbeekina* fauna succeeds the cool-water *Stanoviella–Lytvolasma* fauna. Simultaneous succession of the two faunas occurred along the Eurasian margin of Tethys, for example Jalaid Qi of northern Inner Mongolia. This can be explained by a broadening of the tropical–subtropical belt.

1.3.4 MIXED BIOTAS DUE TO PLATE TECTONICS

The modern mixture of Asian and Australian faunas between the Wallace Line and the Weber Line in the Indonesian Archipelago is evidently caused by collision of the two continental areas during the Middle Miocene. Introduction of organisms from other biogeographical units into the endemic biota through accretion of plates and terrains carrying these organisms is called Noah's ark, as mentioned on p. 15. After the collision of India with Eurasia during the Tertiary, the freshwater snail subfamily, Triculinae, carried by India entered the Mekong River drainage system and radiated into 92 species of 11 genera, which became mixed up with the autochthonous Eurasian fauna. Also, the introduction of the Gondwanan pinacean *Podocarpus* into South-East Asia and Central America in the Tertiary can be explained by northward drift of Australia and South America. Gondwanan tetrapods *(Lystrosaurus)* and flora *(Glossopteris indica, Schizoneura gondwanensis)* have been discovered adjacent to the indigenous Triassic biota in the Lamprabang region of Upper Laos (Fontaine and Workman 1978). One of the interpretations is that they were carried by the Qiangtang–Indosinian Noah's ark which collided with Eurasia during the Triassic. Some have even suggested that the lystrosaurians, kannemeyerians, and cynognathids of North China and the Junggar Basin were introduced from this plate.

It should be noted that such mixing is not a one-shot migration from the Noah's ark. Following the successive collisions of first Qiangtang (end-Triassic), then Gangdise (end-Jurassic), and finally India (end-Eocene) with Eurasia, the Eurasian terrestrial fauna invaded these newly collided lands: first the freshwater Yangtze-type unioniids in Qiangtang (Middle Jurassic), then the Asian *TPN* (*Trigonioides*, *Plicatunio*, *Nippononaia*) bivalves (see Chapter 11) into Gangdise (early Cretaceous), and finally *Hipparion guichongensis* in Jilong, southern Tibet (Pliocene).

1.4 DATA-BASE AND APPLICATION OF CLUSTER ANALYSIS

1.4.1 DATA-BASE

From the beginning of the compilation of this book, it was demanded that the authors collect all the available distributional data of the major fossil groups of the Phanerozoic periods. For each geological period, 3–8 major groups were selected. Data from all published literature dated up to 1984 were collected. Doubtful names listed in over-generalistic or somewhat outdated publications had to be identified by inspections of figures or through personal communication with specialists, failing this they were deleted. The taxonomic unit was genus or species, varying in different periods and different fossil groups, usually genus for benthic animals and species for plankton and plants.

The statistics were based on stratigraphical regions and subregions delineated in the *National Chart of Stratigraphic Regionalization.*[2] Because there are too many subregions (more than 300) and many of them may not have the strata of the period concerned, it was permitted to incorporate adjacent subregions if the similarity of their fossil contents was ascertained, or simply to count according to regions. Those lacking the strata of relevant periods were deleted. Thus arranged, the number of areas dealt with statistically was usually decreased to around 30. Furthermore, in cluster analysis of a certain fossil group, a few areas may have been uncounted due to the lack of a fossil record, or merged with their apparently similar neighbouring areas, so usually a cluster analysis only dealt with a dozen or more operational areas. At the end of the

[2] The *National Chart of Stratigraphic Regionalization* is attached to the first page of every issue of *National Charts of Regional Stratigraphy*. The latter was published in many volumes during the 1960s and 70s; each volume covers a province of China, giving in detail the stratigraphic column of stratigraphic regions and subregions.

procedure the data cards containing details of the regional distribution of genera and species, or bed-by-bed listings of stratigraphical sections, were accumulated and filed. Based on these cards, distribution tables were made by filling in forms with the region and subregion data in columns and the genera and species data in lines. The total database thus arranged consisted of 10 170 cards, 557 tables, and statistics of 12 468 genera and species. They provided about 217 000 items of data required for the cluster analysis. From these tedious but necessary statistics we have obtained a more-or-less quantitative knowledge of the fossil biotas in the geological history of China, as indicated in the following chapters.

1.4.2 APPLICATION OF CLUSTER ANALYSIS IN THE PALAEOBIOGEOGRAPHICAL PROVINCIALIZATION OF CHINA

Q-mode binary cluster analysis

This has been used for most Phanerozoic epochs of China. After consideration of various kinds of similarity coefficients, the Otsuka coefficient, or generalized angular cosine coefficient, was chosen, because it resulted in more agreement with those conclusions reached using traditional methods of analysis:

$$\text{Otc} = \frac{C}{\sqrt{N_1 \cdot N_2}}$$

where C is the number of common taxa (genera or species) between the two areas under comparison, and N_1 and N_2 are total numbers of taxa of the two areas under comparison.

The coefficient data thus obtained were first arranged in similar matrices, then categorized according to their similarities, and presented in hierarchical dendrograms showing regional relationships.

Tests of various methods of forming dendrograms gave quite different results. In binary clustering we used direct formation and mean continuation methods, because these methods use that information about the primary similar matrices fully, and do not introduce distortions. The direct formation method is the simpler procedure, by which a dendrogram can be formed directly from the primary matrix. However, it cannot differentiate poorly delineated groups of points. Non-correlated samples cannot be discriminated from a certain category once they have been mixed up. Therefore, the mean continuation method was applied more often.

Example, Early Denovian brachiopod provincialization. Data-set consisted of 10 samples (areas) — 21, 19, 17, 14, 11, 10, 7, 6, 4, 2 — and 126 variables (genera). Area 19 provided too few data and was deleted; areas 14 and 17 were combined because they were adjacent and apparently similar. So only eight samples were utilized. The common characteristic matrix was as follows:

	21	17 + 14	11	10	7	6	4	2
21	30	13	2	1	4	5	7	4
17 + 14		61	7	6	6	15	15	6
11			14	3	4	4	7	2
10				10	3	5	6	1
7					24	16	8	5
6						39	16	7
4							31	10
2								12

Then the similar matrix:

	21	17 + 14	11	10	7	6	4	2
21	1.00							
17 + 14	0.304	1.00						
11	0.098	0.24	1.00					
10	0.058	0.243	0.254	1.00				
7	0.149	0.157	0.273	0.194	1.00			
6	0.146	0.308	0.171	0.253	0.523	1.00		
4	0.230	0.345	0.336	0.341	0.293	0.460	1.00	
2	0.211	0.222	0.154	0.091	0.295	0.324	0.518	1.00

Figures 1.1 and 1.2 show the dendrograms that resulted from the direct formation method and mean continuation method, respectively. Figure 1.1 was unreasonable because the areas 2, 4 and 6, 7 all belonging to the same Junggar–Xingan province of the Boreal realm, were categorized in two groups. Figure 1.2 was accepted because these four areas were categorized together in accordance with the traditional synthesis. The eight areas were grouped into three biogeographical provinces at the 0.28 similarity level, namely southern (10, 11),

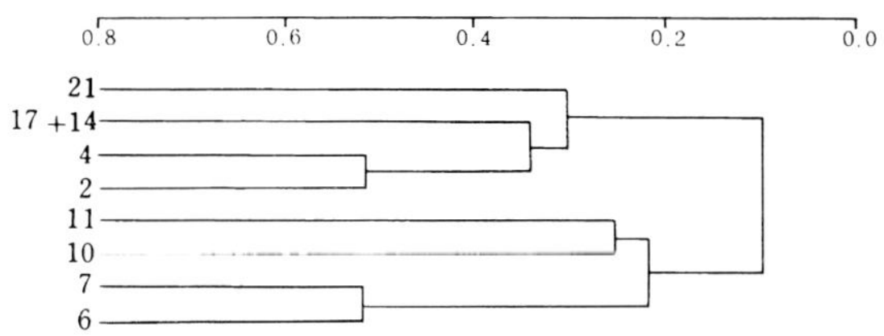

Fig. 1.1 Dendrogram resulting from the direct formation method.

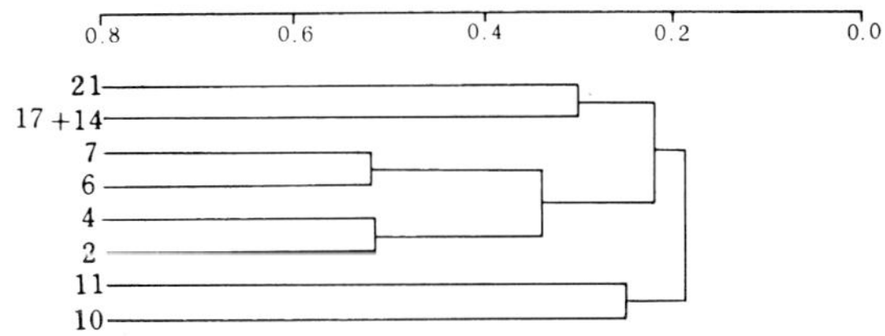

1.2 Dendrogram resulting from the mean continuation method.

Junggar–Xingan (2, 4, 6, 7), and North China–Tarim provinces (14 + 17, 21), and each of these consisted of several biomes.

Blurred cluster analysis

Because the variation of biotic similarity among areas is more-or-less gradual, blurred cluster analysis is required to determine the level at which a biogeographical unit should be demarcated. The first step of its basic procedure is to choose the appropriate blurred similarity coefficient for calculation. There are about a dozen common types of blurred similarity coefficients, among which the suitable ones for biogeographical purposes are the same as those used in binary clustering. The second step is to form the blurred compatibility matrix, R. However, an initial blurred compatibility matrix cannot be used directly in categorizing because, according to blurred mathematical theory, to ensure that samples are categorized into definite groups, avoiding the possibility of equivocally attributing them to one or another group, the blurred matrix to be used in categorizing must satisfy three conditions: reflexivity, symmetry, and transitivity. Those satisfying the three conditions are called blurred equivalent matrices. In fact, only transitivity is needed because an initial blurred compatibility matrix has already satisfied reflexivity and symmetry. For this purpose the third step, fold product operation, is taken.

Fold product is the multiplication in blurred matrix, using '•' as its multiplying mark, to be distinguished from ordinary multiplication. The rule is to take the minor eigenvalue first. One-place fold product of R results in the second-level blurred compatibility matrix R_2:

$$R_2 = R \cdot R.$$

Let $r_{ij}^{(2)}$ denote the eigenvalue of row i and column j of R_2, so we have:

$$r_{ij}^{(2)} = \max_k \min (r_{ik}, r_{jk}) = \bigvee_k (r_{ik} \wedge r_{jk}).$$

Here '$\wedge$' is the minor one of the two eigenvalues, such as $0.8 \wedge 0.5 = 0.5$. '$\vee$' denotes the maximum of the K eigenvalues, such as $0.5 \vee 0.8 \vee 0.6 \vee 0.3 \vee 0.1 = 0.8$. The second-level matrix, R_2, is put under fold product operation again, and results in a third-level blurred compatibility matrix, R_3:

$$R_3 = R_2 \cdot R = R \cdot R \cdot R.$$

and so on, until we get the nth-level blurred compatibility matrix:

$$R_n = R_{n-1} \cdot R = \underbrace{R \cdot R \cdots\cdots R}_{n}.$$

The operation stops when $R_n = R_{n-1}$. It can be proved that the R_n derived in this way is a blurred equivalent matrix suitable for clustering. Given different values for blurred similarity coefficient (or confidence level), F, those values of $r_{ij}^{(n)} \geqslant F$ are categorized in one group and the rest of them in other groups. If $F = 1$, each sample constitutes a separate group. Blurred clusters of various confidence levels can be obtained by gradually decreasing the F value. Based on these results, a blurred dendrogram can be drawn, from which we can observe the clustering of samples visually.

Example. For the Early Triassic bivalves, 107 taxa (variables) and 10 areas (samples) are counted, symbols of the latter being: VI_2, VI, IV_5, IV_1, VII_4, VII_5, VII_6, VII_7, VIII, IX. Generalized angular cosine was used as the blurred similarity coefficient in blurred cluster analysis.

	VI_2	VI	IV_5	IV_1	VII_4	VII_5	VII_6	VII_7	VIII	IX
VI_2	10	1	2	0	3	2	2	2	3	2
VI		15	4	0	10	6	3	5	7	7
IV_5			12	1	8	5	6	7	7	8
IV_1				12	2	2	0	1	1	3
VII_4					53	13	6	14	17	18
VII_5						13	4	6	8	8
VII_6							11	9	7	7
VII_7								28	11	14
VIII									18	12
IX										29

First we arranged the common characteristic matrix $C(10, 10)$ of the samples:

	VI_2	VI	IV_5	IV_1	VII_4	VII_5	VII_6	VII_7	VIII	IX
VI_2	1.000									
VI	0.082	1.000								
IV_5	0.183	0.298	1.000							
IV_1	0.000	0.000	0.083	1.000						
VII_4	0.130	0.355	0.317	0.079	1.000					
VII_5	0.175	0.430	0.400	0.160	0.495	1.000				
VII_6	0.191	0.234	0.522	0.000	0.248	0.334	1.000			
VII_7	0.120	0.244	0.382	0.055	0.363	0.314	0.513	1.000		
VIII	0.224	0.426	0.476	0.068	0.550	0.523	0.497	0.490	1.000	
IX	0.117	0.336	0.429	0.161	0.459	0.412	0.392	0.491	0.525	1.000

Then the initial blurred compatibility matrix $R(10, 10)$ was obtained:

	VI_2	VI	IV_5	IV_1	VII_4	VII_5	VII_6	VII_7	VIII	IX
VI_2	1.000	0.224	0.224	0.161	0.224	0.224	0.224	0.224	0.224	0.224
VI	0.224	1.000	0.430	0.161	0.430	0.430	0.430	0.430	0.430	0.430
IV_5	0.224	0.430	1.000	0.161	0.497	0.497	0.522	0.513	0.497	0.497
IV_1	0.161	0.161	0.161	1.000	0.161	0.161	0.161	0.161	0.161	0.161
VII_4	0.224	0.430	0.479	0.161	1.000	0.523	0.497	0.497	0.550	0.525
VII_5	0.224	0.430	0.497	0.161	0.523	1.000	0.497	0.497	0.523	0.526
VII_6	0.224	0.430	0.522	0.161	0.497	0.497	1.000	0.513	0.497	0.497
VII_7	0.224	0.430	0.513	0.161	0.497	0.497	0.513	1.000	0.497	0.497
VIII	0.224	0.430	0.497	0.161	0.550	0.523	0.497	0.497	1.000	0.525
IX	0.224	0.130	0.497	0.101	0.525	0.523	0.497	0.497	0.525	1.000

We then operated the fold product of R to R_4 because $R_3 = R_4 \bullet R_4$ thus obtained was:
Clustering based on R_4:
Given $F = 0.497$, there resulted for clusters: VI_2; VI; IV_5, VII_4, VII_5, VII_6, VII_7, VIII, IX; IV_1.
Given $F = 0.430$, there resulted three clusters: VI_2; VI, IV_5, VII_4, VII_5, VII_6, VII_7, VIII, IX; IV_1.
Given $F = 0.224$, there resulted two clusters: VI_2, VI, IV_5, VII_4, VII_5, VII_6, VII_7, VIII, IX; IV_1.
Given $F = 0.161$, all samples group into one cluster.

The blurred dendrogram (Fig. 1.3, left) thus obtained was basically coincident with the dendrogram resulting from binary cluster analysis (Fig. 1.3, right), with IV_1, VI_2, and the rest constituting three basic clusters.

The result of traditional synthesis was a subdivision of three provinces, i.e. IV_1, VI_2, and the rest of all other areas, therefore the results of the three methods agree.

1.4.3 PRELIMINARY SUMMARY ON THE APPLICATION OF NUMERICAL CLASSIFICATION TO THE PALAEOBIOGEOGRAPHY OF CHINA

This computer exercise on provincialization in China involved a fairly large data-base. The total input comprised about 217 000 items of data, including: Cambrian 54 995, Ordovician 21 624, Devonian 3022, Carboniferous 58 977, Permian 10 959, Triassic 28 581, Jurassic 13 500, Cretaceous 880, and Tertiary 24 813 items. The results were drawn on 58 dendrograms of cluster analysis, of which only 19, about 33 per cent, were used in this book.

Causes of discordant results between cluster analysis and traditional qualitative analysis

Among the unused dendrograms of cluster analysis, a number were the results of different fossil groups resembling each other, and only the more representative ones were chosen; a few others were abandoned because the samples (areas) in the operation were not large enough to give a clear explanation of provincialization. However, quite a lot of the dendograms were not used because of their discordance with the results from traditional analysis. Common causes of discordance were:

1. In some cases, the taxonomic similarity due to similar biofacies belonging to different areas remarkably exceeded the taxonomic dissimilarity caused by biogeographical division. The cluster dendogram thus produced reflects biofacies or blending of biofacies and biogeographical divisions rather than biogeographical provincialization itself. On the other hand, sometimes the influence of biofacial dissimilarity greatly exceeded the influence of biogeographical similarity; again, the outcome does not represent biogeographical provincialization.

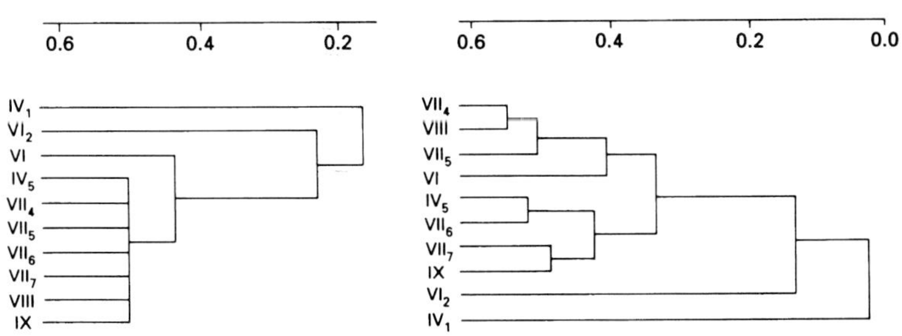

Fig. 1.3 Comparison of a blurred dendrogram (left) and a binary dendrogram (right).

2. The traditional qualitative analysis emphasizes the characteristic taxa which, in spite of their significance in palaeoclimatic and palaeogeographical interpretation, or even their dominance in quantity, were a minority in the whole biota. Due to certain difficulties, the computer operations used in this book were unweighted and each taxon made only one variable. Thus during computer operation, the characteristic minority was diluted by the mass of taxa that were cosmopolitan or common to various provinces. The problem of how to weight the biogeographically characteristic elements in these computer operations is not resolved in this book.

3. There were some small but biogeographically significant fossil groups, such as the Cambrian archaeocyathids, the Late Palaeozoic sphinctozoans, and the Mesozoic scleractinians and solenoporids. However, data on their area distribution were not sufficient to establish a dendrogram, or dendrograms resulting from these data involved too few areas to give a clear explanation. Therefore these groups have not been employed in cluster analysis.

4. A few unsuccessful cases may have been due to the introduction of doubtful data. Fossil lists from the *National Charts of Regional Stratigraphy*, published during the 1960s and 1970s, and other publications contain some taxonomic names that are either too generalized, such as *Productus* and *Unio*, or biogeographically untrustworthy due to lack of understanding of the relevant concept by their compilers. Indiscriminate introduction of these data may have resulted in disturbance to the inferred pattern of provincialization.

Some recommendations on cluster analysis operation

Correctly sift the taxonomic list before adding it to the input.

'The more the better' is not always good for cluster analysis. To avoid the factors causing the above-mentioned failures, a few taxa should not be taken as variables. The following points should be noted:

1. Delete from the data-base those taxa that are indeterminate, too generalized, or biogeographically unreasonable.

2. Choose appropriate groups. Planktonic or other groups apt to be cosmopolitan, such as graptolites and conodonts, are less suitable than benthic groups, especially in regional provincialization. Groups distinctly reflecting certain biofacies, such as green algae, should be discriminately treated or uncounted in order to avoid biofacial influence.

3. It is better not to count taxa that are cosmopolitan or common in most of the areas under operation. Indeterminate species should be excluded.

4. Whether species comformis (cf.) and species affinis (aff.) should be calculated as independent variables or incorporated with the original species should be considered individually. Usually we would delete those sp. cf. and sp. aff. when the originals occur in the same area, while considering those sp. cf. and sp. aff. appearing in different areas from the originals as independent variables.

Establish appropriate areas for clustering

Areas for clustering should not be too finely divided. Areas with the insufficient data should be either deleted or integrated with adjacent biogeographically similar areas. Due to causes other than biogeographical factors, some areas may yield only one or two variables (taxa), among tens or hundreds ready for operation. This will induce unreasonable results. It is better to keep the proportion of variable (taxanomic) numbers between the largest and smallest samples (areas) greater than 15:100. Our computer analysis of the Devonian data achieved relatively satisfactory results with minimal rehandling, because its areas were better established. Cluster analysis of Early Devonian brachiopods, Middle Devonian brachiopods, and Late Devonian corals all basically coincided with the traditional provincialization. In calculating the situation of the Early Devonian brachiopods, we deleted the area 19 which had insufficient data and combined adjacent areas 14 and 17, and obtained more satisfactory results (see p. 18).

Close co-operation between geologists and mathematicians is necessary

The sifting of variables is a difficult problem. According to mathematical theory, variables to be chosen for operation should be reliable, independent, representative, categorically sensitive, and siftable through *R*-mode cluster analysis, factor analysis, and correlation tests, but what do these demands mean for palaeobiogeographical variables? Because we have not been able, so far, to figure out quantitative indices for them, disagreements about choosing variables will certainly arise between geologists and mathematicians. Reasonable solutions can be obtained only through tests for improvement in calculations and through back and forth discussions between specialists. We treated the Triassic data in this way; repeated discussions and sifting of data finally led to relatively satisfactory results; among the nine dendrograms drawn, six showed good coincidence with traditional analysis and were adopted. The same procedure was applied in the treatment of all the data of other geological periods.

Correlation test

A general survey of all the 21 dendrograms considered to be coincident with traditional analysis revealed that the similarity coefficients between provinces were generally below 0.30, only exceptionally reaching 0.31. Provinces can thus be taken as non-correlative biogeographical units. We can test this using another method, i.e. by incorporating all areas within a province into one sample, we can make statistical tests of the correlation coefficient. Take the provincialization of Upper Devonian corals for example:

With 64 variables, the cluster dendrogram of Upper Devonian coral is shown in Fig.1.4 (see also Fig. 6.4). Four groups are formed: (23, 19, 14, 16); 9; (5, 4); 2 (numbers denote area codes). The common characteristic matrix of the four groups (provinces) is as follows:

	(23, 19, 14, 16)	9	(5, 4)	2
(23, 19, 14, 16)	41	4	3	4
9	5	0	1	
(5, 4)		8	2	
2			9	

From the formula for the correlation coefficient:

$$r_{ij} = \frac{NC - N_i N_j}{\sqrt{(N_i N_j)(N - N_i)(N - N_j)}}$$

where N is the number of variables (here $N = 64$); C is the number of common characteristics; and

N_i, N_j are the number of characteristics of samples i and j.

The correlation coefficients among provinces can be calculated as follows:

	(23, 19, 14, 16)	9	(5, 4)	2
(23, 19, 14, 16)	1	−0.165	−0.209	−0.165
9		1	−0.110	0.497
(5, 4)			1	0.119
2				1

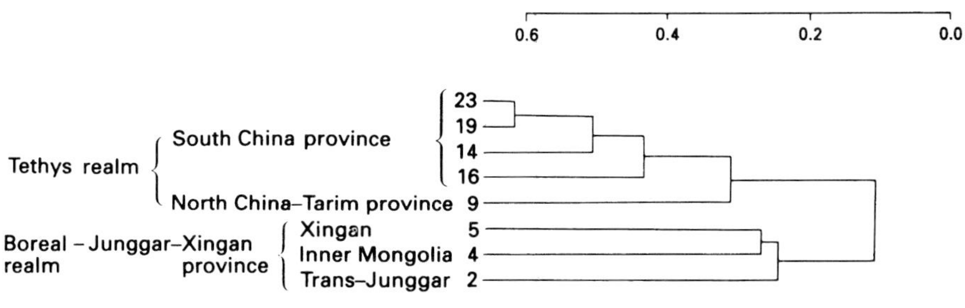

Fig. 1.4 Dendrogram of Upper Devonian corals.

Then check the table of critical values of correlation coefficients for degrees of freedom $f = 64 - 2 = 62$ (check $f = 60$); the critical values of correlation for significance levels $\alpha = 0.10, 0.05, 0.02, 0.01, 0.001$ are:

$\alpha = 0.10$	0.05	0.02	0.01	0.001
$r_a = 0.2108$	0.2500	0.298	0.324	0.408

From the above it is clear that for all significance levels, α, $|r_{ij}| < r_a$, so these provinces are not correlated.

The role of cluster analysis in palaeobiogeographical research in China

1. The 19 adopted dendograms function as a testing and verifying tool, and basically coincide with results from traditional analysis.

2. Based on the results of cluster analysis, a preliminary index of similarity coefficient values adapted to Chinese provinces and realms is summarized, as discussed in Section 1.2.

3. In many cases these dendrograms reflect the division of biomes within a province and serve as a criterion of biome delineation (see, for example, Figs 6.2, 6.3, 6.4, and 9.7).

4. In a few cases, the results of cluster analysis called to attention and helped to correct mistakes in provincialization made by traditional analysis.

Attempts to apply numerical classification methods in palaeobiogeography have, however, left many problems unresolved. Establishment of a quantitative index for a reasonable sifting of variables without deviation from subjective choice is one problem remaining to be solved. Variables — fossil taxa — do not possess the same biogeographical importance; some are index taxa, others not. Logically, different levels of weighting coefficients should be given to variables of different importance, but how to define these levels is a problem unresolved in this book. In short, despite the hard work done on the application of mathematical methods, they function, for the time being, mainly as testing and verifying methods to complement the traditional analysis. Further improvements to make such methods an important, independent, and quantitative criterion in palaeobiogeographical research should be one of our future goals.

REFERENCES

Bé, A. W. H. (1977). An ecological, zoogeographic and taxonomic review of recent planktonic foraminifera. In *Oceanic micropaleontology*, Vol. 1 (ed. A.T.S. Ramsay), pp. 1–100. Academic Press, London.

Beliaev, G. M. (1974). *Ultra-abyssal benthic faunas of the Pacific Ocean*. Nauka, Moscow (in Russian).

Boblinski, I. G. (1956). *Zoogeography*. Nauka, Moscow (in Russian).

Bostwick, D. A. and Nestell, M. K. (1967). Permian Tethyan fusulinid faunas of the Northwestern United States. In *Aspects of Tethyan biogeography*, (ed. C. Adams and D. V. Ager), Systematics Association Publication 7, pp. 93–102. Systematics Association, London.

Boucot, A. J. (1974). Silurian and Devonian biogeography. In Paleogeographic provinces and provinciality, (ed. C. A. Ross). *Journal of Economical Paleontologists and Mineralists* (Special paper) **21**, 165–70.

Boucot, A. J. (1975). *Evolution and extinction rate controls*. Elsevier, Amsterdam.

Boucot, A. J. (1981). *Principles of benthic marine palaeoecology*. Academic Press, New York.

Cao Zhengyao, Li Baoxian, and Guo Shuangxing (1982). The Cretaceous plant-bearing strata in China. In *Stratigraphic correlation chart in China* (ed. Nanjing Institute of Geology and Palaeontology, Academia Sinica), pp. 254–69. Science Press, Beijing (in Chinese).

Cox, C. B. (1973). Triassic tetrapods. In *Atlas of paleobiogeography*, (ed. A. Hallam), pp. 213–25. Elsevier, Amsterdam.

Cox, C. B., Healey, I. N., and Moore, P. D. (1976). *Biogeography an ecological and evolutionary approach*. Blackwell Scientific Publications, Oxford.

ECTBTP (Editorial Committee of Treatise on Basic Theories of Palaeontology) (1983). *Palaeobiogeographic provinces of China*. Science Press, Beijing (in Chinese).

Ekman, I. (1967). *Zoogeography of the sea*, (2nd edn). Sidgwick and Jackson, London.

Fairbridge, R. W. (1979). *Encyclopedia of paleontology*. Dowden, Hutchinson and Ross, Strasberg.

Fisher, A. G. (1960). Latitudinal variation in organic diversity. *Evolution* **14**, (1), 64–81.

Fontaine, H. and Workman, D. R. (1978). Review of the geology and mineral resources of Kampuchea, Laos and Vietnam. *Proceedings of GEOSEA, Bangkok*, Vol. 3, pp. 541–603.

Gou Zhonghai and Li Xiaochi (1982). Preliminary research on the Jurassic bivalve fauna of Nyalam, Tibet. *Contribution to the Geology of the Qinghai–Xizang (Tibet) Plateau* (ed. Editorial Committee of CGQXP), Vol. 11, 43–71. Geological Publishing House, Beijing (in Chinese).

Grabau, A. (1932). Evolution and dispersal center of the Paleozoic faunas. *Bulletin of the Geological Society of China* **11**, (2), 227–41.

Gurianova, E. F. (1964). Zoogeographic regionalization of the faunas of Pacific Ocean. *Physico-geographic atlas of the world*, chart 686 and its explanatory sheet. Izdatelistvo AN USSR and GUGK GGK, Moscow (in Russian).

Hedgpeth, J. W. (ed.) (1957). Treatise on marine ecology and palaeoecology. *Geological Society of America, Memoir* **67** **(1)**.

Kauffman, E. C. (1973). Cretaceous bivalvia. In *Atlas of paleobiogeography*, (ed. A. Hallam), pp. 351–83. Elsevier, Amsterdam.

Kendeigh, S. C. (1961). *Animal ecology*. Prentice-Hall, Englewood Cliffs, New Jersey.

Kiparisova, L. D., Radchenko, G. P., and Gorskii, V. P. (ed.) (1973). *Triassic system*. Stratigraphy of USSR, Vol. 7. Nedra, Moscow (in Russian).

Kusakin, O. G. (1971). Distribution and some characteristics of the vertical distribution of isopod crustaceans in the cold and temperate waters of Pacific Ocean. Unpublished doctorate dissertation, Leningrad University (in Russian).

Li Wenben (1983). Microfloral areas of Early Cretaceous in China. In *Palaeobiogeographic provinces of China* (ed. ECTBTP), pp. 142–51. Science Press, Beijing (in Chinese).

Lu Yanhao, Chu Chaoling, Chien Yiyuan, Ling Huanlin, Zhou Zhiyi, and Yuan Kexin (1974). Bio-environmental control hypotheses and its application to the Cambrian biostratigraphy and palaeobiogeography. *Memoirs of Nanjing Institute of Geology and Palaeontology, Academic Sinica* **5**, 27–116 (in Chinese).

McKenna, M. C. (1973). Sweepstakes, filters, corridors, Noah's Arks and beached Viking funeral ships in palaeogeography. In *Implications of continental drift to the earth sciences*, (ed. D. H. Tarling and S. K. Runcorn), Vol. 1, pp. 295–308. Academic Press, New York.

Menzies, R. J., George, R. Y., and Rowe, G. T. (1973). *Abyssal environment and ecology of the world oceans*. Wiley, New York.

Mu Anzhi (1983). Ordovician ecostratigraphical types and biogeographical regions of China. In *Palaeobiogeographic provinces of China* (ed. ECTBTP), pp. 16–31. Science Press, Beijing (in Chinese).

Palmer, A. R. (1969). Cambrian trilobite distributions in North America and their bearing on the Cambrian paleogeography of Newfoundland. *American Association of Petroleum Geologists, Memoir* **12**, 139–44.

Parker, R. H. (1964). Zoogeography and ecology of some macroinvertebrates, particularly mollusks in the Gulf of California and the continental slope off Mexico. *Videnskabelige Meddelelser fra Dansk Naturhistorisk Forening*, **126**, 12–35.

Raup, D. M. and Stanley, S. M. (1979). *Principles of paleontology*. Freeman, San Fransisco.

Ross, C. A. (ed.) (1974*a*). *Paleobiogeography*, Benmark Papers in Geology. Dowden, Hutchinson and Ross, Strasberg.

Ross, C. A. (ed.) (1974*b*). *Palaeobiogeographic province and provinciality, based on a symposium*, Special Publication of the Society of Economic Palaeontologists and Mineralogists, No. 21. Tulsa, Oklahoma.

Simpson, C. G. (1953). *Evolution and geography*. Oregon State System of Higher Education, Eugen.

Stehli, F. G. (1973). Permian brachiopods. In *Atlas of paleobiogeography*, (ed. A. Hallam), pp. 143–9. Elsevier, Amsterdam.

Stehli, F. G., McAlester, A. L., and Helsley, C. E. (1967). Taxonomic diversity of recent bivalves and some implications for geology. *Geological Society of America, Bulletin* **78** **(4)**, 455–66.

Sun Ge (1981). Discovery of fossil dipteriaccans in the Upper Triassic of eastern Jilin. *Acta Palaeontologica Sinica* **20** **(5)**, 459–67.

Sun Yunchu (1963). Fundamental concept and problems of the transgressions with special discussion on the Palaeozoic faunas and their paleogeographical provinces of China. *Acta Geologica Sinica* **43**, (2), 99–115 (in Chinese).

Sverdrup, H. U., Johnson, M. W., and Fleming, R. H. (1946). *The oceans, their physics, chemistry and general biology* (3rd edn). Prentice-Hall, New York.

Terengiev, P. V. (1963). Tests of variance analysis on quantitative abundance of the terrestrial vertebrate faunas of USSR. *Vestnik of National University of Leningrad, Ser. biology* **4**, 19–26 (in Russian).

Thorson, G. (1957). Bottom communities. *Geological Society of America, Memoir* **67** **(1)**, 461–534.

Tozer, E. T. (1981). Triassic Ammonoidea: geographic and stratigraphic distribution. In *The Ammonoidea*, (ed. M. R. House and J. R. Senior). Systematics Association, special volume No. 18, pp. 397–432. Systematics Association, London.

Valentine, J. W. (1973). *Evolutionary paleoecology of the marine biosphere*. Prentice-Hall, Englewood Cliffs, New Jersey.

Valentine, J. W. and Meade, R. F. (1981). *Californian Pleistocene paleotemperatures*. University of California Publications of Geological Sciences, vol. 40(1), pp. 1–46. University of California, Berkeley, Los Angeles, California.

Vinogradov, M. E. (ed.) (1977). *Oceanology, biology of oceans*, 1, biological structure of oceans. Nauka, Moscow.

Vinogradova, N. G. (1956). Zoogeographic regionalization of abyssal Pacific Ocean. *Doklady (Reports) of Akademia Nauka, USSR* **3 (1)** (in Russian).

Walter, H. (1985). *Vegetation of the Earth,* (3rd edn). Springer-Verlag, Berlin.

Wang Hongzhen, Li Zhimin, and Wang Zhiping (1984). Silurian and Early Devonian rugose coral biogeography of China. *Palaeontographica Americana* **54**.

Wang Naiwen (1984). The rise of Gondwana research in China and some key problems of paleobiogeography related to plate tectonics. *Bulletin of the Chinese Academy of Geological Sciences* **10**, 103–15 (in Chinese with English abstract).

Waterhouse, J. B. (1976). World correlation for Permian marine faunas. *University of Queensland, Department of Geology, Paper* **7 (2)**.

Wei Mochao (1988). *Concise textbook on micropalaeontology*. Geological Publishing House, Beijing.

Wells, J. W. (1957). Corals. In Treatise on marine ecology and paleoecology, Vol. 1, Ecology (ed. J. W. Hedgpeth). *Geological Society of America, Memoir* **67(1)**, 1087–104.

Yin Jixiang and Wen Chuanfen (1988). Diamictites of the Carboniferous and Lower Permian, their stratigraphical characteristic and origin. *Bulletin of the Institute of Geology, Academia Sinica* **3**, 26–55.

Yin Hongfu, Yang Fengqing, Lai Xulong, and Yang Hengshu (1988). Triassic belts and Indosinian development of the Qinling Mountains. *Geoscience — Journal of Graduate School, China University of Geosciences* **2 (3)**, 364–72 (in Chinese with English abstract).

Zhen Chenkui and Zhang Junfu (1959). On the algae provincialisation of western North Pacific. *Haiyang yu Huzao (Seas and Lakes)* 2(**4**), 244–67 (in Chinese).

Ziegler, A. M. (1965). Silurian marine communities and their environmental significance. *Nature* **207**, 270–2.

Ziegler, A. M. (1990). Phytogeographic patterns and continental configurations during the Permian Period. In *Palaeozoic palaeogeography and biogeography*, (ed. W. S. McKerrow and S. R. Scotese). Geological Society of London Memoir No. 12, pp. 363–82.

2 SINIAN

Chen Yiyuan

2.1 GENERAL REVIEW

Despite the complete sequence and widespread distribution of diversified Sinian sediments within China, the research on Sinian metazoan fossils in China has been a weak link all along, which partly hinders the study of the Sinian system.

The finding of hyolithids in the Dengying Formation in Emei Mountain, Sichuan[1] (Liao Shifan 1958) has drawn attention to the existence of Pre-Cambrian metazoans in China. However, if *Anabarites*, an important representative of the first assemblage of small-shelled fossils in China, is referred to the Sinian, the earliest report should date back to 1941

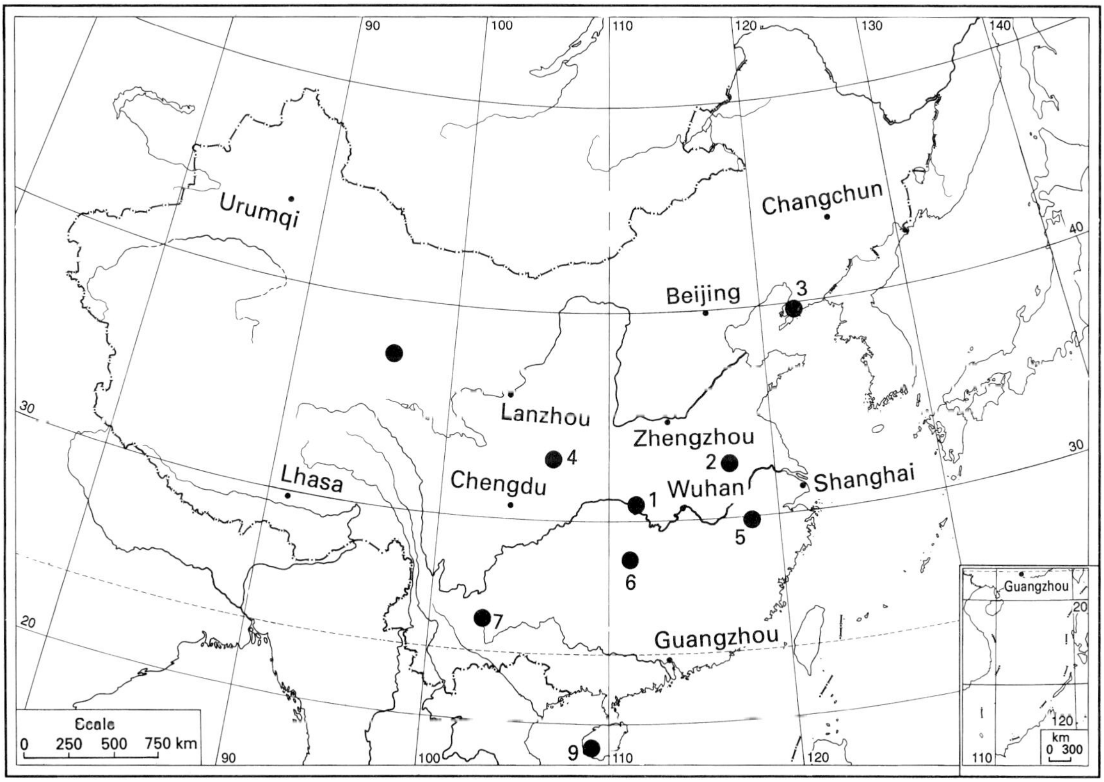

Fig. 2.1 Occurrence of the Sinian metazoans and macroscopic algae in China. ●, fossil localities: 1, eastern Yangtze Gorges; 2, Xuzhou–Huainan; 3, eastern Liaoning; 4, southern Shaanxi; 5, southern Anhui; 6, western Hunan; 7, eastern Yunnan; 8, Quanji Mountains, Qinghai; 9, Hainan Island.

[1] For a description of Chinese geographical names, and a list of Chinese geographical localities, see p. 345 and pp. 346–56.

Table 2.1 Correlation chart of the Sinian of China

		Eastern Yangtze Gorges	Eastern Yunnan	Southern Shaanxi	Xuzhou, Jiangsu and Huainan, Anhui	Western Hunan	Eastern Liaoning	Southern Anhui	Quanjishan, Qinghai
Sinian	Upper	Dengying Doushantuo	Dengying Wangjiawan	Dengying Kuanchuanpu Member Beiwan Member Gaojiashan Member Algal dolomite Doushantuo	Gouhou Jinshanzhai - - - - - Wangshan Shijia Weiji Zhangqu Jiudingshan Niyuan Jiuliqiao	Liuchapu Doushantuo	Xingmincum Cuijiatun Majiatun Shishanlitai Yingchengzi Ganjingzi Nanguanling	Piyuancun Lantian	Zhoujieshan Hongtiegou Heitupo Hongzaoshan Shiyingliang
	Lower	Nantuo Liantuo	Nantuo - - - - - Cheng-jiang	?	Shouxian	Nantuo - - - - - Fulu Changan	Qiaotou	Leigongwu Xiuning	Kubaimu Mahuanggou

when Professor Wang Hongzheng made a survey of the phosphorous ore in Zhongyicun, Jinning County, Yunnan (i.e. *Hyolithes* sp., type a). In 1975 Chen Menge *et al.* reported finding *Sabellidites* sp. in the Gaojiashan Formation, Ningqiang County, Southern Shaanxi. In the late 1970s new discoveries were reported from the eastern Yangtze Gorges, southern Liaoning, and southern Anhui, etc. According to the incomplete statistics at present, metazoan fossils have already been found in six regions, and macroalgae in five regions (Fig. 2.1, Tables 2.1, 2.2).

2.2 SINIAN PALAEOBIOGEOGRAPHY OF CHINA

2.2.1 YICHANG AND THE EASTERN YANGTZE GORGES

The section along the Xilingxia Gorge of the Yangtze River from Liantuo to Shipai, which lies to the west of Yichang city, is one of the typical sections of the Sinian (Table 2.3). A *Charnia*-like pennatulacean fossil was found in the black, organic-rich, thin-bedded micrite of the Shibanton

Table 2.2 Geographical occurrence of four groups of the Sinian soft-bodied metazoans (after Glaessner 1984 with additions of Chinese data)

		S. Australia	C. Australia	SW Africa	S. America	E. Europe	Siberia	China	England	Sweden	Newfoundland	NW Canada	N. Carolina
Coelenterata (Cnidaria)	Hydrozoa	×				?	?				×		
	Scyphozoa	×	×			×	×						
	Conulata	×				?							
	'Medusoids'	×	×	×		×	×		×	×	×		
	Pteridiniidae	×	×	×		×	×					×	
	Rangeidae			×									
	Charniidae	×	×			×	×	×	×		×		
	Erniettidae		×	×		×	×						
Annelids (and probably related phyla)	Cribricyathea			×	?								
	Dickinsoniidae	×				×							
	Sprigginidae	×		?		?							
	Anabaritidae						×	×				×	
	Sabelliditida						×	×					
	Echiura			×									
Ichnofossils	Other 'vermes'	×	×	×		×	×	×				×	
Arthropoda	Vendomiidae	×				×			?				
	Paravancorinidae	×				?							
Phylum uncertain	Redkinia					×							
	Tribrachidium	×				×							

Table 2.3 Geographical occurrence of the Sinian metazoans and macroscopic algae from China

		Yangtze Gorges	S. Shaanxi	E. Yunnan	W. Hunan	S. Anhui	Quanjishan	Huainan	E. Liaoning	Hainan
Pennatulaceans	*Paracharnia*	×								
Annelids	*Sabellidites*		×		×		?	×	×	
	Paleolina							×	×	
	Pararenicola							×		
	Protoarenicola							×		
	Sinosabellidites							×		
	Calyptrina							×		
	Micronemaites	×								
	Sinotubulites	×								
Poriferans	*? Eospicula*	×								
	? Hazelia	×								
Microscopic algae	*Vendotaenia*	×	?	×				×		
	Tyrasotaenia	×				×				
	Chuaria		?	×		×		×	×	×
	Tawuia			×				×	×	×
	Shouhsienia			×		×		×	×	×
	Epiphyton					×				
	Gordonophyton					×				

Member, Middle Dengying Formation in Shibantan, Shipai, Yichang County (Ding and Chen 1981). Because it is quite different both in form and structure from the *Charnia* found in Charnwood, England, Sun Weiguo (1986) established a new genus, *Paracharnia*, for it, and changed the previous name, *Charnia dengyingensis*, into a new name, *Paracharnia dengyingensis* (Ding and Chen), which is considered a member of the Ediacaran fauna. *Sinotubulites, Micronemaites*, and the trace fossil *Paraplanolites* have also been found (Chen *et al.* 1981; Zhao *et al.* 1985).

Problematic sponge spicules, perhaps of *Eospicula* (calcareous triaxone) and *Hazelia* (siliceous monaxone) have been found in slides of the siliceous dolomite lens at the bottom of the Doushantuo Formation, Wangfenggang, Liantuo (Zhao *et al.* 1985). These tiny fossils are easily confused with volcanic glass, thus their identification remains to be verified. There are also trace fossils in the Doushantuo Formation, in which Chen Menge found the mega-alga *Enteromophites siniansis* in 1978 (Zhu and Chen 1984).

2.2.2 GUIZHOU

Zhang Yun (1989) reported multicellular thallophytes with differentiated tissues, which he named *Thallophyca* and *Wengania*, from the Duoshantuo Formation (*c.* 680 ma) of the Wengan–Fuqukan phosphate mines of central Guizhou. *Thallophyca* may represent a primitive metaphyte; *Wengania* is assigned to the solenoporid algae. Zhang Yun concluded that diverse thallophytes had evolved during the Ediacaran.

2.2.3 XUZHOU–HUAINAN (NORTHERN ANHUI AND NORTH-WESTERN JIANGSU)

This region mainly includes areas to the north and south of the Huaihe River of northern Anhui and around Xuzhou city, where the Sinian strata display a complete sequence with abundant fauna (Sun *et al.* 1986; Chen *et al.* 1986). The shale in Liulaobei Formation at the lower part of the Lower Sinian Series (the west slope of Bagongshan Hill, Shouxian County) bears the mega-algae *Vendotaenia, Chuaria,* and *Tawuia,* and worm-like fossils (*Sinosabellidites*). In the Upper Sinian Series, the Jiuliqiao Formation bears worm-like fossils (*Pararenicola* and *Protoarenicola*) and mega-algae (*Chuaria* and *Tawuia*). The Shijia Formation yields *Pararenicola, Calyptrina* and *Paleolina* have been found at the top of the series, in the Jinshanzhai Formation (Xing 1984).

2.2.4 EASTERN LIAONING

The Sinian strata in this region are similar to those in the northern Anhui–north-western Jiangsu region in stratigraphic sequences and sedimentary characteristics. The Changlingzi Formation, corresponding to the upper part of the Lower Sinian Series, bears annelids (*Paleolina* and *Sabellidites*) and macroalgae (*Chuaria, Tawuia,* and *Shouhsienia*) (Chen *et al.* 1982*a*; Xing *et al.* 1985). The so-called 'medusoid' from the Nanguanling Formation, originally discovered at the bottom of the Upper Series, has now been proved to belong to certain sedimentary structures formed by the escaping of water and vapour during diagenesis, and is thus of inorganic origin and not a fossil (Chen 1984; Chen *et al.* 1982*b*). However, the medusoid fossils found in the Xingmincum Formation and Getun Formation of the Upper Sinian Series in 1987, namely *Liaonanella, Jinxianaella,* and *Daliania*, are new members of the Ediacaran fauna (Hong *et al.* 1987). Recently, Zhang (1988) reconfirmed the Ediacaran biota in eastern Liaoning.

2.2.5 SOUTHERN SHAANXI

In 1975 it was reported that *Sabellidites* was found in pelitic siltstone of the Gaojiashan Formation, the Upper Series of the Sinian system in Xuanjiangping and Luojiahe, Ningqiang County (Chen *et al.* 1975). This was the first report of soft-bodied metazoans in China. Recently, Zhang (1986) discovered another type of worm fossil, another trace fossil, and possibly also *Chuaria?* and *Vendotaenia?*

2.2.6 SOUTHERN ANHUI

In 1981, Bi Zhiguo found number of mega-algae in shales of the Lantian Formation in the lower part of the Upper Series of the Sinian system in Xiuning and Yixian County. They were identified as *Epiphyton, Gordonophyton, Tyrasotaenia, Chuaria,* and *Shouhsienia* (Xing *et al.* 1985).

2.2.7 WESTERN HUNAN

Sabellidites occurs in the flint beds intercalated with shales at the upper part of the Liuchapu Formation, Upper Sinian Series, in the Shengtianping phosphorous ore area, Yuanlin County, Hunan (Tang *et al.* 1978). However, many of the specimens may belong to the Chaetopteridae.

2.2.8 EASTERN YUNNAN

Fossils of macroalgae, including *Chuaria, Shouhsienia, Tawuia* and *Vendotaenia,* have been discovered in the Jiucheng Member, Yuhucun Formation, in the upper part of the Upper Sinian Series in Wangjiawan and Meishucun, Kunyang County (Xing 1984).

2.2.9 QUANJI MOUNTAIN OF QINGHAI

The Sinian strata of Quanji Mountain, at the northern margin of the Qaidam Basin, belong to the Quanji Group. At its top, the thin-layered siltstone bears sabelliditid annelids and trace fossils (Wang *et al.* 1980).

2.2.10 HEILONGJIANG

In Longshantun and Mashan of Jixi city, a number of 'metazoans' have been reported from the Mashan Group, such as *Arumberia, Jixiella, Glaessnerina,* and *Mashania* (Liu 1981). These occur in clastic diopsidites, and were recorded by Glaessner (1984) as belonging to the Erniettidae. However, Chen Menge *et al.* (1982*b*) argued that it is incredible for these fossils to have been preserved under such a strong metamorphism. The same surface structures of the most abundant '*Mashania*' could be seen even after being desquamated layer by layer. Such a phenomenon should be more appropriately explained as a result of metamorphism of the rocks.

2.2.11 HAINAN ISLAND

A very recent discovery of the Pre-Cambrian metazoans and mega-algae has been reported but not yet published.

2.3 SUMMARY

Chinese geologists engaged in the late Pre-Cambrian during the past two decades have found and verified many fossils in Sinian strata, including both soft-bodied metazoans and many micro- and macroalgae. Considerable progress has been made, but the researches on metazoans remain at the descriptive stage. There are still insufficient materials available to allow us to draw up a scheme for palaeobiogeographical provincialization. The following are merely some brief ideas on this subject.

The Sinian macroalgae in our country mostly belong to vendotaenids and chuariids. Vendotaenids have been found in both North and South China (Yangtze Platform). There are different opinions concerning the taxonomic position of the chuariids, with a majority taking them as planktonic algae. In North China they occur mainly in the Liulaobei Formation, central Anhui (the bottom boundary of which is dated 840 ma), and in the Changlingzi Formation, eastern Liaoning (the lower boundary is dated 723 ± 43 ma). On the other hand, in South China, *Chuaria* occurs essentially at the top of the Dengying Formation, eastern Yunnan, and in the Lantian Formation, southern Anhui. Hence the macroalgae in North and South China are different in stratigraphical horizons rather than in taxonomy.

As regards the soft-bodied metazoans, the worm fossils from central Anhui (Huainan) have not been seen in other parts of the world. Chen and Zheng (1986) pointed out that the Huainan fauna is obviously simpler in composition and smaller in size than the Ediacaran fauna and should belong to a more primitive biota. Among them, the metazoans from the Liulaobei Formation are thought to be

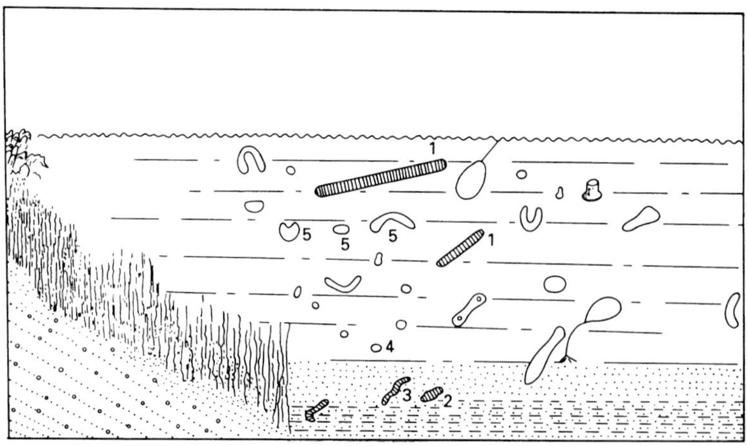

Fig. 2.2 Reconstruction of Xuzhou–Huainan biota (700–900 ma) (after Chen and Zheng 1986). 1, *Sinosabellidites*; 2, *Pararenicola*; 3, *Protoarenicola*; 4, *Chuaria*; 5, *Tawuia*.

inarticulate or primitively articulate, suspension-feeding plankton or nekton; whereas the metazoans from the Jiuliqiao Formation are real worms, beginning to show diversity in life-style, with obviously articulated bodies, deposit-feeding, and with a small proboscis and a coelom.

Consequently, the metazoans of North China are quite different from the Ediacaran fauna in composition and stratigraphical horizon, and belong to a pre-Ediacaran fauna, represented by the *Pararenicola* worm assemblage. The metazoans of the Yangtze Platform in South China belong to the Ediacaran fauna, represented by the assemblage of *Paracharnia*, some worm tubes, and ruganodosariids. The divergence between them seems clear enough to serve as a palaebiogeographical boundary between the North China Platform and Yangtze Platform. However, Zhang (1988) argued that the metazoan-bearing strata of Liulaobei Formation and its equivalents in eastern Liaoning correspond to the Jiuliqiao Formation and other Ediacaran strata both in fauna and in isotopic age, so this problem remains to be solved (Fig. 2.2).

REFERENCES

Chen Menge (1984). A discussion about the "Medusa Fossils" from Wuhanshan Group of Late Precambrian, Lianodong Peninsula, China. *Scientia Geologica Sinica* **1**, 51–7 (in Chinese).

Chen Menge and Zheng Wenwu (1986). On the Pre-ediacaran Huainan biota. *Scientia Geologica Sinica* **3**, 221–31 (in Chinese).

Chen Menge, Chen Xianggao, and Lao Qiuyuan (1975). An introduction to the Metazoa fossil from the Upper Sinian System in Southern Shaanxi and its stratigraphic significance. *Scientia Geologica Sinica* **2**, 181–90 (in Chinese).

Chen Menge, Chen Yiyuan, and Qian Yi (1981). Some tubular fossils from Sinian–Lower Cambrian boundary sequences, Yangtze Gorges. *Bulletin of the Tianjin Institute of Geology and Mineral Resources, Chinese Academy of Geological Sciences* **3**, 117–27 (in Chinese).

Chen Menge, Lin Weixing, Zhang Pifu, Yang Sen, and Bu Dean (1982*a*). Discovery of Sabellinitidae form fossils from Gaojiatan Formation of Sinian system, Liaodong Peninsula. *Scientia Geologica Sinica* **3**, 339–40 (in Chinese).

Chen Menge, Liu Hungyun, Sha Chinan, and Lao Qiuyuna (1982*b*). A discussion of some problems in current Precambrian palaeontologic research of China. *Geological Review* **28** (5), 461–6 (in Chinese).

Ding Qixiu and Chen Yiyuan (1981). Discovery of soft metazoans from the Sinian System along Eastern Yangtze Gorges, Hubei. *Earth Science—Journal of Wuhan College of Geology* **2**, 53–7 (in Chinese).

Glaessner, M. F. (1984). *The dawn of animal life*. Cambridge University Press, Cambridge.

Hong Zhoumin, *et al.* (1987). The discovery of the medusa-like fossils from the Sinian Xingmincun and Getun Formations in Southern Liaoning. In *Abstracts of the International Symposium of Precambrian and Cambrian Geology*, (ed. Wang X., Ma G., Xiang L., Xing Y.), pp. 33–4 (in Chinese).

Liao Shifan (1958). On the edge of the Dengyin Limestone in light of the discovery of hyolithids in it. *Geological Review* **8**, (3), (in Chinese).

Liu Xiaoliang (1981). Metazoa fossils from Mashan Group near Jixi, Heilongjiang. *Bulletin of the Chinese Academy of Geological Sciences* **1**, (1), 71–83 (in Chinese).

Sun Weiguo (1986). Late Precambrian pennatulids (sea pens) from the Eastern Yangtze Gorges, China: *Paracharnia* gen. nov. *Precambrian Research* **31**, 361–75.

Sun Weiguo, Wang Guixiang, and Zhou Benhe (1986). Macroscopic worm-like body fossils from the Upper Precambrian (900–700Ma.) Huainan district, Anhui, China and their stratigraphic and evolutionary significance. *Precambrian Research* **31**, 377–403.

Tang Tianfu, Zhang Junming, and Jiang Xianjian (1978). Discovery and significance of the Late Sinian fauna from western Hunan and Hubei. *Acta Stratigraphica Sinica* **2**, (1), 32–44 (in Chinese).

Wang Hongzhen (1941). Preliminary notes on the Tsongyichun Phosphate Ore, Kunyang County, Yunnan. *Bulletin of the Geological Society of China* **21**, 12–29.

Wang Yunshan, Zhuang Qingxing, Shi Congan, Liu Jiyang, and Zhang Liangshi (1980). Quanji Group along the northern border of Chaidamu Basin. In *Sinian Suberathem in China*, (ed. Tianjin Institute of Geology and Mineral Resources). Tianjin Science and Technology, Tianjin, China, pp. 214–30 (in Chinese).

Xing Yusheng (1984). The Sinian System and its position in the geological time scale. *Scientific papers on Geology for International Exchange* **1**, 37–47 (in Chinese).

Xing Yuxheng, Duan Chenghua, Liang Yuzuo, and Cao Renguan (1985). Late Precambrian palaeontology of

China. *Geological Memoirs, Series B* **2**, 182–92 (in Chinese).

Zhang Luyi (1986). A discovery and preliminary research of late Late Sinian Gaojiashan biota in Ningqiang County, Shanxi. *Bulletin of Xi'an Institute of Geology and Mineral Resources, Chinese Academy of Geological Sciences* **13**, 68–88 (in Chinese).

Zhang Peifu (1988). The Ediacaran biota in Eastern Liaoning Province. *Bulletin of the Tianjin Institute of Geology and Mineral Resuources, Chinese Academy of Geological Sciences* **20**, 110–25 (in Chinese).

Zhang Yun (1989). Multicellular thallophytes with differentiated tissues from Late Proterozoic phosphate rocks of South China. *Lethaia* **22**, (2), 113–32.

Zhao Ziqiang, Xing Yusheng, Ma Guogan, and Chen Yiyuan (1985). *Biostratigraphy of the Yangtze Gorge area, Part 1, The Sinian*. Geological Publishing House, Beijing (in Chinese).

Zhu Weiqing and Chen Menge (1984). On the discovery of macrofossil algae from the Late Sinian in the Eastern Yangtze Gorges, South China. *Acta Botanica Sinica* **26**, 558–60 (in Chinese).

3 CAMBRIAN

Yang Jialu

3.1 CAMBRIAN PALAEO-BIOGEOGRAPHICAL PROVINCIALIZATION OF THE WORLD

The inaccuracy of palaeomagnetic data for the Cambrian period, which is dated about half a billion years ago, means that the knowledge of palaeobiogeography, palaeoclimatology, and geotectonics must be applied to its plate reconstruction. Scientists (Zoninshain *et al.* 1977; Ziegler *et al.* 1979; Scotese *et al.* 1979, 1990; Palmer 1981; Lin Jinlu *et al.* 1985) consider that most Cambrian plates were located in low latitudes and only a few of them may have reached middle latitudes. All the continents, except Gondwana, were small and scattered, although further studies may substantiate the further dismembering of an originally integral plate or reintegration of plates. Figure 3.1 shows the palaeobiogeographical provincialization of the world, using the base map of reconstruction of Franconian age (Late Cambrian) compiled by Scotese *et al.* (1979, 1990) plus the following modifications on the location of some plates:

1. Separation of the Sino-Korean plate from the Yangyze plate, and its removal to lower latitudes.
2. Disintegration of the Tarim–Sino-Korean plate into Sino-Korean plate and Tarim. During the Cambrian, Tarim and its western extension—Kara Kum—were integrated with the southern and western parts of the Kazakhstan plate to the north and Iran plate to the south, thus composing a greater plate. In Fig. 3.1, in order to correspond with the succeeding periods, only the Tarim itself is presented.
3. The Kazakhstan plate is divided into a northern part and a southern plus western part, and its northern part is much nearer to the Siberian plate than to its southern plus western part.

Cambrian plates were sited at similar latitudes, so the global latitudinal climate zonation had little palaeobiogeographical significance during this period. The global realms were separated by oceans and deep seas.

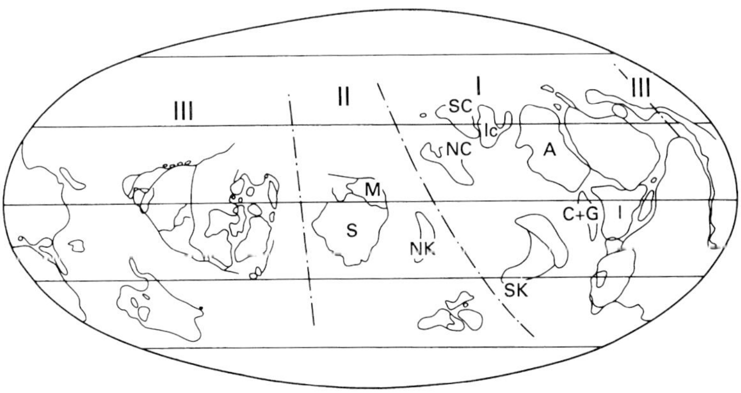

Fig. 3.1 Map of the Cambrian palaeobiogeographical provincialization of the world. I, Asia–Australian realm; II, Siberian realm; III, North America–Atlantic realm; A, Australia; Ic, Indochina; M, Mongolia; NC, North China; NK, North Kazakhstan; C+G, Qiangtang and Gangdise, S, Siberia; SC, South China; SK, South Kazakhstan.

Three realms have been established, mainly based on trilobite distributions (Cowie 1971; Kobayashi 1971; Jell 1974 Lu *et al.* 1974; Ergaliev 1980; Palmer 1981; Wittke 1984). The Asia–Australian realm included the western and southern parts of Kazakhstan, Iran, Chinese territory south of the Ebinur–Gaxunnur–Xar Moron line, South-East Asia, Australasia, and Antarctica. They were scattered along the western and north-western sides of Gondwana. Its Early Cambrian was characterized by the flourishing of Redlichiina and absence of Olenellina. The Middle and Late Cambrian were dominated by Damesellacea, Dolichometopidae, Ordosiidae, and Leiostegiidae, representatives of which include *Chuangia*, *Chuangioides*, *Chuangiella*, *Eochuangia*, *Iranochuangia*, *Chuangina*, *Idamea*, plus *Charchaqia*, *Hedinaspis*, and *Westergaardites*. The North America–Atlantic realm included South and North America, eastern, western, and north-western Europe, distributed along the eastern and south-eastern sides of Gondwana. Its Early Cambrian was characterized by the flourishing of Olenellina and the absence of Redlichiina, and the Middle and Late Cambrian by endemic genera of Pterocephaliidae, Richardsonellidae, and Catillicephalidae, as well as a few nektonic polymers, such as certain genera of Paradoxididae and Olenidae. The Siberian realm occupied a vast region east of the Urals and north of Lake Balkash and the Ebinur–Gaxunnur–Xar Moron line, including north Kazakhstan, Junggar, Mongolia, north north-eastern China, and Siberia itself. It extended eastward to Alaska and westward, inserting between the two previous realms, to Spain and Morocco. It shows mixed features of the above two realms plus a few endemic groups, such as the Early Cambrian Saukiandidae, Resserpsidae, Abadiellidae, Jakutidae, Edelsteinaspidae, and Fallotaspinae.

The territory of China spans two realms. Areas north of the Ebinur line occupy a corner of the Siberian realm, and areas to the south of this line belong to the main part of the Asia–Australian realm, where trilobites, bradoricopids, and 'small shells' are in great abundance and diversity. They occurred in these Asia–Australian areas earlier than in other regions, so these areas are considered to be their source regions.

The replacement of Crypterozoic soft-bodied biota by Phanerozoic shelled ones (trilobites, 'small shells', bradoricopids, archaeocyathids, and inarticulate brachiopods) was a great leap in biological evolution. Most of the Early Cambrian biotas belonged to the benthos and, judging from the sedimentary facies, mainly inhabited the shallow shelf waters. The deep oceans served as barriers to separate the realms. On the other hand, in Middle and Late Cambrian, plankton and nekton trilobites appeared. They were able to disperse from shallow seas to deep water, to occupy new ecological habitats. Because these migratory genera and species blurred the biogeographical boundaries to a certain extent, care should be taken to avoid the influence of ecological effects in provincialization of high-rank palaeobiogeographical units, so it is necessary to emphasize the significance of the benthos instead of plankton and nekton.

3.2 CAMBRIAN PALAEOBIOGEOGRAPHICAL PROVINCIALIZATION OF CHINA

3.2.1 INTRODUCTION

The great event of this period was the sudden occurrence of marine shelly invertebrates. There was a great increase in the diversity and abundance of the Cambrian biota compared with the Sinian. At the beginning of the Early Cambrian, many groups of 'small shells' suddenly appeared and then declined; to be followed by the successive appearance of inarticulate brachiopods, archaeocyathids, and more advanced arthropods, such as bradoricopids, trilobites, and isopods. In the Middle and Late Cambrian, echinoderms, nautiloids, and dendroid graptolites of the advanced phyllum Chordata appeared, as well as conodonts (whose taxonomical position is still in question). Among the numerous groups of organisms, the dominant one is the Trilobita. In its adaptive radiation it occupied the neritic, bathyal, and upper abyssal zones and became the overwhelming group. Because of its swift evolution and widespread distribution, it serves as the basis of both biostratigraphy and palaeobiogeographical provincialization (Zhang *et al.* 1980; Lu *et al.* 1982). In this chapter emphasis is laid on the distribution patterns of trilobites, with a moderate consideration of other organisms.

The base of the Lower Cambrian is delineated at a horizon within the Zhongyicun Member, Meishucun Formation, Kunyang Country, Yunnan (Table 3.1), i.e. between the small shell *Paragloborilus–Siphogonuchites* zone and the underlying *Anabarites–Circotheca* zone (Luo *et al.* 1982). The boundary of the Middle and Lower Series lies between the Xuzhuangian and the Longwangmiaoan stages. The same boundary in eastern Yunnan is set within the Douposi Formation, between the *Sinoptychoparia–Kütsingocephalus* zone (upper) and the *Chittidilla–Kunmingaspis* zone (lower). In North China it is delineated between the Xuzhuang Formation and the Maozhuang Formation, i.e. between the *Hsüchuangia–Ruichengella* zone (upper) and *Shantungaspis* zone (lower). The topmost boundary of the Cambrian is defined as between the end of the trilobite *Mictosaukia* zone and conodont *Cordylodus proavus* zone of the Fengshan Formation in North China, and the beginning of the trilobite *Missiquoia perpetis* zone and conodont *Cordylodus intermedius* zone of the Yeli Formation. Cambrian biogeographical regions are of little significance, as discussed in Chapter 1, the useful palaeobiogeographical units are realm, province, and biome, the latter being further divided into communities. Apart from brachiopods, middle and late biotas are quite different from those in the Early Cambrian. Although Late Cambrian trilobites are different from those of the Middle Cambrian at generic, family, and even superfamily levels, their distribution pattern is inherited from the latter. Thus these two are put together for consideration, and only two palaeobiogeographical maps have been compiled—for the Early Cambrian and for the Middle and Late Cambrian, respectively—to show the palaeobiogeographical provincialization and its evolution.

3.2.2 EARLY CAMBRIAN PALAEOBIOGEOGRAPHICAL PROVINCIALIZATION OF CHINA

Early Cambrian trilobites came into being at the Qiongzhusi Stage in the Yangtze shallow sea. Representatives of four orders occurred in succession. The earliest one is the family Pagetiidae of the Eodiscina (Wuding County, Yunnan; Emei Mountain Sichuan; northern Guizhou and the Yangtze Gorges) which thrived in the Qiongzhusi Stage, during which time Eodiscidae of the same order had just a few representatives, appearing only in the Beishan area of Gansu. Members of the Abadiellidae and Redlichiidae appeared a little later than, or nearly simultaneously with, the Pagetiidae, and *Redlichina* soon began to thrive and expand into other shallow waters (Tarim shallow sea, North China shallow sea), declining at the Longwangmiaoan Stage. The early representatives of Corynexochida (e.g. Balangiidae, Cheiruroididae, and Oryctocephalidae) originated mainly in offshore shallow seas (Balang Formation in the border area of western Hunan and eastern Guizhou) or outer shelf to more-or-less bathyal zone (Yangjiaping Formation in the border of western Hunan and eastern Guizhou; Hetang Formation and Shuijingtuo Formation in Yangtze Gorges, and became the dominant members in the Canglangpuan. *Dinesus* (=*Erbia*) of Dinesidae is an important member of the Early–Middle Cambrian fossil record in Kazakhstan, Siberia, and Queensland (Australia). It has been found in the Beishan area of Gansu, Xichuan County of Henan and Junxian County of Hubei. Another member of the same family, *Tingyuania*, has been found in Anhui and northern Jiangsu. Not until late in the Canglangpuan Stage did the Ptychopariacea appear, and this group began to thrive in the Longwangmiaoan Stage. Subsequently, it replaced the other trilobites and became the dominant group, flourishing not only in Yangtze but also in North China. Archaeocyathids, although a shallow-sea benthic group, can be found not only in shallow seas of continental plates, but also in volcanic arc areas. Many of them were cosmopolitan, and endemic genera were few. While archaeocyathids appeared early in Siberia, in China they occurred mainly in the Canglangpuan Stage and are represented only by Septoidae and Taenioidae. Because of their limited and uneven distribution, they serve as a secondary factor in the analysis of palaeobiogeographical provincialization (Zhang Sengui 1982, 1983; Yuan and Zhang 1980). Small shelly faunas, or the tiny shell-bearing

Table 3.1 Correlation chart of the Cambrian of China

<table>
<tr><th colspan="3" rowspan="2"></th><th colspan="5">South China province</th></tr>
<tr><th>Eastern Yunnan</th><th>Northern Guizhou</th><th>Western Hubei</th><th>Danzhai, Guizhou</th><th>Western Hunan–eastern Guizhou</th></tr>
<tr><td rowspan="11">Cambrian</td><td rowspan="3">Upper</td><td>Fengshanian</td><td rowspan="4"></td><td rowspan="4">Loushanguan Group</td><td rowspan="3">Sanyoudong</td><td rowspan="4">Lushan Group</td><td rowspan="3">Tongren Dolomite Member
Tingziguan</td></tr>
<tr><td>Changshanian</td></tr>
<tr><td>Gushanian</td></tr>
<tr><td rowspan="2">Middle</td><td>Zhangxiaan</td><td rowspan="3">Qinjiamiao</td><td>Laochatian</td></tr>
<tr><td>Xuzhuangian</td><td>Shuanglongtan</td><td>Shilengshui</td><td>Jialao</td><td>Baojing</td></tr>
<tr><td rowspan="6">Lower</td><td rowspan="2">Longwangmiaoan</td><td>Douposi</td><td>Gaotai</td><td>Kaili</td><td>Nadong</td></tr>
<tr><td>Longwangmiao</td><td>Qingxudong</td><td>Shilongdong</td><td>Wuxun</td><td>Qingxudong</td></tr>
<tr><td rowspan="2">Canglangpuan</td><td rowspan="2">Canglangpu</td><td>Jindingshan</td><td>Tianheban</td><td>Palang</td><td>Palang</td></tr>
<tr><td>Mingxinsi</td><td>Shipai</td><td>Bianmachong</td><td>'Mingxinsi'</td></tr>
<tr><td>Quiongzhusian</td><td>Qiongzhusi</td><td>Niutitang</td><td>Shuijingtuo</td><td rowspan="2">Jiumenchong</td><td rowspan="2">Yangjiaping</td></tr>
<tr><td>Meishucunian</td><td>Meishucun</td><td>Dayan Member</td><td>Tianzhushan Member</td></tr>
</table>

metazoan groups which flourished for a very short time, are mainly restricted to the Yangtze epicontinental sea and the Tarim shallow sea, thus playing only a supplementary role in palaeobiogeography.

The trilobites in the Early Cambrian were at their primary stage of evolution. Their ecological patterns were also monotonous; most of them were benthic. There were not yet open-sea plankton and nekton. Of course, further analysis of the benthos will reveal that some of them may have lived in lower parts of shallow seas, and a few may have been able to migrate into an even deeper, slightly reducing environment.

The Chinese part of the Asia–Australian realm can be divided into three provinces, i.e. South China province, Tarim province, and North China province (Fig. 3.2). The South China province is characteristic of abundant and highly diversified trilobites, small shells, and bradoricopids, as well as numerous archaeocyathids and brachiopods (Table 3.2). The important trilobites include *Chengkouia*, *Hunanocephalus*, *Yinites*, *Eoredlichia*, *Diandongaspis*, *Sinodiscus*, and *Tsunyidiscus* (Table 3.3).

The South China province can be subdivided into three biomes (Table 3.2). The first one is the Yangtze biome which is the most representative of the South China province (Zhou and Yuan 1980; Luo 1984*a*; Zhang *et al.* 1980). Besides the abundant 'small shells', bradoricopids, and archaeocyathids, the trilobites especially occur in great abundance and diversity (Fig. 3.3, Tables 3.2, 3.3), and this area is thought to be one of their centres of origin (Lu *et al.* 1974). Representatives include *Eoredlichia*, *Yunnanocephalus*, *Pachyredlichia*, *Wangzishia*, *Maopingia*, *Duotingia*, *Mianxiandiscus*, and *Yinites*. This is a typical neritic benthic biome, except that the eastern part of the Upper

<table>
<tr><th></th><th colspan="3">Tarim province</th><th colspan="3">North China province</th><th>Central Siberia province</th></tr>
<tr><th>Western Zhejiang</th><th>Kalping, Xinjiang</th><th>Kuruktag, Xinjiang</th><th>Beishan, Gansu</th><th>Jinxi, Liaoning</th><th>Northern Qaidam</th><th>Lajishan, Qinghai</th><th>Yichun, Heilongjiang</th></tr>
<tr><td>Xiyangshan</td><td rowspan="5">Awatag Group</td><td rowspan="3">Tursaktag Group</td><td rowspan="3">Mazongshan</td><td>Fengshan</td><td rowspan="3">Upper Olonbulag</td><td rowspan="3">Liudaogou Group</td><td rowspan="9">?
Chenming
?</td></tr>
<tr><td rowspan="2">Huayansi</td><td>Changshan</td></tr>
<tr><td>Gushan</td></tr>
<tr><td rowspan="2">Yangliugang</td><td rowspan="2">Mohe'ershan Group</td><td rowspan="2">Xishuang-yingshan</td><td>Zhangsia</td><td rowspan="2">Lower Olonbulag</td><td rowspan="6">Nidanshan Group
?</td></tr>
<tr><td>Xuzhuang</td></tr>
<tr><td rowspan="2">Dachenling</td><td rowspan="2">Sayilik</td><td rowspan="3">Xidashan</td><td rowspan="4">Shuang-yingshan</td><td>Maozhuang</td><td rowspan="4">Xiaogaolu</td></tr>
<tr><td>Mantou</td></tr>
<tr><td rowspan="2">Hetang</td><td>Usugar</td><td>Jianchang</td></tr>
<tr><td>Xorbulak</td><td>Xishanbulake</td><td></td></tr>
</table>

Table 3.2 Number and distribution of the Early Cambrian fossil genera from China

	Asia–Australian realm										Siberian realm
	S. China province						Tarim province		N. China province		C. Siberia province
	Yangtze biome			W. Hunan–Xichuan biome							
Taxa	UP	SEY	N	WH	X	WZ	K	B	NS	SM	HM
Trilobites	122	8	11	39	11	9	18	8	49	18	3
Archaeocyathids	18	1	0	0	0	0	3	0	0	0	5
Bradoricopids	15	0	0	0	0	0	3	0	0	0	0
Small shell fossils	114	0	2	0	0	0	23	1	4	18	0
Brachiopods	24	0	4	3	1	3	8	0	10	8	0
Total	293	9	17	42	12	12	55	9	63	26	8

UP, Upper Yangtze; SEY, south-eastern Yunnan; N, Nanjing; WH, western Hunan–southern Anhui; X, Xichuan; WZ, western Zhejiang–central Hunan biome; K, Kalping biome; B, Beishan biome; NS, North China–southern north-eastern China biome, SM, southern marginal North China biome; HM, Hinggan–Mangol biome Yichun–Horqin Youyi Qianqi.

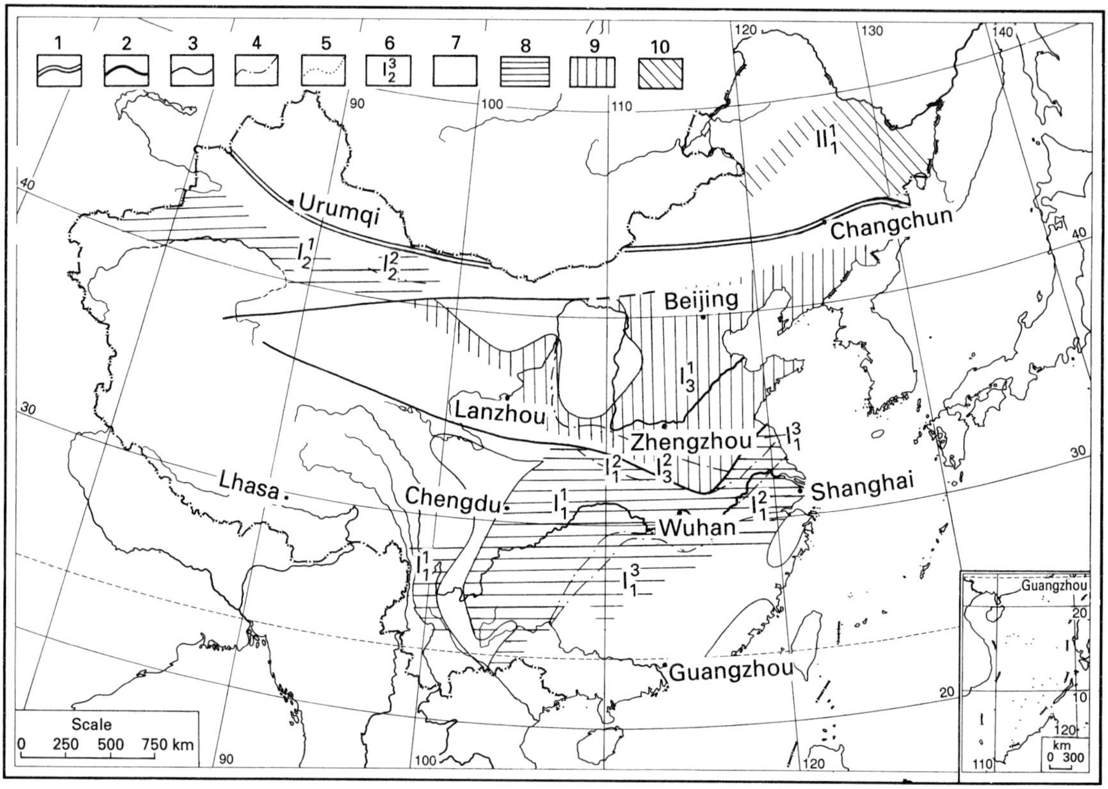

Fig. 3.2 Map of the Early Cambrian palaeobiogeographical provincialization of China. 1, Boundary of realms; 2, boundary of regions; 3, boundary of provinces; 4, boundary of biomes; 5, boundary of communities; 6, roman digit = realm mark, subscript = province mark, superscript = biome mark; 7, Oldland. 8–10, Marine and paralic regions: 8, South China; 9, North China; 10, Junggar–Xingan. I, Asia–Australian realm; I_1, South China province; I_1^1, Yangtze biome; I_1^2, Western Hunan–Xichuan biome; I_1^3, Western Zhejiang–Central Hunan biome; I_2, Tarim province; I_2^1, Keping (Kalpin) biome; I_2^2, Beishan biome; I_3, North China province; I_3^1, North China and Southern North-eastern China biome; I_3^2, south margin of the North China biome; II, Siberian realm; II_1, Central Siberia province; II_1^1, Xingan–Mongolia biome.

Yangtze Sea may have been deeper, with a submerged outer shelf. The second biome is the Western Hunan–Xichuang biome, stretching alongside the southern and northern flanks of the Yangtze biome. The southern flank extends from western Hunan and southern Anhui to the Suzhou–Hangzhou area in the east. The northern flank now only comprises the Zhen'an–Xichuang belt in the Qinling Geosyncline. Diversified triobites, such as *Changaspis*, *Balangia*, *Protoryctocephalus wuxunensis*, and *Hunanocephalus*, are found here, but there are no 'small shells', bradoricopids, or archaeocyathids. The third biome is called the Western Zhejiang–Central Hunan biome, extending from central Hunan to western Zhejiang and Chuxian County, Anhui. Here a deeper and slightly reducing environment resulted in the occurrence of monotonous and less numerous trilobites, characterized by the *Arthricocephalus–Hunanocephalus* assemblage.

Organisms of the Tarim province were much less diversified than those of the South China province. Yet the fossil records of both provinces are quite similar. They yield closely related taxa, and two-

Table 3.3 Comparison of the Early Cambrian trilobite families, subfamilies, and genera from South China province with those from North China province

S. China province	N. China province
Eodiscida: *Hupeidiscus*, *Mianxiandiscus*	
Metaredlichiinae: *Metaredlichia*, *Breviredlichia*, *Pseudoredlichia*, *Metaredlichioides*	
Pararedlichiinae: *Eoredlichia*, *Pachyredlichia*	
Wutingaspinae: *Wutingaspis*, *Chaoaspis*	
Abadiellidae: *Parabadiella*	
Dolerolenidae: *Malungia*, *Paramalungia*	
Yinitidae: *Drepanopyge*, *Drepanuroides*, *Yinites*	
Balangidae: *Balangia*	
Yunnanocephalidae: *Yunnanocephalus*	
Chengkouiidae: *Chengkouia*, *Xiugiella*	
Dolichometopidae: *Hoffetella*	
Oryctocephalidae: *Changaspis*, *Protoryctocephalus*	
Feilongshanidae: *Arthricocephalus*, *Feilongshania*	
Redlichiinae: *Redlichia*, *Syndianella*	*Redlichia*
Palaeolenidae: *Palaeolenus*, *Megapalaeolenus*	*Palaeolenus*, *Megapalaeolenus*
Ptychopariidae: *Yuehsienszella*	*Shantungaspis*, *Probowmania*, *Chengshanaspis*, *Psilostracus*
Protolenidae: *Protolenella, Ichangia*, *Hsüaspis*, *Wangzishia*, *Bergeroniellus*	*Hsüaspis* (*Madianaspis*), *Bergeroniellus*
Cheiruroideidae: *Cheiruroides*, *Shabaella*, *Hunanocephalus*	*Cheiruroides*
Dinesidae: *Dinesus* (= *Erbia*)	*Tingyuania*
	Zacanthoididae: *Prozacanthoides*
	Tengfengdaeii: *Luaspides*

thirds of the genera and species are identical. The difference lies in the smaller diversity of trilobites and a few local members (such as *Kepingaspis* and *Tannudiscus*) in Tarim. The Tarim province can be divided into two biomes. The Kalping biome is typically benthic, as shown by the fossils found in Kalping and Kuruktag. Representatives of the trilobites include *Kepingaspis*, *Tianshanocephalus*, and *Pseudoprotolenella*. In the Beishan biome there are some elements of the Opidiscidae and Pagetiidae, e.g. *Serrodiscus*, *Tannudiscus*, and *Calodiscus*, which show aspects of the North America–Atlantic and Siberian realms. Such an assemblage probably reflects a relatively deep sea environment.

The North China province is characterized by the absence of archaeocyathids and bradoricopids. 'Small shells' are few; the number of trilobite genera and subgenera is less than half of that in the South China province, and there are few identical families and genera between the two (Table 3.3). Typical trilobites are *Shantungaspis*, *Psilostracus*, *Probowmania*, *Luaspides*, *Prozacanthoides*, and *Tingyuania*, reflecting a shallow shelf environment. Two biomes have been identified: one covers North China and the southern part of north-eastern China, the other the southern margin of North China. The former is characterized by '*Bonnia*', *Prozacanthoides* etc. The latter, being the slope of the North China

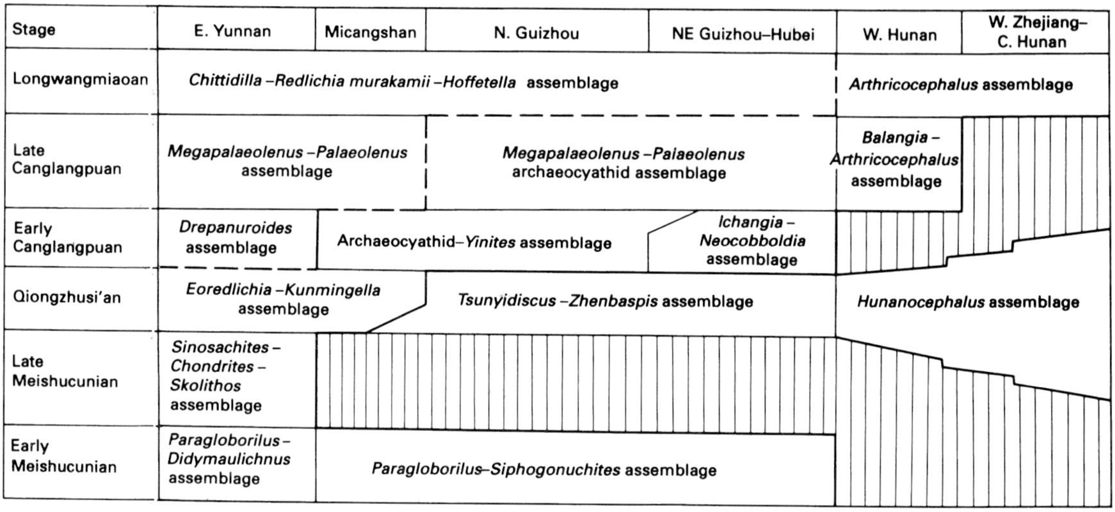

Fig. 3.3 Early Cambrian marine communities (in the appearance of assemblages) of South China. Note: assemblages to the left of the thick dashed line were littoral–neritic and experienced a warm environment; those to the right of the thick solid line experienced deeper and less oxygenated environments; those between the dashed line and the solid line were in normal shallow marine environments.

Platform, hosted a deeper assemblage—*Bergeroniellus*, *Hsüaspis* (*Madianaspis*), and *H.* (*Yinshanaspis*).

The Siberian realm occupies the so-called Northern Geosynclines in China, or areas to the north of Ebinur–Gaxunnur–Xar Moron line. There are very few fossil localities. Archaeocyathid assemblages have been found in Jilin and Inner Mongolia, containing *Ajacicyathus*, *Archaeocyathus*, *Archaeopharetra*, *Ethmophyllum*, *Protopharetra*, and *Syringocyathus*(?), and comparable with those of the Lenian Stage in Siberia. Trilobites, such as *Proerbia* and *Inouyina* found in Wuxinzhen, Yichun County, Heilongjiang, compare closely with those of the Siberian realm.

The cluster analysis of Early Cambrian trilobites (Fig. 3.4) shows that the similarity ratio of the Yichun assemblage to those of the Asia–Australian realm is approximately zero, so that it belongs to another realm—the Siberian realm. Within the Asia–Australian realm, the similarity ratio of the biota in North China with those in South China and Tarim is very low (0.06); they should be attributed to at least different provinces. These results are consistent with the observations mentioned above. On the other hand, the similarity ratio between the Tarim and South China biotas is so high that they could not be separated into two groups.The diversity difference of these two biotas cannot be shown using the clustering methods applied in this book.

In the Yangtze epicontinental sea east of the Xikang–Yunnan Oldland[1], and the open sea southeast of Yangtze, the Early Cambrian palaeogeographical changes in time and space caused regular alterations of the distribution pattern of sediments and biota (Fig. 3.3, Table 3.4). Thus 14 biotic communities (in the appearance of assemblages) are identified. As denoted in Chapter 1, here the word 'community' refers to a succession of interrelated fossil communities found *in situ* in nearby horizons. Detailed discussion of each community cannot be given here; instead, Table 3.4 shows the name of each community, its age and distribution area, its country rock type, ecological type, and environmental interpretation.

[1] Oldland means a terrestrial area in geological times. It is a term used widely in Chinese geology.

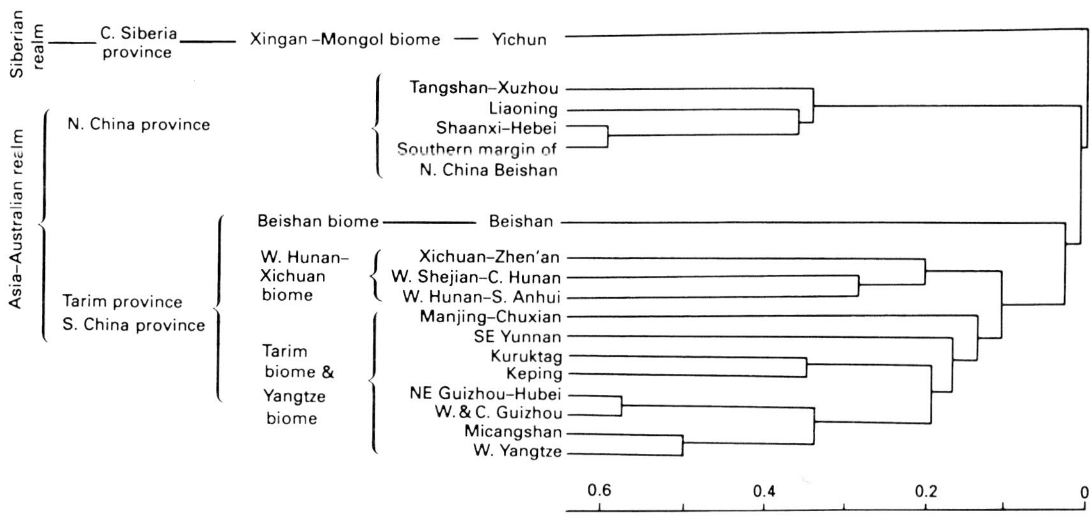

Fig. 3.4 Cluster analysis of the Early Cambrian trilobites of China.

Four preliminary conclusions can be drawn from the features of Early Cambrian communities and from their changes in time and space.

1. From the late Meishucunian Stage to the Longwangmiaoan Stage the normal epicontinental benthic communities extended their distribution, accompanying the normalization of sea-bottom physical and chemical conditions. The main environmental factors influencing this distribution were changes of water depth and the relative content of dissolved oxygen and hydrogen sulphide. Although in the Qiongzhusian Stage benthic communities also spread widely, only the *Eoredlichia– Kunmingella* community to the east of the Xikang–Yunnan Oldland belongs to the normal shallow benthic community; the others, such as the *Tsunyidiscus–Zhenbaspis* and *Hunanocephalus* communities are attributed to a weakly reducing environment, which represents the lower part of the continental shelf or the upper part of the slope.

2. The general relief of the sea bottom was, for a long period, higher in the west and lower in the east, so from northwest to southeast there occurred successive terrestrial, littoral, and neritic zones, the lower part of the shelf and upper part of the slope. The almost identical communities over the vast area of the central Yangtze shallow sea (Guizhou, eastern Sichuan, and Hubei) indicate a flat sea bottom. However, the relief southeastward from western Hunan became steeper, and there the shelf turned into a slope.

3. Judging from the small archaeocyathid reefs in Shaanxi, northern Sichuang, western Hubei, and northern Guizhou of the Yangtze sea basin during the Canglangpuan Stage, the depth of water was about 20–30 m and there was a small amount of terrigenous debris. The climate was probably rather warm.

4. The Early Cambrian of China used to be considered as a period of transgression. While this is true for North China, the situation of Yangtze is more complicated. A tendency toward sea-level lowering can be detected after the diachronous transgression at the beginning of the Early Cambrian. This phenomenon can be testified by the following:

(a) Distribution areas of the normal shallow benthic communities were enlarged from the late Meishucunian to the Longwangmiaoan.

(b) Environmental shifts from the lower part of the shelf or the upper part of the slope into the

Table 3.4 List of the Early Cambrian assemblages (representing communities) of South China

Assemblage Name	Main ecological type	Main lithological type	Environment	Age	Geographical distribution
1. *Paragloborilus–Didymaulichnus*	Free benthics	Sandy dolomite, phosphorite	Carbonate-phosphate platform	E. Mei.	Eastern Yunnan
2. *Paragloborilus–Siphogonuchites*	Free benthics	Collophane dolomite, phosphatic silicalite, siliceous phosphorite, limestone, shale	Tidal flat–stagnant shallow sea	E. Mei.	Upper Yangtze (not including eastern Yunnan)
3. *Sinosachites–Chondrites–Skolithos*	Free benthics	Black siltstone or silty mudrock	Near–shore shelf	L. Mei.	Eastern Yunnan–Emei
4. *Eoredlichia–Kunmingella*	Free benthics	Yellow-green shale and siltstone	Inner shelf	Qiong.	Eastern Yunnan, Emei, NE Sichuan
5. *Tsunyidiscus–Zhenbaspis*	Planktonics, free benthics	Black shale	Outer shelf	Qiong.	Guizhou, Hubei, NE Sichuan
6. *Hunanocephalus*	?	Black carbonaceous shale	Outer shelf–inner slope	Qiong.	C. Hunan–W. Zhejiang, Lower Yangtze, W. Hunan–S. Anhui
7. *Drepanuroides*	Free benthics	Grey-green, yellow-green sandy shale, interbedded by sandstone	Littoral sandy beach	E. Cang.	Eastern Yunnan, NE Yunnan
8. *Archaeocyathid–Yinites*	Attached benthics, free benthics	Muddy archaeocyathid limestone and yellow-green shale	Inner shelf–littoral sandy beach	E. Cang.	Northern Sichuan, northern Guizhou
9. *Ichangia–Neocobboldia*	Free benthics	Grey-green siltstone, shale	Continental shelf	E. Cang.	Hubei
10. *Megapalaeolenus–Palaeolenus*	Free benthics	Grey-yellow sandstone, shale	Littoral sandy beach	L. Cang.	Eastern Yunnan, SE Yunnan, Emei, Micang Hills
11. *Megapalaeolenus–Palaeolenus–Archaeocyathid* (Fig. 3.5)	Free benthics, attached benthics	Archaeocyathid limestone, shale	Smooth continental shelf	L. Cang.	Guizhou, SE Sichuan, NE Sichuan, Hubei
12. *Balangia–Arthricocephalus*	Free benthics	Grey-green, yellow-green calcareous sandstone, shale	Marginal shelf	L. Cang.	Eastern Yunnan, western Hunan
13. *Chittidilla–Redlichia murakamii–Hoffetella*	Free benthics	Dolomite and limestone sandy shale	Tidal flat	Long.	Yangtze
14. *Arthricocephalus*	Free benthics	Limestone, dolomitic limestone	Continental shelf, inner slope	Long.	W. Hunan–S. Anhui, C. Hunan–W. Zhejiang

Mei., Meishucunian; Cang., Canglangpuan; Long., Longwangmiaoan; Qiong., Qiongzhusian; E., Early; L., Late.

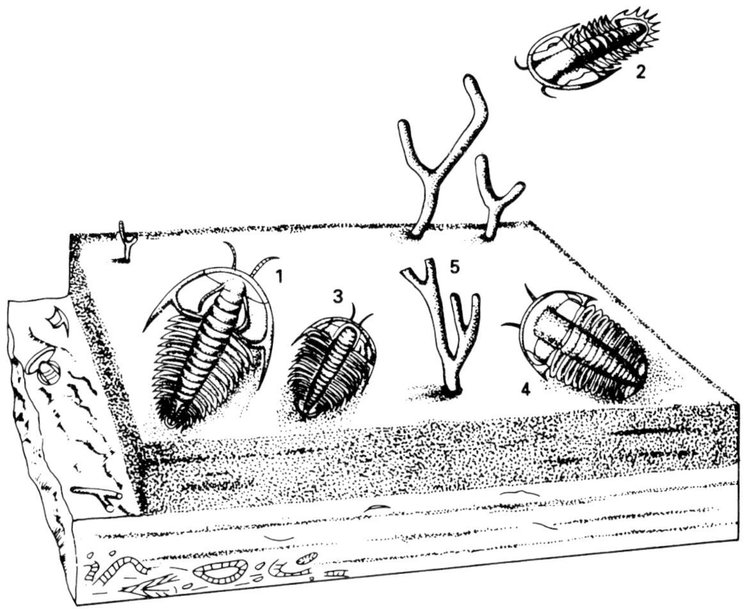

Fig. 3.5 Reconstruction of the late Canglangpuan *Megapalaeolenus–Palaeolenus*–archaeocyathid community of the Early Cambrian of Guizhou, Yangtze biome. 1, *Redlichia*; 2, *Kaotenia*; 3, *Yuehsienszeella*; 4, *Megapalaeolenus*; 5, archaeocyathids.

upper part of the shelf are displayed in a number of localities from the late Meishucunian to the Longwangmiaoan.

(c) The Hannan Oldland in southern Shaanxi was enlarged from the Qiongzhusian to the Longwangmiaoan.

(d) Regressive sequences appeared in the Early Cambrian in eastern Sichuan and western Hunan.

It could be argued that a global ice age might have caused the lowering of sea level. However, according to present data, global ice ages have been recognized only in the Sinian and late Ordovician. There was no such large-scale glaciation in the Cambrian. Another interpretation is the uplift of the Yangtze plate as a whole, which is more plausible.

3.2.3 MIDDLE AND LATE CAMBRIAN PALAEOBIOGEOGRAPHICAL PROVINCIALIZATION OF CHINA.

The general features of the biosphere in this period (Fig. 3.6, Table 3.5) are quite different from those in the Early Cambrian. 'Small shells', bradoricopids, and archaeocyathids became greatly reduced or had become extinct, while dendroid graptolites and nautiloids appeared, as did new trilobites groups. Besides the still abundant benthic trilobites, many pelagic nektonic and planktonic trilobites appeared. At an early stage of the Middle Cambrian the Ptychopariida diversified rapidly, with more than ten new families replacing the Early Cambrian Redlichiina as the dominant shallow benthic group. The Dolichometopidae and Dorypygidae (Corynexochida) were also important. Miomerid planktonic Agnostina replaced the benthic Eodiscina,

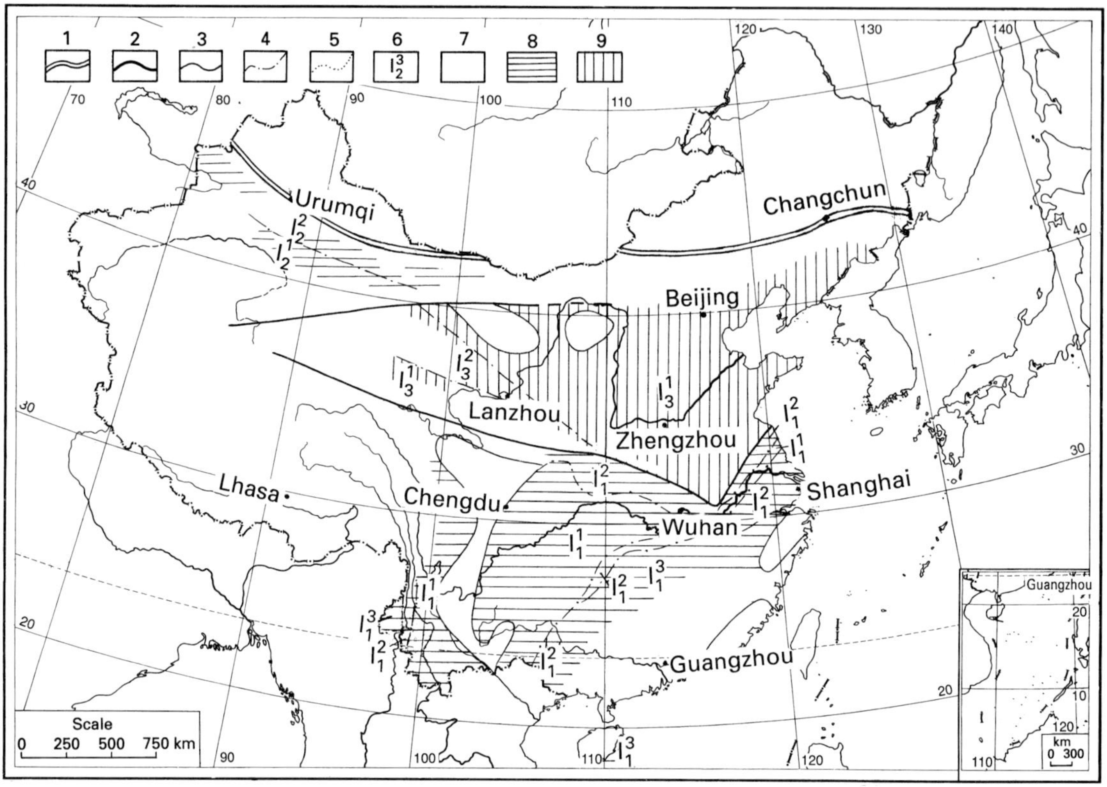

Fig. 3.6 Map of the Middle and Late Cambrian palaeobiogeographical provincialization of China. 1, Boundary of realms; 2, boundary of regions; 3, boundary of provinces; 4, boundary of biomes; 5, boundary of communities; 6, roman digit = realm mark, subscript = province mark, superscript = biome mark; 7, Oldland. 8–9, Marine and paralic regions: 8, South China; 9, North China. I, Asia–Australian realm; I_1, South China province; I_1^1, Yangtze biome; I_1^2, Platform Margin (of Yangtze) biome; I_1^3, Extra-platform (of Yangtze) biome; I_2, Tarim province; I_2^1, Kuruktag–Ejin Qi biome; I_2^2, Huocheng, Beishan, Hang'ula biome; I_3, North China province; I_3^1, North China–Qaidam biome; I_3^2, Qillian biome.

and spread out into deeper waters. From the late Middle Cambrian to the early Late Cambrian important families, such as the Lisaniidae, Ordosiidae, and Damesellidae, occurred. In the middle Late Cambrian the Leiostegiidae, Changshaniidae, Kaolishaniidae, and Pterocephaliidae were important representatives of benthic trilobites, as were the Asaphidae, Olenidae, and Ceratopygidae for pelagic nekton. In the late Late Cambrian the Saukiidae, Ptychaspididae, and Tsinaniidae were important benthic trilobites.

Junggar and the northern part of north-east China lack data for middle and Late Cambrian trilobites and are tentatively attributed to the Siberian realm on the basis of data derived from adjacent areas, notably Kazakhstan, Mongolia, and Xingkai Lake. The rest of China belonged to the Asia–Australian realm, which can be further divided into three provinces, the South China province, Tarim province, and North China province.

A comparison of the trilobite suborders, families, and genera in the provinces of South and North China

Table 3.5 Number and distribution of the Middle–Late Cambrian fossil genera from China

		Oligomerian trilobites	Pliomerian trilobites	Nautiloids	Brachiopods	Others	Total
S. China province							
Yangtze biome	Yangtze-Zhongdian	70	1	1	7	1	80
	SE Yunnan	36	2	0	7	0	45
	Danzhai, Guizhou	26	1	0	0	0	27
Platform Margin biome	S. marginal Yangtze	140	40	6	8	1	195
	N. marginal Yangtze	24	9	0	0	0	33
	W. marginal Yangtze	50	5	0	4	1	60
Extra-platform biome	Sandu–C. Hunan–W. Zhejiang	56	36	12	11	13	128
Tarim province							
Kuruktag-Ejin Qi biome		58	18	0	2	2	80
Huocheng–Beishan–Hangula biome		46	14	1	9	2	72
N. China							
N. China–Qaidam biome	N. China–S. NE China	279	14	32	14	8	347
	Qaidam	22	0	0	0	0	22
Qilian biome		63	18	0	10	0	91

is listed in Table 3.6. There are altogether 30 suborders and families plus numerous genera in common, but each province also has its own unique and dominant families. South China province has 14 endemic and dominant families (the first 14 in the left-hand column of Table 3.6). North China province has eight dominant families (the last eight in the right-hand column of Table 3.6). Each has a large number of endemic genera. Such a prominent difference verifies the existence of two provinces. In the early Middle Cambrian, *Kaotaia* and *Kütsingocephalus* were the characteristic members of the South China province, while *Hsüchuangia*, *Inouyia*, *Lorenzella*, *Metagraulos*, *Poriagraulos*, and *Sunaspis* represented the North China province. In the Middle Cambrian and early Late Cambrian the South China province was characterized by *Bergeronites (Palaeadotes)*, *Chatiania*, *Eoshengia*, *Fenghuangella*, *Jiangnania*, *Neoanomocarella*, *Paracoosia*, *Paradamesella*, and *Paradamesops*, while the North China province was characterized by *Amphoton*, *Damesella*, *Bergeronites*, *Crepicephalina*, *Liaoyangaspis*, *Stephanocare*, and *Taitzuia*. In the middle and late Late Cambrian the difference in trilobites between the two provinces diminished and both were characterized by elements of the Ptychaspididae and Saukiidae. Of course, each province still had a few local members, e.g. *Changia* and *Eochuangia* for the North China province and *Lonchopygella*, *Eosaukia*, and *Shengia* for the South China province. In addition, the latter had some nektonic trilobites, such as *Niobella*, *Acrocephalina*, *Charchaqia*, *Yuepingia*, and *Pseudoyuepingia*.

The South China province consists of three biomes: the shallow benthic Yangtze biome, the mixed-type Platform Margin (of Yangtze) biome, and the nektonic–planktonic Extra-platform (of Yangtze) biome (Table 3.7, Fig. 3.7).

Table 3.6 Comparison of the middle-late Cambrian trilobite families, subfamilies, and genera from South China province with those from North China province

S. China province	N. China province
Elrathiellidae: Elrathiella	
Bolaspididae: *Jialaopsis*	
Protohediniidae: *Protohedinia*	
Burlingiidae: *Burlingia, Schmalenseeia*	
Papyriaspididae: *Hedinaspis, Pianaspis*	
Guizhoucephalinidae: *Guizhoucephalina*	
Eoacidaspididae: *Paraacidaspis*	
Macropygiidae: *Promacropyge, Macropygites*	
Hungaiidae: *Chencunia*	
Hapalopleuridae: *Jegorovaia, Huamiaocephalus, Araiopleura*	
Palaeoharpidae: *Loganopeltoides*	
Asaphidae: *Yuepingia, Pseudoyuepingia, Niobella, Charchaqia, Psiloyuepingia, Pseudogriphasaphus*	*Parayuepingia*
Ceratopygidae: *Proceratopyge, Wannania, Diceratopyge*	
Dokimocephalidae: *Wuhuia, Lorrtetina*	*Wuhuia*
Agnostina: many genera	more genera
Dorygygidae: *Dorypyge, Olenoides, Kootenia*	*Dorypyge*
Corynexochidae: *Corynexochus, Chatiania*	*Corynexochus*
Dolichometopidae: *Paramphoton, Fuchouia*	*Amphoton, Paramphoton, Pseudamphoton, Amphotonella, Sunia, Fuchouia*
Ptychopariidae: *Kaotaia, Eosoptychoparia, Sinoptychoparia, Sanduspis*	*Emmrichella, Eosoptychoparia*
Proasaphiscidae: *Kütsingocephalus, Manchuriella Szeaspis, Jiubaspis, Eymekops*	*Manchuriella, Proasaphiscus, Honanaspis, Proasaphiscina, Zhongtiaoshanaspis, Shanxiella, Szeaspis, Eymekops*
Anomocarellidae: *Neoanomocarella*	*Anomocarella, Liaoyangaspis, Peishania*
Solenopleuridae: *Pseudosolenoparia, Modocia, Pseudosolenopleura, Neosolenopleurella, Solenoparia, Acrocephalina*	*Solenoparia, Solenopleura, Plesisolenoparia, Changingia, Squarrosoella, Protrachoparia, Catasolenopleura, Hyperoparia, Solenoparina, Jiagouia, Eilura, Eujinnamia*
Agraulidae: *Cyclolorenzella, Fenghuangella*	*Cyclolorenzella, Poriagraulos, Parachittidilla, Metagraulos, Lorenzella, Inouyia, Inouyops, Paralorenzella, Ptyctolorenzella*
Catillicephalidae: *Distazeris*	*Qilianshania*
Damesellidae: *Fengduia, Ariaspis, Paradamesops, Blackwelderia, Paradamesella, Teinislion, Palaeadotes*	*Damesella, prodamesella, Drepanura, Stephanocare, Blackwelderia, Chiawangella, Teinistion, Shantungia, Bergeronites (B.), Damesops*
Elviniidae: *Enshia, Yunlingia, Paraenshia*	*Taishania, Maladioides*
Shumardiidae: *Eoshumardia, Koldinioidia, Shumardia, Liriamnica*	*Koldinioidia, Shumardia*

Table 3.6 (*cont.*)

Monkaspidae: *Monkaspis*, *Liaoningaspis*, *Metaliopareua*	*Monkaspis*, *Lianingaspis*, *Parakoptura*, *Koptura*, *Walcottaspidella*, *Paraliaoningaspis*
Alokistocaridae: *Elrathia*, *Nangaops*, *Waergangia*	*Houmengia*, *Parapachyaspis*
Pterocephaliidae: *Maladioidella*, *Pseudaphelaspis*, *Aphelaspis*, *Eugonocare*, *Paracoosia*, *Cilia*	*Maladioidella*, *Changshanocephalus*, *Eugonocare*, *Aphelaspis*
Olenidae: *Olenus*, *Westergaardites*	*Shihuigouia*
Shirakiellidae: *Shirakiella*	*Shirakiella*
Lisaniidae: *Aojia*, *Lisania*, *Shengia*, *Extrania*	*Aojia*, *Lisania*, *Metalisania*, *Lisaniella*, *Paralisaniella*
Anomocaridae: *Hunanaspis*	*Datongites*, *Haniwa*, *Haniwoides*, *Lioparia*, *Parakotuia*, *Wentsuia*, *Wutingshania*, *Callaspis*
Tsinaniidae: *Tsinania*, *Dictyites*, *Zhujia*	*Tsinania*, *Dictyites*, *Sunaspis*, *Leiaspis*, *Sunaspidella*
Komaspidae: *Irvingella*, *Parakomaspis*, *Komaspis*	*Irvingella*, *Parakomaspis*, *Komaspis*
Leiostegiidae: *Chuangia*, *Chuangiella*, *Pagodia*	*Chuangia*, *Eochuangia*, *Plethopeltella*, *Pagodia*
Kaolishaniidae: *Kaolishania*, *Peichiashania*, *Mansuyia*, *Anhuiaspis*, *Tangjiaella*, *Prochuangia*, *Kaolishaniella*	*Kaolishania*, *Peichiashania*, *Mansuyia*, *Prochuangia*, *Kaolishaniella*, *Taianocephalus*, *Tingocephalus*, *Hamikaolishania*
Ptychaspididae: *Quadraticephalus*, *Prosaukia*, *Jingxiania*	*Quadraticephalus*, *Ptychaspis*, *Anderssonella*, *Changia*, *Saukioides*, *Keithia*
Saukiidae: *Saukia*, *Eosaukia*, *Prosaukia*, *Calvinella*, *Tellerina*, *Lophosaukia*	*Saukia*, *Sinosaukia*, *Prosaukia*, *Calvinella*, *Lichengia*, *Tellerina*
Changshaniidae: *Parachangshania*	*Changshania*, *Prochangshania*, *Parachangshania*
Liostracinidae: *Liostracina*	*Liostracina*
Ordosiidae: *Wanshania*	*Ordosia*, *Taitzuia*, *Poshania*, *Inouyella*, *Levisia*, *Plesioinouyella*, *Pseudotaitzuia*, *Eotaitzuia*, *Menocephadites*
	Conocoryphacea: *Bailiella*
	Crepicephalidae: *Crepicephalina*, *Hsüchuangia*, *Bagoshania*, *Idioura*, *Beikuangaspis*
	Tengfengiidae: *Tengfengia*
	Kingstoniidae: *Kingstonia*, *Wanwanaspis*
	Holanshaniidae: *Holanshania*, *Acanthocephalus*
	Plagiuridae: *Ruichengella*, *Paraplagiura*
	Polycyrtaspididae: *Oxygonaspis*, *Hebeia*

The sea was somewhat hypersaline in the Yangtze biome area, so the benthic elements were less diverse; of the trilobite genera and subgenera found (more than 100 in total), half are in common with those in North China and Qaidam; there are still some endemic members, such as *Kaotaia*, *Kütsingocephalus*, *Douposiella*, *Sinoptychoparia*, *Sanhuangshania*, *Paranomocare*, and *Fengduia*. Miomere trilobites and other fossils are very few (Zhang *et al.* 1980). This biome area includes south-eastern Yunnan because there is a typical member of the Yangtze biome, *Kaotaia*, and the area was

Table 3.7 List of the Middle–Late Cambrian assemblages (representing communities) of South China

Age	Assemblage name	Main ecological type	Main lithological type	Environment	Geographical distribution
Early Xuzhuangian	1. *Kütsingocephalus*	ben.	Grey-purple, grey-green marl, argillaceous dolomite interbedding sandstone and shale	Near-shore shallow sea	E. Yunnan, Emei, Michangshan
	2. *Kütsingocephalus–Kaotaia*	ben.	Dolomitic limestone, sandstone and shale	Shallow sea	Guizhou, SE Sichuan, Hubei
	3. *Kütsingocephalus–Kaotaia–Pagetia* (Fig. 3.8)	ben., plan., nek.	Grey-green sandstone and shale Hubei	Marginal shallow sea	Danzhai, Guizhou, Pugi,
Late Xuzhuangian	1. *Protohedinia –Szeaspis*	ben.	Grey limestone, argillaceous dolomite interbedding sandstone and shale	Near-shore shallow sea	E. Yunnan
	2. *Shilengshuia–Jiubaspis*	ben.	Thin-bedding dalomite interbedding breccia-dolomite	Tidal flat	N. Guizhou
	3. *Ehmaniella–Kootenia*	ben., nek.	Shale, sandstone	Marginal epicontinental sea	Danzhai, Guizhou
	4. *Kootenia–Pagetia*	mainly nek. and plan.	Thick-bedding limestone, laminated dolomite	Upper slope	W. Hunan, E. Yunnan
	5. *Wuania–Proasaphiscus*	ben.	Dolomite, feldspar-quartz siltstone	Near-shore shallow sea	SE Yunnan
Zhangxiaan	1. *Paranomocare–Peishania*	ben.	Dolomitic limestone, limestone	Epeiric shelf	NE Guizhou, SW Hubei
	2. *Lisania–Lejopyge*	ben., plan.	Dark grey limestone	Marginal platform-upper slope	W. & N. Hunan, E. Guizhou, Suzhou-Hanzhou
	3. *Lejopyge–Phalagnostus*	plan., ben.	Dark grey thin-bedding limestone and shale	Slightly reduced quiet basin	C. Hunan, W. Zhejiang
	4. *Pagetia–Peronopsis*	plan.	Black shale interbedding limestone	Stagnant deeper sea	Sandu, Guizhou
	5. *Solenoparops–Prilorenzella*	ben.	Dolomite, dolomitic limestone	Limited platform	SE Yunnan
Gushanian	1. *Fengduia–Liaoningaspis*	ben.	Limestone	Epeiric shelf	SE Sichuan, SW Hubei
	2. *Chatiania–Glyptagnostus stolidotus*	ben., plan.	Limestone, dolomitic limestone	Marginal platform-upper slope	E. Guizhou, W. Hunan, N. Hunan, S. Anhui, Suzhou-Hanzhou

Gushanian	3. *Blackwelderia–Paralorenzella*	ben.	Mudstone interbedding limestone	Epeiric shelf	SE Yunnan
	4. *Bergeronites (P.)–Cyclolorenzella*	ben.	Mudstone, siltstone	Epeiric shelf	Baoshan, Yunnan
	5. *Bergeronites (P.)–Proceratopyge*	ben., nek.	Argillaceous limestone, thick-bedding limestone	Continental slope	Chuxian and Quanjiao, Anhui
	6. *Glyptagnostus stolidotus–Pseudagnostus*	plan.	Argillaceous limestone	Continental slope	W. Zhejiang
Changshanian	1. *Enshia*	ben.	Limestone interbedding dolomite	Epeiric shelf (epicontinental sea)	SE Sichuan, SW Hubei
	2. *Chuangia–Billingsella*	ben.	Limestone interbedding sandstone and mudstone	Epeiric shelf	SE Yunnan
	3. *Prochuangia–Proceratopyge*	ben., nek., plan.	Limestone, dolomite	Marginal platform-upper slope	W. Hunan, E. Guizhou, N. Hunan, S. Anhui
	4. *Chuangia–Parachangshania*	ben.	Shale interbedding limestone	Marginal shelf	W. Yunnan
	5. *Proceratopyge–Pseudaphelaspis*	nek., plan.	Dolomitic limestone	Continental slope	Chuxian, Anhui
	6. *Proceratopyge–Pseudo-glyptagnostus*	nek., plan.	Limestone	Continental slope	W. Zhejiang
	7. *Proceratopyge–Pseudo-yuepingia*	nek.	Mark Shale	Continental slope	Sandu, Guizhou
Fengshanian	1. *Prosaukia*	ben.	Dolomite, limestone	Epeiric shelf (epicontinental sea)	SE Sichuan, W. Hubei, SW. Hubei
	2. *Prosaukia–Billingsella*	ben.	Mudstone, siltstone, limestone	Epeiric shelf	SE Yunnan, W. Yunnan
	3. *Mictosaukia–Cilioceras*	ben., nek., plan.	Limestone	Marginal platform-upper slope	W. Hunan, N. Hunan
	4. *Jingxiania–Saukia*	ben.	Limestone interbedding shale	Marginal platform-upper slope	S. Anhui
	5. *Lotagnostus–Acaroceras* (Fig. 3.9)	plan., nek.	Shale	Continental slope	W. Zhejiang
	6. *Niobella–Guizhoucephalina*	nek.	Shale	Continental slope	Sandu, Guizhou

ben., benthics; nek., nektonics; plan., planktonics.

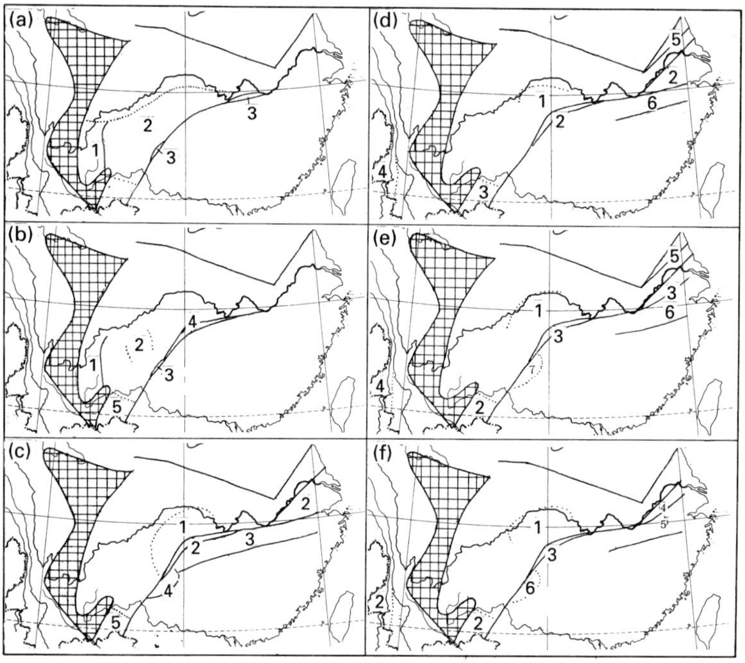

Fig. 3.7 Distribution of the Middle and Late Cambrian communities (assemblages) of South China. (a) early Xuzhuangian (Middle Cambrian); (b) late Xuzhuangian (Middle Cambrian); (c) Zhangxiaan (Middle Cambrian); (d) Gushanian (Late Cambrian); (e) Changshanian (Late Cambrian); (f) Fengshanian (Late Cambrian); the numbers on each figure correspond to the numbered communities in Table 3.7.

connected with the Yangtze shallow sea. However, in Funing, Wenshan, Mengzi, and other countries many trilobite genera similar to those in North China and Qaidam occur. Few trilobites have been found on the western flank of the Xikang–Yunnan Oldland. In addition to *Calvinella* and *Haniwa* found in the Songdagou Formation (Late Cambrian) in Batang, western Sichuan, *Kunmingaspis*, *Kailiella*, *Probowmania*?, *Pseudosolenoparia*, *Wudanggia*, *Douposiella*, *Xingrenaspis*, and *Wuxunaspis*? (Leng 1983) have been discovered in the lower member of Yinchangyou Formation, or the lower Middle Cambrian, in Zhongdian, Yunnan (possibly the upper Lower Cambrian is also included). These findings provide evidence to attribute the western flank to the same biome.

The biome of the platform margin of Yangtze surrounds the Yangtze biome. Its southern fringe includes Tunpan (Guangxi), Yuping (Guizhou), Fenghuang, Baojing, and Taoyuan (Hunan), Qingyang (southern Anhui), and the Jiangsu–Zhejiang border (Yang 1978; Yang *et al.* 1984; Qian and Zhou 1984). The northern fringe includes Chuxian (Anhui), Zhenping, and Ziyang (southern Shaanxi). The western fringe is stretched to Baoshan (western Yunnan) (Sun and Xiang 1979; Luo 1982, 1984*b*). This biome consists of abundant trilobites, including large benthic and many plankton agnostids and other nektonic trilobites. Many of them are local members, such as *Chatiania*, *Fenghuangella*, *Dingxiangaspis*, *Neoanomocarella*, *Paracidaspis*, *Wanshania*, *Yuepingia*, *Pseudoyuepingia*, *Stigmatoa*, *Paradamesella*, *Pseudaphelaspis*, *Rhyssometopus*, and *Olenus austriacus*.

The Basinal biome area extends from western Zhejiang, central Hunan, Hainan Island, to Mangsi of Luxi County, western Yunnan. Among the relatively fewer trilobite genera, the nekton take up half of the total number. Most of the trilobites are widespread, but a few are restricted, such as *Erixanium*, *Fuyangia*, *Zhajia*, *Wujiajiania*, *Metalizania*, *Charchaqia subquadrata*, *Westergaardites tachenensis*, *Olenus sinensis*, *Xystridura yaxianensis*, *X. hainanensis*, and *X. orientalis*.

The North China province includes two biomes: the benthic North China–Qaidam biome and the mixed-type Qilian biome. There is no planktonic biome. The North China–Qaidam biome is rich in benthic trilobites, especially in North China and the southern part of north-eastern China. They include *Hsüchuangia*, *Changia*, *Drepanura*, *Bergeronites*, *Liaoyangaspis*, *Delinghaspis*, *Parakomaspis*, and *Damesella*. There are fewer trilobites in the Qilian biome, two-thirds of them are benthic. The representative members are *Pseudamphoton*, *Lajishanaspis*, *Qilianaspis*, *Qilianshania*, *Datongites*, and *Parakotuia*. An (1982) and Wang Zhihao (1984) recorded abundant conodont fauna from North and north-eastern China. They are closely related to those of the south-western interior of the United States (coefficient of similarity 0.69), and are related to a lesser degree to Kazakhstan (0.48) and Baltica (0.25) (Bergstrom 1990).

The Tarim province actually extends westward to incorporate Karatau of southern Kazakhstan and Iran, and southward to the Salt Range of Pakistan and Spiti, India. However, the situation in other countries is not considered here. It is quite different from the other two provinces in its lower abundance of trilobites (Table 3.5) and lack of shallow benthic fauna due to hypersaline conditions. There is a mixed-type biome in its northern margin (the Kuruktag–Ejina Qi biome), and a planktonic biome (the Huocheng–Beishan–Hangula biome) that inhabited a more offshore setting (Lu *et al.* 1981). Nearly 60 genera of polymerid trilobites have been discovered in the former biome. Half of them are similar to those in the biome of the platform margin

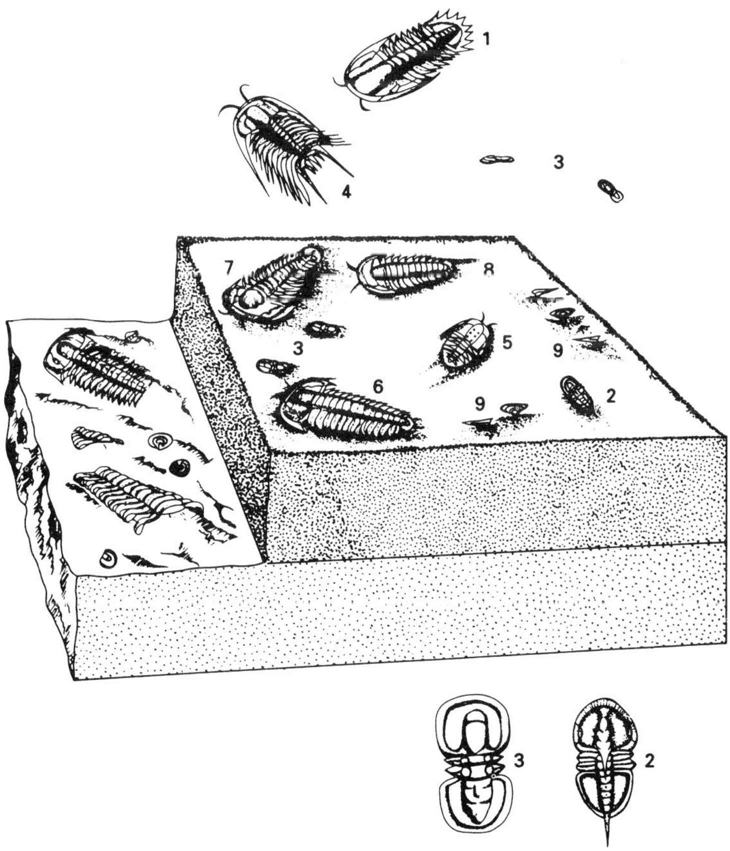

Fig. 3.8 Reconstruction of the early Xuzhuangian *Kütsingocephalus–Kaotaia–Pagetia* community of the Middle Cambrian at Danzhai, Guizhou, South China. 1, *Kootenia*; 2, *Pagetia*; 3, *Peronopsis*; 4, *Oryctocephalus*; 5, *Metabalangia*; 6, *Kaotaia*; 7, *Kütsingocephalus*; 8, *Xingrenaspis*; 9, *Homotreta*.

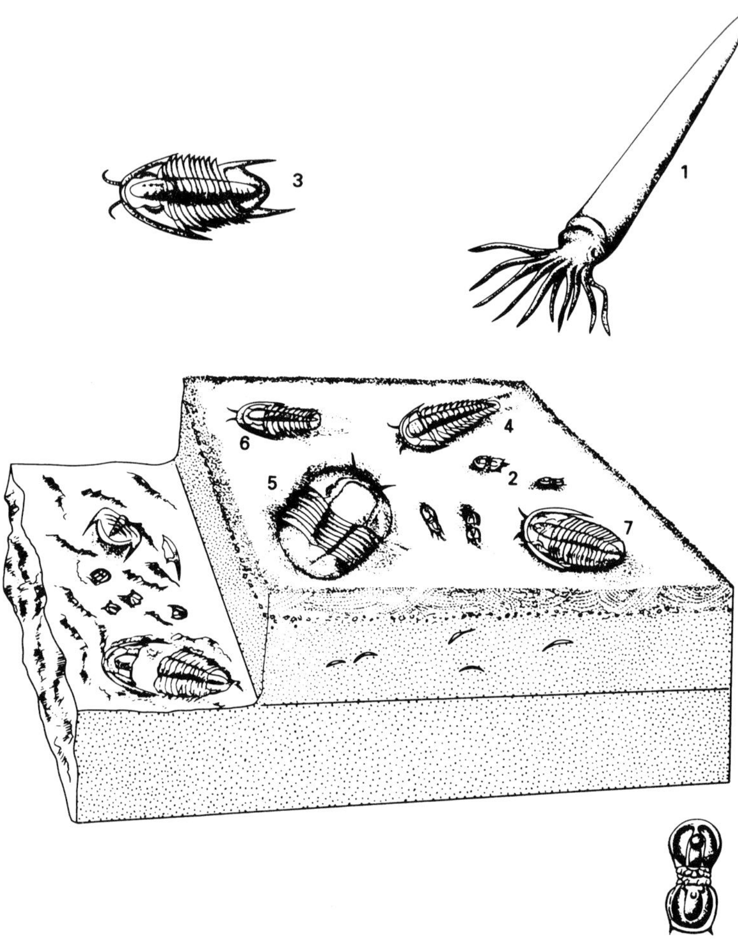

Fig. 3.9 Reconstruction of the Fengshanian *Lotagnostus–Acaroceras* community of the Late Cambrian in western Zhejiang. 1, *Acaroceras*; 2, *Lotagnostus*; 3, *Proceratopyge*; 4, *Hedinaspis*; 5, *Niobella*; 6, *Duibianaspis*; 7, *Pseudaphelaspis*.

of Yangtze, and so are the miomerid agnostids, except for a few local elements such as *Kuruktagaspis*, *Kuruktagella*, *Tamdaspis*, *Rhadinopleura*, *Euptychoparia*, *Westergaardites formis*, *W. pelturae*, and *Diceratopyge mobergi*. The Huocheng–Beishan–Hangula biome is similar to the Basinal biome. Most genera are the same but the species are different, e.g. *Xystridura tianshanensis*, *Centropleura shuangyingshanensis*, *C. gansuensis*, *C. beishanensis*, *C. vaulta*, *C. loveni*, *Charchaqia subeiensis*, *Ch. rustrica*, *Ch. spinicaudata*, *Hedinaspis tianshanensis*, *H. obesa*, *H. brevica*, *Diceratopyge distincta*, and *D. mobergi*. In addition, there are several endemic polymerid trilobites, such as *Teratokoptura*, *Keguqinia*, *Huochengella*, *Sayramaspis*, *Bulbolenus*, and *Guozia*.

Biomes are subdivided on the basis of biofacies, palaeogeographical framework, and palaeotectonic features. The Middle and Late Cambrian contains three types—benthic, planktonic, and mixed type (Figs 3.8, 3.9). The Yangtze biome and North China–Qaidam biome are of benthic type, which was located in epicontinental shelf environments of the platform, with benthos as the dominant organisms; the Basinal biome and Huocheng–Beishan–Hangula biome are of planktonic type, which was located in the miogeosynclinal or continental slope environment, with the plankton and nekton as the major organisms. The Platform Margin (of Yangtze) biome, the Kuruktag biome, and the Qilian biome are attributed to the mixed type, with both benthos and plankton; both may coexist in one bed, may occur in successive strata, or appear synchronously in different localities. Tectonically these biomes are located in the marginal areas between platform and geosyncline, or within the geosyncline; palaeogeographically they are located in the lower part of the continental shelf or the upper part of the continental slope, and they may occur in minor areas with alternating shallower and deeper water outside an immense continental shelf.

Thirty-two communities have been recognized through detailed research on their distinctive environments and taxonomy (Table 3.7, Fig. 3.7). Five preliminary conclusions have been drawn from the geographical distribution and evolution of the communities.

1. The number of communities and the area of their distribution continued to increase through time (Fig. 3.7). There were only three communities in the early Xuzhuangian, with their distribution area restricted to the triangle covering eastern Yunnan, Guizhou, northern Hunan, and south-eastern Hubei. The benthic trilobites were dominant; there were very few plankton and nekton, and the swimming capabilities of the latter were weak. The number of communities grew to five in the late Xuzhuangian and Zhangxiaan. The increased salinity in Yunnan and Guizhou was unfavourable for benthic trilobites; however, pelagic planktonic trilobites occurred in great abundance and actually extended the distribution area of the communities. During these ages the organisms were concentrated on the southern margin of the Yangtze shallow sea and the continental slope, i.e. from south-eastern Yunnan, through Guizhou, Hubei, and Hunan to southern Anhui and western Zhejiang. The number of communities in the Late Cambrian went up to 6–7 due to a further enlarged distribution area of the trilobites, and also due to the greatly increased number of pelagic nektonic trilobites; the distribution area extended to northern Anhui and western Yunnan. Compared with the features of the Early Cambrian, this tendency is very obvious.

2. Communities of different types show tendencies of inheritance and migration (Figs 3.10, 3.11). The fact is true for most areas, that successive

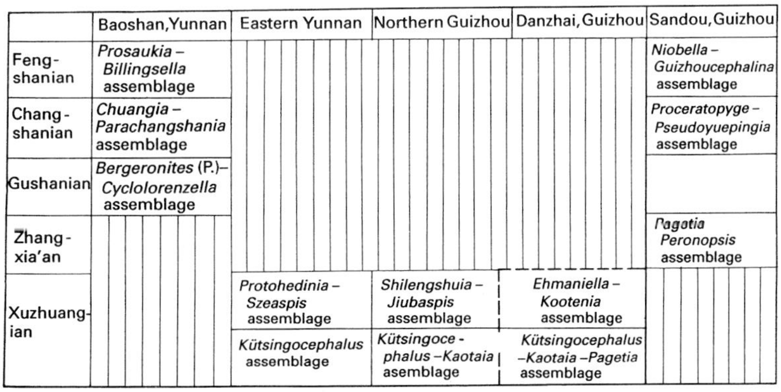

Fig. 3.10 Middle and Late Cambrian communities (in the appearance of assemblages) of South China, profile 1. Note: assemblages to the left of the dashed line were benthos, those to the right of the thick solid line were plankton or nekton, those between both these lines were mixed type, with both plankton and benthos.

	N.E. Guizhou –S.W. Hubei	E. Guizhou– W. Hunan–N. Hunan	W. Zhejiang
Fengshan-ian	*Prosaukia* assemblage	*Mictosaukia* – *Cilioceras* assemblage	*Lotagnostus* –*Acaroceras* assemblage
Changshan-ian	*Enshia* assemblage	*Prochuangia* – *Proceratopyge* assemblage	*Proceratopyge* – *Pseudoglyptagnostus* assemblage
Gushanian	*Fengduia* –*Liaoningaspis* assemblage	*Chatiania* –*Glyptagnostus stolidotus* assemblage	*Glyptagnostus stolidotus* – *Pseudagnostus* assemblage
Zhangxiaan	*Paranomocare* –*Peishania* assemblage	*Lisania* –*Lejopyge* assemblage	*Lejopyge* –*Phalagnostus* assemblage
Xuzhangian		*Kootenia* –*Pagetia* assemblage	

Fig. 3.11 Middle and Late Cambrian communities (in the appearance of assemblages) of South China, profile 2. See also the note to Fig. 3.10.

communities in a certain area show features of inheritance. The benthic, planktonic, and mixed communities were each developed and successively evolved in their own geographical distributions. This, in fact, is closely related with the palaeogeographical environments and palaeotectonic framework. Of course, there are some subtile changes. Eastern Guizhou, western Hunan, and northern Hunan were located in the intermediate belt of two types of communities (benthic and planktonic), so that they became the most variable zone; for example, in this zone during the late Xuzhuangian there developed the planktonic communities, which were replaced by mixed type, and partially even by benthic communities in the Fengshanian (such detailed change is not shown in Fig. 3.11). Such temporal change of ecological types actually represents geographical migration.

3. The salinity of sea water was the essential factor controlling the distribution of the benthic communities. Whenever the salinity approximated to the normal value, the benthic communities developed in the Yangtze shallow sea, e.g. in Table 3.7, early Xuzhuangian (also Fig. 3.7) and late Zhangxiaan.

4. Ecological types, palaeogeographical environments, and the palaeotectonic framework are interrelated. It is known that origination, development, and extinction are closely related to the

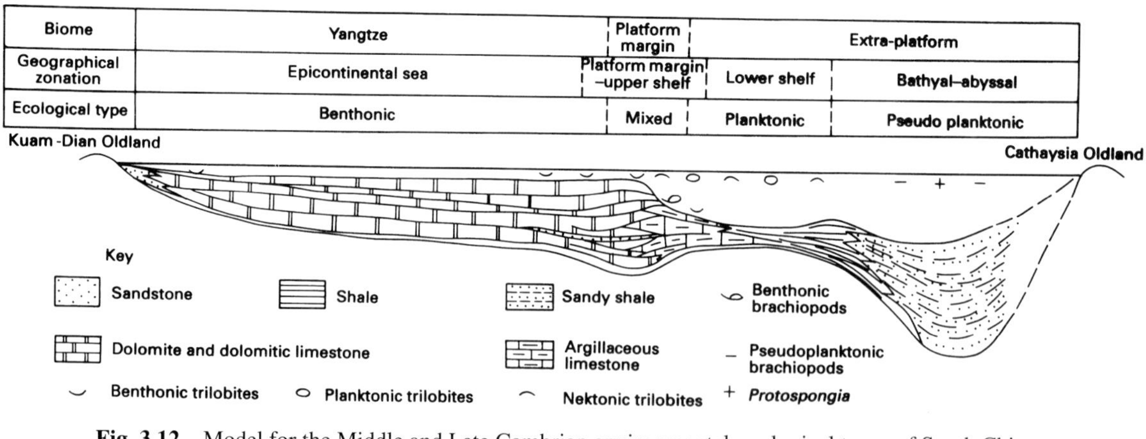

Fig. 3.12 Model for the Middle and Late Cambrian environmental-ecological types of South China.

environment as well as the biology of the organisms themselves. After the Middle and Late Cambrian, the ecological differentiation of trilobites became more and more distinctive, and each type occupied a certain area (Fig. 3.12). The Yangtze biome was a benthic community occupying the epicontinental sea, which was a stabilized platform. The Basinal biome was occupied by planktonic communities in the lower part of the slope and deeper waters, in a zone which belonged tectonically to the transitional and active miogeosyncline. The Platform Margin biome occupied the zone between the two above-mentioned biomes, being transitional in ecological type, palaeogeographical environment, and palaeotectonic setting; there were both benthos and plankton, i.e. mixed type of organisms. This biome occupied the carbonate shelf margin and the upper part of the continental slope; the relief was uneven, and the sea-level fluctuated constantly. Tectonically this was the intermediate area between the platform and the geosyncline. Detailed study of the distribution and stratigraphical succession of the mixed-type communities will help us to define the limits of the Yangtze shallow sea and Yangtze platform.

5. Mixed-type communities have greater diversity and abundance. Figure 3.13 shows statistics of the trilobite genera and subgenera for the three

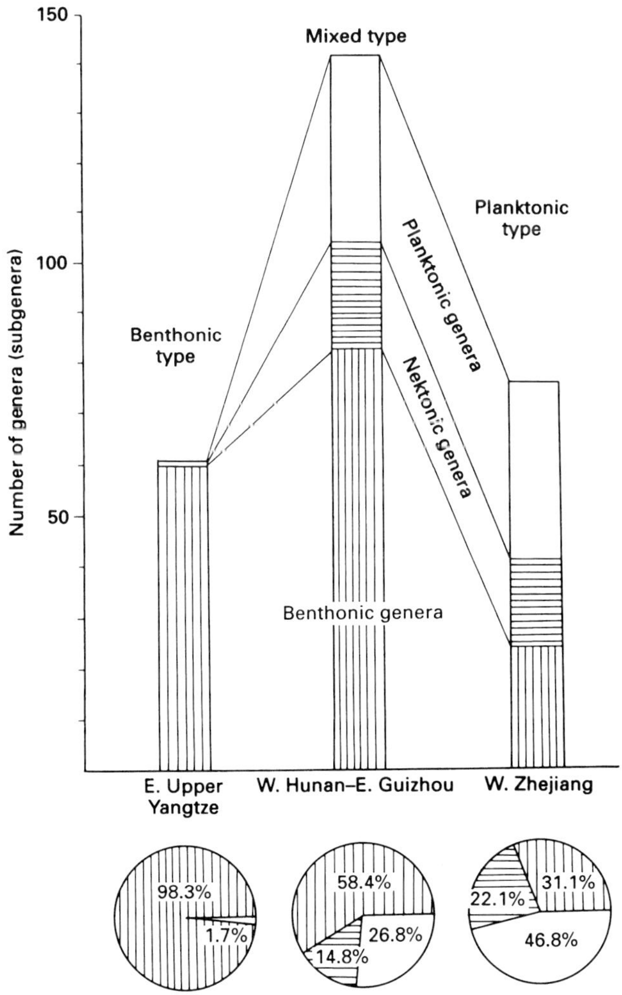

Fig. 3.13 Statistics of the three ecological types of Middle and Late Cambrian trilobites (genera and subgenera).

types of Middle and Late Cambrian communities. The mixed type has the greatest diversity with a total number of 142 genera and subgenera, double that in each adjacent area. Diversities of each of the three different habitats of trilobites are also the highest in the mixed type. This is because the mixed type lived in a more advantageous environment, i.e. the transition zone from the shallow sea to the slope, with fluctuating relief and changing depth of sea, also connected with the open sea. This zone provided all the conditions of a normal shallow sea beneficial to the benthic trilobites, including illumination, salinity, dissolved oxygen, and microbial nourishment, and meanwhile avoided the unfavourable conditions found in the Yangtze shallow sea itself which, as an epicontinental sea, underwent frequent salinity increases (marked by gypsum and salt deposits) that greatly threatened the benthic trilobites. The transition zone was also adjacent to the deeper sea of the continental slope, which the nekton and plankton could reach without difficulty. In contrast, it would have been difficult for the benthic trilobites to survive in the deep-sea environment. In short, this biome had the advantageous environmental conditions of different zones so that organisms of both two ecological types could flourish.

3.3 RECONSTRUCTION OF THE SINO-KOREAN, YANGTZE, AND TARIM PLATES

The Sino-Korean plate embraces Qaidam, Qilian, North China, north Korea, and the southern part of north-eastern China. The Yangze plate covers the wide area extending westward to Longmenshan, northward to southern Qinling, and southward to north Vietnam. The Tarim plate covers Tarim and Karakum of the CIS (until recently, USSR), and adjacent areas to the north and south (Beishan, Huocheng, Minor Karatau of the CIS, Iran) (Li and Tang 1982, 1983). During the Cambrian they were all incorporated with, or affiliated to, the Gondwanan palaeocontinent (Wang and Liu 1979). Their mutual relationship and palaeogeographical locations can be studied using palaeobiogeography, palaeoclimatology, crustal movement, and palaeomagnetism, with the premise that the plates linked together, or situated close to each other, should have similar biotic, climatic, and geological features.

3.3.1 PALAEOBIOGEOGRAPHY

In terms of palaeobiogeography, the Sino-Korean plate belongs to the North China province, the Yangtze plate to the South China province, and Tarim plate to Tarim province (Wang *et al.* 1982, 1985). Although as different provinces they have quite obviously differentiated genera and species, yet they yield similar benthic trilobite groups (Table 3.3, Table 3.6), so they are attributed to the same Asia–Australian realm. The North China biota differs from the other two. Moreover, the biotic diversity of the Tarim plate is quite different from that of the Yangtze plate, although they have similar kinds of organisms (Table 3.2, Table 3.5). Hence from the biological information these three plates are likely to have been separate in the Cambrian (Yang 1987*b*).

3.3.2 PALAEOCLIMATE

The Yangtze shallow sea in the Early Cambrian contained abundant archaeocyathids, or even small archaeocyathid reefs. These archeocyathid fossils are preserved in grey to greyish-green marls and limestones, which may pass stratigraphically and geographically into greyish-green clastics and grey limestones. The facies denotes a warm, humid climate. They are quite different from the abundant archeocyathids in the Siberian realm, mainly found in brick-red to purple clayey dolomitic limestones. The Siberian plate is regarded as being under a tropical marine environment and is placed near the equator. The Tarim shallow sea had only a few Archaeocyatha, and the North China shallow sea was almost devoid of them. These suggest different palaeoclimatic features.

To a certain extent, the diversity of organisms reflects their relative distance from the equator. The

Yangtze shallow sea showed a great diversity of organisms in the Early Cambrian (Table 3.2), while such a diversity appeared in the North China shallow sea in the Middle and Late Cambrian (Table 3.5). These facts seem to indicate a warmer climate and lower latitude for these two plates during the mentioned epoch.

It is well known that at the beginning of the Early Cambrian the Yangtze shallow sea was rich in phosphate (Lu 1984; Wang Shenglong 1984). Many large phosphorous deposits were formed. They were basically chemogenic in origin, although biogenesis has not been excluded. These deposits were formed in an arid climate, and seas of the low–middle latitudes are usually considered to have been most favourable. In modern seas, 24–28° latitude is the best area for the deposition of phosphorus, and 16–40° latitudes are considered to be favourable (Liu and Zheng 1985). There are few phosphorous deposits in the shallow-sea areas of Tarim and Karakum, CIS, but many more are to be found on their northern slopes, indicating a similar latitudinal and climatic situation to that of the Yangtze shallow sea. In contrast, there was very little phosphorous deposition in the North China shallow sea, the latitude and climate of which were obviously quite different from those of the Yangtze and Tarim shallow seas.

On the Yangtze and Tarim plates a great deal of dolomite, and even gypsum and salt deposition, formed from the late Early Cambrian to the Late Cambrian, indicating an arid, hot climate. Heterochronously, in the North China shallow sea, purple clastics with sparsely scattered salt pseudomorphs were deposited from the late Early Cambrian to the early Middle Cambrian, also suggesting an arid, hot climate. From the late Middle Cambrian to the Late Cambrian in North China, deposits were completely of limestone with a great abundance of trilobites, which clearly suggests a warm climate.

To sum up (Table 3.8), Yangtze and Tarim were more-or-less alike in climate change, but both were quite different from North China. The former two changed from warm to hot and arid, while the latter changed from hot and arid to warm. This is significant for inferred plate movement.

Table 3.8 The Cambrian palaeoclimates of China

	N. China shallow sea region		Yangtze shallow-sea region		Tarim shallow-sea region	
Late Cambrian	Warm	Limestone High trilobite diversity	Hot, arid	Dolomite and limey dolomite Gypsum No trilobites, or rare		Hot, arid
Middle Cambrian	Hot, arid	Purplish-red sandstone and shale Halite pseudomorphs High trilobite diversity	Damp, hot	Yellow-green standstone and shale Archaeocyathid reefs More trilobites	Limestone, dolomitic limestone Rare archaeocyathids	Damp, hot
Early Cambrian			Warm, arid	Dolomite, phosphatic and dolomitic limestone Abundant phosphatic small shelly fossils		Arid, warm

3.3.3 PALAEOGEOGRAPHY

Judging from lithofacies and the characteristic communities of mixed type and planktonic type, as well as the palaeotectonic pattern, it could be inferred that continental slopes and deep basins surrounded the North China shallow sea, Yangtze shallow sea, and Tarim shallow sea (Wang *et al.* 1985; Yang 1987*a*).

Around the Yangtze shallow sea a relatively complete sequence of slope deposition is preserved. This was a passive continental margin; its south-eastern part extends from western Guangxi, eastern Guizhou, northern Hunan, central Hunan, northern Jiangxi to southern Anhui, western Zhejiang, and the Jiangsu–Zhejiang border; its northern flank consists of the bordering area of Sichuang and Shaanxi and Chuxian county in eastern Anhui; its south-western flank stretches in Luxi County in western Yunnan. No data for the western flank and north-western flank are available.

The slope deposition is also well preserved on the northern flank of the Tarim shallow sea, which covers the Hocheng–Kuruktag–Kawaburak–Beishan–Erjina Qi area.

The southern and south-western parts of the North China shallow sea had a volcanic island arc, probably facing a vast sea to the south.

These slope seas and basins in fact separated the plates.

3.3.4 CRUSTAL MOVEMENT

Disregarding the fact that different parts of a plate might display dissimilar tectonic roles, most parts of the same plate usually underwent vertical movements in the same manner.

As has already been described above, the Yangtze sea became shallower after the Qiongzhusian; oldlands extended gradually, indicating crustal elevation. This epeirogeny was, of course, uneven; the Hannan and the Xikang–Yunnan oldlands were uplifted more slowly than those to their east.

The vertical movement of the crust under North China displayed contrasting situations. After the late Canglangpuan, the sea gradually transgressed onto its southern, eastern, and north-eastern flanks. In the Late Cambrian only the north-western corner was above sea-level. This fact suggests a tendency of crustal subsidence. However, the descending altitude was unequal, with the southern and eastern flanks of the North China sea subsiding in a more pronounced manner.

The Tarim shallow sea was probably similar to the Yangtze shallow sea, and thus also belonged to the crustal ascending type.

In brief, we came to the conclusion that the North China plate and the Yangtze plate were separated. The Tarim plate has some common features with the Yangtze plate. Obviously there was some kind of connection between the two.

3.3.5 PALAEOMAGNETISM

The Cambrian palaeomagnetic data are few and comparatively unreliable, yet some general ideas can be formed from them (Lin *et al.* 1985; Zhu and He 1987). According to the palaeomagnetic data obtained by Li Pu and Lian Qizhong from the lower part of the Lower Cambrian in Meishucun, Kunyang County, eastern Yunnan (see Luo *et al.* 1982), at the Zhongyicun the palaeolatitude was 31.9°– 45.7°; at the Dahai time, 18.2°– 44.2°; at the Badaowan, 48.7°– 57.3°; and at the Yuanshan, 55.7°– 44.9°. These data show a tendency for migration from mid-low latitude to middle or mid-high latitude. Gao *et al.* (1983) offered palaeomagnetic data for the north-eastern part of the Sino-Korean plate. In the Early Cambrian Nanpiao (western Liaoning) had a palaeolatitude of 13°S; in the Middle Cambrian 15.7°S; in the Late Cambrian, 9.8°S, 11.6°N, 21°N. In the Early Cambrian, Laozhuanghu had a palaeolatitude 16.7°S; in the Middle Cambrian, 19.8°N, which suggests a tendency of the northern part of the North China plate to shift from southern low latitude to northern low latitude. Zu Hong's synthesis produces results in broad agreement (Fig. 14.8). However, Lin *et al.* (1985) gave different data for the Early Cambrian: palaeolatitudes of *c.* 37° for

North China and *c*. 10° for South China, which are not compatible with the palaeoclimatic data. Despite these differences, all data indicate a considerable distance between North and South China, so the available palaeomagnetic data confirm the conclusion that Yangtze was separated from the Sino-Korean plate in the Cambrian.

Conclusions:

1. The Sino-Korean plate is believed not to have been connected to other plates and to have been situated at a low latitude, on the basis of:

(a) it being an independent bioprovince;
(b) it being separated from other plates by its continental margin;
(c) it having features of palaeoclimate and crustal movement different from those of Yangtze and Tarim; and
(d) the palaeomagnetic difference.

Figure 3.1 represents the locations of the palaeocontinents in the Middle Cambrian; palaeomagnetic data of Nanpiao and Laozhuanghu in western Liaoning suggest quite a different picture from the one shown here. Taking into consideration that from the early to the Late Cambrian the Sino-Korean plate migrated northward from the equator, the northeastern part of the Sino-Korean plate is tentatively laid on a northern latitude of 10°.

2. The palaeobiogeography,palaeogeographical framework, crustal movements, and palaeoclimate of Yangtze demonstrate its isolation from the other plates. Considering the palaeomagnetic data and the abundant phosphorous deposits, Kunyang (eastern Yunnan) is placed in a northern mid-low latitude. Because of many features similar to those in Australia, Yangtze is laid near Australia.

3. The biota, crustal movements, and palaeoclimate in the Tarim plate are quite similar to those in the Yangtze plate. Nevertheless, considering that the diversity and abundance of organisms are far inferior to those in Yangtze plate, the Tarim shallow sea yielded fewer phosphorous deposits (not comparable with those in the Yangtze plate), and the Tarim plate has its own marginal slopes, it should be separated from other plates. Then how can its similarities and differences with the Yangtze plate be explained? In our opinion, Tarim and Yangtze were very close to Gandwana, which accounts for their similarities in biotas and crustal movements. Both of them lay at the corresponding latitudes to the south and north of the equator, so the palaeoclimatic features could have been very similar. The disparity in biotic diversity and abundance, and in the phosphorous content, reflect that they were separate, independent plates with different regional environments; the former disparity may also be due to different level of research. Of course, this inference should be further explored by the palaeomagnetic data.

4. The traditional view of taking Kazakhstan as a whole plate should be reconsidered. Here only a brief comment is presented because the major part of the plate lies beyond the Chinese border. In the Minor Karatau in southern Kazakhstan and in Beishan Mountain, Gansu, the Late Cambrian trilobites (Elgaliev 1980) bear the features of the Asia–Australian realm, but the Early Cambrian trilobites bear the features of the Siberian realm. There could be two explanations for these data:

(a) The southern Kazakhstan plate in the Early Cambrian was joined with northern Kazakhstan to make a part of the Siberian realm, and then separated from it, due to the Xingkai orogeny, and joined with the Tarim plate; or
(b) Southern Kazakhstan was separated from northern Kazakhstan since the Early Cambrian because, viewing from the palaeogeographical framework, southern Kazakhstan seems to be a part of the continental margin of the Tarim plate. In the Early Cambrian a deep shelf environment existed. Such an environment may have enabled the Siberian organisms to migrate to southern Kazakhstan which gave its Early Cambrian trilobites some Siberian features.

The second explanation is favoured in this book.

REFERENCES

An Taixiang (1982). Study on the Cambrian conodontens from North and Northeast China. *Science Reports, Institute of Geosciences University of Tsukuba* Sec. 2, **3**, 113–59, pls. 1–17.

An Taixiang, *et al.* (1983). *The conodonts of North China and the adjacent regions*. Geological Publishing House, Beijing (in Chinese).

Bergstrom, S. M. (1990). Relations between conodont provincialism and the changing palaeogeography during the Early Palaeozoic. *Geological Society of London, Memoir* **12**, 105–23.

Chang Wentang (1980). A review of the Cambrian of China. *Journal of the Geological Society of Australia* **27**, 137–50.

Cowie, J. W. (1971). Lower Cambrian faunal provinces. *Geology Journal* Special Issue **(4)**, 31–46.

Ergaliev, G. H. (1980). *Middle and Upper Cambrian trilobites of Lesser Karatay, Alma-Ata*. Nauka, Kazakhstan (in Russian).

Gao Rongfan, Hang Heling, Zhu Zhiwen, Liu Haishan, and Fan Yiqing (1983). The study of palaeomagnetism in Northeastern Massif during pre-Late Paleozoic. *Contribution of the Project of Plate Tectonics of Northern China* **1**, 264–74 (in Chinese).

Jell, P. A. (1974). Faunal provinces and possible planetary reconstruction of the Middle Cambrian, *Journal of Geology* **82**, 315–50.

Kobayashi, T. (1971). The Cambro-Ordovician faunal provinces and the interprovincial correlation discussed with special reference to the trilobites in eastern Asia. *Journal of the Faculty of Science, University of Tokyo* Sec. 2, **18**, pt. 1, 129–299.

Leng Chonglin (1983). A discovery of the Cambro-Ordovician in Jingjiang, Zhongdian, Yunnan. *Journal of Stratigraphy* **11**, (1), 77 (in Chinese).

Li Chunyu and Tang Yaoqing (1982). *Explanatory notes to the tectonic map of Asia*. Cartographic Publishing House, Beijing (in Chinese).

Li Chunyu and Tang Yaoqing (1983). Some problems on subdivision of Palaeo-Plates in Asia. *Acta Geologica Sinica* **57**, (1), 1–10 (in Chinese).

Lin Jinlu, Fuller, M., and Zhang Wenyou (1985). Preliminary Phanerozoic polar wander paths of the North and South China blocks. *Nature* **313**, 444–9.

Liu Baojun and Zeng Yunfu (1985). *Bases and methods of lithofacies and paleogeography*. Geol. Publishing House, Beijing (in Chinese).

Lu Yanhao (1984). Genesis, geologic and geographic distribution of the Cambrian phosphate deposits in China–a biostratigraphic view. In *Symposium of the 5th International field Workshop and Seminar on Phosphorite*, pp. 69–89. Geological Publishing House, Beijing, (in Chinese).

Lu Yanhao, Zhu Zhaoling, Qian Yiyuan, Lin Huanling, Zhou Zhiyi, and Yuan Kexing (1974). Bio-environmental central hypothesis and its application to the Cambrian biostratigraphy and paleozoogeography. *Memoirs of the Nanjing Institute of Geology and Palaeontology, Academia Sinica* **5**, 27–116 (in Chinese).

Lu Yanhao, Zhou Zhiyi, and Zhou Zhiqiang (1981). Cambrian–Ordovician boundary and their related trilobites in the Hangula region, West Nei Monggol. *Bulletin of the Xi'an Institute of Geology and Mineral Resources, Chinese Academy of Geological Sciences* **2**, (1), 1–22 (in Chinese).

Lu Yanhao, Zhu Zhaoling, Qian Yiyuan, Lin Huanling, and Yuan Jinliang (1982). Correlation chart of Cambrian in China with explanatory text. In *Stratigraphic correlation chart in China with explanatory text*, Nanjing Institute of Geology and Palaeontology, Academia Sinica, pp. 28–54. Science Press, Beijing (in Chinese).

Luo Huilin (1982). On the occurrence of Late Cambrian Gushan trilobite fauna in Western Yunnan. *Contribution to the Geology of Qinghai-Xizang (Tibet) plateau* **10**, 1–12 (in Chinese).

Luo Huilin (1984*a*). Subdivision and correlation of the Cambrian System in south-eastern Yunnan. *Acta Geologica Sinica* **58** (2), 87–96 (in Chinese).

Luo Huilin (1984*b*). The discovery of the Late Cambrian Gushan-aged strata in western Yunnan and their significance. *Geological Review* **30**, (5), 425–9 (in Chinese).

Luo Huilin, Jiang Zhiwen, Wu Xiche, Song Xueliang, and Quyang Lin (1982). *The Sinian–Cambrian Boundary in eastern Yunnan, China*. People's Publishing House of Yunnan (in Chinese).

Palmer, A. R. (1981). Cambrian. In *Treatise on invertebrate paleontology*, pt. A, Introduction (ed. R. C. Moor), pp. 119–35. Geological Society of America and University of Kansas Press.

Qian Yiyuan and Zhou Zemin (1984). Middle and early Upper Cambrian trilobites from Kushan, Jiangsu with reference to their distribution in the lower Yangtze region. *Acta Palaeontologica Sinica* **23**, (2), 170–84 (in Chinese).

Scotese, Ch. R., *et al.* (1979). Paleozoic base maps. *Journal of Geology* **87 (3)**, 217–77.

Scotese, Ch. R. and Mckerrow, W. S. (1990). Revised World maps and introduction. Palaeozoic, Palaeogeography and Biogeography, *Geological Society of London, Memoir*, **12**, 1–21.

Sun Yangzhu and Xiang Liwen (1979). Late Upper Cambrian trilobite fauna from western Yunnan. *Bulletin of the Chinese Academy of Geological Sciences* **1**, (1), 1–133 (in Chinese).

Wang Hongzhen (1985). *Atlas of the palaeogeography of China*. Cartographic Publishing House, Beijing (in Chinese).

Wang Hongzhen, Xu Chengyan, and Zhou Zhengguo (1982). Tectonic development of the continental margins on both sides of the Palaeo-Qinling marine realm. *Acta Geologica Sinica* **56**, (3), 270–80 (in Chinese).

Wang Quan and Liu Xueya (1979). The drift of Yangtze continental segment and the early action of Gondwana Oldland–an important tendency of dynamic action of the Phanerozoic globe. *Journal of Changchun College of Geology* **33**, (1), 1–7 (in Chinese).

Wang Shenglong (1984). On the palaeoclimate condition in the formation of marine phosphorite deposits. In *Symposium of the 5th International field Workshop and Seminar on Phosphorite*, pp. 307–18. Geological Publishing House. Beijing (in Chinese).

Wang Zhihao (1984). Late Cambrian and Early Ordovician conodonts from North and Northeast China with comments on the Cambrian-Ordovician boundary. In *Stratigraphy and Palaeontology of Systemic Boundaries in China* **2**, 195–258. Anhui Science and Technology Publishing House, Wuhu (in Chinese).

Wittke, H. W. (1984). Middle and Upper Cambrian trilobites from Iran: their taxonomy, stratigraphy and significance for provincialism. *Palaeontographica*, Abt. A, Bd. **183**, 91–161, pls. 1–11.

Yang Jialu (1978). Middle and Upper Cambrian trilobites of western Hunan and eastern Guizhou. *Professional Papers of Stratigraphy and Palaeontology* **4**, 1–82 (in Chinese).

Yang Jialu (1987*a*). The Cambrian biogeography and the palaeocontinent reconstruction in China and its adjacent areas. In *Palaeo-tectonogeography and palaeobiogeography of China and adjacent regions, Abstracts of papers in treatise contributed to the third all-China congress on tectonics* (ed. Wang Hongzhen). China University of Geoscience Press, Wuhan (in Chinese).

Yang Jialu (1987*b*). An attempt at the distribution of Cambrian biogeographical provinces in China. *Collected Papers of Lithofacies and Palaeogeography* **3**, 99–113 (in Chinese).

Yang Jialu, Hu Chunsheng, and Jiang Xinsheng (1984). On new trilobite zones of the early Late Cambrian Laochatian Formation. *Earth Science-Journal of Wuhan College of Geology* **25**, (2), 23–31 (in Chinese).

Yuan Kexing and Zhang Sengui (1980). Lower Cambrian Archaeocyatha of central and southwestern China. *Acta Palaeontologica Sinica* **19**, (5), 380–92 (in Chinese).

Zhang Quanzhong and Zhou Zemin (1985). Early Cambrian trilobite fauna of Kunshan, Jiangsu. *Acta Palaeontologica Sinica* **24**, (2), 259–66 (in Chinese).

Zhang Wentang and Zhu Zhaoling (1979). Notes on some trilobites from the Lower Cambrian Houjiashan Formation in southern and southwestern parts of North China. *Acta Palaeontologica Sinica* **18**, (6), 513–25 (in Chinese).

Zhang Wentang, Lu Yanhao, Zhu Zhaoling, Qian Yiyuan, Lin Huanling, Zhou Zhiyi, Zhang Sengui, and Yuan Jinliang (1980). Cambrian trilobite faunas of southwestern China. *Palaeontologia Sinica*. Whole Number **159**, New Series B, 16. Science Press, Beijing (in Chinese with summary in English)

Zhang Sengui (1982). Oldest trilobite in the world and correlation of some archaeocyathid assemblages. *Acta Palaeontologica Sinica* **21**, (5), 517–24 (in Chinese).

Zhang Sengui (1983). Early Cambrian archaeocyathids from Kuruktag, Xinjiang. *Acta Palaeontologica Sinica* **22**, (1), 9–19 (in Chinese).

Zhou Zhiyi and Yuan Jinliang (1980). Lower Cambrian trilobite succession in southwest China. *Acta Palaeontologica Sinica* **19**, (4), 331–9 (in Chinese).

Zhu Hong and He Zhengxuan (1987). Study on the paleomagnetism of the Palaeozoic in the northern margin of the Alxa Massif and discussion on the evolution of plate tectonics. In *Palaeozoic biostratigraphy and tectonic evolution of the Alxa Massif margin* (ed. Zhu Hong, Zheng Zhanchang, and H Xinyi), pp. 154–9. China University of Geosciences Press (in Chinese).

Ziegler, A. M., *et al.* (1979). Paleozoic paleogeography. *Annals of Earth Planetary Science*, **(7)**, 473–502.

Zonenshayn, L. P., and Gorodnitskiy, A. M. (1977). Paleozoic and Mesozoic reconstruction of the continents and oceans. Article 1. Early and middle Paleozoic restructions. *Geotectonics* **11**, 83–94.

4 ORDOVICIAN

Li Zhiming

4.1 ORDOVICIAN PALAEO-BIOGEOGRAPHICAL PROVIN-CIALIZATION OF THE WORLD

A number of new marine invertebrate faunas came into being in Ordovician times, and the generally rather monotonous features of the Cambrian biota were changed. Among the fauna were trilobites, graptolites, articulate brachiopods, nautiloids, and rugose and tabulate corals, which are of great significance in stratigraphical correlation and subdivision, biofacies analysis, and palaeobiogeographical provincialization. Many scholars, such as Whittington and Hughes (1972), Jaanusson (1973), Berry (1974, 1979), Lu Yanhao *et al.* (1976), Chen Junyuan (1980), Sweet and Bergstrom (1984), Cocks and Fortey (1988, 1990), Bergstrom (1990), and Finney and Chen (1990), have studied Ordovician palaeobiogeographical provincialization based on the groups of organisms that they specialized in. Although the principles on which their studies were based and the boundaries that they proposed are different from author to author, most of them have mentioned latitudinal control over the global-scale distribution of fauna, especially over the distribution of planktonic graptolites and nektonic nautiloids. Cocks and Fortey (1990) indicated the latitudinal control especially on benthic cratonic faunas. Evidently, the latitude-controlled global-scale climate zonation played a role in Ordovician palaeobiogeographical provincialization, because the southward shift of Gondwana placed a large part of the continent in medium and high latitudes. The rest were still scattered plates, situated mostly in low latitudes, but a few of them were also distributed in both medium and high latitudes (Fig. 4.1). The situation in the Cambrian Period, when all the land blocks were near the equator, was changed.

As for China, a few palaeomagnetic data and the fact that gypsum occurred in the Middle Ordovician of North China (Fengfeng Formation) show that North China was an arid region in low latitude, whereas the fauna in the vast area of central–southern and south-western China was highly diversified, indicating a near-equator situation. Many corals occurred in the Late Ordovician of western Zhejiang, eastern Qinling, the Shaanxi–Gansu–Ningxia Basin, and Xinjiang, indicating that all these regions were

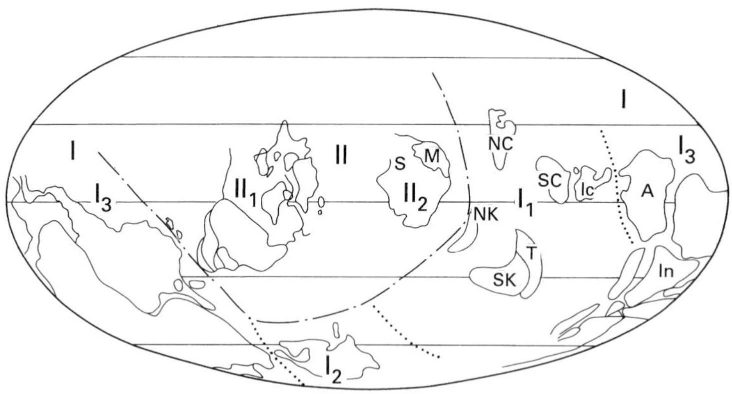

Fig. 4.1 Map of the Ordovician palaeobiogeographical provincialization of the world. I, Austral realm; I_1, China-European region; I_2, Baltic–Scandinavian region; I_3, South America–Australian region; II, North America–Siberian realm; II_1, North American region; II_2, Siberian region; A, Australia; In, India; Ic, Indochina; M, Mongolia; NC, North China; NK, North Kazakhstan; S, Siberia; SC, South China; SK, South Kazakhstan; T, Tarim.

located in tropical areas. To sum up, all parts of China were located in the middle-lower latitudes in the Ordovician.

The Ordovician palaeobiogeographical provincialization is quite controversial, based on studies of different taxa, among which the trilobites have been studied in detail. It also differs from Early to Late Ordovician, with the Early Ordovician (excluding Tremadoc) biogeographical provincialization being the most prominent. For the Early Ordovician, the world can be divided into a North America–Siberian realm and a Meridional realm. The North America–Siberian realm can be subdivided into the North American and Siberian regions. The Meridional realm can be subdivided into the Baltic–Scandinavian region, the Sino-European region, and the South America–Australian region. The Middle and Late Ordovician palaeobiogeographical pattern is generally similar to that of the Early Ordovician, except that the dispersion of fauna blurred the distinction of each region.

During the Ordovician, China extended across two regions. Xingan province and Altay province, to the north of the line from the Ertix River to the Xar Moron River, belong to the Siberian region. The remaining parts belong to the Sino–European region, where many endemic organisms occurred, including trilobites, such as trinucleids, taihungshaniids, hommatocnermids, and *Cyclopyge*; cephalopods, such as Manchuroceratidae, Lituitidae, and *Sinoceras*; and Late Ordovician tabulate corals, such as agetolitids. In comparison with the Cambrian Period, the western section of the northern boundary of the realm moved northward from the Ebinur–Gaxunnur line to the Ertix River line.

4.2 ORDOVICIAN PALAEOBIOGEOGRAPHICAL PROVINCIALIZATION OF CHINA

4.2.1 INTRODUCTION

Ordovician palaeogeography

There were various kinds of sedimentary belts, reflecting different marine environmental regimes, in the Ordovician of China (Shen 1974; Lu *et al.* 1976; Wang 1980; Lai *et al.* 1982), including epicontinental sea (North China and southern Tibet), shelf (Yangtze and the rest of Tibet), shelf margin

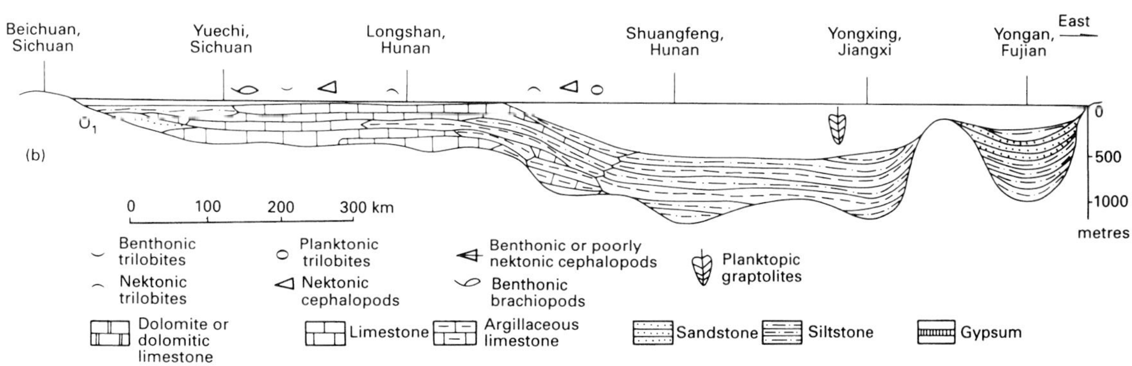

Fig. 4.2 Model for the Ordovician environmental-ecological types of (a) North China and (b) South China.

(Jiangnan, see Fig. 4.3; Shaanxi–Gansu–Ningxia Basin), and deeper sea (South China, Tianshan, and Xingan). These belts were distributed regularly in each small plate from centre to edge, signifying gradual changes from shallow water to deep water. For example, the major part of North China was epicontinental sea, which deepened northward and westward, and gradually turned into shelf and marginal sea (Fig. 4.2a); and the major part of the Yangtze region was shelf sea, and changed gradually south-eastward into shelf margin and deeper waters (Fig. 4.2b). Such changes can be identified in each land block, but in some cases the change is clear, while in other cases it is not easy to distinguish.

Relationship between biotas and environments

The main Ordovician groups were trilobites, graptolites, brachiopods, cephalopods, corals, stromatoporoids, bryozoans, echinoderms, and conodonts. Different fauna were adapted to different sedimentary environments, as expressed by facies (Fig. 4.2). The North China Platform has mainly shallow carbonate deposits, in which the principle fossils are nautiloids with poor swimming ability, and benthic gastropods, but in the early stage of the Early Ordovician (the base of the Yeli Formation) some planktonic graptolites occurred due to deepening of the sea. Lithofacies are varied in the main body of the Yangtze Platform from Lower to Upper Ordovician, and rocks of each series are variable in space, but generally two main lithofacies may be identified: one is an argillaceous carbonate (including the Fenxiang Formation, part of the Dawan Formation, the Guniutan Formation, Baota Formation and Linxiang Formation, part of the Miaopo Formation, and part of the Tongzi Formation); the other is a fine clastic arenaceous–pelitic facies (including part of the Dawan Formation, part of the Miaopo Formation, part of the Tongzi Formation, and the Meitan Formation). The argillaceous carbonate facies is dominated by shelly organisms, mainly benthos (trilobites, brachiopods, stromatoporoids, and bryozoans), and also some nektonic cephalopods. The fine clastic arenaceous–pelitic facies yields planktonic organisms such as graptolites. Thus they formed a distinct biological assemblage, different from that in North China, and belonging to a shallow shelf assemblage with extremely abundant organisms. The same and similar assemblages are found in the Middle Ordovician in Tibet, on the western border of the North China platform, and the northern fringe of Tarim. In Jiangnan, the Early Ordovician and the lower part of the Middle Ordovician (Ningguo Formation, Hulo Formation) are composed of argillosiliceous facies; the upper part of the Middle Ordovician and the lower part of the Upper Ordovician are of calcareous pelitic deposits (Yenwashan Formation, Huangnigang Formation); in the upper Upper Ordovician, it changes into carbonaceous–siliceous facies. The main organisms accompanying such sedimentary facies are plankton (graptolites) in the Early Ordovician; benthic brachiopods, trilobites, and nektonic cephalopods in the late Middle Ordovician and the early Late Ordovicians; and planktonic graptolites again in the late Late Ordovician. The Ordovician as a whole in south-eastern China consists of flysch formations dominated by planktonic graptolites.

From the above, it can be concluded that the distribution of organisms was closely related to the sedimentary facies, and different ecological groups occurred in different tectono-sedimentary belts of the sea; put another way, the environment controlled the biotic type. Cocks and Fortey (1990) showed a clear Early Ordovician (Arenigian) ecological differentiation from platform trilobites, pelagic trilobites, to marginal deep-water isograptids in Yangtze. They have recognized a dikelokephalinid platform trilobite assemblage, a deeper pelagic *Carolinites* assemblage, and a deep-water isograptid and olenid assemblage. Mu Anzhi (1983) made a summary of the ecological types of the Ordovician in which he distinguished three types: North China type (epicontinental shallow-sea, benthic, carbonate deposits), Central China type (shelf sea, mixed benthic and planktonic organisms, carbonate and sandshale deposits), and South China type (deep sea planktonic organisms, sandshale and carbonate deposits). In fact, this classification is a reflection of the ecological differentiation of organisms from near-shore to offshore on a block. Such a model is suitable for the distribution of organisms in each block of China.

Geographical distribution of the main taxa

In the Ordovician palaeobiogeographical provincialization, stress is placed on the geographical distribution of cephalopods, trilobites, corals, and, to a lesser extent, brachiopods and graptolites. The available data from the fossil record show that the Early Ordovician contains 153 genera of trilobites, 104 of genera of brachiopods, 107 genera of cephalopods, and 65 genera of graptolites; the Middle Ordovician contains 142 genera of trilobites, 106 genera of brachiopods, 94 genera of cephalopods, 57 genera of graptolites, and 17 genera of corals (of which five genera are rugose corals); the Late Ordovician has 90 genera of trilobites, 89 genera of brachiopods, 22 genera of cephalopods, 49 genera of tabulate corals, 32 genera of rugose corals, and 22 genera of graptolites. The distribution of these organisms was correlated with water depth, nature of the sea-bottom, and depositional conditions.

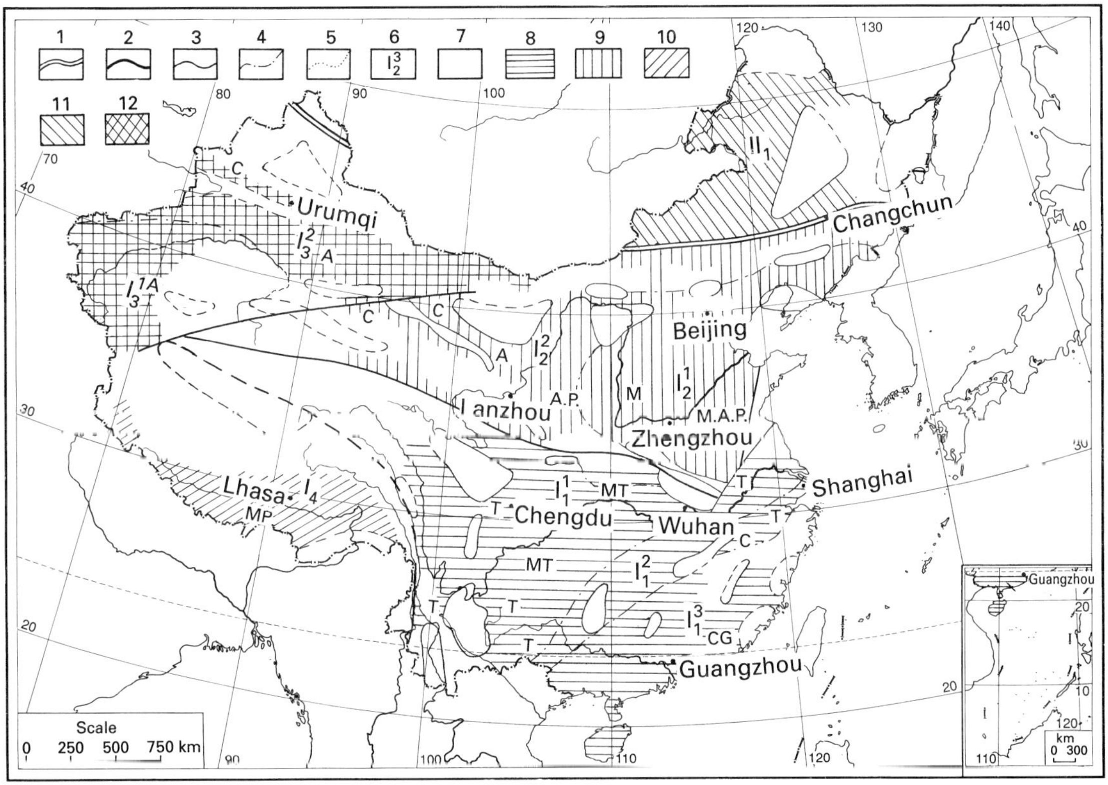

Fig. 4.3 Map of the Early Ordovician palaeobiogeographical provincialization of China. 1, Boundary of realms; 2, boundary of regions; 3, boundary of provinces; 4, boundary of biomes; 5, boundary of communities; 6, roman digit = realm mark, subscript = province mark, superscript = biome mark; 7, Oldland. 8–12, Marine and paralic regions: 8, South China; 9, North China; 10, Gondwana; 11, Junggar–Xingan; 12, transitional. I, Austral realm, China–European region; I_1, South China province; I_1^1, Yangtze biome; I_1^2, Jiangnan biome; I_1^3, South-eastern China biome; I_2, North China–Qilian province; I_2^1, North China biome; I_2^2, Ordos–Qilian biome; I_3, Tarim province; I_3^1, Keping biome; I_3^2, Kuruktag biome; I_4, Tibet province; II, North America–Siberian realm, Siberian region; II_1, Xingan province; M, Manchuroceratidae; A, Armenoceratidae; P, Polydesmidae; T, Taihungshanidae; C, *Cardiograptus*; G, *Glossograptus*.

Table 4.1 Correlation chart of the Ordovician of China

<table>
<tr><td colspan="3"></td><th colspan="6">S. China province</th><td></td></tr>
<tr><td colspan="3" rowspan="2"></td><th colspan="3">Yangtze subprovince</th><th rowspan="2">Jiangnan Western Zhejiang</th><th colspan="2">Southern China</th><th rowspan="2">Northern China</th></tr>
<tr><th>Western Huber</th><th>Northern Sichuan</th><th>Northern Guizhou</th><th>Western Jiangxi</th><th>Hainan</th></tr>
<tr><td rowspan="14">Ordovician</td><td>Upper</td><td>Ashgill</td><td>Guanyinqiao
Wufeng</td><td>Wufeng</td><td>Wufeng</td><td>Sanjushan
Wufeng
Changwu</td><td>Wufeng</td><td>Gangoucun</td><td rowspan="2"></td></tr>
<tr><td rowspan="3">Middle</td><td rowspan="2">Caradoc</td><td>Linxiang</td><td>Jiancaogou</td><td>Linxiang</td><td>Huangnigang</td><td>Shikou</td><td>Jianling</td></tr>
<tr><td>Baota</td><td>Baota</td><td rowspan="2">Baota</td><td>Yenwashan</td><td>Hanjiang</td><td rowspan="2">Shatang</td><td rowspan="2">Fengfeng</td></tr>
<tr><td>Llandeilo</td><td>Miaopo</td><td rowspan="2">Shizipu</td><td>Hulo</td><td>Hulo</td></tr>
<tr><td rowspan="10">Lower</td><td rowspan="2">Llanvirn</td><td rowspan="2">Guniutan</td><td rowspan="5">Meitan</td><td rowspan="6">Ningguo</td><td rowspan="8">Ningguo</td><td rowspan="10">Yahua</td><td rowspan="3">Upper Majiagou</td></tr>
<tr><td rowspan="3">Xiliangsi</td></tr>
<tr><td rowspan="5">Arenig</td><td rowspan="2">Dawan</td></tr>
<tr><td rowspan="4">Lower Majiagou</td></tr>
<tr><td rowspan="3">Honghuayuan</td><td rowspan="3">Zhaojiaba</td></tr>
<tr><td rowspan="2">Honghuayuan</td></tr>
<tr><td rowspan="4">Yinzhubu</td></tr>
<tr><td rowspan="3">Tremadoc</td><td rowspan="2">Fenxiang</td><td rowspan="3">Tongzi</td><td rowspan="3">Tongzi</td><td rowspan="2">Liangjiashan</td></tr>
<tr><td rowspan="2">Yinzhubu</td></tr>
<tr><td>Nanjinguan</td><td>Yeli</td></tr>
</table>

The cephalopods were distributed mainly in the stable regions (platforms) (Chen 1980). Relatively abundant cephalopods have been found in the Early Ordovician in Yangtze, the Keping district of Tarim, Tibet, North China, and the Shaanxi–Gansu–Ningxia region. They were not abundant in the early Early Ordovician (Xinchang or Yeli age), but a large number of new families and genera appeared in the late Early Ordovician (Ningguo, or Liangjiashan plus Majiagou ages), which formed the actinoceratid fauna in North China and endoceratid fauna in Yangtze. In the Middle Ordovician, owing to the shallowing of pre-existing deep-water areas, the distribution of cephalopods extended from the previous areas to Zhejiang– Jiangxi, northern Tibet, western Yunnan, and Guanniaohe (Heilongjiang). However, the distribution area of cephalopods in the Late Ordovician diminished significantly due to uplift of the North China Platform and the spread of reducing conditions on the sea bottom in the late Late Ordovician, which caused the deposition of black shale in many areas (Wufeng shale).

The graptolites were distributed mainly in argillaceous facies (in normal and abnormal environments) and fine clastic rock facies, and usually did not coexist with cephalopods. The South-east type (Pacific type) of graptolites was distributed relatively extensively, while the Yangtze type (Atlantic type) was limited to a small area (Mu 1980, 1983). The Wufengian graptolite fauna was the most abundant in the Late Ordovician throughout the world, and was distributed extensively (in Yangtze, Jiangnan, the eastern part of northern Qilian, and western Yunnan).

N. China–Qilian province			Tarim-Junggar province			Tibet province		Hinggan
Ordos–Qilian			Tarim	Middle Tianshan	Junggar	Shanzha	Nyalam	
Shaanxi–Gansu–Ningxia	Zhouzishan, Nei Mongol	Yumen, Gansu						
Beiguoshan	Baiyanhuashan	Nanshimenzi	Yingan	Hudukedaben	Basitawukuduke	Gangmusang	Hongshantou	Aihui
	Taoqupo				Jiabosa'er			
	Sheshan							
Chedao	Gongwusu	Yaomoshan	Qilang	Aketashi	Keziletesikela	Ke'erduo	Jiacun Group (upper part)	Luohe
Pingliang	Lashizhong		Kanling	Kekesaleixi				Duobaoshan
	Wulaike							
Sandaogou	Kelimoli	Yingou	Sa'ergan	Talejihe		Mende'ema	Jiacun Group (lower part)	Xiqiuhe
Shuiquanling	Zhuozishan		Qiulitag	Fenggou				
Machuan	Sandaokan			Xin'ertai			Rouqiecun (upper part)	

The trilobites were distributed mainly in shales and argillaceous limestones, but very scarcely in pure limestone, dolomite, and argillaceous rocks that were formed in abnormal environments (Lu 1975). Corals in China began to occur in the Early Ordovician (Majiagou Formation); were sparsely scattered in the Middle Ordovician, with a few genera; and developed widely during the Late Ordovician in eastern and western Junggar, central Tianshan, the northern fringe of Qaidam, Qilian, the Shaanxi–Gansu–Ningxia Basin, southern Qinling, southern Guizhou of the Yangtze region, and western Zhejiang of the Jiangnan region (Yu 1960; Yu *et al.* 1963). Most of these areas are distributed along the margins of stable blocks.

In his palaeobiogeographical work, Bergstrom (1990) made comparisons of South and North China with other areas. For the Arenigian, South China was similar to the Baltic and North American interior, while North China (including southern north-east China) was an independent province. For the late Llanvirnian to Llandeilian, South China was basically of Baltic type. For the late Caradocian to Early Ashigillian, South China remained of Baltic type. There are no Middle and Late Ordovician data for North China.

The stratigraphical scheme (Table 4.1) is adopted from that proposed by the Nanjing Paleontological Institute (1981), but the Guanyinqiao horizon and its corresponding horizon were put in the Upper Ordovician. We have only drawn two palaeobiogeographical maps, i.e. Early and Late Ordovician (Figs 4.3 and 4.8, respectively), because of the fact that the outlines of the provincialization of Middle and Late Ordovician are roughly the same, except for the indistinctness of the latter. For the Middle Ordovician biogeographical provincialization please refer to the Late Ordovician provincialization map (Fig. 4.8).

4.2.2 EARLY ORDOVICIAN PALAEOBIOGEOGRAPHICAL PROVINCIAZATION OF CHINA

Palaeobiogeographical provincialization of Early Ordovician cephalopods

From the Early Ordovician times, organisms became much more diverse compared with the Late Cambrian. The biogeographical provincializations are different between the early Early Ordovician (Tremadocian) and the late Early Ordovician (Arenigian). The cephalopods from the Nanjinguan Formation, Fenxiang Formation, and Honghuayuan Formation in Yangtze belong mainly to the Ellesmeroceratidae, Proterocameroceratidae, and Manchuroceratidae, many taxa of which were in common with those in North China, such as *Manchuroceras*, *Coreanoceras*, and *Hopeiceras* (Table 4.2). In the Dawan Formation, the cephalopods changed greatly, and such families as the Endoceratidae, Protocycloceratidae, and Lituitidae became dominant and typical, and consequently formed the distinctive feature of the Yangtze region. It is generally called the *Endoceras–Protocycloceras* fauna, disparate from those in North China but relatively close to those in the Baltic region.

The cephalopods in Jiangnan occured in the Hangzhou and Wulingshan areas, and were

Table 4.2 Occurrence of the main Early Ordovician cephalopods in the regions of China

Genus	Yangtze province			N. China province			
	Upper Yangtze	Lower Yangtze	Jiangnan	N. China	Ordos–Qilian	Tarim	Tibet
Ellesmeroceras	×	×	×	×		×	
Manchuroceras	×			×			×
Coreanoceras	×			×			
Hopeiceras	×			×			×
Protocycoloceroides	×	×	×		×	×	
Vaginoceras	×	×	×				
Chisiloceras	×		×			×	
Meiltanoceras	×				×		×
Trifucatoceras	×						
Bothoceras	×						
Armenoceras				×	×	×	
Wutinoceras				×	×		×
Nybyoceras				×			
Cyrtonybyoceras				×			
Ordosoceras				×	×		×
Polydesmia				×	×		
Stolbovoceras				×	×		
Tianshanoceras						×	
Kalpinoceras						×	
Qiulitagoceras						×	
Ceorgina							×
Ancistroceras	×				×		

mainly *Endoceras*, *Vaginoceras*, *Cameroceras*, *Chisiloceras*, *Cyclostomiceras*, *Richardsonoceras*, *Dideroceras*, *Protocycloceras*, *Bactroceras*, *Eosomichelinoceras*, and *Centronoceras*, dominated by the family Endoceratidae. Compared with those in the Yangtze region, there were fewer genera and a lack of members of the family Lituitidae. Almost all the genera of cephalopods also occurred in Yangtze.

The cephalopods in North China prior to the Liangjiashan age were rather abundant and characterized by Ellesmeroceratidae, Bassleroceratidae, and Manchuroceratidae, many genera of which were the same as those in Yangtze. The Lower Majiagou Formation was typical for its abundant armenoceratoids, which was a distinctly different cephalopod fauna from that in the Yangtze province (dominated by endoceratids) and quite similar to those in Siberia and central parts of North America. Here the armenoceratoid family, Polydesmidae, is only found in North China. Limited cephalopods occured in the Shaanxi–Gansu–Ningxia Basin margin and Qilian region, mainly in the late Early Ordovician (corresponding to the Lower Majiagou Formation) including *Wutinoceras*, *Armenoceras*, *Endoceras*, *Meitanoceras*, *Kirkoceras*, *Protocycloceroides*, *Shanxiceras*, and *Ancistroceras*. The occurence of armenoceratoids shows that a formation should belong to the North China type, but endoceratids were also present in a certain proportion. Besides, *Ancistroceras* (Baltic–Scandinavian region) and *Protocycloceroides* of the Lituitidae family, which are characteristic of Yangtze, indicate some relation with this region.

According to Chen Ting'en (1981), the cephalopods in Tibet, mainly found in southern Tibet (Jiacun Group), consist of 12 genera, five of which were found in Yangtze and many of which were typical members of North China. Therefore, he held that they belonged to the same biota as North China.

The cephalopods in Tarim occured mainly in the Keping district (Qiulitage Group), with many endemic genera (such as *Tianshanoceras*, *Quilitagoceras*, *Kalpinoceras*, and *Subaxiceras*), also with some members such as *Stereoplasmocerina* and *Murryoceras* found in North China, and some common members of the Yangtze region, such as *Protocycloceroides*, *Vaginoceras*, and *Chisiloceras*. The other areas yielding cephalopods include Kuruktag, Yidun–Zhongdian, and southern Qinling, but with fewer fossils. The Kuruktag fauna is similar to that of Keping, while those of Yidun-Zhongdian and southern Qinling are similar to that of the Yangtze region. Thus, the whole of China seems to be within one biogeographical region in the early Early Ordovician, based on cephalopods, characterized by manchuroceratoids. In the late Early Ordovician, the southern and northern parts were evidently differentiated: North China was dominated by armenoceratoids, which could be called a armenoceratoid fauna, while the Yangtze region was dominated by the Protocycloceratidae and Endoceratidae, with the appearance of lituitoid members, where the faunal characteristics were very similar to those in Europe. North China, although greatly different from Yangtze, still yielded Yangtze members along the margin of the platform. The Tarim fauna was closely related to that of the Yangtze, but showed a strong endemic flavour. The Tibetan fauna was similar to that of North China, but with fewer genera and species. Thus, based on cephalopods, four provinces can be established: Yangtze, Tibet, Tarim, and North China.

Palaeobiogeographical provincialization of Early Ordovician trilobites

Early Ordovician trilobites were distributed widely except in Tibet, Junggar, and south-eastern China. The trilobites found in North China were limited in number, and distributed sporadically, mainly occurring in the basal part of the Yeli Formation and Majiagou Formation, and in the Kelimoli Formation of the western margin of the Shaanxi–Gansu–Ningxia Basin. The trilobites in North China include many North American members, such as *Hystricurus*, *Bathyurus*, *Lelosteum*, *Jujuyaspis*, and *Symphysurus*, in addition to many endemic genera, such as *Eoisotelus*, *Tienshifuia*, *Penchiopsis*, and *Laishuiia* (Table 4.3). These two types took up a dominant proportion, but the common genera in the Yangtze province, such as *Dikelokephalina*,

Table 4.3 Occurrence of the main Early Ordovician trilobites in the regions of China

	Yangtze province		N. China province			Tarim	
Genus	Yangtze	Jiangnan	N. China	Zhouzishan Hills	Qilian	Keping	Tianshan–Kuruktag
Asaphopsis	×	×					×
Dikelokephalina	×	×	×				
Loshanella	×						
Psilocephalina	×	×					
Symphysurus	×	×	×	×		×	
Ptychopyge	×	×	×	×		×	
Eoisotelus		×?	×				
Norinia		×				×	×
Taihungshania	×	×					
Tungtzuella	×						
Ceratopyge		×			×		
Hysterolenus		×					×
Hanchungolithus	×						
Ningkianolithus	×						
Triarthrus		×			×		
Inkouia		×			×		×
Hystricurus	×	×	×				
Szechuanella	×	×			×		
Bathyurus			×				
Onychopyge			×		×		
Tienshifuia			×				
Illaenus	×			×		×	
Geragnostus	×	×					×
Harpides	×					×	×
Endymionia				×	×		

Wanliangtingia, *Apatokephalus*, and *Ptychopyge*, were also found in this region. Most of the trilobites found on the western margin of the North China platform, except a few endemic genera (*Ordosaspis*) occur in Yangtze.

The Yangtze province has the most abundant trilobites in China, which had begun to show a strong endemic aspect since the early Early Ordovician. Quantitative work on the Early Ordovician trilobites found in the region, amounting to 109 genera, has shown that 42 of them are endemic genera (38.5 per cent of the total number of genera). The early Early Ordovician (Nanjinguan Formation, Fengxiang Formation, and Hunghuayuan Formation) was dominated by families of Taihungshaniidae (*Taihungshania*, *Omeiopsis*, *Tungtzuella*), Dikelocephalinidae, and Loshanellidae, as well as by *Lohanpopsis*, *Chungkingaspis*, *Sichuanella*, and *Psilocephalina*, which were unique to Yangtze, and also yielded some common genera of Eurasia;

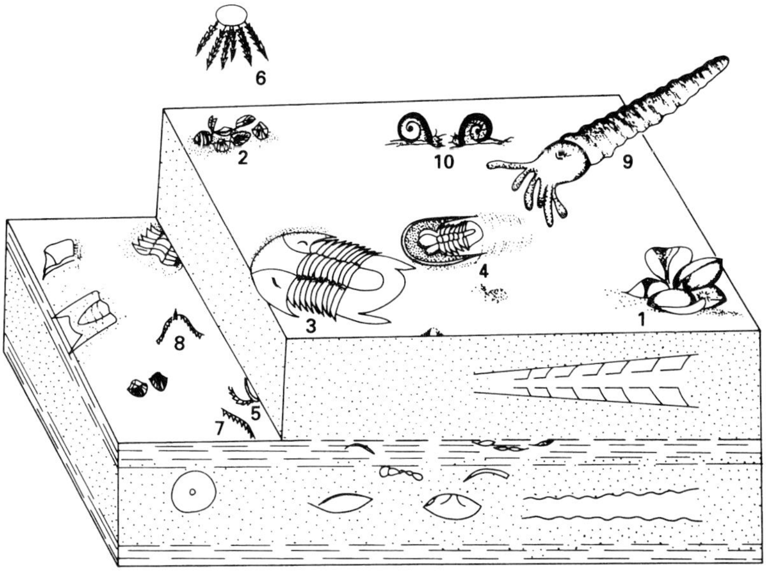

Fig. 4.4 Reconstruction of the *Yangtzeella–Taihungshania–Sinorthis* community of the Early Ordovician Dawan Formation in Yangtze province. 1, *Yangtzeella*; 2, *Sinorthis*; 3, *Taihungshania*; 4, *Ningkianolithus*; 5, *Tetragraptus*; 6, *Glyptograptus;* 7, *Azygograptus*; 8, *Didymograptus*; 9, *Protocycloceras*; 10, *Ecculiomphalus*.

Taihungshania was also found in southern Europe and Turkey. In the late Early Ordovician (Dawan Formation, Guniutan Formation), many trinucleids appeared, which were mainly found in the Middle Ordovician in Eurasia. The Hanchungolithinae were found only in southern China (with four genera, i.e. *Hanchungolithus*, *Ningkianolithus*, *Yinpanolithus*, and *Ceratolithus*). Meanwhile, members of the *Taihungshaniidae* family still survived. If we were to name the province after trilobites, it could be called Taihungshaniidae–Hanchungolithinae province (Fig. 4.4).

More Early Ordovician trilobites were found in the Jiangnan district (Lu Yanhao *et al.* 1976); they belonged to the Olenidae (such as *Triarthrus*, *Inkouia*, and *Parabolinella*), Ceratopygidae (such as *Hysterolenus*, *Ceratopyge*), and the genus *Symphysurus*. The former two families were the most important non-agnostid trilobites of Jiangnan type in the Late Cambrian, and were absent or scarce in North China and Yangtze. However, these genera, except *Inkouia* and *Parabolinella*, although scarce, are still found in Yangtze, and some Yangtze members, such as *Taihungshania*, *Shumardia*, *Psilocephalina*, and *Dikelokephalina*, occurred in the Jiangnan region. Hence there is a difference between the two regions, but their relationship is still quite close.

In the Keping district of Tarim, trilobites were found in the upper subgroup of the Quilitag Group, the major members of which were *Hammatocneus*, *Illaenus*, *Ptychopyge*, *Pseudocalymene*, *Nileus*, *Norinia*, and *Symphysurus*. The former five genera are common members of the Dawan Formation in Yangtze, while the latter two are mainly distributed in Jiangnan. Trilobites in Kuruktag (north-eastern Tarim) and central Tianshan differ from those in the Keping district. In Tianshan–Kuruktag a lot of Jiangnan members occur, such as *Geragnostus*, *Leiagnostus*, *Hysterolenus*, *Inkouia*, *Symphysurus*, *Trinodus*, and *Norinia*, of which planktonic trilobites account for a considerable proportion.

The trilobites found in Qilian include *Onychopyge*, *Inkouia*, *Ceratopyge*, *Triarthrus*, *Shumardia*, *Symphysurus*, *Apatokephalus*, *Szechuanella*, *Leiostegium*, *Pseudostricurus*, *Parabolinella*, *Harpides*, *Nileus*, *Yinaspis*, *Kainella*, and *Geragnostus*, of which many members are in common with Jiangnan, also including members typical of Yangtze and North China. Other areas yielding trilobites are western

Yunnan and north-eastern China. Trilobites are scarce in north-eastern China, and the nature of the fauna here is yet not known. Western Yunnan yields Hanchungolithinae and Taihungshaniidae, as found in the Yangtze province.

From the main trilobites mentioned above, we know that the Early Ordovician trilobites in all parts of China are related but still distinguishable among different provinces. The Yangtze province yields the largest number of genera and species, of which endemic genera take up a large proportion, and is most similar to Europe. The trilobite fauna in Jiangnan differs greatly from that in Yangtze, but the typical Yangtze members can be found in Jiangnan and the common members shared by both regions takes up nearly half of the total genus number in Jiangnan. The difference between the two regions is caused by the different biofacies and environments, hence they can be considered as two different biomes in one province. The trilobite fauna in western Yunnan shows no difference from that in Yangtze. North China is distinguished from Yangtze by more North American elements and the lack of a few typical Yangtze elements, such as taihungshaniids, triuncleids, and loshanellids. All the trilobite genera reported from the western margin of the North China Platform, except one endemic genus (*Ordosaspis*), can be found in Yangtze. Thus it may be said that although they belong to different provinces, the features of their biotas and environments are similar. The trilobites in Qilian reflect the intermixing of elements from various regions, but are mainly of North China type. The trilobites in the Keping district of Tarim are close to those in Kazakhstan and, to a lesser degree, to those in Yangtze, although a few typical members are lacking; the marginal parts, such as central Tianshan and Kuruktag, yield trilobites close to those in Jiangnan. In Junggar there has not yet been a detailed report on Early Ordovician trilobites. Thus, according to the trilobite data available, three provinces are established: North China province, including North China and the Ordos–Qilian biomes; South China province, including the Yangtze and Jiangnan biomes; and Tarim province, including the Keping and Tianshan–Kuruktag biomes.

Palaeobiogeographical provincialization of Early Ordovician graptolites and brachiopods

The graptolites in China have always been provincialized into an Atlantic or Central China (Yangtze) type, and a Pacific or South China (Pearl River, or isograptid) type. Such provincialization has evoked two major schools of interpretation. Cocks and Fortey (1990), among others, attributed the two types to the adaptation of graptolites to different water depths: the former type being epiplankton, thus restricted to shallow epicratons, while the latter lived below the thermocline, and was thus extracratonic. On the other hand, Finney and Chen (1990) and Berry and Wilde (1990) preferred the influence of latitude and temperature or water-mass specificity, combining the influence of ocean currents and latitudinal temperature gradient. According to them, the Atlantic type was temperate while the Pacific type was tropical.

In the Yangtze type (or the Atlantic graptolite fauna), the Early Ordovician was characterized by a great number of pendent and deflexed *Didymograptus* as well as *Azigograptus*, and aspects of the South-China-type graptolites, *Cardiograptus* and *Didymograptus*. Thus the Yangtze region is considered to be of a transitional type. The Jiangnan region yielded South-China-type graptolites, which were rather abundant, and some special members such as *Oncograptus*, *Cardiograptus*, *Paracardiograptus*, *Glossograptus*, *Skiagraptus*, and Sinograptidae. This fauna is also called the Pacific graptolite fauna. The Pearl River region (south-east region) also yielded South-China-type graptolites, but the sediments were all graptolite shale facies, and its graptolites were not so abundant as in Jiangnan; moreover, a few typical members of Jiangnan, such as Sinograptidae, *Pterograptus*, and *Nicholsonograptus* have not been found. The Early Ordovician graptolites in western Yunnan were of Central China type. In central Tianshan and Qilian, they were South China type. The graptolites in North China mainly occurred in the bottom part of the Lower Ordovician; the dendroids were the most abundant, with many endemic genera and species. It was called the North China type by Wang Xiaofeng (1980).

The brachiopods were distributed mainly in the Yangtze region, where many endemic members occurred, such as *Yangtzeela*, *Sinorthis*, *Lepidorthis*, *Metorthis*, *Xinanorthis*, *Martella*, and *Pseudomimella* (see Fig. 4.4). The branchiopods in North China are mainly found in the Yeli Formation; the major representatives are *Finkelnburgia*, *Eosatrematorthis*, *Inbricatia*, *Plectorthis*, and *Tritoechia*, very similar to those in Yangtze. In addition, brachiopods are found in Qilian, Xingan, and Ordos, but are rare and scattered.

Cluster analysis of Early Ordovician cephalopods

Cluster analysis has been undertaken for the Early Ordovician cephalopods (Fig. 4.5). Data of 107 genera have been collected throughout China, and their localities grouped into 20 areas, some of which are very large because very few cephalopod genera have been found. This must have affected the accuracy of the statistics; nevertheless, a palaeobiogeographical pattern, similar to that described above, can still been seen (Fig. 4.5). The influence of biomes is even clearer. In the Middle and lower parts of Fig. 4.5, areas of North China are grouped into a cluster; areas of Lower and Upper Yangtze are grouped into another cluster; Himalaya is close to Yangtze; and North China is to some extent related to Yangtze. Other clusters indicate different biomes; for example, areas around the Yangtze Platform form one cluster: south Qingling to its north, Yanyuan to its west, Wuling to its south-east. The Ordos margin and Keping district are placed next to South China (Yangtze), which indicates that they were similar in ecology. Therefore, the cluster distribution of Early Ordovician cephalopods was evidently controlled by biogeographical factors and ecological differentiation.

Summary of the Early Ordovician

Summing up the above metioned information about fossils of various taxa, in the Early Ordovician the whole of China, except Xingan, can be included in the Sino-European region (a part of the Meridional realm), which is further divided into four provinces: North China–Qilian province (North China biome

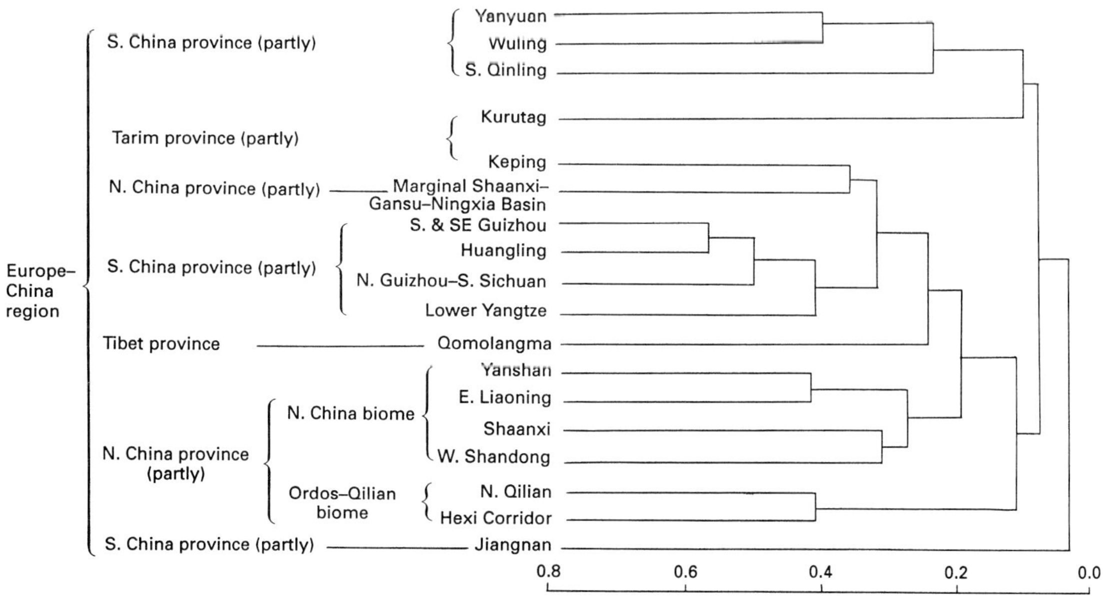

Fig. 4.5 Cluster analysis of the Early Ordovician cephalopods of China.

and Ordos–Qilian biome), South China province (Yangtze biome, Jiangnan biome, and South-east biome), Tarim province (Keping biome, Tianshan–Kuruktag biome), and Tibet province (Fig. 4.3). In Junggar and Altay (northern Xinjiang), there is, as yet, no exact record of the Lower Ordovician.

4.2.3 MIDDLE ORDOVICIAN PALAEOBIOGEOGRAPHICAL PROVINCIALIZATION OF CHINA

In the Middle Ordovician, the land area of China expanded continuously; small land blocks were enlarged or joined together and large land blocks were uplifted gradually, especially in North China, where the sea was limited to its western margin in the late Middle Ordovician. It has been suggested that, in accordance with such changes, biogeographical provincialization was not so distinctive as it had been in the Early Ordovician. In particular, the boundaries of subprovinces were displaced greatly, in accordance with the changes of sea area (which was obviously different from that in the Early Ordovician epoch). The other explanation for diminishing provincialism is that South and North China were approaching each other; this could be caused by the separation of South China from Gondwana and the disappearance of the small ocean basin along central Qinling between South and North China.

Palaeobiogeographical provincialization of Middle Ordovician cephalopods

The cephalopods in the Yangtze region consisted predominately of Lituitidae and Michelinoceratidae (Table 4.4, Fig. 4.6). Members of these two families had occurred in great numbers in the late Early Ordovician, so it can be said that the main feature of the Middle Ordovician in the Yangtze region was similar to that of the Early Ordovician. During this epoch, the principal genera of cephalopods were

Table 4.4 Occurrence of the main Middle Ordovician cephalopods in the regions of China

Genus	Yangtze province		N. China province		Tarim	Tibet	
	Yangtze	Jiangnan	N. China	Marginal N. China		South	North
Lituites	×		×				×
Cyclolituites	×				×		
Trilacinoceras	×				×		
Sinoceras	×	×		×	×	×	×
Jiangshanoceras	×	×		×			
Discoceras	×	×		×	×		
Euasiaticoceras	×	×		×			×
Fengfengoceras			×				
Tofangoceras			×	×			
Gonioceras			×				
Badouceras			×				
Protocycloceras	×		×				
Parormoceras	×		×				
Stereoplasmocerina			×	×			
Protocycloceroides	×				×		
Richardsonoceras	×		×	×	×		
Kalpinoceras					×		
Eosomichelinoceras	×					×	×

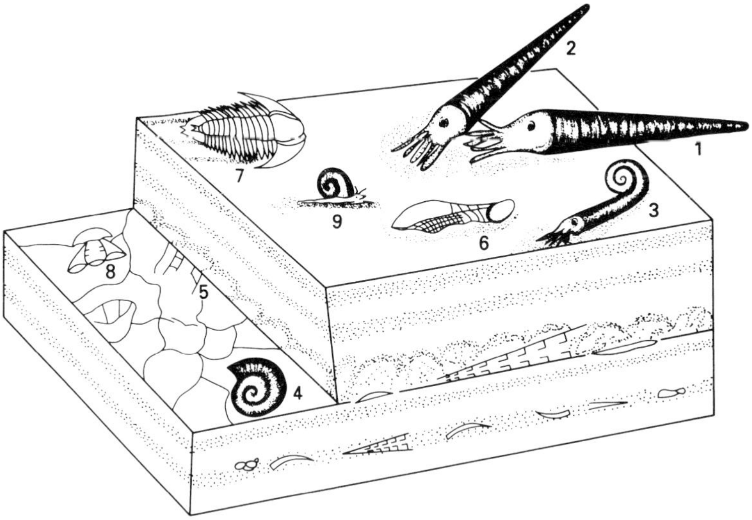

Fig. 4.6 Reconstruction of the *Sinoceras–Cyclopyge* community of the Middle Ordovician Baota Formation in the Yangtze biome. 1, *Sinoceras*; 2, *Orthoceras*; 3, *Lituites*; 4, *Discoceras*; 5, *Vaginoceras*; 6, *Cyclopyge*; 7, *Remopleurides*; 8, *Neseuretus*; 9, *Ecculiomphalus*.

Lituites, *Cyclolituites*, *Sinoceras*, *Asymmetrocerus*, *Chunnanoceras*, *Estonioceras*, *Dideroceras*, *Dongkalaceras*, *Fanggangoceras*, *Protocycloceroides*, *Parormoceras*, *Richardsonoceras*, *Jiangshanoceras*, and *Troedssonella*, of which the Lituitidae, *Estonioceras*, and *Troedssonella* were distributed mainly in the Baltic–Scandinavian region. *Sinoceras* was unique to China; *Parormoceras* occurred in East Asia; *Dideroceras* occurred in northern Europe and East Asia. There were other genera, not listed here, which were also widely distributed, while endemic genera accounted for about one-quarter of the total. In general, Eurasian members and endemic Yangtze members are dominant in the fossil record.

The Middle Ordovician deposits in the Jiangnan region are of argillaceous carbonate (Nianwashan Formation), quite similar to the deposits of the Baota Formation in Yangtze. Hence the cephalopod assemblages of both regions are nearly the same. The cephalopods occurring in Jiangnan were almost all found in Yangtze, the only difference being that the genera and species of cephalopods in Jiangnan were not so diverse as those in Yangtze.

In North China, cephalopods have been found in the Fengfeng Formation and the lower part of the Taoqupo Formation (Shaanxi), which continued to bear aspects of the Siberian and North American fauna, including *Actinoceras*, *Gonioceras*, *Tofangoceras*, *Tofangocerina*, *Badouceras*, *Vaginoceras*, *Protocycloceras*, *Fenfengoceras*, *Lituites*, and *Stereoplasmocerina* (Chen *et al.* 1984). Compared with the Early Ordovician, it shared more common members with the Yangtze region. *Sinoceras*, *Dideroceras*, and *Jiangshanoceras* have also been found in its western margin, but still more fossils were northern-type cephalopods.

In Tarim, cephalopods have been found mainly in the Keping district and only a few were found in Kuruktag, with the same features as the Keping fauna. The cephalopods occurring in Keping include *Cyclolituites*, *Ancistroceras*, *Trilacinoceras*, *Ctenoceras*, *Protocycloceroides*, *Michelinoceras*, *Sinoceras*, *Richardsonoceras*, *Kalpinoceras*, and *Allumettoceras*. The characters of this assemblage are very close to those in the Yangtze region, the only difference being the occurrence of very abundant lituitids in the latter. Many *Sinoceras* fossils have been found in the Beishan region, Gansu.

In northern Tibet, cephalopods occur in the Ke'erduo Formation of the Xainza district; and in southern Tibet, in the Jiacun Group of the Nyalam district. Both cephalopod faunas are very similar. They include *Sinoceras*, *Michelinoceras*, *Eosomichelinoceras*, *Euasiaticoceras*, *Allumettoceras*, and *Beloitoceras*. All these cephalopods have also been found in Yangtze, so they should belong to the same province as Yangtze. Moreover, the Middle Ordovician in northern Tibet is composed of argillo-calcareous rocks, similar to the limestone of the Baota Formation of Yangtze, hence the environments were rather similar too. However, in southern Tibet *Orthoceras* was absolutely dominant, while in northern Tibet, relatively abundant planispiral and cyrtocone elements occur.

In north-eastern China, there are a few cephalopods from the northern part of Inner Mongolia: *Kionoceras*, *Dideroceras*, and *Sabwinnipegoceras*, the last of which is a new genus; the others are widespread genera but seldom found in North China.

To conclude, Middle Ordovician members of the Lituitidae and *Sinoceras* have been found almost everywhere apart from Xingan. If the area was to be named after the cephalopod fauna, it could be called the Lituitidae–*Sinoceras* biogeographical region. Within this region, North China had more southern members than it did in the Early Ordovician, but it was still dominated by actinoceratids, thus forming the transitional feature of the Siberian and Eurasian faunas. The other regions are quite different from the Siberian realm. Meanwhile, the Jiangnan subprovince and the Tibet province of the Early Ordovician epoch were no longer easy to identify, and the differentiation between the Tarim and Yangtze provinces was also not clear.

Palaeobiogeographical provincialization of Middle Ordovician trilobites

Middle Ordovician trilobites (Table 4.5) show similar changes to the cephalopods; again, provincialization was not so distinct as it was in the Early Ordovician. The Yangtze region was still the most abundant area for trilobites, their representatives being trinucleids and asaphids. Endemic genera became less abundant than in the Early Ordovician, taking up approximately 17 per cent, while the cosmopolitan genera increased correspondingly. The main typical members in this region include: *Birmanites*, *Reedocalymene*, *Calymenesun*, *Prosopiscus*, *Phorocephala*, *Xenocyclopyge*, *Paraphillipsinella*, and *Hammatocnemis*, most of which are endemic to Asia.

Trilobites found in Jiangnan (Yenwashan Formation) were not abundant. Most of these were common Yangtze genera, while some were endemic genera, such as *Xiushuilithus* and *Jianxilithus* of the Trinucleidae. *Lonchodomas*, *Neseuretus*, *Paraceraurus*, and *Sphaerexochus* are found in the Xainza district of northern Tibet; all were cosmopolitan genera, also found frequently in the Yangtze region. There is, as yet, no report of Middle Ordovician trilobites from southern Tibet. In western Yunnan, trilobites have been found in the lower part of the Lower Pupiao Formation of the Baoshan district, including *Ampyxinella*, *Basiliella*, *Birmanites*, *Calymenesun*, *Illaenus*, and *Lonchodomas*; *Calymenesun* was a typical member of the Yangtze region; the other fossils were also common in the Yangtze region; thus it belongs to the same bioprovince.

In North China, trilobites have been found mainly on the Shaanxi–Gansu–Ningxia Basin margin (Gongwusu Formation) in the western part of the North China Platform. They are few in number. Besides genera common in the Miaopo Formation and Baota Formation of the Yangtze region, there are a small number of endemic ones (*Lamproscutellum*, *Pseudostygina*, and *Tongxinaspis*).

The distinction between northern and southern provinces tended to be even more vague for trilobites than for cephalopods. In northern Qilian, as in the Early Ordovician, many planktonic trilobites occurred, a situation similar to that of the Jiangnan region. The trilobites found in southern Qilian and the Lajishan district of central Qilian were almost the same as those in northern Qilian. Many trilobites have been found in the Keping district of Tarim (Upper Sa'ergan Formation, Kanling Formation, and Qilang Formation). Of the 33 genera, only *Ampyxinella* was found in Kazakhstan, *Calymenesun* is typical of Yangtze, and *Cyclopyge* typical of the Mediterranean region. Approximately 50 per cent of the trilobites found here were also found in the Yangtze region, so the overall character is similar to Yangtze. The trilobites occurring in Kuruktag include *Ampyx*, *Asaphus*, *Basiliella*, *Parabulbaspis*, *Cekovia*,

Table 4.5 Occurrence of the main Middle Ordovician trilobites in the regions of China

	Yangtze province			N. China province			
Genus	Yangtze	Jiangnan	W Yunnan	Marginal N. China	Qilian	Tarim	Xingan
Hammatocnemis	×	×		×		×	
Lonchodomas	×		×	×		×	×
Miaopopsis	×						
Tangyaia	×		×	×	×	×	
Birmanites	×		×	×		×	
Reedocalymene	×						
Calymenesun	×		×			×	
Cyclopyge	×	×			×	×	
Xenocyclopyge	×						
Prosopiscus	×						
Phorocephala	×						
Remopleurides	×	×		×		×	
Paraphillipsinella	×						
Quyuania	×						
Jianxilithus		×					
Xiushuilithus		×					
Pseudostygina				×			
Tongxinaspis				×			
Yumenaspis					×		
Nanshanaspis					×		
Xinjiangia						×	
Kanlingia						×	
Bothycheilus							×

Illaenus, *Lonchodomas*, *Ogygites*, and *Paracybeloides*; apart from *Paracybeloides*, other genera could be found in Yangtze.

The data given above indicate that the southern and northern division of trilobites continued to exist, reflected by the absence of some typical Yangtze members in the northern region, but this division is more obscure and there is very little difference between the biomes within the same province. A limited number of trilobites have been found in Xingan, occurring in the lower part of the Luohe Formation, a stratigraphical unit spanning Middle and Late Ordovician; its Middle Ordovician trilobites include *Sphaerexochus*, *Ampyx*, *Lonchodomas*, *Stenoparia*, and *Platylichas*, mainly widespread genera.

Palaeobiogeographical provincialization of Middle Ordovician graptolites and corals

Two types of Middle Ordovician graptolites can still be distinguished—the Central China type and the South China type. The graptolites in Yangtze region (Central China type) and Jiangnan region (South China type), both mainly found in the early Middle Ordovician (Miaopo Formation, Hulo Formation), exhibited little difference from each other. Central China has yielded *Gymnograptus*, which was found elsewhere only in northern Europe, while *Abrograptus* and *Jianxigraptus* in the Jiangnan region have not been found in Central China. Meanwhile, the Pearl River region (South China type) has exclusively graptolitic shale facies, characterized by *Glossograptus* and *Paraglossograptus*. Other regions with graptolite fauna similar to those of Central China are Zhouzishan of Inner Mongolia and Keping, whereas the South-China-type graptolites occur in Qilian, Kuruktag, and central Tianshan.

There have been only a few reports about Middle Ordovician corals. The areas concerned are southern Guizhou, Funing (Yunnan), Quxian (western Zhejiang), Longxian (Shaanxi), Qilian, Kuruktag, western Junggar, and Beishan. Lin Baoyu (1974) divided the Ordovician corals of China into southern and northern types. The southern region was represented by *Plasmoporella*, *Reuschia*, *Calostylis*, *Brachylasma*, and *Yohophyllum*, and can be further subdivided into south-eastern and south-western districts. The northern region was characterized by the absence of the common genera of the southern region and the presence of *Favistella* and *Lichenaria*, thus displaying similarity with corals in the Siberia–North American region. Liao (1990) also concluded that during the Middle and Late Ordovician the corals of North China were most closely related to those of Siberia and North America while South China occupied an independent province.

Cluster analysis of the Middle Ordovician trilobites

We have calculated the similarity index of Middle Ordovician trilobites (142 genera so far published) which fall into 18 districts; with those localities that obviously belonged to same biogeographical region being incorporated before computer analysis. It can be seen from the results (Fig. 4.7) that western Junggar, Kuruktag, Beishan, Keping, and Qilian in the north can be merged into one biogeographical province; Jiangnan and Wuling are somewhat different, thus should be united into one group but still related to the Yangtze region to some extent, which is in agreement with the analysis of ecological differentiation in the text. Lower Yangtze, Upper Yangtze, Baoshan, Lijiang–Kunming and south-eastern Yunnan belong to one biogeographical province. Strangely, the Shaanxi–Gansu–Ningxia Basin margin and Yangtze region are very similar, reflecting the closeness of the two in ecological terms and the vagueness of provincial boundaries mentioned above. Most of the trilobites in the Xingan region were widespread genera but still show closeness to those in the northern region.

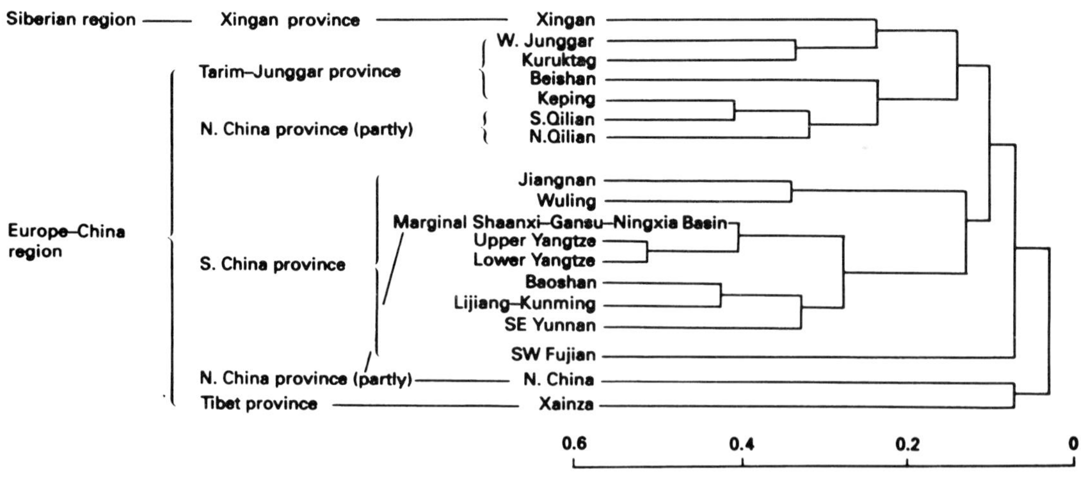

Fig. 4.7 Cluster analysis of the Middle Ordovician trilobites of China.

Summary of the Middle Ordovician

Systematic analysis of provincialization plus clustering of the trilobites shows that the general biogeographical pattern of the Middle Ordovician did not change significantly from that of the Early Ordovician, but that the biomes had to be regrouped to some extent. The North China–Qilian province can still be divided into a North China biome and Qilian biome, whereas in the South China province the Jiangnan biome became so similar with the Yangtze that they should be merged together. In Tibet, differentiation between north and south appeared; southern Tibet was essentially dominated by orthoceratid nautiloids, while in northern Tibet there occurred many planispiral and cyrtocone genera and species, but both parts still belonged to the same biogeographical province. The Tarim province retained its previous status.

4.2.4 LATE ORDOVICIAN PALAEOBIOGEOGRAPHICAL PROVINCIALIZATION OF CHINA

Palaeobiogeographical provincialization of Late Ordovician trilobites and brachiopods

In the Late Ordovician (Fig. 4.8), many blocks of China continued to be subjected to epeirogenic uplift. Most prominently, the North China Platform had already elevated above sea-level, and deposition occurred only in the Yaoxian area of Shaanxi along its western margin. The Xikang–Yunnan Oldland expanded, the Yunnan–Guizhou Oldland emerged, and most parts of south-eastern China and the major part of Tarim uplifted at the end of the Late Ordovician. Such large-scale regression may have been caused by a global temperature decrease and subsequent glaciation. All these changes affected the distribution of fauna.

In Yangtze, the Late Ordovician includes the Linxiang Formation, Wufeng Formation, and Guanyinqiao Formation, belonging to shelly facies or graptolite–shelly mixed facies. A reducing event occurred during the Ashgillian, causing the deposition of the Wufeng black graptolitic shale in Yangtze as well as in Jiangnan and western Yunnan. The benthic shelly fauna occurred in the Linxiang Formation (Caradocian) and Guanyinqiao Formation (Ashgillian). Most genera of the Linxiangian trilobites had survived from the Middle Ordovician, of which the most widespread were *Nankinolithus*, *Cyclopyge*, and *Hommatocnemis*. *Nankinolithus* was a typical genus distributed in Yangtze, western Yunnan, Jiangnan, the western margin of North China, and Beishan. Later, an early to middle Ashgillian (Wufengian) *Foliomena* assemblage, consisting mainly of thin-shelled and small brachiopods, occurred in Yangtze. This assemblage is known from the Laurentia, Baltic–Avalonia, North Gondwana, and Shan–Thai block (Cocks and Rong 1988), apparently occupying deep-water outer-margin sites. The most common trilobite and brachiopod in the late Ashgillian were, respectively, *Dalmanitina* and *Hirnantia*, forming the famous *Dalmanitina–Hirnantia* fauna (Rong and Harper 1988). This fauna can be traced from Jiangnan, western Yunnan, and northern Tibet all the way to western Europe around the margin of a glaciation area in Africa, and has thus been regarded as a peri-Gondwana fauna, although *Dalmanitina* has also been reported from Qilian and the Daxingan Mountains. In addition, there were some endemic genera, such as *Encrinuroides*, *Kweichowilla*, and *Scutellum* (*Planiscutellum*). To the north of Qinling, a number of trilobites were discovered in Beishan, plus a few in Tarim (*Ampyx*, *Ampyxinella*, and *Xinjiangia*), eastern Junggar (*Calliope*, *Enerinuroides*, *Fuyuia*, *Junggarella*, *Parisoceraurus*, and *Remopleurides*), and Xingan (*Cheirurus*, *Encrinuroides*, and *Isoteloides*).

It can be concluded from this list that most were widespread genera apart from a few endemic ones, and the Yangtze members were seldom found in these regions. The trilobites in Tarim are much closer to those in Kazakhstan; other regions were also different from Yangtze.

Palaeobiogeographical provincialization of Late Ordovician corals

The impoverished situation of Middle Ordovician corals changed in the Late Ordovician, when either

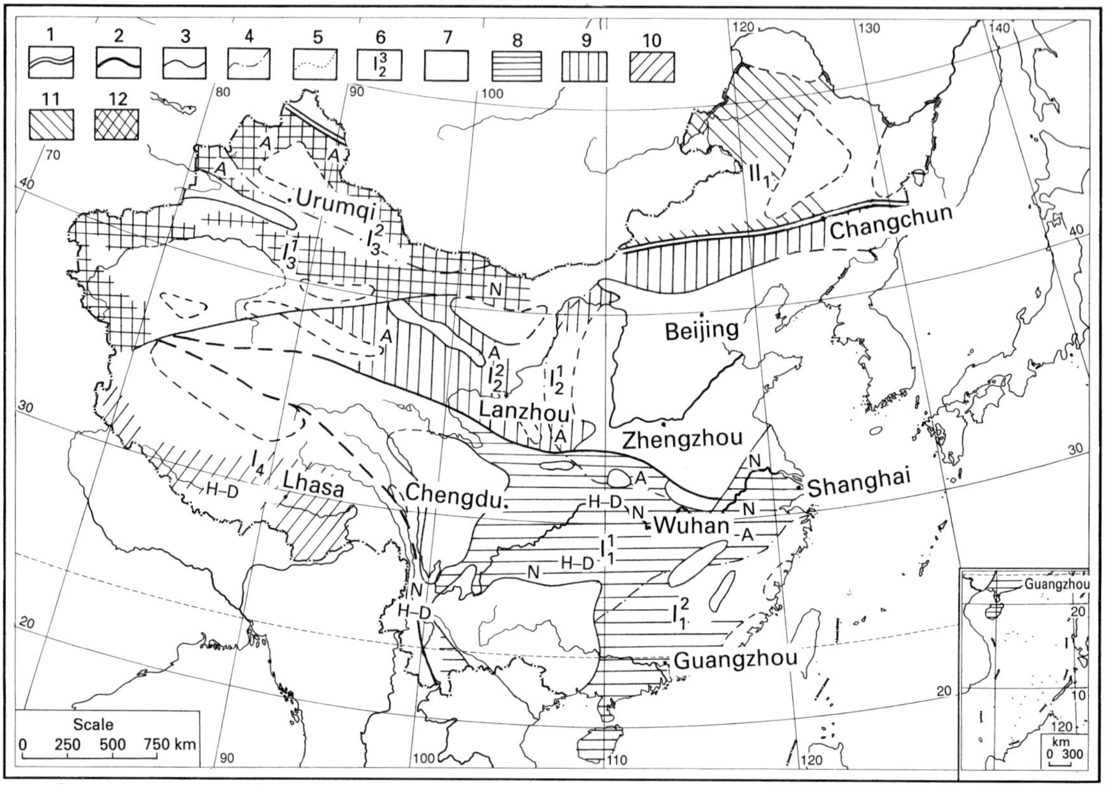

Fig. 4.8 Map of the Late Ordovician palaeobiogeographical provincialization of China. 1, Boundary of realms; 2, boundary of regions; 3, boundary of provinces; 4, boundary of biomes; 5, boundary of communities; 6, roman digit = realm mark, subscript = province mark, superscript = biome mark; 7, Oldland. 8–12, Marine and paralic regions: 8, South China; 9, North China; 10, Gondwana; 11, Junggar–Xingan; 12, transitional. I, Europe–Asia–Australian realm, Europe–China region; I_1, South China province; I_1^1, Yangtze biome; I_1^2, Jiangnan–South-eastern China biome; I_2, North China–Qilian province; I_2^1, Ordos biome; I_2^2, Qilian biome; I_3, Tarim province; I_3^1, Tarim biome; I_3^2, Junggar biome; I_4, Tibet province; II, North America–Siberian realm, Siberian region; II_1, Xingan province; N, *Nankinolithus*; A, Agetolitidae; H–D, *Hirnantia–Dalmanitina*.

tabulate or rugose corals developed and spread to much wider areas (Table 4.6). In the south, they have been discovered in Zhejiang, eastern Jiangxi, Batan, and Zhongzan (western Sichuan), Jinping (western Yunnan), and central and southern Qinling; and in the north, in Xingan, eastern and western Junggar, central Tianshan, Beishan, the northern margin of Qaidam, Qilian and the western margin of the North China Platform. The distributional pattern shows that the corals were then principally distributed along the margins of the platforms, and not many occurred on the platforms proper.

In Yangtze, corals were found mainly in the Guanyinqiao Formation of Wufengian age (He 1978) (Fig. 4.9). Leleshus (1970) subdivided the Late Ordovician coral faunas into four provinces, i.e. Baltic, Central Asian, Siberian, and Antarctic, among which the Central Asian province was characterized by agetolitid tabulates. The agetolitids were extremely widespread in China. The present data show that these occurred in almost every coral locality, except Xingan, suggesting that China (except Xingan) and Central Asia should belong to the same province—the Agetolitid coral province.

Table 4.6 Occurrence of the main Late Ordovician tabulate corals in the regions of China

	S. China province					N. China–Qilian province			
Genus	W. Zhejiang	Upper Yangtze	Yidun–Zhongdian	Qamdo–Simao	S. Qinling	Ordos	Qilian	Junggar	Xingan
Agetolites	×		×	×	×	×	×	×	
Agetolitella	×				×	×			
Subagetolites						×			
Calapoecia	×						×		×
Sarcinula	×					×			
Nyctopora	×						×	×	
Plasmoporella	×	×	×		×	×	×	×	
Neoplasmoporella								×	
Sibiriolites									×
Neosibiliolites	×							×	
Toeniolites	×					×		×	
Foestephyllum						×		×	
Casmiolithus								×	
Protrochiscalithus								×	
Proheliolites									×

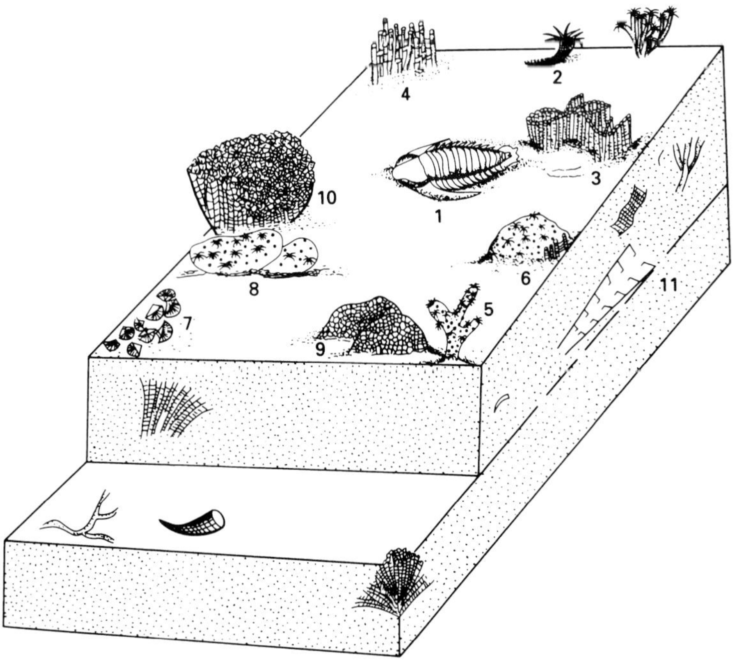

Fig. 4.9 Reconstruction of the *Agetolites–Plasmoporella* community of the Late Ordovician Sanjushan Formation in Yangtze province. 1, *Remopleurides*; 2, *Tryplasma*; 3, *Catenipora*; 4, *Sarcinula*; 5, *Sibiriolites*; 6, *Stelliporella*; 7, *Sowerbyella*; 8, *Plasmoporella*; 9, *Palaeofavosites*; 10, *Agetolites*; 11, *Orthoceras*.

Because the agetolitids were most diversified in China (*Agetolites*, *Agetolitella*, *Subagetolites*, and *Paragetolites*), Lin Baoyu (1983) suggested that South China was the origination centre of the Agetolitidae. The frequent tabulate corals associated with them include mainly *Calapoecia*, *Diploepora*, *Eofletcheria*, *Taenilites*, and *Wormsipora*. In the Late Ordovician, rugose corals were also rather abundant and did not differ much between South and North China, because some major members, such as *Sinkiangolasma*, *Brachylasma*, *Streptelasma*, and *Tryplasma*, were found in both parts. However, South and North China can still be separated on the basis of the occurrence of the tabulate *Amsassia* and rugose *Favistella* in the north and their absence in the south. Such genera were found mainly in the Siberia–North American realm, so more Boreal members were mixed in the north. Based on corals, two biogeographical regions can be distinguished: Xingan region, which was a part of the Siberian realm (Lin Baoyu (1983) named it as *Calapoecia–Sibiriolites* province), and Agetolitid Fauna region, which can be further subdivided into a South China province and a North China–Qilian province (*Agetolites–Favistella* Fauna province).

As far as the Junggar area is concerned, Lin and Chou (1977) and Lin and Wang (1985) suggested a faunal subprovince, considering that the heliolitids were dominant while tabulate representatives were few. It should be noted that *Agetolites* was distributed as far as Junggar but disappeared north of this area.

The distribution of Late Ordovician cephalopods and graptolites

Late Ordovician cephalopods were not widely distributed in China. Small, sparsely distributed cephalopods occur in the Lingxiang Formation in Yangtze. More cephalopods are found in western Zhejiang (Sanjushan Formation) and the Yaoxian–Dongchuan–Longxian area of Shaanxi (the upper part of the Taoqupo Formation). The cephalopods in western Zhejiang, along the south-eastern margin of the Yangtze Platform, include *Yushanoceras*, *Jiangshanoceras*, *Orthonybyoceras*, *Beloitoceras*, *Diestocrina*, and *Trocholites* (Chen *et al.* 1984). This faunal assemblage has many new genera and species with strong endemism. The Yaoxian–Dongchuan–Longxian area along the south-western margin of the North China Platform yields principally *Rizoceras*, *Kiaeroceras*, *Temperoceras*, *Diestocrina*, *Yaoxianoceras*, and *Beloitoceras*. The above two areas display some similarity, which favours the argument that after the disappearance of the small Early Palaeozoic ocean basin in central Qinling, North China and Yangtze were approaching each other.

The graptolites occurred mainly in the late Late Ordovician (Wufengian age), and were widespread in South China (Yangtze, western Yunnan, Jiangnan) and eastern Qilian (North China). During this period the complications of graptolite rhabdosomes (e.g. the increase of graptolite twigs; *Tangyagraptus*), and their attachments (the increase of appendage spines from one to six pairs (Li and Li 1985)), and the net-like reticulation of the thecal wall, were the three main tendencies in graptolite evolution. These tendencies were side-branches from the mainstream of evolution, reflecting morphological adaptation to the environment. The complicated graptolites thrived most in Yangtze and Jiangnan, thus bearing a strong endemic character distinctive from those of other regions.

Cluster analysis of the Late Ordovician trilobites

We have again undertaken statistical analysis of the Late Ordovician trilobites, and have obtained 14 regions and 90 genera after some deletion and incorporation. The results (Fig. 4.10) show that Xingan and Junggar can be merged into one province; Kuruktag, northern Qilian, and the Ordos Basin margin can be merged into another province, while Upper and Lower Yangtze, Kunming–Zhaojue, and Jiangnan belong to one province. However, why Beishan appears to be similar to Yangtze remains a problem to be explored further. Xainza in Tibet and Longmenshan in Sichuan are quite distinctive from the above regions.

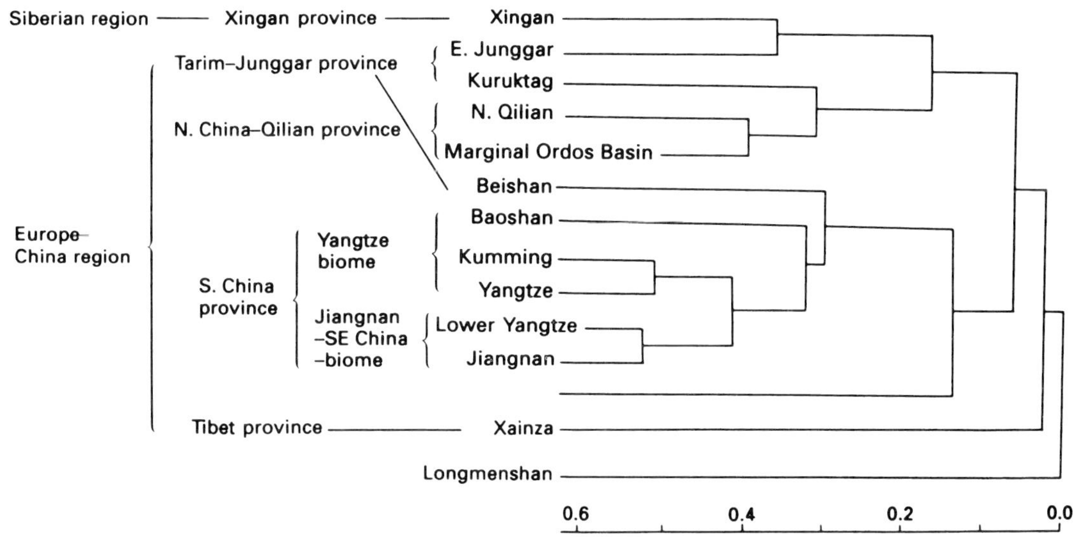

Fig. 4.10 Cluster analysis of the Late Ordovician trilobites of China.

4.3 IMPLICATIONS FOR PLATE TECTONICS

From the above statements it is clear that palaeobiogeographical provincialization was most obvious in the Early Ordovician and became obscure after the Middle Ordovician, especially between South and North China. Whether it was due to epicontinental uplift or the approaching of the southern and northern blocks is of interest to plate tectonics. The situation is different in Tibet; the regions north and south of the Yarlung Zangbo River had no significant difference in the Early Ordovician, but became differentiated in the Middle Ordovician. Cyrtocone and planispiral cephalopods were hardly found to the south of the Yarlung Zangbo River but became abundant to the north of it, indicating some linkage of the northern part to Eurasia. This feature became even more obvious in the Late Ordovician, when the *Dalmanitina–Hirnantia* fauna were present in northern Tibet but absent in southern Tibet, which suggests that the northern block, as a whole, probably had begun to separate from southern Tibet or Gondwana since the Middle Ordovician. There were basically no Early Ordovician sediments in Junggar, where the faunal characteristics in Middle Ordovician times were close to those in Tarim. However, in the Late Ordovician, heliolitids became extremely abundant, and the trilobites were close to that of Xingan, indicating that Junggar was gradually moving away from Tarim and approaching Siberia.

REFERENCES

Bergstrom, S. M. (1990). Relations between conodont provincialism and the changing palaeogeography during the Early Palaeozoic. *Geological Society of London, Memoir* **12**, 105–22.

Berry, W. B. N. (1974). Facis distribution patterns of some marine benthic faunas in Early Paleozoic platform environments. *Paleogeography, Paleoclimatology, Palaeoecology* **15**, 155–68.

Berry, W. B. N. (1979). Graptolites biogeography: a biogeography of some Lower Palaeozoic plankton. In *Historical biogeography, plate tectonics and changing environment*, (ed. J. Gray and A. J. Boucot), pp. 106–16. Oregon State University Press, Corvalis.

Berry, W. B. N. and Wilde, P. (1990). Graptolite biogeography: implications for palaeogeography and palaeoceanography. *Geological Society of London, Memoir* **12**, 129–38.

Chen Junyuan (1980). Character of Ordovician cephalopod provincialism, palaeoclimate, palaeogeography. *Scientific Papers on Geology for International Exchange* **1**, 27–32. Geological Publishing House, Beijing (in Chinese).

Chen Junyuan, *et al.* (1984). Ordovician biostratigraphy of Western Ordos. *Memoirs of Nanjing Institute of Geology and Palaeontology, Academia Sinica* **20**, 1–32 (in Chinese).

Chen Ting-en (1981). Ordovician and Silvrian cephalopod provincialism in Xizang with comments on the continental drift. In *Proceedings of Symposium on Qinghai-Xizang (Tibet) Plateau* (Beijing China) (ed. Liu Dongsheng), Vol. 1, pp. 159–64. Science Press, Beijing.

Cocks, L. R. M. and Fortey, R. A. (1988). Lower Palaeozoic facies and faunas around Gondwana. In *Gondwana and Tethys*, (ed. M. G. Audley-Charles and A. Hallam), pp. 183–200. Geological Society of London and Oxford University Press.

Cocks, L. R. M. and Fortey, R. A. (1990). Biogeography of Ordovician and Silurian faunas. *Geological Society of London, Memoir* **12**, 97–104.

Cocks, L. R. M. and Rong Jiayu (1988). A review of the Late Ordovician *Foliomena* brachiopod fauna with new data from China, Wales and Poland. *Palaeontology* **31**, 53–67.

Finney, S. C. and Chen Xu (1990). The relationship of Ordovician graptolite provincialism to palaeogeography. *Geological Society of London, Memoir* **12**, 123–8.

He Xinyi (1978). Tetracoral Fauna of the Late Ordovician Guanyinqiao Formation, Bijie, Guizhou province. *Professional Papers of Stratigraphy and Palaeontology* **6**, 1–45 (in Chinese).

Jaanusson, V. (1973). Ordovician articulate brachiopods. In *Atlas of palaeobiogeography*, (ed. A. Hallam), pp. 19–25. Elsevier, Amsterdam.

Lai Caigen, *et al.* (1982). *Stratigraphy of China*, Vol. 5, *The Ordovician System of China*. Geological Publishing House, Beijing (in Chinese).

Leleshus, V. L. (1970). Paleozoogeografiya ordovilea, silura i rannego devona po tabulyatomorfnym korallam i granitsy siluriyskoy sistemy. *Akad. Nauk SSSR. Izvestiya, ser. geol.*, 184–92

Li Guangcen and Lin Baoyu (1982). On some geological problems in Eastern Kunlun Mountain. *Contribution to the Geology of the Qinghai–Xizang (Tibet) Plateau* **1**, 28–48 (in Chinese).

Li Yaoxi, Song Lishing, Zhou Zhiqiang, and Yang Jingyao (1973). Ordovician. In *Early Palaeozoic stratigraphy in Western Daba Mountain*, pp. 62–84. Geological Publishing House, Beijing (in Chinese).

Li Zhiming and Li Daqing (1985). *Appendispinograptus*, a new subgenus of *Climacograptus*. *Earth Science – Journal of Wuhan College of Geology* **10**, 35–41 (in Chinese).

Liao Weihua (1990). The biogeographic affinities of East Asian corals. *Geological Society of London, Memoir* **12**, 175–80.

Lin Baoyu (1974). A preliminary study on stratigraphical distribution and zoogeographical provinces of China. *Acta Geologica Sinica* **48**, (1), 5–22 (in Chinese).

Lin Baoyu (1983). Ordovician tabulate corals of China. *Acta Palaeontologica Sinica*, **22**, (2), 487–91 (in Chinese).

Lin Baoyu and Chou Xinghu (1977). Tabulata and Heloilitida of the Upper Ordovician of the Chejiang and Jiangxi provinces. *Professional Papers of Stratigraphy and Palaeontology* **3**, 108–208 (in Chinese).

Lin Baoyu and Wang Baoyu (1985). Some Ordovician heliolitid corals from Jiabosar Formation of Jiabosar district, Xinjiang. *Acta Palaeontologica Sinica* **24**, (4), 351–6 (in Chinese).

Lu Yanhao (1975). Ordovician trilobite faunas of central and southwestern China. *Palaeontologica Sinica* **B11**. Sciences Press, Beijing (in Chinese).

Lu Yanhao, *et al.* (1976). Ordovician biostratigraphy and palaeozoogeography. *Memoirs of Nanjing Institute of Geology and Palaeontology* **7**, 1–83.

Mu Anzhi (1980). Ordovician graptolite and biogeographical regions of China. *Scientific Papers on Geology for International Exchange* **1**, 33–40. Geological Publishing House, Beijing (in Chinese).

Mu Anzhi (1983). Ordovician ecostratigraphical types and biogeographical regions of China. In *Palaeobiogeographic provinces of China* (ed. Geol. Soc. China), pp. 16–31. Science Press, Beijing (in Chinese).

Rong Jiayu and Harper, D. A. T. (1988). A global synthesis of the latest Ordovician Hirnantian brachiopod faunas. *Transactions of the Royal Society of Edinburgh: Earth Sciences* **79**, 383–402.

Shen Xinfu (1974). *Subdivision and correlation of Ordovician of China*. Geological Publishing House, Beijing (in Chinese).

Sweet, W. C. and Bergstrom, S. M. (1984). Conodont provinces and biofacies of the Late Ordovician. *Geological Society of America, Special Paper* **196**, 69–87.

Wang Xiaofeng (1980). The Ordovician system in China. *Acta Geologica Sinica* **54**, (1), 1–8; (2), 1–8 (in Chinese).

Whittington, H. B. and Hughes, C. P. (1972). Ordovician geography and faunal provinces deduced from trilobite distribution. *Philosophical Transactions of the Royal Society of London, Series B* **263**, 235–78.

Yu Changming (1960). Late Ordovician corals of China. *Acta Palaeontologica Sinica* **8**, (2), 65–90 (in Chinese).

Yu Changming, *et al.* (1963). *The coral fossils of China*. Science Press, Beijing (in Chinese).

5 SILURIAN

He Xinyi

5.1 SILURIAN PALAEOBIOGEOGRAPHICAL PROVINCIALIZATION OF THE WORLD

5.1.1 AN OUTLINE

The Early–Middle Llandoverian was similar to the end of the Late Ordovician in the general features of palaeogeography and biotas. In the southern hemisphere, Gondwana was still integral, except for a number of small separate plates along its margins, and it was situated mostly in the middle and high latitudes south to 30°S. Present-day northern Africa and southern South America were situated in the polar circle. The late Late Ordovician is noted for widespread glaciation, especially in Gondwana. Under the influence of the immense glaciation, global sea-level underwent a sharp drop of up to more than 100 m during the late Ashgill of Late Ordovician (Rong Jiayu *et al.* 1984). The environmental changes were so dramatic as to cause the decline or extinction of many marine groups. The most obvious example by which to illustrate this phenomenon is the almost complete extinction of the widespread *Dalmanitina–Hirnantia* fauna during the late Ashgill. Furthermore, up to the 90 per cent of the Late Ordovician trilobite genera disappeared during late Ashgillian. Nautiloids also showed a sharp decline.

Beginning in the Early Silurian, the global climate warmed and the glaciers melted, leading to global transgression and a significant alternation of biotas. Laurentian, Baltic, Kazakhstan and Sino-Korean plates were all between the latitudes 30°N and 30°S, while most of the Siberian plate (including Mongolia) was in the range 20–50°N. The dominant form of the Early–Middle Llandoverian deposits on the margins of all continents, especially the Eurasian, was graptolite shale facies. The expansion of transgression since late Llandoverian up to the Wenlockian (Middle Silurian) caused the flattening of land surfaces, the development of the shelly facies in shallow seas, and the widespread distribution of various marine groups in areas of the northern hemisphere, such as Europe, North America, Urals, Central Asia, Siberia, and Yangtze, with genera and species endemic to each region. This provides important data for the biogeographical provincialization of the Silurian.

Based on the research of many specialists of Silurian benthic faunas (such as brachiopods and corals), four palaeobiogeographical realms are

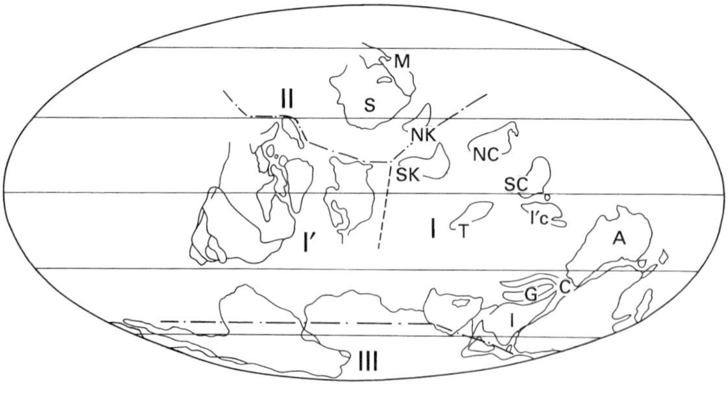

Fig. 5.1 Map of the Silurian palaeobiogeographical provincialization of the world. I, East Proto-Tethyan realm; I′, West Proto-Tethyan realm; II, Boreal realm; III, Malvinokaffric realm; A, Australia; G, Gangdise; I, India; Ic, Indochina; M, Mongolia; NC, North China; C, Qiangtang; S, Siberia; SC, South China; SK, South Kazakhstan; T, Tarim.

recognizable in the Silurian: the Boreal realm, East and West Proto-Tethyan realms, and the Malvinokaffric realm. (Fig. 5.1). For details please refer to Ivanovsky (1965), Kaljo and Klaamann (1973), Boucot (1974, 1990), Pickett (1975), Ziegler *et al.* (1977), Oliver (1980), Hill (1981), Lin Baoyu *et al.* (1984), and Wang Hongzhen *et al.* (1984, 1989).

5.2 SILURIAN BIOGEOGRAPHICAL PROVINCIALIZATION OF CHINA

5.2.1 GENERAL INTRODUCTION

During the Silurian, China spanned two palaeobiogeographical realms, the Boreal and the East Proto-Tethyan. The Junggar–Xingan region belonged to the former. It ranged from Altay of northern Xinjiang, eastern and western Junggar, through northern Tianshan, north-eastern Inner Mongolia to Daxingan and the Xiaoxingan Mountains. This realm is characterized by the presence of brachiopod *Tuvaella* fauna in the Middle–Late Silurian, and the scarcity of coral fossils, with *Altaja* and *Diplophyllum* (elements typical of the Boreal Realm) found only in a few localities.

The East Proto-Tethyan realm was to the south of the Ebinur–Gaxunnur–Xar Moron River line. The East Proto-Tethyan realm in the Early–Middle Silurian of China can be subdivided into Tianshan–Qilian province, Qinghai–Tibet–Western Sichuan province, Yangtze province, and Southern Tibet–Western Yunnan province in terms of the biotas, depositional patterns, and structural features. The Yangtze province covered a great area with various depositional and ecological patterns, and thus can be further subdivided into three biomes; similarly, the Qinghai–Tibet–Western Sichuan province can be subdivided into four biomes, and Tianshan–Qilian province and Southern Tibet–Western Yunnan province into two biomes each. The Late Silurian was noted for changes in palaeogeographical environments resulting from the regression of (see Fig. 5.6). Junggar–Xingan province remained within the Boreal realm, while some parts of the Tethyan realm were exposed due to regression, leading to the subsequent reduction in shelly facies. Hence, only three Late Silurian provinces are identifiable in the Proto-Tethyan realm: Tianshan–Inner Mongolia province, Qinling–Yangtze province (containing three biomes), and Southern Tibet–Western Yunnan province. The demarcation line between Southern Tibet–Western Yunnan province and Qinling–Yangtze province is the fault zone of the Yarlung Zangbo River. In the literature, southern Tibet used to be called the 'Himalayan province', where all three Silurian series are displayed completely.

The Early Silurian was dominated by graptolite facies, while the Middle and Late Silurian were dominantly shelly facies. The Early and Middle Silurian biotas were somewhat similar to those of Yangtze; nektonic organisms, such as nautiloids and conodonts, of the Middle and Late Silurian were very similar to contemporary faunas in western Europe.

The Silurian in China is developed extensively and displays various facies, including shelly, graptolite, and mixed. In addition to corals, brachiopods, nautiloids, graptolites, and conodonts there were trilobites, bryozoans, stromatoporoids, bivalves, gastropods, echinoderms, and primitive fish. The most abundant were brachiopods (up to about 160 genera), corals (more than 100 genera), trilobites (55 genera), nautiloids (44 genera), graptolites (approximately 30 genera), and conodonts (about 30 genera and many species). They can serve as the basis for division of both province and biome.

Table 5.1 shows the subdivision and correlation of the Chinese Silurian. Tables 5.2–5.4 show the major Silurian biotas and their ecological types and sedimentary facies for the different biomes of the Junggar–Xingan province and Yangtze province.

5.2.2 EARLY AND MIDDLE SILURIAN PALAEOBIOGEOGRAPHY OF CHINA

Some of the Silurian marine benthic organisms (e.g. brachiopods and corals) exhibit distinctive biogeographical provincialization because of the climatic zonation and prolonged separation by deep-sea basins, high mountain ranges, and deep rifts. Biome formation was obviously related to the

Table 5.1 Correlation chart of the Silurian of China

<table>
<tr><th colspan="3"></th><th>Xingan region</th><th>Inner Mongolia region</th><th>Eastern Junggar region</th><th>Yangtze region</th><th>Qilian region</th><th>Western qinling</th><th>Batang Sichuan</th><th colspan="2">Himalayan region</th></tr>
<tr><td rowspan="6">Silurian</td><td rowspan="2">Upper</td><td>Pridolian</td><td>Gulanhe</td><td rowspan="2">Xibiehe</td><td rowspan="4">Hongliuxia</td><td>Yulongshi</td><td rowspan="2">Hanxia Group</td><td rowspan="2">Beilongjiang Group</td><td rowspan="2">Yongren</td><td colspan="2" rowspan="2">?</td></tr>
<tr><td>Ludlowian</td><td>Woduhe</td><td>Miaogao</td></tr>
<tr><td rowspan="2">Middle</td><td>Upper Wenlockian</td><td rowspan="2">Bashilixiaohe</td><td rowspan="2">Bate’aobao</td><td>Guandi</td><td rowspan="2">Quannaogoushan Group</td><td rowspan="2">Zhouqu</td><td rowspan="2">Sanze</td><td rowspan="4">Shiqipo Group</td><td rowspan="2">Upper Formation</td></tr>
<tr><td>Lower Wenlockian</td><td>Xiushan</td></tr>
<tr><td rowspan="2">Lower</td><td>Upper Llandoverian</td><td rowspan="2">Huanghuagou</td><td colspan="2" rowspan="2"></td><td>Hanjiatien Group Shiniulan</td><td>Angzanggou Group</td><td>Jiannigou</td><td rowspan="2"></td><td rowspan="2">Lower Formation</td></tr>
<tr><td>Lower and middle Llandoverian</td><td>Longmaxi</td><td>Zhaohuajing</td><td>Anzigou</td></tr>
</table>

Table 5.2 Comparison of the main Silurian taxa among some provinces in the East Tethyan realm

Age	Yangtze province	Qinghai–Tibet –W. Sichuan province	Tiangshan–Qilian province
Early Silurian	cor. *Halmophyllum, Mucophyllum, Crassilasma, Phaulactis, Pilophyllum, Squameofavosites* bra. *Schizophoria, Protathyrisina, Nikiforovaena, Prothyris* naut. *Euthyoceras, Yunnanoceras* con. *Dentacodina multidentata, Spagnathodus crispus, Ozarkodina typica*	cor. *Squameof avosites, Phaulactis, Nipponophyllum, Altaja, Schlotheimophyllum, Holmophyllum* bra. *Howellella, Atrypella, Protathyris* naut. *Ophioceras–Yunnanoceras* con. *Spathognathodus crispus*	cor. *Mesof avosites, Palaeofavosites* bra. *Eospirifer tingi* tri. *Scutellum*
Middle Silurian	cor. *Holmophyllum primzevum, Kyphophyllum, Mucophyllum, Miculiella, Somphoporella, Carnegiea, Pilophyllum, Ketophyllum equitabulatum* bra. *Howellella shiqianensis, Nalivkinia, Atrypoidea, Nalivkinia, Atrypoidea, Nucleospira* tri. *Coronocephalus–Kaillia* *naut. Schuanoceras–Heyuncunoceras* gra. *Stomatograptus sinense, Monograptus chuchlenois* con. *Panderodus unicostatus, Ozarkodina edithae*	cor. *Holmophyllum, Mucophyllum, Stereoxylodes* bra. *Ferganella borealis* gra. *Crytograptus sakmaricus, Monograptus priodon, Monoclimacis*	cor. *Ptychophyllum, Mucophyllum, Nanshanophyllum, Kyphophyllum* bra. *Coelospira, Eospirifer, Mutationella* tri. *Encrinus* naut. *Sichuanoceras*
Late Silurian	Cor. *Amplexoides, Ceriaster, Idiophyllum, Shaansiphyllum, Kodono phyllum, Holophragma, Gyalo phylloides, Pseudophaulactis, Kaikottia Maitanopora* bra. *Borealis, Striispirifer, Zygospiraella, Pentazmerus, Stricklandia* tri. *Ptillillaenus–Gaotania, Encrinuroides, Sonkania, Coronocephalus* naut *Yichangoceras, Sonkanoceras* con. *Spathognathodus obesus, Spathognathodus parahassi, Pterospathodus celloni* gra. *G. persculptus Oktavites communis, Monoclimacis griestoniensis*	gra. *Monograptus, Rastrites, Oktavites, Streptograplus nodifer* gra. *Glyptograptus persculptus, Oktavites spiralis–Stomatograptus gandis*	Angzanggou Formation gra. *Spirograptus turriculatus, Streptograptus crispus–Monoclimacis griestonensis* Ningxia; Zhaohuajing Formation cor. *Pseudophaulactis, Porfirieviella, Tungussop, Brachyelasma, Grewingkia, Halysites, Palaeofavosites* bra. *Holorhynchus, Striispirifer*

cor., corals; bra., brachiopods; con., conodonts; gra., graptolites; naut., nautiloids; tri., trilobites

Table 5.3 Comparison of the main Silurian taxa among the biomes in Junggar–Xingan province

Age	Xingan biome	Inner Mongolian biome	Junggar biome
Late Silurian	Upper: Gulanhe Formation bra. *Howellella elegans, Ferganella* Lower: Woduhe Formation brachiopods bra. *Tuvaella gigantea, Tanuspirifer sinicus, Protoleptostrophia* cor. *Enterolasma*	Xibiehe Formation cor. *Spongophyllum, Ptychophyllum, Weissermelia* bra. *Conchidium, Atrypella*	Hongliuxia Formation (S_2– S_3) bra. *Tuvaella gigantea, Tanuspirifer Leptostrophia, Stegerhynchus* cor. *Syringaxon*
Middle Silurian	Bashilixiaohe Formation bra. *Tuvaella reckovskii, Dalmanella initalensis*	Bateaobao Formation (M. Sil. – early L. Sil.) anth. *Altaja, Mucophyllum, Diplophyllum, Squameofavosites* bra. *Tuvaella, Conchidium kirkidium*	
Early Silurian	Huanghuagou Formation bra. *Chonetoidea, Hindella*		

cor., corals; bra., brachiopods; anth., anthozoans; M. Sil., Middle Silurian; L. Sil., Late Silurian.

differentiation of ecological environments and sedimentary settings. Two different kinds of psychrophilic brachiopods proliferated in Boreal and Malvinokaffric realms. The Boreal realm was predominated by *Tuvaella*, and the Malvinokaffric realm by *Clarkeia*. The *Tuvaella* fauna originated in the late Llandoverian and survived into the Late Silurian. It was first found in the Late Llandovery strata of Central Asia, and then from south-eastern Kazakhstan, through Altay, the western Sayan Mountains, Tuva of Russia, the Far East, western Mongolia, to east Junggar of Xinjiang, north-eastern Inner Mongolia and Da Xingan, and the Xiao Xingan Mountains. Whether the *Tuvaella* fauna should be regarded as a distinctive biogeographical unit (or province) has been much debated in recent years. Su Yangzheng (1981) and Zhang Zixin *et al.* (1983) have reviewed the problem. According to Zhang, the percentage of the endemic genera for the *Tuvaella* fauna is too low (11.1 per cent) to form a separate biogeographical province, so it should be included in the Uralian–Codilleran region proposed by Boucot (1974). Wang Yu *et al.* (1984) argued that the *Tuvaella* fauna of Central to eastern Asia can only be taken as a biogeographical subprovince or endemic centre. However, we do not think that the endemism percentage is an appropriate criterion for non-tropical provincialization, as discussed in Chapter 1. Taking into account the fact that the *Tuvaella* fauna was distributed widely in northern Xinjiang (Junggar, northern Tianshan), north-eastern Inner Mongolia and Xingan, as well as in Tuva, USSR, and Mongolia, and that it represents a distinctive cool-water fauna, the present author suggests that it should be the criterion for the Junggar–Xingan province, which in turn belongs to the Boreal realm. (Fig. 5.2).

Table 5.4 Comparison of the main Silurian taxa and sedimentary facies among the biomes in Yangtze province

Age	Qinling (Batang, W. Sichuan)	Upper Yangtze (semi-graptolite facies – semi-shelly facies)	Hunan–Guangxi–Guangdong (all graptolite facies)	Lower Yantze (clastic facies)
Lower Silurian	L–U. Sil. of Batang are mostly carbonate facies cor. *Entelophyllum, Stortophyllum, Spinolasma, Squameofavosites* bra. *Conchidium* West Qinling cor. *Entelophyllia, Pilophyllum*	U. Sil. of E. Yunnan are shelly facies cor. *Holmophyllum, Phaulactis, Squameofavosites* bra. *Protathyris* naut. *Yunnanoceras* con. *Spathognathodus crispus*	L–U. Sil. of Qinfang, Guangxi are all graptolite facies *Prist. tumesceus, P. bohemicus, P. nilssoni*	Fossil fish *Ningualepis, Sinocanthus faucunensis*
Middle Silurian	cor. *Microplasma gotlandica, Aphyllum, Zelophyllum, Syringopora bifurcata* bra. *Leptostrophia sp., Pentamerus* tri *Coronocephalus* (West Qinling) cor. *Holmophyllum cf. primavum, Mucophyllum, Mesofavosites obiliquns*	M. Sil. of E. Yunnan are shelly facies cor. *Holmophyllum, Ketophyllum, Micula* bra. *Howellella, Atryp* naut. *Sichuanoceras, Heyuncunoceras* tri. *Proetus, Acanthopyge* con. *Hideodella cf. equidentata* M. Sil. of NE Guizhou grap. *Monograptus, riccartonensis, Stomatograptus* bra. *Xinanospirifer, Salopina minuta*	M. Sil. are partly clastics containing trilobites and brachiopoda gra. *Monog. riccartonensis* tri. *Coronocephalus* bra. *Eospirifer*	bra. *Nalivkinia, Striispirifer* naut. *Sichuanoceras* tri. *Coronocephalus*
Upper Silurian	cor. *Cysticonophyllum, Diplochone, Halysites, Mesofavosites* (West Qinling) cor. *Amplexoides appendiculatus, Kodonophyllum, Ceriaster* bra *Leptaena depressa* gra. *Oktavites spiralis, Streptograptus crispus*	Upper L. Sil. of Shaanxi are shelly facies and semi-shelly facies. gra. *Monoclimacis griestoniensis* cor. *Idiophyllum, Maikottia, Paraceriaster, Amplexoides, Pycnactis, Pseudophaulactis, Meitanopora* Lower parts are graptolite facies *Monog. sedgwickii–G. persculptus*	Upper graptolites *Oktavites spiralis, Sprograp. turriculatus* Lower graptolites *G. persculptus, Monog. sedgwickii*	gra. *Z. G. persculptus–Pristiog. leei*

cor., corals; bra., brachiopods; tri., trilobites; gra., graptolites; naut., nautiloids; con., conodonts; L–U. Sil., Late–Upper Silurian; U. Sil., Upper Silurian; M. Sil., Middle Silurian

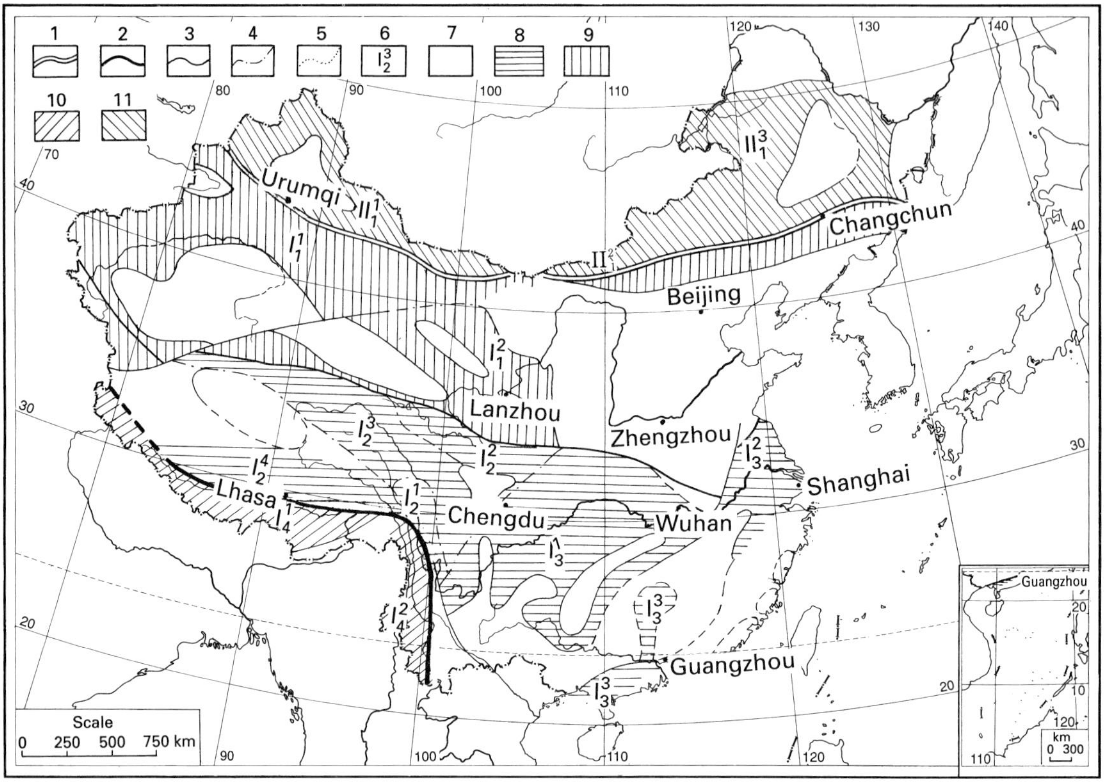

Fig. 5.2 Map of the Early and Middle Silurian palaeobiogeographical provincialization of China. 1, Boundary of realms; 2, boundary of regions; 3, boundary of provinces; 4, boundary of biomes; 5, boundary of communities; 6, roman digit = realm mark, subscript = province mark, superscript = biome mark; 7, Oldland; 8–11, Marine and paralic regions: 8, South China; 9, North China; 10, Gondwana; 11, Junggar–Xingan. I, East Proto-Tethyan realm; I_1, Tianshan–Qilian province; I_1^1, Southern Tianshan–Beishan biome; I_1^2, Qilian biome; I_2, Tibet–Qinghai–Western Sichuan province; I_2^1, Yidun–Batang biome; I_2^2, Western Qinling biome; I_2^3, Markam biome; I_2^4, Northern Tibet biome; I_3, Yangtze province; I_3^1, Upper Yangtze biome; I_3^2, Lower Yangtze biome; I_3^3, Central Hunan–Qinfang biome; I_4, Southern Tibet–Western Yunnan province; I_4^1, Southern Tibet biome; I_4^2, Western Yunnan biome; II, Boreal realm; II_1, Junggar–Xingan province; II_1^1, Junggar biome; II_1^2, Northern Inner Mongolia biome; II_1^3, Xingan biome.

Early and Middle Silurian provincialization of major groups

Early and Middle Silurian brachiopods

Represented by *Nalivkinia* and *Xinanospirifer*, the Silurian brachiopods of Yangtze province are rich and varied, and have been well studied. Wang Yu *et al.* (1984) reported that so far 12 genera of early Llandoverian brachiopods (including *Dalmanella*, *Eospirifer*, and *Hindella*) have been described in southern China (including Yangtze). However, all of them occurred widely over the world. Seventeen genera of brachiopods proliferated in southern China during the middle Llandoverian. Some of them were newcomers, such as *Borealis*, *Atrypina*, *Nalivkinia*, *Zygospirifer*, *Qianomena*, *Beitaia*, and *Kritorhynchia*. The last three genera have been found only in Yangtze, while *Nalivkinia* originated in the Uralian–Codilleran region. *Borealis* first occurred during the middle Llandoverian in the Baltic.

According to the published data, the late Llandoverian–early Wenlockian strata in China yield more than 60 genera of articulate brachiopods, of which 26 genera, including *Atrypinopsis*, *Atrypoidea*, *Nikiforovaena*, and *Pleurodium*, came from the Ural-Codillera. The Lower Silurian series yields *Nalivkinia*, *Striispirifer*, and *Antigaleatella* in the Qilian Mountains. It is of interest that the Zhongning–Tongxin area of Ningxia yielded brachiopods such as *Holorhynchus*, *Eospirifer*, and *Nalivkinia* during the Early–Middle Llandoverian, while, in contrast, *Holorhynchus* occurred mainly in the late Late Ordovician, in Central Asia.

Spiriferids and atrypids were predominant among Middle Silurian brachiopods. In Yangtze, *Eospirifer*, *Striispirifer*, and *Howellella* were the most widespread genera. Furthermore, some newcomers, such as *Xinanospirifer* and *Nalivkinia*, are also often found in the Middle Silurian of Yangtze, Qilian, and Altay. *Nucleospira* was a major and geographically widespread member of the nucleospirids. *Spinochonetes*, a strophomenid common in Yangtze, spanned the time from late Early to Middle Silurian.

According to Rong and Yang (1981) and Rong *et al.* (1984), there were five benthic assemblages amongst the Silurian brachiopods in the Sichuan–Guizhou–Hubei area of Upper Yangtze, summarized as follows:

Benthic assemblage 1. This contains numerous *Lingula*, abundant bivalves, and some rhynchonellids. Interbedding with ripplemarks characterizes the sediments, e.g. the upper part of the Rongxi Formation in Leijiatun, Shiqian Co., north-eastern Guizhou. The assemblage is assumed to have lived in a water depth of 0–10 m.

Benthic assemblage 2. This is represented by the abundant *Eocoelia* community. In addition, monotonous *Nalivkinia* and *Nucleospira* can also be included in this assemblage, which occurs in sections of the Majiaochong and Xiushan Formations in the borderland of Sichuan–Guizhou–Hunan–Hubei. The assemblage was distributed down to about 30 m below sea-level.

Benthic assemblage 3. This assemblage is represented by the *Pentamerus* community, which occurs with the greatest numbers among all five assemblages. The upper Luorepin Formation, Yichang City, western Hubei, yields a great number of *Pentamerus* preserved *in situ*, in static water. The uppermost part of the Xiangshuyuan Formation in Shiqian Co. has provided an abundance of *Paraconchidium* with their dorsal and ventral valves usually separated, indicating that they were deposited under turbulent-flow environments. This assemblage usually lived down to a depth of to 60 m below sea-level.

Benthic assemblage 4. This assemblage lived in quiet water, and is noted for the *Stricklandia* community. Also in this assemblage are *Pleurodium*, (which occurs in the upper Xiangshuyuan Formation of the Lower Silurian in north-eastern Guizhou) and *Zygospiraella*, (found in some sections of the Luorepin Formation of western Hubei, in which *Stricklandia* also occurs). This assemblage lived at a greater depth than benthic assemblage 3, but less than 100 m below sea-level.

Benthic assemblage 5. An assemblage called the *Clorinda* community by Ziegler, it occurs in the Luorepin Formation of Yichang City, and the Xiangshuyuan Formation of north-eastern Guizhou. However, large numbers of *Stricklandia* or *Pentamerus* are associated with this genus. Therefore there seems to be no reason to establish the *Clorinda* community on the basis of only the presence of *Clorinda*. In fact this community has also been found in assemblages 3 and 4. Assemblage 5 lived down to a maximum depth of 100–200 m, below sea-level.

Early–Middle Silurian corals

Fossil beds of two localities have proved to be the lowest stratigraphic horizons yielding Early Silurian corals (Wang and He 1981). One of them is located in Leijiatun, Shiqian Co., north-eastern Guizhou, with corals occurring in the limestone at the bottom of the Longmaxi Formation. These include the rugose corals *Brachyelasma* and *Parastauria*, the tabulate coral *Palaeofavosites rugosus*, and heliolitoids *Propora conferta guizhouensis*. The other locality is the Zhongning–Tongxin area in Ningxia, yielding *Paramplexoides*, *Grewingkia*, *Dinophyllum*, *Pseudophaulactis*, and *Brachyelasma* in the Zhaohuajing Formation, which lies below beds containing the Early–Middle Llandoverian graptolite community—*Climacograptus minutus*, *C. miserabilis* (He Xinyi *et al.* 1987). In addition, Early Silurian corals also flourished in Beishan area, Gansu. They consist mainly of tabulate and heliolitoid corals, such as *Palaeofavosites*, *Halysites*, and *Hexismia*. The latter has been found mostly in the Lower–Middle Silurian of North America and Kazakhstan.

The Middle Silurian corals flourished mostly in the Guandi Formation of eastern Yunnan in Yangtze province, and in the upper part of the Xiushan Formation of north-eastern Guizhou in Yangtze province. The Middle Silurian corals in the Qilian–Qinling area were found in the Quannaogoushan Formation of north Qilian and in the Zhouqu Formation of western Qinling, which differed in general features from those in Yangtze. Besides, a great number of Middle Silurian corals also proliferated in Tianshan, mainly central–southern Tianshan, and in the Baiyun'obo area of northern Inner Mongolia. In recent years, an abundance of corals, especially *Favosites* and other tabulate corals, has been found in the Middle Silurian Sanze Formation of Batang in western Sichuan.

The Himalayan area, including Zanda, has been found to yield Early–Middle Silurian corals, such as *Amplexoides*, *Palaeofavosites*, and *Heliolites repora*. Their general features are not yet completely known. They may be related to the Yangtze coral community.

Now we turn to discussion of the provincialization of the Early–Middle Silurian coral fauna.

Yangtze coral province

As pointed out above, the shelly facies was well developed in this province from late Early to Middle Silurian. It was typical of epicontinental platform, with rich and varied fossils. Wang Hongzhen and He Xinyi (1981) reported on the coral-yielding horizons and the sequence of coral assemblages. Statistics by Li Ziming (see Table 1.4) show high endemism of this province both in corals and in other groups. Liao (1990) also reported that during the early and late Llandoverian, Yangtze showed an coral endemism of 35 per cent, with a high level of similarity with eastern Australia (58 per cent).

The Early Silurian corals found in the Shiqian–Sinan area of north-eastern Guizhou were the best developed and most complete in sequence in the Yangtze region. Palaeogeographically, this area was near the margin of the central Guizhou Oldland, and was also near the Silurian deposition centre in Yangtze. As stated above, the coral community at the lowest horizon of the Lower Silurian of southern China (corresponding to the Early–Middle Llandoverian) is that found in the limestone at the bottom of the Longmaxi Formation at Leijiatun, Shiqian. Above this community there is Longmaxi graptolithic shale, 5 m thick, followed upwards by the Xiangshuyuan Formation of shelly facies (corresponding to middle Llandoverian according to the brachiopod and coral communities). This formation contains rugose and tabulate corals in large numbers, such as *Brachyelasma*, *Crassilasma*, *Dinophyllum*, *Pycnactis*, *Amplexoides*, *Palaeophyllum*, *Ceriaster*, *Rhabdocyclus*, *Cantrillia*, *Cystiphyllum*, *Rhizophyllum*, *Troedssonites*, *Syringoporinus*, *Schedohalysites*, and *Helioplasmolites*. The rugose corals are represented by the *Dinophyllum–Rhabdocyclus* assemblage. Apart from such endemic members as *Amplexoides* and *Ceriaster*, the other members are those frequently discovered in the Early Silurian of North Europe, Central Asia, and Australia. Recently, *Amplexoides* has also been found in the Lower Silurian of New South Wales, Australia. It is noteworthy that, in China, *Rhizophyllum* is present in a very low stratigraphical horizon (middle Llandoverian), because this genus has not been found until the Middle Silurian in Europe and almost all other parts of the world, except Australia in which it has been found in the Lower Silurian.

The Late Llandoverian of north-eastern Guizhou is noted for a diversified coral fauna in the Leijiatun Formation, with abundant *Kodonophyllum*, *Tabularia*, *Zelophyllum*, *Cystostauria*, *Pilophyllia*, *Cysticonophyllum*, *Maikottia*, *Palaeofavosites paulus*, *Meitanopora*, and *Thecostegites*, known as the *Kodonophyllum–Maikottia* assemblage. The so called Shiniulan Formation (in northern Guizhou, southern Sichuan) and Luorepin Formation (in western Hubei) are equivalent in stratigraphic horizon to the Leijiatun Formation. The formations yield members of the same coral assemblage, such as *Kodonophyllum* and *Maikottia*, despite respective endemic elements typical of each formation. Corals with dissepiments, such as *Strombodes*, *Yassia*, and *Entelophyllum*, occur in large numbers in the upper part of the Shiniulan Formation of southern Sichuan. Apart from *Amplexoides*, *Brachyelasma*, *Eostauria*, *Kodonophyllum*, and *Palaeophyllum*, i.e. the members often found in Yangtze, the Luorepin Formation in western Hubei is noted for the full development of *Onychophyllum* (the typical member of the Lower Silurian in western Europe), *Pseudophaulactis* (prevailing in the Lower Silurian of Europe and Siberia), and the abundance of slipper-like or semi-slipper-like *Rhizophyllum* and *Calceocystiphyllum*. Early Silurian corals of northern Guizhou and north-eastern Yunnan have a number of genera and species in common, such as *Dinophyllum yunnanense*, *Lindstroemophyllum*, and *Holmophyllum conicum*, and should be regarded as one biome, despite the fact that *Kodonophyllum* and *Maikottia* have not yet found in north-eastern Yunnan.

Although many of the Early Silurian corals in Yangtze were cosmopolitan, there were still quite a few endemic genera, such as *Amplexoides*, *Paramplexoides* (a survivor from the late Ordovician), *Eostauria*, *Dinophylloids*, *Yangtzephyllum*, and *Meitanopora*. A change of situation took place after the Telychian, Early Silurian. A reduction of carbonate deposits in the area resulted in movement of the flourishing centre of corals to southern Shaanxi and northern Sichuan. For example, the Ningqiang Formation of southern Shaanxi yields rugose corals in extreme abundance. The important representatives are *Idiophyllum*, *Gyalophylloides*, *Chonophyllum*, *Pilophyllia*, *Miculiella*, *Shensiphyllum*, and *Kyphophyllum*, constituting the *Idiophyllum–Kyphophyllum* assemblage. Besides the rugose corals, this formation also yields tabulate corals in great number, especially *Somphoporella*, *Carnegiea*, *Palaeofavosites*, *Subalveolites*, and *Erlangbapora* the *Somphoporella–Carnegiea* assemblage (Lin Baoyu *et al.* 1984).

In comparison with southern Shaanxi, the upper part of the Xiushan Formation in northern and south-eastern Guizhou and its corresponding strata have far fewer corals. Those found include *Ketophyllum*, *Zelophyllum*, *Tabularia*, *Pseudoamplexus*, *Rhizophyllum*, and *Palaeofavosites*. *Idiophyllum*, (from the upper part of the Xiushan Formation of north-western Hunan) and *Shensiphyllum* (from the upper part of the Shamao Formation of western Hubei) are comparable to the corals of the Lower Silurian in southern Shaanxi. Therefore, these areas may belong to the same province. In summary, the Early Silurian corals of Yangtze province, notably rugose, have many members in common with those in the Baltic, Kazakhstan, and Australia. No doubt they should be regarded as belonging to the east Proto-Tethyan realm. On the other hand, the Telychian (Early Silurian) many corals, such as *Chonophyllum*, *Miculiella*, *Cysticonophyllum* originated from Europe. Many of them are similar to those of the same age in Kazakhstan but have not much in common with those seen in Australia. Thus it may be concluded that the Early Silurian coral fauna of Yangtze province was more closely related to that of the Urals and Kazakhstan in the East Proto-Tethyan realm, apart from some endemic members.

According to Yu Changmin *et al.* (1988), due to the discovery of the index graptolites *Monoclimacis griestoniensis*, *Oktavites tullbergi*, and others, the Ningqiang Formation, formerly referred to Middle Silurian, should be attributed to the Telychian Stage, Lower Silurian.

Tianshan–Qilian coral province

The Early Silurian strata of this province are predominated by graptolithic facies with few coral interbeds. The Middle Silurian corals developed in the southern Tianshan-Kuqa region, with representatives such as *Microplasma gotlandicum*, *Holmophyllum*, *Amplexoides*, *Palaeofavosites*, and *Favosites*.

In addition, the Lower Silurian in Keping of Xinjiang yields *Streptelasma*, *Kodonophyllum*, *Plasmopora*, and *Propora*, besides the associated brachiopods, such as *Dalmanella* and *Eospirifer*. The Early–Middle Silurian corals in the Tianshan area are represented by those of the Kifuke Formation including *Gyalophyllum*, *Microplasma*, *Holmophyllum*, *Zelophyllum*, *Strombodes*, *Yassia*, and *Kyphophyllum*.

The most abundant corals in the Lower–Middle Silurian of Qilian area are represented by those found in the Early Silurian Zhaohuajing Formation in the Tongxin–Zhongning area, Ningxia, and the Middle Silurian Quannaogoushan Formation in the Qilian Mountains. The Early Silurian corals of Tongxin area have been described briefly. As far as their features and taxa are concerned, they contain members of European type, such as *Dinophyllum* and *Grewingkia*, on the one hand, and Early Silurian members seen in Siberia (of the Boreal realm), such as *Tungussophyllum* and *Profieviella*, on the other. It is of interest that *Paramplexoides*, the member thought to be unique to Yangtze province, has also been found in the Tongxin area, reflecting the intermixing between provinces. Coral fauna from the Quannaogoushan Formation is represented mainly by *Nanshanophyllum* and *Ptychophyllum*, having few taxa in common with those in the Ningqiang Formation of Yangtze. Although the fauna is relatively monotonous, it has a relatively close afinity with those found in Europe and Kazakhstan.

Qinghai–Tibet–Western Sichuan coral province

Corals of the Early and Middle Silurian have a limited distribution in this province. In the Yidun–Batan area of western Sichuan, the lower part of the Lower Silurian consists of graptolites, while the upper part is of shelly facies, yielding such corals as *Cysticonophyllum*, *Tryplasma*, *Diplochone*, *Mesofavosites*, and *Halysites*. The Sanze Formation of Middle Silurian yields varied corals, notably tabulate corals, typical elements of which include *Acanthohalysites*, *Favosites*, *Mesofavosites*, *Schedohalysites*, *Syringopora bifurcata*, *Squameofavosites*; typical rugose corals are *Microplasma gotlandica asiatica*, *Zelophyllum*, and *Tryplasma*. Other organisms in this formation include brachiopods (*Pentamerus* and *Pleurodium*) and trilobites (*Coronocephalus*). The Batan area differs much from Yangtze province in the general features of the Early–Middle Silurian corals. The most developed rugose corals are cystiphylloid. The members typical of Yangtze province, such as *Amplexoides*, *Eostauria*, *Maikottia*, *Idiophyllum*, and *Erlangbapora* are scarce in the Batan area. It is of interest to note that *Schedohalysites* has been found only in the Middle Silurian of Australia, and has not yet been discovered in Asia.

The Silurian in western Qinling is of geosynclinal depositional type. The Lower Silurian strata consist of graptolite facies or mixed facies. A number of coral fossils have been discovered (Yang Hengshu *et al.* 1990), such as *Amplexoides*, *Ceriaster*, and *Kodonophyllum*. Corals of the Middle Silurian Zhouqu (or Mar) Formation flourish in Diebu and Zhouqu of Gansu, and in Ruoergai of north-western Sichuan. Typical corals include *Stereoxylodes*, *Holmophyllum*, *Gyalophyllum*, *Hedenstroemophyllum*, *Favosites gotlandicus*, *Mesofavosites obiliqunus*, and *Falsicatenipora dazhubaensis*. These formations have some corals in common with the Ningqiang Formation in southern Shaanxi and northern Sichuan of the Yangtze province, although the genera typical of the Early Silurian in Yangtze province are absent here (except *Amplexoides*).

Southern Tibet–Western Yunnan coral province

Silurian corals have been found in only two localities in this province. One of them is the lower part of the Siqipo Group of the Lower Silurian at north Jiacun, Nielamu of Tibet, yielding *Triplophyllum nyalamense* and *Favosites* sp. The other is Zanda County of southern Tibet, where the scientific exploration team of Wuhan College of Geology discovered some rugose and tabulate corals of Early–Middle Silurian. These corals include *Kodonophyllum*, *Amplexoides*, *Thamnopora zandaensis*, *Cladopora*, and *Heliolites* (He Xinyi 1990). This coral community, similar to Yangtze type, obviously has a close afinity with that found in the Proto-Tethyan realm in the presence of such genera as *Kodonophyllum* and *Heliolites*, and thus is tentatively incorporated in the Proto-Tethyan realm.

Early and Middle Silurian cephalopods

Silurian nautiloids have a wide distribution in China, except south-eastern China. They are commonly found in Yangtze, Qinling, Tibet, western Yunnan, and the Himalayan region. According to Chen Junyuan (1983) three biogeographical provinces of Silurian cephalopods are recognizable in China.

Yangtze cephalopod province

This region (Chen *et al.* 1981) yields an abundance of nautiloid fossils, with 63 known genera and more than 300 species. Known as the *Sonkanoceras* fauna, they began to thrive during the Shiniulan age of the Early Silurian, but were mainly restricted to northern Guizhou. Almost at the same time, or a little later, a widespread fauna, *Yichangoceras*, occurred in western Hubei. They belong to the Luorepin age.

During the Xiushan age (probably Middle Silurian; a stage of the progressive enlargement of transgression), the Yangtze region was dominated by the widespread *Sichuanoceras* fauna, which has also been found in southern Jiangsu, Qinling Mountains and eastern Yunnan.

The Dongkacuo area of Bango in central Tibet yields, in addition to *Sichuanoceras*, nautiloids quite similar to that in Russia, Europe, and America.

Southern Tibet–Western Yunnan cephalopod province

Graptolite facies essentially predominated in this region during the Lower Silurian; no nautiloids have been found in these strata. Although western Yunnan yields nautiloids in strata of the Middle and Upper Silurian, the species and genera are monotonous, with *Michelinoceras* and its subgenus *Kopanioceras* in full flourish. The fauna differ from those of Yangtze, but have a similarity with western Europe. The Himalayan region also provides *Michelinoceras*, its subgenus *Kopanioceras*, and *Columenoceras* in Middle and Upper Silurian strata, and should therefore be included in the Southern Tibet–Western Yunnan province. These genera and subgenera are directly comparable to the nautiloids of the same stage in the Middle–Upper Silurian of Bohemia (Chen 1975).

North China cephalopod province

Up to now, nautiloids of the Silurian have been found only in the lower Bate'aobao Group (now known as the Bate'aobao Formation) in a few localities, including Baiyun'ebo of Inner Mongolia, where they are noted for the abundance of *Trocholites* (Middle Ordovician to Middle Silurian), and, to a lesser extent, *Phragmoceras*, which is often found from the upper part of the Middle Silurian to the lower part of the Upper Silurian in Europe and North America. The Silurian cephalopod provincialization of this region and relation to adjacent regions are uncertain because of insuficient data.

Early and Middle Silurian trilobites

Silurian trilobites were on the decline in China as in the whole world. The major families were the Encrinuridae, Proetidae, Cheirnuridae, and Hapridae. According to Xiang Liwen (1984), Chinese Silurian trilobites may be divided roughly into three regions, namely, North China, Yangtze, and western Yunnan. Geographically, they correspond roughly to the three biogeographical provinces of cephalopods.

North China trilobite province

This fauna ranges from Xinjiang, western Junggar, through Tianshan and Inner Mongolia to central Jilin. *Encrinurus* occurred in this region in larger numbers during the Middle–Late Silurian. The Bate'aobao Group (Middle–Late Silurian) in northern Inner Mongolia yields trilobites in large numbers, such as *Cheirurus*, *Gravicalymene*, *Cyphoproetus*, *Scutellum*, and *Phacops*. However, these genera persisted for a long time. Some of them range from the Silurian to the Devonian.

Yangtze trilobite province

There are trilobites at all three series of Silurian in this region, with a greater concentration in certain horizons. The trilobite *Ptilillaenus–Gaotania* assemblage contained many genera of late Early Silurian in Sichuan, Guizhou, and western Hubei. Most of them were endemic, apart from European members such as *Odontpleura*. The late Lower Silurian trilobites belong to the so-called *Coronocephalus–Kailia* assemblage. *Coronocephalus*, often occurring in the Lower–Middle Silurian of the Yangtze region, had the widest distribution in the upper and lower Yangtze and Jiangnan regions and also in southern Tianshan and the Bayan Har Mountains. Some members of this assemblage descended from the Early Silurian or persisted into the Late Silurian.

Western Yunnan trilobite province

Published data of the distribution of this trilobite fauna are very limited. Only the yellow shale of the Renheqiao Formation (Early and Middle Silurian) in the Shidian area has been found to yield trilobites

(*Leonaspis* and *Primaspis*). They are very different from the Yangtze trilobite fauna.

Early and Middle Silurian palaeobiogeographical provincialization of China

On the basis of the development and distribution of major Early–Middle Silurian organisms (brachiopods, corals, cephalopods, and trilobites), and with the sedimentary facies, palaeogeography, palaeotectonics, palaeomagnetism, and ecological environments in mind, the Early–Middle Silurian palaeobiogeographical provinces in China have been distinguished, as in Fig. 5.2.

Junggar–Xingan province

Due to the of absence of carbonate sediments in the Silurian, only a few solitary small rugose and tabulate corals occurred in the Junggar–Xingan province of the Boreal realm. This province can further be subdivided into three biomes (Table 5.3) mainly on the basis of the distribution of brachiopods and corals, as well as sedimental characteristics.

The Chinese part of East Proto-Tethyan realm in the Early–Middle Silurian can be further subdivided as follows:

Yangtze province

Upper Yangtze biome

This includes Sichuan, Guizhou, western Hubei, north-eastern Yunnan, part of western Yunnan (Lijiang River), southern Shaanxi, and north-western Hunan, and is a typical carbonate sedimentary area of platform type. Both the Lower and Middle Silurian have predominantly carbonate facies (and mixed facies or clastic facies on occasion). but the lower part of the Lower Silurian is composed mainly of graptolite shale facies. This region is rich in fossils of many types (Fig. 5.3), with abundant brachiopods, corals, nautiloids, trilobites, and sometimes even reefs.

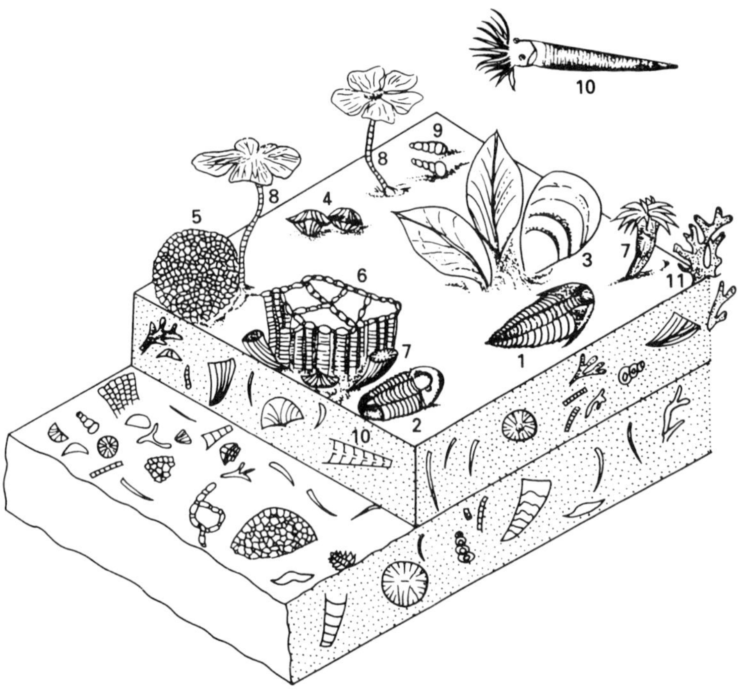

Fig. 5.3 Reconstruction of the late Early Silurian Upper Yangtze biome (The Gorges). Trilobites: 1, *Coronocephalus*; 2, *Proetus*. Brachiopods: 3, *Pentamerus*; 4, *Striispirifer*. Tabulates: 5, *Favosites*; 6, *Halysites*. 7, Rugose coral, streptelasmatid; 8, crinoid *Petalocrinus*; 9, gastropod *Hormotoma*; 10, nautiloid *Orthoceras*.

Lower Yangtze biome

South-eastern Hubei, central Anhui, southern Jiangsu, and central Jiangxi fall within this region. Silurian series are complete here. The Lower Silurian consists mainly of graptolite facies, and the Middle–Upper Silurian of clastic facies. Despite the absence of coral fossils, other benthic organisms, such as brachiopods, trilobites, bivalves, and gastropods, were well developed and formed a transitional assemblage (Fig. 5.4). The biotas and accompanying deposits suggest that this region was located on the platform margin near the open sea.

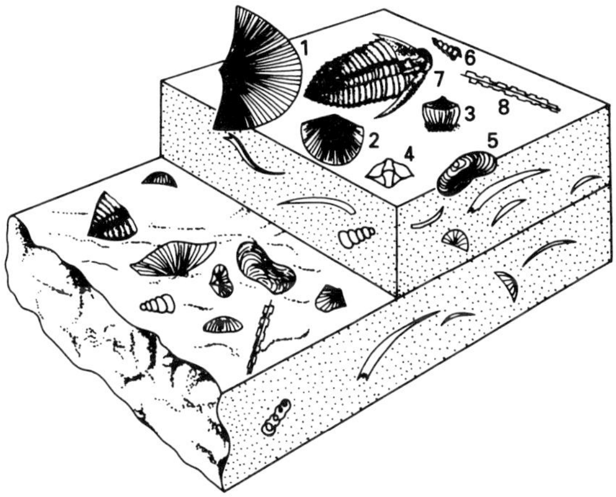

Fig. 5.4 Reconstruction of the transitional Lower Yangtze biome community of the Early Silurian Dabaidi Formation at Siaofeng, Zhejiang. 1–4, Brachiopods: 1, *Strophomena*; 2, *Fardenia*; 3, *Resseralla*; 4, *Eospirifer*; 5, bivalve *Modiolopsis*; 6, gastropod *Hormotoma*; 7, trilobite *Coronocephalus*; 8, graptolite *Climacograptus*.

Central Hunan–Qinfang biome

This region includes southern Hunan, northern Guangxi, Liantan (Guangdong), and Qinzhou–Fangcheng (Guangxi). In Middle and Lower Silurian strata graptolite facies dominate. Early Silurian graptolites flourished at Liantan, forming abundant graptolite communities. In the Qinfang area, Early Silurian sediments reach their maximum thickness of more than 2000 m. In the Middle Silurian, the Hepu Group still contained such graptolites as *Cyrtograptus* and *Monograptus priodon*; however, their distribution area tended to decrease.

Tianshan–Qilian province

Southern Tianshan–Beishan biome

This region ranges from Kuqa of Xinjiang, southern Tianshan, through Beishan in northern Gansu, and eastwards to Ejina Qi of Inner Mongolia. In Lower Silurian strata graptolite facies dominates, whereas in the Middle and Upper Silurian, shelly facies prevails. In the Beishan area, the Silurian strata, notably Lower–Middle ones, yield abundant corals; a few Late Silurian corals are also found in the northern part of the Badan Jaran desert of Ejina Qi.

Qilian biome

This extends from the western corridor of the Yellow River, Liupanshan, through the border areas within Shaanxi, Gansu, and Ningxia to the Tongxin–Zhongning area of Ningxia. This biome is seen in all Silurian series, but various areas differ in the completeness of the sequence. The lower part of the Lower Silurian is usually dominated by graptolite facies, with mixed facies in some areas (e.g. the Zhongning area of Ningxia; Li 1980). The Middle Silurian consists of shelly facies, rich in varied fossils, and the Upper Silurian comprises purple clastic flysch deposits with little sign of fossils.

Qinghai–Tibet–Western Sichuan province

Yidun–Batong biome

This is a narrow north–south zone including Yidun, Muli, and Batang of western Sichuan. The lower part of the Lower Silurian is graptolite shale, while the upper part of the Lower Silurian and Middle Silurian are strata of shelly facies where corals developed. As mentioned above, the best developed are tabulate corals, followed by brachiopods, trilobites, and bivalves.

Western Qinling biome

Diebu, Luqu, Zhouqu, Wudu, Lueyang of Gansu, north of Zoige in Sichuan, fall within this region. The Lower Silurian consists of graptolite facies; the Middle Silurian Zhouqu Group comprises mixed facies, in which corals, (predominated by *Stereoxylodes*) and brachiopods are present in larger numbers by comparison with the Lower Silurian strata, and show some distinguishing features.

Markang biome

This ranges from Qinchuang, Guangyuan of northern Sichuan to Pinwu and Baoxing in the south. The Silurian sediments are composed mainly of clastic rocks of great thickness, known as the Maoxian Group. The lower subgroup is considered to be Middle and Late Silurian mainly yielding corals, brachiopods, and trilobites. The corals include *Pilophyllum*, *Spongophyllum cf. shearshii*, *Tryplasma*, *Squameofavosites*, and *Somphopora*; the trilobites include *Coronocephalus* sp.

Northern Tibet biome

This covers a vast area including the north of the Yarlung Zangbo fault zone. Silurian strata are exposed only in Bango, Xainza, and Rutog-Lazhulong of the Ali area. The Lower Silurian consists of graptolite facies and mixed facies. The corals, such as *Amplexoides* and *Palaeofavosites*, are similiar to those seen in Yangtze. However, they are monotonous in species and genera. The Middle and Upper Silurian are fragmentary in distribution, with the remains of a few planktonic organisms, especially conodonts, but few megafossils.

Southern Tibet–Western Yunnan province

This falls within the Gondwana side of the realm. The biotas were quite different from those in the three preceding provinces and thus should be attributed to an independent region.

Western Yunnan biome

This is a narrow, longitudinal zone, ranging from the north of Baoshan, southwards to Zhenkang, westwards to Luxi, and eastwards to Changning. There is general development of graptolite facies in the Lower and Middle Silurian, with plankton in the majority, and the graptolite assemblages can often be found in strata (Fig. 5.5).

Southern Tibet biome

This includes the Himalayan area. Silurian strata outcrop in Nielamu and Zanda, they consist of all series, although incomplete in exposure. The lower series consists of mixed facies containing graptolites, corals, and brachiopods, or shelly facies, while the Middle–Upper Silurian are of shelly facies, with nautiloids and conodonts in larger numbers. This biome has a close affinity with those seen in western Europe, but is very different from those found in Yangtze.

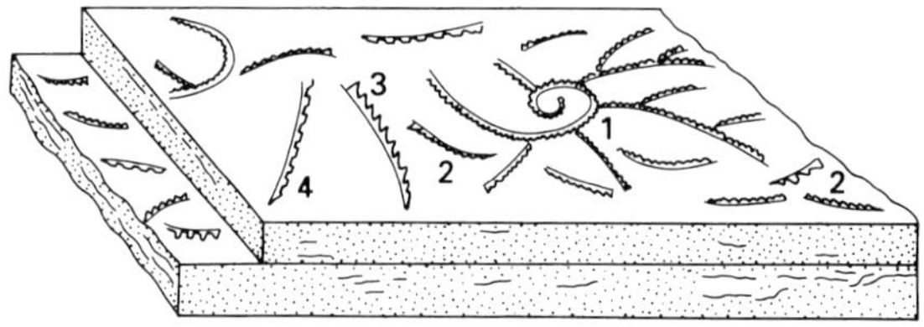

Fig. 5.5 Reconstruction of the graptolite community of the Middle Silurian Planktonic facies in the western Yunnan biome. 1, *Cyrtograptus*; 2, *Pristiograptus*; 3, *Monograptus*; 4, *Monoclimacis*.

5.2.3 LATE SILURIAN PALAEOBIOGEOGRAPHICAL PROVINCIALIZATION OF CHINA

A great regression is recorded in China, especially in southern China, from the late Late Silurian to the early Early Devonian. Most of the Yangtze area was exposed. The flooded areas include the Qilian–Qinling area (Qilian–Qinling trough) and part of the upper Yangtze area (eastern Yunnan, part of western Sichuan). In addition, the Tianshan and Inner Mongolian troughs (including Xingan) continued to receive sediments after the Middle Silurian epoch. The Tibetan sea was connected to the western Yunnan sea, and there was continuous sedimentation during the Middle and Late Silurian, with a consequential similarity in biotas.

Generally speaking, the shelly strata of the Late Silurian (both for the geosynclinal and platformal types) show a high degree of development of corals and brachiopods, particularly those seen in Tianshan, Inner Mongolia, western Qinling, and eastern Yunnan of the Yangtze province. Upper Silurian graptolite facies is mainly seen in the Qinzhou–Fangcheng area of Guangxi. Furthermore, there are graptolite localities in the Upper Silurian of western Junggar, western Yunnan, and the Maoxian Group of north-western Sichuan.

On the basis of the distribution of Late Silurian corals, brachiopods, nautiloids, and graptolites, and considering the tectonic background and sedimentary facies, the Late Silurian provincialization of China is given as follows (Fig. 5.6).

(1) Boreal realm; Junggar–Xingan province is noted for the *Tuvaella* fauna.

(2) East Proto-Tethyan realm.

(a) Tianshan–Inner Mongolia province. Central and southern Tianshan, northern Inner Mongolia, and central Jilin fall within this province. There was general development of corals and brachiopods in the Late Silurian.

(b) Qinling–Yangtze province. This is mainly composed of western Qinling, Upper Yangtze (eastern Yunnan included), and the Jinshajiang valley of western Sichuan. There was a higher degree of development of Late Silurian shelly organisms in this province. In the Lower Yangtze area there are mainly clastic rocks lacking fossils.

(c) Southern Tibet–Western Yunnan province. The Upper Silurian in western Yunnan is

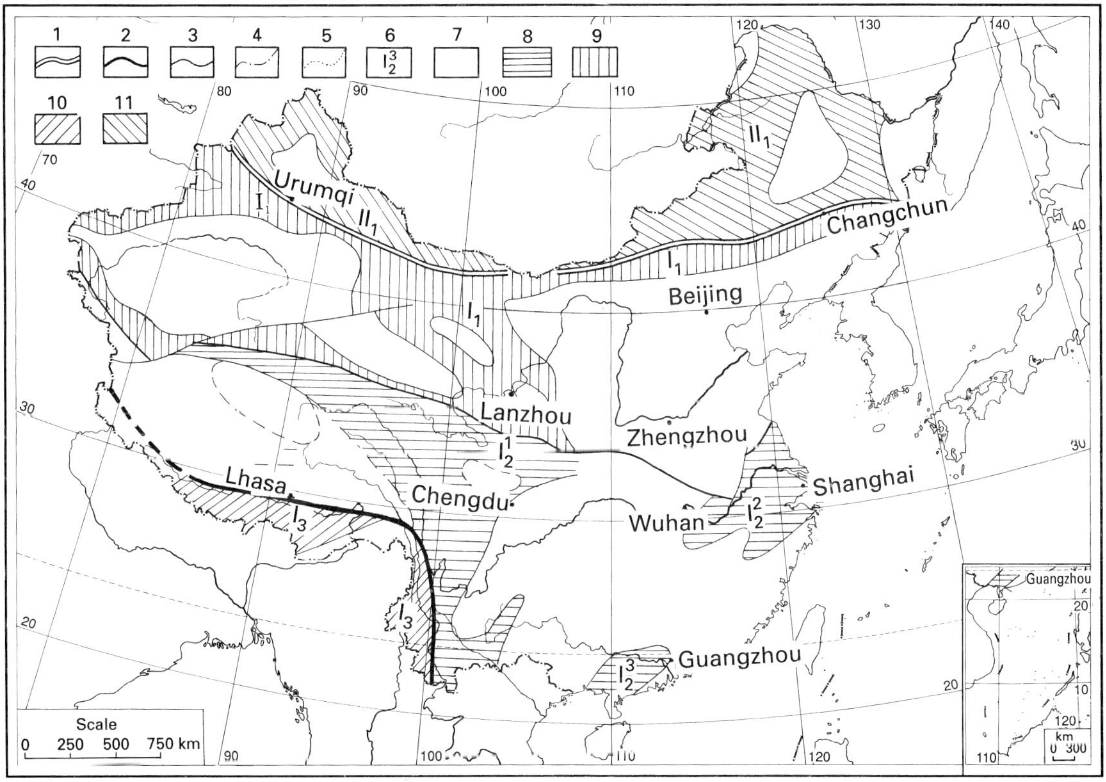

Fig. 5.6 Map of the Late Silurian palaeobiogeographical provincialization of China. 1, Boundary of realms; 2, boundary of regions; 3, boundary of provinces; 4, boundary of biomes; 5, boundary of communities; 6, roman digit = realm mark, subscript = province mark, superscript = biome mark; 7, Oldland. 8–11, Marine and paralic regions: 8, South China; 9, North China; 10, Gondwana; 11, Junggar–Xingan. I, East Proto-Tethyan realm; I_1, Tianshan–Inner Mongolia province (including Liaoning–Jilin biome); I_2, Qinling–Yangtze province; I_2^1, Qinling–Upper Yangtze biome; I_2^2, Lower Yangtze biome; I_2^3, Guangdong–Guangxi–Qinfang biome; I_3, Southern Tibet–Western Yunnan province; II, Boreal realm; II_1, Junggar–Xingan province.

mainly composed of shelly facies containing cephalopods and conodonts, and southern Tibet is similar. There may also be graptolite facies in western Yunnan. The abundant nautiloids and conodonts of this province have similarities to those seen in the Bohemian area, which was then part of Gondwana.

Main groups of Late Silurian organisms in China

Late Silurian brachiopods

Late Silurian brachiopods flourished in the northern geosyncline, including the Junggar–Xingan province and Tianshan–Inner Mongolia province. There were also rich brachiopod faunas in the Bailongjiang Group of western Qinling. However, they had a limited distribution in Upper Yangtze. Opinions are varied on the age of the upper part of the Guandi Formation in eastern Yunnan. It might be included in the early Ludlow series of the Late Silurian. Brachiopod genera such as *Atrypoidea*, *Nikiforovaena*, *Striispirifer*, and *Eospirifer* are found in this upper Guandi Formation, above which are the Miaogao and Yulongshi Formations, yielding Late Silurian brachiopods, such as *Howellella tingi*, *Protathyris*, and *Spirinella*. The Miaogao Formation belongs to the Middle and Late Ludlow Series, while the Yulongshi Formation should be attributed to the Pridolian, according to the published data on conodonts.

The Chinese Ludlowian to Pridolian strata yield articulate brachiopods with a total of 71 genera, of which 20 genera (28 per cent of the total) are members of the Ural–Cordillera type, according to Wang Yu *et al.* (1984). This shows that these brachiopods exhibited a distinctive provincialization.

The Late Silurian brachiopods in the Xingan area of the Boreal realm are represented by the fauna of the Woduhe Formation, the most common members being *Tuvaella*, *Tanuspirifer*, *Stegerhynchella*, *Protocortezorthis*, and *Leptostrophia*. The first three of these genera are indices of the biogeographical region, including Junggar, Altay, Tuva, and the Far East of Russia as well as this area.

The upper Bate'aobao Group, or the Upper Silurian Xibiehe Formation in northern Inner Mongolia yields brachiopods such as *Morlongia*, *Atrypoidea*, and *Howellella*. According to Su Yangzheng *et al.* (1983) many new genera have been found, such as *Atrypina*, *Spirigerena*, *Tenellodermis*, *Conchidium*, and *Protathyrisina gushaomiaoensis*. The brachiopod fauna as a whole should be of Ludlowian age. More than half the genera were cosmopolitan. However, *Atrypoidea* mainly flourished in the Uralian–Cordilleran area. The brachiopod assemblages have a close affinity with those seen in the Boreal realm.

Late Silurian corals

By comparison with the flourishing areas of Early and Middle Silurian corals, the main area of Late Silurian corals in China seems to have been shifted northwards and concentrated in the area extending from Tianshan, Beishan, through Inner Mongolia to Liaoning–Jilin. The following is an outline of the coral distribution in various provinces.

Junggar–Xingan province

The coral fauna of this province is roughly similar in geographical distribution to brachiopod *Tuvaella* fauna, except that corals occur essentially in the carbonate strata. Western and eastern Junggar of Xinjiang yield an abundance of Late Silurian corals. The Kaokesaiergai Formation in eastern Junggar contains, for example, *Kodonophyllum junggarense*, *Chlamydophyllum*, *Rhizophyllum*, *Kyphophyllum*, *Entelophyllum*, *Mesofavosites*, *Squameolites*, and *Heliolites*. There are also many rugose corals in the Upper Silurian Kekexiongkeduke Formation of western Junggar. They include *Kodonophyllum*, *Chlamydophyllum*, *Cystiphyllum*, *Rhizophyllum*, *Holmophyllum*, and *Kyphophyllum*. The coral fauna of eastern Junggar is comparable to that seen in western Junggar; it is similar to that found in the Proto-Tethyan realm in Europe, and has fewer genera in common with the Late Silurian Boreal realm (Siberian) fauna.

Late Silurian corals are very limited in the Xingan area, only streptelasmatid corals, such as *Streptelasma* and *Enterelasma*, have been found. The latter of these often occurs in the Middle–Late Silurian strata of North America and Siberia, and is one of the typical representatives of the Boreal realm. Therefore, the coral fauna of the Xingan area should belong to the Boreal realm.

Tianshan–Inner Mongolian province

In this province Late Silurian corals are well developed, especially in central and southern Tianshan, northern Inner Mongolia, and the western Batan–Jilin area. The Kurul Formation of central–southern Tianshan generates rugose corals (such as *Cystiphyllum, Holmophyllum, Holmophylloides, Microplasma, Kyphophyllum, Pilophyllum, and Ketophyllum*), and tabulate corals (such as *Favosites, Squameofavosites,* and *Heliolites*).

In northern Inner Mongolia, Late Silurian corals flourished, notably in the Xibiehe Formation, which yields, for example, *Spongophyllum, Entelophyllum, Mucophyllum, Kyphophyllum, Ptychophyllum, Weissermelia,* and *Holmophyllum*. The Bate'aobao Formation, which is upper Middle to lower Upper Silurian and underlying the Xibiehe Formation, yields *Altaja, Weissermelia, Diplophyllum,* and *Strombodes*. However, when describing the Late Silurian corals in the same area, Guo Shengzhe (1978) held that *Altaja* and *Weissermelia* occur in the Xibiehe Formation. In Siberia, Altay, and Tuva, *Altaja* is restricted to the Lower–Middle Silurian series. It seems that in China this genus occurs at a slightly higher stratigraphic horizon. In northern Inner Mongolia, most of the Late Silurian corals are members of European type in the Proto-Tethyan realm, e.g. *Kyphophyllum, Holmophyllum, Weissermelia* and *Spongophyllum*, despite the presence of members, such as *Altaja* and *Diplophyllum*, typical of the Boreal realm (Siberia and North America), and the presence of an Australian member, *Mucophyllum*. It is noteworthy that in the Zhusilenhairhan area of western Inner Mongolia, the conglomerate at the bottom of the Lower Devonian series yields Late Silurian coral rubble, such as *Wintunastrea, Prohexagonaria, Evenkiella,* and *Schlotheimophyllum*. The first two of these are common members of the Upper Silurian of western North America and northern Canada. *Prohexagonaria* also occurs in the Upper Silurian strata of Tuva in Russia, but *Schlotheimophyllum* is a typical element on the Proto-Tethyan. These facts show that, on the one hand, these faunas can be regarded as representatives of the Boreal realm, but, on the other hand, they contain typical members of the Proto-Tethyan realm. It seems from the above analysis that there was an intermixing of organisms, probably because northern Inner Mongolia was situated between the Boreal and Proto-Tethyan plates, and southern and northern seas were connected in this area.

Qinling–Yangtze province

In this province, Late Silurian corals developed fully in the Beilongjiang Group of Diebu in western Qinling, where 32 genera of rugose corals have been identified. The lower part of this group yields the following major genera: *Cystiphyllum, Nipponophyllum, Ketophyllum, Columnolasma, Columnaria, Kyphophyllum, Neocystiphyllum, Zelophyllum, Strombodes,* and *Rhizophyllum*, and the upper assemblages consist mainly of phaceloid to massive corals, such as *Entelophyllia, Spongophyllum, Altaja*, and, to a lesser degree, *Ptychophyllum, Micula, Schlotheimophyllum, Holmophyllum, Cyathactis,* and *Gyalophyllum*. Furthermore, the upper part of this group yields tabulate corals and heliolitids, such as *Squameofavosites, Dictyofavosites, Favosites kogulaensis,* and *Heliolites tcherganensis*.

The Qujing area of eastern Yunnan is the area where the Upper Silurian strata containing corals and brachiopods are best developed in Upper Yangtze. The Miaogao Formation contains corals such as *Crassilasma, Phaulactis, Pilophyllum, Kyphophyllum, Cystiphyllum, Microplasma, Squameofavosites,* and *Favosites*. The overlying Yulongshi Formation also yields corals in small numbers, such as *Pseudocystiphyllum* and *Squameofavosites*. Of the preceding Late Silurian corals, only *Phaulactis, Pilophyllum, Squameofavosites,* and *Favosites* occur in the Beilongjiang Group of western Qinling, other genera are very different because of the difference in biome (platform type for Yangtze, and geosynclinal type for Qinling) and province (the Beilongjiang Group has a closer affinity with the Urals and Central Asia in terms of Late Silurian corals).

In Yidun and Batan of western Sichuan the Late Silurian Yongren Formation yields an abundance of corals, such as *Stortophyllum*, *Entelophyllum*, *Tryplasma*, *Spinolasma*, *Squameofavosites* cf. *bohemicus*, and *Favosites*. They differ much from those of the Beilongjiang Group in general features. However, they are similar to the Late Silurian corals of Alengchu, western Yunnan of the Upper Yangtze, probably because this area was in the marginal depression of the Xikang–Yunnan Oldland and was connected to the western Yunnan sea trough. It is interest of note that, so far, *Stortophyllum* has been found only in the Upper Silurian of western Yunnan and Yidun of western Sichuan.

Late Silurian cephalopods

Up to now little data has been published on Chinese Late Silurian cephalopods. In the Yangtze region, only the Miaogao Formation of eastern Yunnan has yielded an *Enthyocycloceras* fauna. At the same time the pioneering species of *Yunnanoceras* occurs in this region, indicating the beginning of a further evolution of nautiloids. The overlying Yulongshi Formation of late Late Silurian is noted for the flourish of a *Yunnanoceras* fauna, the major genera being *Yunnanoceras*, *Eostereotoceras*, *Euryrizoceras*, *Ecotocyrtoceras*, *Byronoceras*, and *Kionoceras*.

In the southern Tibet–Western Yunnan region, the Middle–Late Silurian nautiloid fauna were relatively monotonous in diversity, but were characterized by the thriving of *Michelinoceras* and its subgenus *Kopaninoceras*. The Middle to Late Silurian Keya Formation of the Qomolangma area enlists *Michelinoceras transiens*, *M.(Kopaninoceras) capax*, *M.(Kopaninoceras) dorsatum*, *Harrisoceras*, *Columenoceras*, and *Geisonoceras*. Because both species of *Kopaninoceras* occur in the Middle and Late Silurian strata of Bohemia, this fauna is merely regarded as Middle–Late Silurian but not definitely Late Silurian (Chen Junyuan 1983).

In the Baoshan area of western Yunnan, the Middle–Upper Silurian also yields cephalopods in larger numbers, and is noted for the flourishing of *Michelinoceras* and its subgenus *Kopaninoceras*. They are closely related to those of the same stage in Tibet.

In Tianshan–Inner Mongolia, Late Silurian cephalopods have been found in two localities. One is the Beishanbao Formation of eastern Junggar, yielding *Plagiostomoceras*. The other is Bate'aobao in Darhan–Mumingan Lianhe Qi of Inner Mongolia, where some genera and species have been found. However, it is uncertain whether they should belong to the Late Silurian. The stratigraphic horizon in which cephalopods occur is in the Bate'aobao Formation. Zou Xiping (1983) has reported the discovery of some genera, such as *Grafronoceras*, *Beteroboceras*, *Protophragmoceras*, and *Phragmoceras*. They are similar in general features to Late Silurian members in North America, and some of them, (e.g. *Protophragmoceras*) are distributed mainly in the Middle Silurian of Bohemia. Therefore, the Bate'aobao Formation of Inner Mongolia corresponds to late Middle and early Late Silurian.

Late Silurian trilobites

Few Late Silurian trilobites have been found in China. In the Yangtze region, only the upper part of the Guandi, Miaogao, and Yulongshi Formations of eastern Yunnan yield *Acanthopyge*. The highest horizon bearing Silurian trilobites in China is the Yulongshi Formation which yields *Encrinurides* and *Warburgella* as well.

In Tianshan, Inner Mongolia, and central Jilin, the Middle–Late Silurian trilobites are present at the Daban Group, Bate'aobao Group, and Erdaogou Group. All of these yield *Encrinurus*.

In Darhan–Mumingan Lianhe Qi, Inner Mongolia, the Middle–Late Silurian Bate'aobao Group generates trilobites in larger numbers: *Encrinurus*, *Cheirurus*, *Calymene*, and proetids (Wu Hongji 1983).

In summary, the Late Silurian trilobite faunas in China are roughly distributed in two major regions. One of them is the Yangtze region, especially eastern Yunnan, noted for *Acanthopyge* in the Late Silurian. The other is the Boreal realm, including Xinjiang, Inner Mongolia, and the Liaoning–Jilin area, and is represented by *Encrinurus*.

Biomes of the Qinling–Yangtze province

Qinling–Upper Yangtze biome

This covers mainly western Qinling in southern Gansu, north-western Sichuan, eastern Qinling, north-western Hubei, as well as the Yidun and Batang area, western Sichuan. In western Qinling, Upper Silurian strata are of geosynclinal deposits, consisting of very thick carbonate facies. The benthic organisms, such as corals and brachiopods, developed in these strata. In the Batan area the Upper Silurian Yongren Formation is also composed of shelly facies, with corals in large num bers. During the Late Silurian epoch, the remarkable regression resulted in exposure of the Yangtze region except the eastern Yunnan gulf, with the Miaogao and Yulongshi Formations formed in the Qujing area. These are localized deposits, with brachiopods, predominating.

Lower Yangtze biome

This ranges from south-eastern Hubei, through southern Jiangsu to part of both Zhejiang and Anhui. Represented by the Maoshan Group, the Late Silurian strata consist of a series of nearshore epicontinental sediments, having red beds in the middle part, with scarce fossils, including antiarchid bones and bivalves, in small numbers.

Guangxi–Guangdong–Qinzhou–Fangcheng biome

This covers mainly the area of the Yunkai Mountains and the Qinzhou–Fangcheng area of south-eastern Guangxi. The Upper Silurian Fangcheng Group is a series of thick clastic deposits with no other fossils except rich graptolites, and thus reflects a special ecological environment.

Northern Tibet biome

So far, available data on this region are very scarce. There may be a few fossils in the Kilmu Group near Bango; the detailed description cannot be given here.

5.3 SOME NOTES ON THE BIOGEOGRAPHICAL PROVINCIALIZATION OF THE SILURIAN PERIOD

It is well known that palaeobiogeographical provincialization is related to the palaeoclimate and sea-water palaeotemperature, which are, in turn, affected by palaeolatitude and ocean currents, on the one hand, and by the palaeogeographical position of the continental plates and prolonged geographical separation and isolation (e.g. high mountains, deep sea) on the other hand. The Silurian palaeobiogeographical provincialization is, at present, mainly based on benthic brachiopods and corals, because their distribution was correlated closely to palaeoclimatic, palaeogeographical, and ecological conditions. For example, according to his studies on brachiopods, Boucot (1974) has distinguished two Silurian biogeographical realms: the Boreal realm (covering the greatest area) and the Malvinokaffric realm. This division is made on the basis of palaeoclimatic zonation and palaeogeographical position, as well as the presence of brachiopod communities unique to each realm.

Great changes took place in the Silurian. With the plates migrating towards high latitudes, the climatic zonation became much more conspicuous than in the Ordovician. Consequently, bipolarity of organisms became remarkable. In this respect we suggest that Boucot's Boreal realm can be further subdivided. The area near the high latitudes reflects a cool climate which covered Siberia, Tuva, Mongolia, and Xingan, and the deposits of which yielded the cold-water *Tuvaella* fauna; this area is appropriately called the Boreal. Sytova and Ulitina (1983) reported a fauna of Early Silurian solitary streptelasmatids, such as *Tungussolasma*, *Densiphyllum*, *Pterophrentis*, *Pycnactis*, *Holophragma*, in Tuva and Mongolia; these also occur in the Lower Silurian of Siberia.

Other parts of Boucot's 'Boreal realm' yield completely different faunas. In our Proto-Tethyan areas, massive rugose and tabulate corals flourished.

Table 5.5 Paleobiogeographical provincialization of late Silurian brachiopods, showing bipolarity of relationship (from unpublished materials of Fu Lipu)

Age		Cold south temperate zone (or south frigid zone)	Warm south temperate zone			Torrid zone	North temperate zone			
		Malvinokaffric province	Amazon province	S. Asia–Australia province	N. Africa province		China subprovince	Boreal subprovince	Appalachian subprovince	Tuva–Hinggan subprovince
							5 Coelospira			
Late Silurian	Ludlo-Pridolian	*Clarkeia* *Australina*	*Isorthis (Arcuella)* *Coelospira* ⑤	*Atrypoidea* ① *Capeliniella* ② *Molongia* ③	*Dubaria* ④	*Loilemia*	*Atrypoidea* ① *Molongia* ③ *Gannania*	*Atrypodia* *Capeliniella* ② *Dubaria* ①	*Pectorhyncha* *Eatanioides*	*Tuvaella* *Tannuspirifer*

1–5 mark the bipolar distribution of five genera, with *Clarkeia* and *Tuvaella* faunas in higher north and south latitudes, respectively.

Brachiopods had a wide distribution in the tropical zone immediately south and north of the palaeo-equator. During the Early Silurian *Eocoelia*, *Pentamerus*, (notably *P. borealis* and *P. oblongus*), and *Stricklandia* prevailed in low latitudes. The latter two genera were abundant, no matter whether in limestone facies of platform type or clastics of geosynclines and continental shelf, while *Eocoelia* seems to be limited to clastic facies and was distributed over a smaller area compared with *Pentamerus*.

The brachiopods of the Malvinokaffric realm in the southern hemisphere were represented by the *Clarkeia* fauna, and were characterized by a low diversity and absence of limestone deposits, the main associated elements being a few solitary corals. Therefore, this fauna is usually considered as a cold-water fauna, similar to the *Tuvaella* fauna seen in the Boreal realm.

A provincialization of Late Silurian brachiopods, showing bipolarity, is given in Table 5.5.

REFERENCES

Boucot, A. J. (1974). Early Palaeozoic evidence of continental drift. In *Plate tectonics–assessments and reassessments*, (ed. C. F. Kahle). Memoirs of the American Association of Petroleum Geologists, Vol. 23, pp. 273–94. American Association of Petroleum Geologists.

Boucot, A. J. (1990). Silurian Biogeography, *Geological Society of London, Memoir* **12**, 191–6.

Chen Junyuan (1975). Nautiloid fossils from Qomolangma region. In *A report of scientific expedition in the Mount Jolmo Lungma region, Palaeontology* **1**, 267–94 (in Chinese).

Chen Junyuan (1983). Silurian cephalopod provincialism. In *Palaeobiogeographic provinces of China*, pp. 43–52. Science Press, Beijing (in Chinese).

Chen Junyuan, Liu Gengwu, and Chen Tingen (1981). Silurian nautiloid faunas of Central and Southwestern China. *Memoirs of Nanjing Institute of Geology and Palaeontology, Academia Sinica* **13**, 1–104 (in Chinese).

Guo Shengzhe (1978). Late Silurian rugose corals from northern Bailingmiao area in Inner Mongolian

Autonomous Region. *Professional Papers of Stratigraphy and Palaeontology* **6**, 101–43. Geological Publishing House, Beijing (in Chinese).

He Xinyi, Liu Shumei, and Gao Jianguo (1987). The discovery of Lower Silurian graptolites fauna from the Zhongning area of Ningxia and its geological significance. In *Palaeozoic biostratigraphy and tectonic evolution of the Alxa Massif Margin*, (ed. Zhu Hong *et al.*), pp. 144–53. China University of Geosciences Press (in Chinese).

He Xinyi (1990). Silurian and Devonian rugose corals and tabulates of the Ngari area. In *Palaeontology of Ngari, Tibet (Xizang)*, (ed. Yang Zunyi *et al.*), China University of Geosciences Press.

Hill, D. (1981). *Treatise on invertebrate paleontology, Part F. Coelenterata, Supplement 1, Rugosa and tabulata*. Geological Society of America and the University of Kansas Press.

Ivanovskiy, A. B. (1965). Stratigraphical and paleobiogeographical review of the Ordovician and Silurian Rugosa. *Akad. Nauk. SSSR, Sibirskoe otd., Inst. Geol. Geofiz., Nauka* (Moscow), 119 (in Russian).

Kaljo, D. and Klaamann, E. (1973). Ordovician and Silurian corals. In *Atlas of Palaeobiogeography*, (ed. A. Hallam), pp. 37–45. Elsevier, Amsterdam.

Li Yaoxi (1980). Shelly facies and mixed facies of Lower Silurian of Shaanxi, Gansu and Ningxia. *Geological Review* **26**, (6), 479–85 (in Chinese).

Liao Weihua (1990). The biogeographic affinities of East Asian corals. *Geological Society of London, Memoir* **12**, 175–9.

Lin Baoyu (1979). The Silurian System of China. *Acta Geologica Sinica*, **53**, (3), 173–91 (in Chinese).

Lin Baoyu, Dong Deyuan, Wang Xiaofeng, and Guo Dianheng (1984). *Stratigraphy of China*. Vol. 6, *The Silurian System of China*. Geological Publishing House, Beijing (in Chinese).

Oliver, W. A., Jr, (1980). Coral in the Malvino-kaffric Realm. *Muensterische Forschungen zür Geologie und Palaeontologie* **52**, 13–27.

Oliver, W. A., Jr., and Pedder, A. E. H. (1979). Biogeography of Late Silurian and Devonian rugose corals in North America. In *Historical biogeography, plate tectonics and the changing environment*, (ed. J. Gray and A. J. Boucot), pp. 131–45. Oregon State Univ. Press.

Pickett, J. (1975). Continental reconstructions and the distribution of coral faunas during the Silurian. *Journal and Proceedings of the Royal Society of New South Wales* **108**, 147–56.

Rong Jiayu and Yang Xuechang (1981). Middle and late Early Silurian brachiopods faunas in Southwest China. *Memoirs of Nanjing Institute of Geology and Palaeontology, Academia Sinica* **13**, 173–270 (in Chinese).

Rong Jiayu, Johnson, M. E., and Yang Xuechang (1984). Early Silurian (Llandovery) sea-level changes in the Upper Yangtze region of Southwestern China. *Acta Palaeontologica Sinica*, **23**, (6), 147–65 (in Chinese).

Su Yangzheng (1981). On the geological and geographical distribution of *Tuvaella* with reference to its habitat. *Acta Palaeontologica Sinica* **20**, (1), 567–76 (in Chinese).

Su Yangzheng *et al.* (1983). Brachiopods of the Bateaobao Formation (early Late Silurian) from Bateaobao, Darhan Mumingan Joint Banner, Inner Mongolia. In *Silurian and Devonian rocks and faunas of the Bateaobao area in Darhan Mumingan Joint Banner, Inner Mongolia*, (ed. Li Wenguo, Rong Jiayu, and Dong Deyuan), pp. 79–102. The People's Publishing House of Inner Mongolia (in Chinese).

Sytova, V. A. and Ulitina, L. M. (1983). *The Early Palaeozoic rugosa of Mongolia and Tuva*. Nauka, Moscow (in Russian).

Wang Hongzhen and He Xinyi (1981). Silurian rugose coral assemblages and paleobiogeography of China. *Geological Society of America* **187**, 55–63.

Wang Hongzhen, Li Zhiming, and Wang Zhiping (1984). Silurian and Early Devonian rugose coral biogeography of China. *Palaeontographica Americana* **54**, 423–6.

Wang Hongzhen, *et al.* (1989). *Classification, evolution and biogeography of the Palaeozoic corals of China*. Science Press, Beijing (in Chinese with English summary).

Wang Yu, Boucot, A. J., Rong Jiayu, and Yang Xuechang (1984). Silurian and Devonian biogeography of China. *Geological Society of America, Bulletin* **95**, 265–79.

Wu Hongji (1983). Silurian trilobites from Darhan Mumingan Joint Banner, Inner Mongolia. In *Silurian and Devonian rocks and faunas of the Bateaobao area in Darhan Mumingan Joint Banner, Inner Mongolia*, (ed. Li Wenguo, Rong Jiayu, and Dong Deyuan) pp. 175–84. The People's Publishing House of Inner Mongolia (in Chinese).

Xiang Liwen (1984). Trilobite. In *Stratigraphy of China*, No. 6, The Silurian System of China (ed. Lin Baoyu *et al.*), pp. 207–9. Geological Publishing House, Beijing.

Yang Xuechang and Rong Jiayu (1982). Brachiopods from the Upper Xiushan Formation (Silurian) in the Sichuan–Guizhou–Hunan–Hubei border region. *Acta Palaeontologica Sinica* **21**, (4), 417–35 (in Chinese).

Yang Hengshu, He Xinyi, and Duan Lilan (1990). Advance on study of Silurian in Zoige and Tewo regions, west Qinling Mountains, China. *Geoscience—Journal of the Graduate School, China University of Geosciences* **4**, (3), 13–26 (in Chinese).

Yu Changmin, *et al.* (1988). Review of the age of the Ningqiang Formation. *Journal of Stratigraphy* **12**, (3), 210–24 (in Chinese).

Zhang Zixin, Rong Jiayu, and Di Qiaoling (1983). Silurian *Tuvaeella gigantea* faunule (Brachiopoda) of the Barkol area, Northeastern Xinjiang. *Acta Palaeontologica Sinica* **22**, (3), 278–94 (in Chinese).

Ziegler, A. M., Hansen, K. S., Johnson, M. E., Kelly, M. A., Scotese, C. R., and Van der Voo, R. (1977). Silurian continental distribution, paleogeography, climatology and biogeography, *Tectonophysics* **40**, (1–2), 13–43.

Zou Xiping (1983). Silurian nautiloids from Bateaobao, Darhan Mumingan Joint Banner, Inner Mongolia. In *Silurian and Devonian rocks and faunas of the Bateaobao area in Darhan Mumingan Joint Banner, Inner Mongolia*, (ed. Li Wenguo, Rong Jiayu, and Dong Deyuan) pp. 165–73, The People's Publishing House of Inner Mongolia (in Chinese).

6 DEVONIAN

Zhao Xiwen

6.1 DEVONIAN PALAEO-BIOGEOGRAPHICAL PROVINCIALIZATION OF THE WORLD

The Devonian succeeded the Caledonian orogeny, which had resulted in great changes in plate framework and tectonogeography in the world. Smith *et al.* (1981), Ziegler (1981), and Scotese and McKerrow (1990) have attempted to reconstruct the Devonian palaeocontinents. It is clear that the relative positions of continents and their latitudinal distribution have provided the palaeotectonic and palaeogeographical background for palaeobiogeography.

Many scholars (House 1971; Cocks and McKerrow 1973; Johnson and Boucot 1973; Oliver and Pedder 1976; House 1979; Norris 1979; Savage *et al.* 1976; Ziegler 1981; Pedder and Oliver 1990; Young 1990) have contributed a great deal on Devonian palaeobiogeography. This valuable information enables us to draw a brief sketch of the Devonian biosphere. Here the whole world is divided into three realms (Fig. 6.1), according to fossil (mainly brachiopod and coral) distribution at the generic level and taking into consideration the tectonopalaeogeographical framework and palaeoclimate zonation (Zhao 1985).

1. The Boreal realm comprises western and eastern North America, the Urals, Siberia, Mongolia and the Chinese territory to the north of the Ebinur–Gaxunnur–Xar Moron suture, e.g. Xinjiang and north-eastern China.

2. The Proto-Tethyan realm is subdivided into a northern region and a southern region. The former consists of the Appalachians, western Europe, Kazakhstan, the main part of China to the south of the Ebinur–Gaxunnur–Xar Moron suture and to the north and east of the Bangong Lake–Lancangjiang suture, and South-East Asia. The latter contains northern South America, North Africa, southern Europe, the Middle East, part of Tibet and western Yunnan to the south of the Bangong Lake–Lancangjiang suture, Australia, and New Zealand.

3. The Malvinokaffric realm includes southern South America, South Africa, and Antarctica.

Some of the boundaries between these realms are clear and distinctive, but others are vague and transitional. For example, the Urals evidently belong to

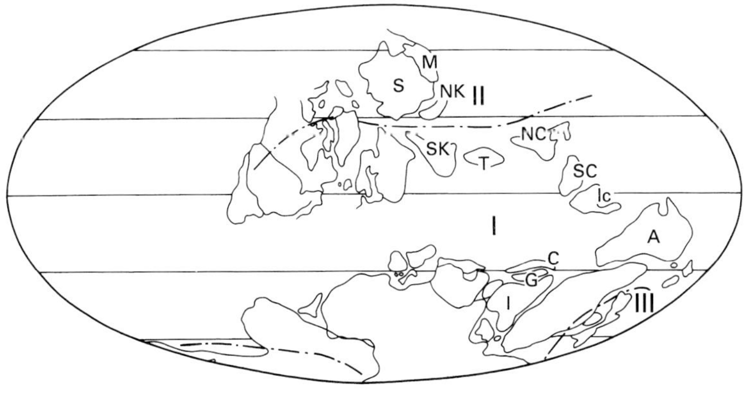

Fig. 6.1 Map of the Early Devonian palaeobiogeographical provincialization of the world. I, Proto-Tethyan realm; II, Boreal realm; III, Malvinokaffric realm; A, Australia; G, Gangdise; I, India; Ic, Indochina; M, Mongolia; NC, North China; NK, North Kazakhstan; C, Qiangtang; S, Siberia; SC, South China; SK, South Kazakhstan; T, Tarim.

the Boreal realm, but the area shows a rather close relationship with the Tethyan realm.

Temporally, the provincialization was most obvious in the Early Devonian, because after the Caledonian orogeny great topographical differentiation and weak transgression created severe migration barriers for marine organisms. Subsequently, following the increase of transgression and the planation of topography, migration of organisms became easier, so the provincialization was not as clear as that in the earlier period, but it still persisted in the Middle and Late Devonian.

6.2 DEVONIAN TECTONO-GEOGRAPHICAL FRAMEWORK OF CHINA

The tectono-palaeogeographical framework of China experienced considerable changes in the Devonian because of the influence of the Caledonian orogeny (Wang Hongzhen 1981). North China, Tarim, and Qaidam blocks were amalgamated into a larger plate, while the transitional crust along Fujian, Zhejiang, and Jiangxi, or the south-eastern continental margin of the Yangtze plate and adjacent island arc, were transformed into continental crust incorporated along the south-eastern margin of the Yangtze plate. As a result, the Yangtze plate was enlarged into the South China plate. Similarly, new 'rims' of Caledonian folding belts were added along some marginal areas of the North China plate.

However, these changes did not alter the essential framework in the Devonian of China. The Central Asia–Mongolia Palaeo-ocean will lay between the Siberian plate and the North China–Tarim plate, occupying great a width. The Xingan–Mongolia area in China was still one part of the Siberian plate, which was probably of temperate–cool climate with biotas different from those in North China. The North China–Tarim plate was an oldland lacking Devonian sediments except for some intermontane basins, although its margins were generally covered with sea water. The biota of its northern margin, i.e. southern Tianshan, Badan Jaran, and southern Jilin, was basically similar to that in South China, but mixed with a small number of Boreal and North American elements.

The width of the Kunlun–Qinling Palaeo-ocean has been a subject of dispute due to the lack of palaeomagnetic data. It is believed by some to have been a wide ocean at that time, but by others to be a small-scale ocean basin. However, Devonian biogeographical data show that this palaeo-ocean did not hinder biotic exchange between its northern and southern sides.

The tectonic meaning of the Hohxil–Jinshajiang fracture zone in the Devonian is not quite clear, since there are few palaeontological data from Qiangtang. As the biota in Qamdu, west of the Jinshajiang River, is close to that of the Hunan–Guizhou–Guangxi region, it may be postulated that both areas should not be separated by a wide ocean.

It is evident that the subducted oceanic crust represented by the Bangong Lake–Lancangjiang fracture zone was responsible for the biotic isolation in the Devonian. As is shown in the world palaeogeographic map (Fig. 6.1), the Gangdise–Nyanqingtanghula–Himalaya–western Yunnan area to the south of the fracture zone was likely to be located on the southern coast of the Proto-Tethyan realm. As for its northern boundary, some suggest that its western section, west to Dingqing Country, should be to the north of Lazhugung in Ritog County.

6.3 DEVONIAN PALAEO-BIOGEOGRAPHICAL PROVINCIALIZATION OF CHINA

6.3.1 THE BASIS AND PATTERN OF DEVONIAN PROVINCIALIZATION IN CHINA

Devonian biosphere

The Devonian biota is mostly characterized by the organic colonization of the continents and by the domination of fish. It is known at present that Devonian fish lived in fresh or brackish water in littoral deltas, and it was impossible for them to across the vast ocean. Therefore, they can be used as a basis for provincialization (Pan *et al.* 1982; Liu

Table 6.1 Correlation chart of the Devonian of China

			Tethyan realm						Boreal realm			
			Central China region						Tibet–Western Yunnan region		Junggar–Xingan region	
			South. China province				N. China–Tarim province				Junggar–Xingan province	
			Hunan–Guizhou–Guangxi	Qinling	Longmen-shan	Western Sichuan	Southern Tianshan	Badan Jaran	Gangdise–Nianqing-tanggula province	Qomolangma–W. Yunan province	Junggar	Xingan
Devonian	Upper	Famennian	Xikuang-shan	Doushishan	Maoba	Talipe	Keziletage	?	Chakuo-luoma	?	Honggu-leleng	Upper Daminshan
		Frasnian	Shetienqiao	Cakuote	Shawotze		Yimugang-tawu	Xiping-shan			?	Lower Daminshan
	Middle	Givetian	Dunggang-lin	Gudaoling	Kuanwushan	Congna	Tataaiertage	Wuotuo-shan Ikewusu	Dardong		Hujiersite	
		Eifelian	Yingtang	Lure Dangduogou	Yangmaba	Qiongcuo		Zhusileng			Mongkelu	Huoboshan Beikuang
	Lower	Emsian	Sipai Yujiang	Gala	Xiejiawan Ganxi	Gerong			Langmagou	Banmandaodi	Mangeer	Onor
		Siegenian	Nakaoling	Upper Putunggou	Bailiupin Pingyipu					Alengchu		
		Gedinnian	Lianhua-shan	Lower Putunggou						Shanjiang	Wutubu-lake	Luotuoshan

Shifan 1983, 1984). For example, Young (1981) has divided the Early Devonian fish into five provinces, one of which is the South China unit, the *Galeaspis*–Yunnanolepid province. The endemism of the Devonian fish of South China has been emphasized by many, e.g. Pan and Dineley (1988). Wang Nianzhong (1984) reported a fish assemblage from the Xitun Member, Cuifengshan Formation of eastern Yunnan, among the 187 genera of which only one is non-endemic — the thelodontid *Turinia*. Liu (1983, 1984) reported that from a total 45 genera of Early Devonian fish in South China, the only non-endemic one is the petalichthyid *Lunaspis*, which is also known in Germany and Australia. On this basis, Young (1990) suggested an isolated tropical position for South China, which is in accordance with our viewpoint.

Devonian palaeobotanic provincialization has not been studied in as much detail as that of marine invertebrates, but some researchers have produced quite fine work on it. Gierlowski (1978) divided Early Devonian plants into three floras:

(1) Gondwana flora in the southern cool temperate zone;

(2) Northwest American flora in the warms or subtropical zone; and

(3) Tropical flora. The common elements of the tropical flora were *Drepanophycus Zosterophyllum*, and *Taeniocrada*. Most parts of China were attributed to the tropical flora.

In addition, Kimura (1984) had discussed the flora province of East and South East Asia.

In the Devonian, there were many marine invertebrates such as brachiopods, corals, stromatoporoids, bryozoans, ammonoids, bivalves, gastropods, conodonts, tentaculites, trilobites, and graptolites (only in the Early Devonian).

Brachiopods were very abundant, including strophomenids, spiriferids, atrypids, dalmanellids, rhynchonellids, terebratulids, athyrids, pentamerids, and productids. Most of them were sessile benthos, so they were separated by continents as well as by abyssal oceans. Because of this, they can be key animals for geographical provincialization. Furthermore, Devonian brachiopods are well studied and widely distributed in China. Much has been published about this group, so these are probably the most important organisms for the study of palaeobiogeography in China (Wang and Rong 1983).

Corals were also abundant and diversified (Yu *et al.* 1979; Wang Zhiping 1985). Moreover, they are sensitive to climate. colonial hermatypic corals are distributed in tropical and semitropical regions, while the small solitary corals appear mainly in warm or cool climates. Corals have been discovered almost everywhere in China, so they provide a very rich source of information for study. According to Liao and Ruan (1988), the Devonian corals can be subdivided into four provinces, namely: Junggar–Xingan, Uralo–Tianshan, South China, and 'Palaeo-Tethys' (roughly our Tibet–Western Yunnan).

Devonian ammonoids are dominated by goniatitids. Although nektonic, they do have some significance for provincialization.

Bivalves may sometimes be used as a reference in provincialization (Liu 1979). Zhang Renjie and Pojeta (in Pojeta 1986) reported that from the diversified bivalve assemblage of the Early and Middle Devonian of Guangxi, at least half of the many new taxa described are endemic.

As to trilobites, Yuan (1988) noted that of the 33 species of proetid Famenian trilobites from Guangxi, 23 species are endemic.

Other groups, such as bryozoans, stromatoporoids, conodonts, graptolites, and tentaculites, have not been well studied. Thus, they are better used only exceptionally in the study of palaeobiogeographical provincialization.

According to ecological differentiation, there are three chief types in the Devonian: reef facies, mainly composed of corals and stromatoporoids; shelly facies, mainly of brachiopods, (represented by the 'Xiangzhou-type' Devonian in South China); and pelagic facies, consisting of ammonoids, graptolites, tentaculites, and conodonts (examples are the 'Nandan type' and the Devonian in the Qinzhou–Fangcheng region, Guangxi province). Ecological differentiation is the basis for the community association or biome classification.

Results of cluster analysis of Devonian biotas

Cluster analysis has been undertaken at generic level using Otsuka coefficients. The analysis is presented in Figs 6.2–6.4. The results suggest the following conclusions:

1. Similarity coefficients were smallest in the Early Devonian. For instance, the similarity coefficients between biomes in South China province in the Middle and Late Devonian were 0.544 and 0.43, respectively; while the value for the Early Devonian was only 0.225. This verifies the fact that in the

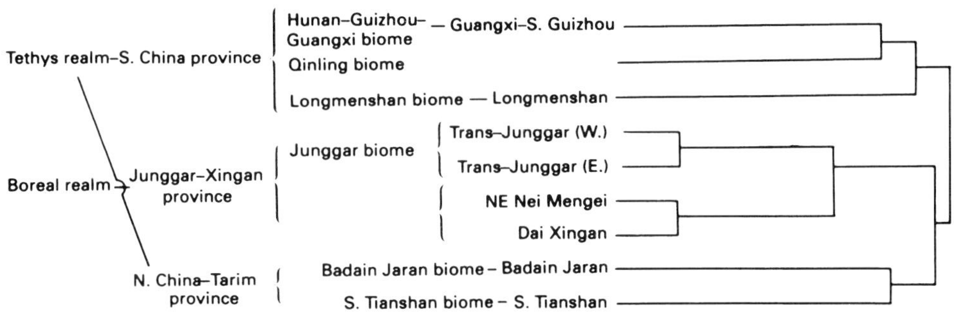

Fig. 6.2 Cluster analysis of the Early Devonian brachiopods of China.

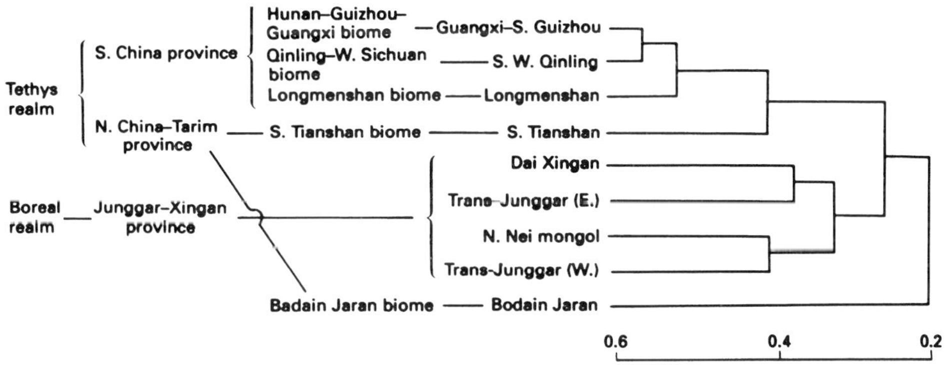

Fig. 6.3 Cluster analysis of the Middle Devonian brachiopods of China.

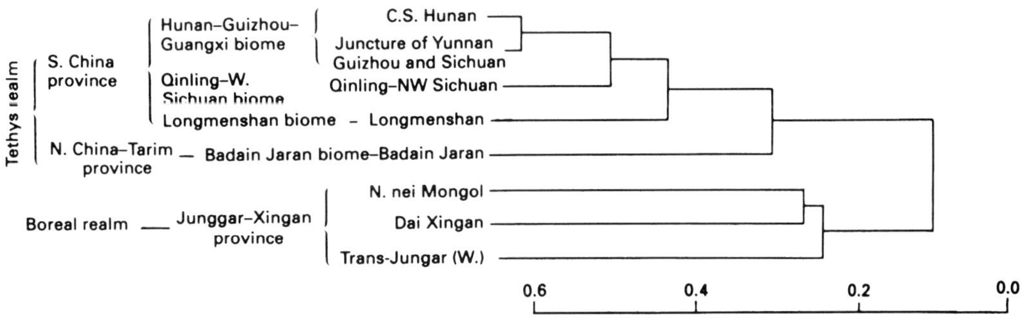

Fig. 6.4 Cluster analysis of the Late Devonian anthozoans of China.

Early Devonian the biotas were clearly different in each area, the endemic elements formed a great proportion of the organisms, and provincialization was strong.

2. The Hunan–Guizhou–Guangxi biome, Qinling biome, and Longmenshan biome shared very high similarity coefficients in the Early, Middle and Late Devonian. This tells us that they were closely related and can be combined to form one biotic province. The same applies to the Junggar and Xingan biomes.

3. The results of cluster analysis differ for the Southern Tianshan and Badan Jaran biomes (on the northern margin of the North China–Tarim province) in each epoch. In the Early Devonian these biomes were more similar to the Junggar–Xingan province. Both biomes were more closely related to the Junggar–Xingan province than to the South China province, since the South China province was only associated with the Southern Tianshan biome in the Middle Devonian and with the Badain Jaran biome in the Late Devonian. Also it can be seen from the world palaeogeographic map (Fig. 6.1) that they are indeed closer to the Boreal realm.

Scheme of provincialization

Taking into account palaeotectonics, palaeogeography, climate zonation, and the presence and absence of typical genera, and based on the results of cluster analysis, the provincialization of the Devonian in China can be defined as follows (see also Figs 6.5–6.7):

I, Tethyan realm: Central China region, consisting of the area south of the Ebinur–Gaxunnur–Xar Moron River suture and north of the Bangong Lake–Lancangjiang suture; I_1, South China province; I_1^1, Hunan–Guizhou–Guangxi biome; I_1^2, Qinling biome; I_1^3, Longmenshan biome; I_1^4, Western Sichuan biome; I_2, North China–Tarim province; I_2^1, South Tianshan biome, I_2^2, Badan Jaran biome.

I′, Proto-Tethyan realm, Tibet–Western Yunnan region, including areas to the south and west of the Bangong Lake–Lancangjiang suture; I'_1, Gangdise–Nianqingtanngula province; I'_2, Qomolangma–Western Yunnan province.

II, Boreal realm, Central Asia–Mongolia region, consisting of areas to the north of the Ebinur–Gaxunnur–Xar Moron River suture, i.e. Xinjiang, north-eastern China, and part of the People's Republic of Mongolia: II_1, Junggar–Xingan province; II_1^1, Junggar biome; II_1^2, Xingan biome.

This is not quite the same as the associations given by cluster analysis, largely because cluster analyses are made without weighting genera or generic groups of palaeobiogeographical significance.

The temporal development of Chinese provincialization is in step with that of the rest of the world. It was most noticeable in the Early Devonian, when endemism was high, but was not as noticeable in the Middle and Late Devonian when endemism decreased.

6.3.2 EARLY DEVONIAN PALAEOBIOGEOGRAPHICAL PROVINCIALIZATION OF CHINA

Early Devonian strata in China are not evenly developed, Gedinnian strata are limited and Siegenian and Emsian strata are more widespread. Another characteristic of Early Devonian stratigraphy is the development of terrestrial deposits, usually with fossils of fish, showing the particular importance of fish in Early Devonian provincialization.

Central China region

The importance of polybranchiaspids, eugaleaspids, and yunnanspids in Early Devonian provincialization is acknowledged by scholars both in China and elsewhere (Wang Junqing 1984). Polybranchiaspids include *Polybranchiaspids*, *Diandonggaspis*, *Dongfangaspis*. *Laxaspis*, and *Sanchaspis* in the Cuifengshan Formation in Yunnan, *Neoduyunaspis* in the Wudang Formation in Guizhou, and *Lunmenshanaspis*. *Szechuanaspis*, *Sanqiaspis*, and *Dongfangaspis* in the Pingyipu Formation in Longmenshan, Sichuan. Eugaleaspids consist of *Eugaleaspis*. *Yunnanogaleaspis*, and *Nanpanaspis* in the Cuifengshan Formation in Yunnan. Yunnanaspids include *Yunnanolepis*. *Phymolepis*, *Qujinolepis*, and *Zhanyilepis* in the Cuifengshan Formation in Yunnan, and *Yunnannolepis* and

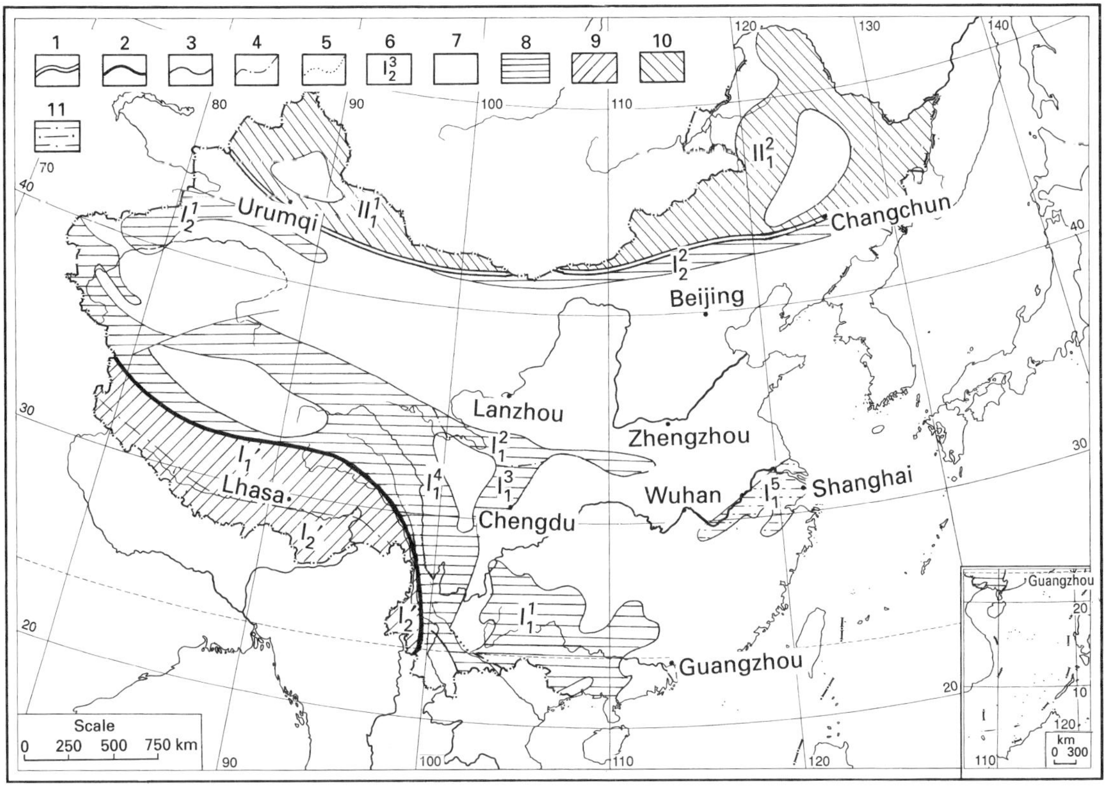

Fig. 6.5 Map of the Early Devonian palaeobiogeographical provincialization of China. 1, Boundary of realms; 2, boundary of regions; 3, boundary of provinces; 4, boundary of biomes; 5, boundary of communities; 6, roman digit = realm mark, subscript = province mark, superscript − biome mark; 7, Oldland. 8–10, Marine and paralic regions: 8, South China; 9, Gondwana; 10, Junggar–Xingan. I, Proto-Tethyan realm, Central China region; I_1, South China province; I_1^1, Hunan–Guizhou–Guangxi biome; I_1^2, Qinling biome; I_1^3, Longmenshan biome; I_1^4, Western Sichuan biome; I_1^5, Lower Yangtze biome; I_2, North China–Tarim province; I_2^1, Southern Tianshan biome; I_2^2, Badan Jaran biome; I′ Proto-Tethyan realm, Tibet–Western Yunnan region; I'_1, Gangdise–Nianqingtanngula province; I'_2, Qomolangma–Western Yunnan province; II, Boreal realm, Central Asia–Mongolia region; II_1, Junggar–Xingan province; II_1^1, Junggar biome; II_1^2, Xingan biome.

Qujinolepis in the Lianhuashan Formation in Guangxi. These polybrachiaspids and primitive antiarchs are very peculiar either in internal structure or in external features, and are totally different from the agnathans and antiarchs in other parts of the world. What is more, they only occur in southern China, and represent an independent fish region. Unfortunately, the fish fossils in northern China have not been well studied, so the northern limit of this realm is not known.

The Early Devonian flora in southern China is represented by *Zosterophyllum* and *Drepanophycus*, which are widely distributed in Yunnan, Guizhou and Guangxi.

The brachiopods in this region are characterized by the *Orientospirifer–Dicoelostrophia* fauna, which includes many endemic genera (Wang and Rong 1983). The Emsian brachiopods in southern China amount to 70 genera, of which 19 (27 per cent) are endemic. As a result of the less widespread marine strata of the early and middle Early Devonian in southern China, the biotas are not well understood, so it is possible that there may be similar proportions of endemic elements of other groups. The brachiopods in Qinling were closely associated with the Tethyan realm. The study of rhynchonellids from northern Qinling reveals that *Linguopugnoides* are seen in Czechoslovakia, the Carnic Alps, Central Asia, the Urals, North America, and Australia; *Phoenicitoechia* is also seen in Czechoslovakia; *Mongolirhynchia* in Inner Mongolia (China); *Lisopleura* in eastern North America; and *Septalaria* mostly in North America. The Early Devonian fauna in southern Tianshan is similar to that of South China. However, the significance of the presence of the Ural element, *Karpinskia* is disputed. Some researchers (Hou Hongfei and Wang Shitao 1985) consider it of significance in provincialization, but others (Wang and Rong 1983) consider it to be only of environmental significance. Badain Jaran, on the northern margin of the North China Oldland, still belongs to the Central China region. Some common elements of the Boreal realm, *Fallaxispirifer* for instance, are found in this area, in addition to endemic genera such as *Badanjarania*.

The Devonian rugose corals in this region are also very characteristic (Yu *et al.* 1982), mainly Disphyllidae, Ptenophyllidae, and the cystoid Goniophyllidae and Cystiphyllidae. In the South China region, 28 per cent of elements are endemic, while the endemic elements are greatly reduced in Longmenshan, western Qinling and southern Tianshan, and Zhusilenghaierhan on the northern margin of the Tarim–North China Oldland, although these areas have much in common with South China. It is interesting that affinities with other parts of the world show a regular change from south to north. The coral assemblage in South China is closely related to that in Australia; it contains the typical Australian elements *Australophyllum*, *Trapezophyllum*, and *Psydracophyllum*. Northwards to western Qinling the typical Appalachian elements *Neophyma*. *Aulacophyllum*, and *Bethanyphyllum* are found beside the typical Australian genera *Embolophyllum*, *Pseudoconophyllum*, and *Vepresiphyllum*. Further northwards to Zhusilenghaierhan on the northern margin of the Tarim–North China Oldland, North American elements increase, forming up to one-quarter of the total generic number of the coral assemblage, and some important western European elements are usually seen. These distributions reveal the positions of the plates in the Early Devonian, i.e. South China closer to Australia, and Tarim–North China Oldland margins closer to Europe and eastern North America (Fig. 6.1).

It should be noted that the coral assemblages in the Alengchu and Banmandaodi Formations in Lijiang, north-western Yunnan were small, solitary corals, closely related to Australian ones. It is worth studying whether they have a special biogeographical meaning.

Tabulate corals also show obvious provincialism, which is supported by the endemic genera *Natalophyllum, Cystitrypanopora*, *Neotroedssonites*, and *Gansupora*. Other genera, such as *Corolites* and *Oculipora*, were associated with Central Asia and the Urals.

Bivalves existed in littoral to shallow sea environments, showing very strong provincialism. However, they have not been studied in as much detail as brachiopods and corals. It is known so far that *Dysodonta* occurs in the Cuifengshan Formation, eastern Yunnan, and in the Lianhuashan Formation, Guangxi, and it is an endemic genus in eastern Yunnan, Guangxi, and north Vietnam. In addition, *Actinodonta* of the late Early Devonian in Yunnan and *Guangxiconcha* of the late Early Devonian in Guangxi were endemic genera in these areas. Other genera, such as *Modionorpha*, *Sphenotomorpha*, *Cypricardinia* and *Sanguinolites*, have not been found yet in the Junggar–Xingan region, although they were not endemic. These gaps are likely to be due to poor collection and study.

South China province

This was rich in endemic elements and distinctive in biotic assemblages. It can be divided into five biomes.

Hunan–Guizhou–Guangxi biome

This is the most representative biome in the Central China region. Most groups, whether fish, brachiopods, corals, or bivalves, show very strong provincialism, which is largely the consequence of palaeogeography—a large-scale epicontinental embayment into the continent. In the Qinzhou–Fangcheng area, Lower Devonian strata overlie Upper Silurian, consisting of sandstones and shales with graptolites (e.g. *Monograptus uniformis*) representing pelagic facies.

Qinling biome

The brachiopods were similar to these in the Hunan–Guizhou–Guangxi biome (Fu Lipu 1983) and the rugose corals were less endemic but mixed with many eastern North American and Australian elements.

Longmenshan biome

Fish and brachiopods were very much like those in the Hunan–Guizhou–Guangxi biome. *Orientospirifer*, *Rostrospirifer*, *Eosophragmophora*, *Howellella*, *Dicoelostrophia*, and *Uncinulus* are found in this biome and are also common in Hunan–Guizhou–Guangxi. There is only one endemic genus of rugose corals in this biome, comprising only 6 per cent of the whole coral assemblage. There are only five common coral genera in the two biomes, indicating some variation between them.

Western Sichuan biome

This also includes northern Yunnan. It lacks endemic elements of brachiopods and corals and is closely related to Australia. It represents a palaeogeographical environment which communicated well with the outer sea (RGTSNIGP 1982).

Lower Yangtze biome

This is characterized by non-marine fish, including agnathans, such as *Ningguolepis* and *Sinacanthus*. In addition, there were a small number of antiarchs.

Tarim–North China province

This contains some elements from other regions apart from the specific endemic elements of the Central China region.

Southern Tianshan biome

Brachiopods are basically similar to those of the Hunan–Guizhou–Guangxi biome except for the existence of the special genus *Karpinskia*. Australian elements of rugose corals decreased but eastern North American elements markedly increased.

Badan Jaran biome

Eastern North American elements of rugose corals increased and brachiopods were mixed with Boreal realm elements and contained a few endemic elements. Therefore, it is very complex in composition.

Tibet–Western Yunnan region

Fossils have not been well studied in this region and Early Devonian strata are absent in some areas. Some areas,such as Qomolangma and Baoshan, Yunnan, have yielded only graptolites and lack brachiopods and corals, as the Lower Devonian is composed of strata of deep-water facies.

Relatively rich fossil brachiopod and coral faunas are found in Xainza, northern Tibet. Most of the brachiopods found there belong to cosmopolitan genera, such as *Atrypina*, *Cryptatrypa*, and *Gypidula*. In addition, there is a typical Australian element, *Notoleptaena*, which indicates the close relationship of this area with Australia and the location of both in the southern belt of the Tethyan realm.

Moderate coral assemblages are found only in Xainza, northern Tibet (Xu *et al.* 1981; Yu and Liao 1982), including eight genera: *Aknisophyllum*, *Embolophyllum*, *Gurieskiella*, *Loylophyllum*, *Martinophyllum*, *Metriophyllum*, *Neaxonella*, and *Pseudamplexus*. This coral assemblage is very similar to that found in Lijiang, Yunnan, with an Otsuka coefficient of 0.34, and most of its genera are related with Australia. Hence, the conclusion is the same whether it is based on brachiopods or on corals.

Gangdise–Nianqingtanggula province

This is rich in benthos, such as corals and brachiopods.

Qomolangma–Western Yunnan province.

This is dominated by pelagic groups, for example the graptolites *Neomonographus* and *Monograptus*, and tentaculites found in Qomolangma. The graptolites in western Yunnan (Ni and Jiao 1983) are similar to those in north-eastern Thailand, southern Tibet, and Malaysia but contain an endemic species, *Monograptus fangensis*.

Central Asia–Mongolia region

Only one of the provinces, Junggar–Xingan province, exists in China. According to palaeotectonic-palaeogeographical reconstructions, this region lay in the broad marine area of the plate margin. It communicated freely with other seas, and thus contained relatively few endemic elements. Palaeoclimatically, it belonged to the northern warm temperate zone, and the organisms were very different from those in the Tethyan realm.

The Early Devonian flora in this region is not known as well as that in the South China region. It is only found in the Wutubulake Formation in western Junggar, and is known as the *Sciadophyton pristinum* association.

Among brachiopods, the *Orientospirifer–Dicoelostrophia* fauna (including *Otospirifer*, *Douvilina*, *Protathyris*, and *Sulcuspidella*), which is common in South China, has not been found in this region. On the contrary, common genera in this region, such as *Leptaenopyxis*, *Paraspirifer*, *Rhytistrophia*, *Resserella*, *Brachyprion*, *Coelospira*, *Chonostrophia*, and *Fallaxispirifer*, do not occur in the Central China region. Of these genera, *Brachyprion* and *Coelospira* are distributed elsewhere, mainly in northern Canada, the Cordillera and Nevada in the western United States, and other areas in the Boreal realm. In addition, there are many cosmopolitan genera in this region (also common in southern and northern China), such as *Acrospirifer*, *Atrypa*, *Chonetes*, *Howellella*, and *Leptaena*. Recently, Hou and Boucot (1990) suggested a Balkhash–Mongolia–Okhotsk region in the Old World realm, based on Devonian, especially Early Devonian, brachiopods. Its southern boundary in Chinese territory coincides with our Central Asia–Mongolia boundary. The criteria on which they establish the region are:

(1) its biogeographical mixture, or the presence of a number of genera restricted elsewhere to the East American realm (9–10 genera) and two genera from the Tasman region

(2) a relatively high percentage (> 30 per cent) of endemism (more than 12 genera); and

(3) the absence of purely Uralian and South China taxa.

Their analysis is consistent with our opinion.

Rugose corals are less abundant in the western part of this region (Junggar) and more abundant in its eastern part (Inner Mongolia and Xingan). The important genera are *Siphonophrentis*, *Schlotheimophyllum*, *Keriophyllum*, and *Chlamydophyllum*. This coral assemblage is typified by small size, lack of dissepimental zone, and development of a zone of marginal thickening, probably characteristics of the warm temperate climate. Larger individuals with wide dissepiment zone and complicated structure, such as *Embolophyllum* and *Phacellophyllum* are rarely found in this region. Another characteristic of rugose corals in this region is the small number of endemic elements; *Sinochlamydophyllum* is the only endemic one in 48 genera. This indicates a good connection of the region with the open sea, and such palaeogeography provides the answer to the occurrence of North American and European elements in this region.

The tabulate corals in the Junggar–Xingan province are also very characteristic, represented by abundant *Bogimbailites*, *Cleistopora*, *Pleurodictyum*, and *Squamites*. The general aspect is more like that in the Lake Balkash region of Kazakhstan but different from that in South China, and, moreover, the number of genera and species is much lower.

The bivalves in this province are far from well studied. It is known at present that in north-eastern China and Inner Mongolia *Nuculoidea*, *Nuculites*, *Leptodesma* (*Leptodesma*), *Megambonia*, *Linoptera*, *Modiomorpha*, and *Prosocoelus* exist. In northern

Xinjiang *Palaeonielo*, *Leptodesma* (*Leptodesma*), *Pterinopecten* (*Newellipecten*), *Lyriopecten*, *Goniophora*, *Eoschzodus*, and *Paracyclas* have been reported. Among these, *Megambonia*, *Prosocoelus*, and *Pterinopetcen* are not found in South China, giving some hints for provincialization.

Junggar biome

This is rare in rugose corals and lacks the *Orientospirifer–Dicoelostrophia* brachiopod fauna, but contains many elements from western European and eastern North-American provinces of the Tethyan realm.

Xingan biome

This is distinctive in rugose coral assemblages and also lacks the *Orientospirifer–Dicoelostrophia* brachiopod fauna.

6.3.3 MIDDLE DEVONIAN PALAEOBIOGEOGRAPHICAL PROVINCIALIZATION OF CHINA

Generally speaking, provincialization in the Middle Devonian is not as noticeable as that in the Early Devonian, due to increasing transgression. The endemic elements were fewer and the cosmopolitan

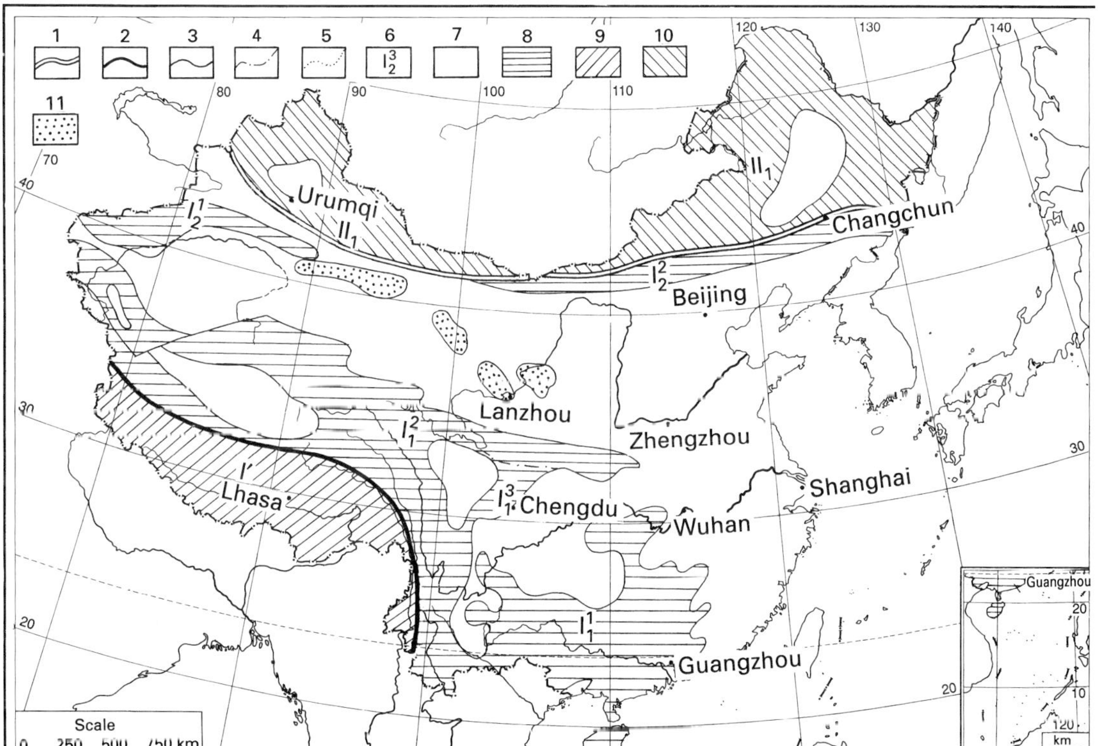

Fig. 6.6 Map of the Middle Devonian palaeobiogeographical provincialization of China. 1, Boundary of realms; 2, boundary of regions; 3, boundary of provinces; 4, boundary of biomes; 5, boundary of communities; 6, roman digit = realm mark, subscript = province mark, superscript = biome mark; 7, Oldland. 8–10, Marine and paralic regions: 8, South China; 9, Gondwana; 10, Junggar–Xingan. Terrestrial region, intermontane basins. I, Proto-Tethyan realm, Central China Region; I_1, South China province, I_1^1, Hunan–Guizhou–Guangxe biome; I_1^2, Qinling–Western Sichuan biome; I_1^3, Longmenshan biome; I_2, North China–Tarim province; I_2^1, Southern Tianshan biome; I_2^2, Badan Jaran biome; I′, Proto-Tethyan realm, Tibet–Western Yunnan region; II, Boreal realm, Central Asia–Mongolia region; II_1, Junggar–Xingan province.

ones became more abundant. However, the basic palaeobiogeographical framework is similar to that of the Early Devonian.

Central China region

Fish provincialism in the Middle Devonian was less striking than that of the Early Devonian, with the extinction of Early Devonian agnathans and primitive antiarchs. Some specialized antiarchs were flourishing and widely distributed. In the late Middle Devonian they spread to Yumen (Gansu), western Jiangxi, and Zhongning (Ningxia). The endemic elements were *Wudinolepis*, *Xichonolepis*, *Dianolepis*, and *Hohsienolepis*. The fish assemblage at this time was mostly characterized by the great increase of cosmopolitan elements, especially *Bothriolepis*, which was not only widely distributed (in Yunnan, Hunan, Guangdong, Jiangxi, and Ningxia) but also had high species diversity. This represented the acme of the genus.

In discussing Middle Devonian fish assemblages, one should not neglect to mention the discovery of *Microbrachium* in the Qujing Formation of Taoyuan-Dongshan, Zhanyi Country, Yunnan, in 1976–77 (Pan *et al.* 1982). This fossil had not been reported in the world for about 100 years (apart from its occurrence in northern Scotland and adjacent islands along this northern coast), ever since it was found in the Middle Old Red Sandstone of England in 1888. Therefore, the discovery of *Microbrachium* in Zhanyi, Yunnan, seems to verify the suggestion that South China may have had some biotic relationship with England, with both regions belonging to one biotic realm. Another discovery of interest is the Givetian thelodont (Agnatha) ichthyo-assemblage from Shidian, western Yunnan (Wang *et al.* 1986). This assemblage includes *Turinia pagoda*, which has well-developed winglets projecting from around the circumference of the crown. According to Burrett *et al.* (1990) this remarkable feature is only seen in turinids from Australia.

Middle Devonian floras are represented by *Protolepidodendron scharyanum–Barrandeina dusliana*, located in Guangxi, Yunnan, and Guangdong (also in Tibet).

As regards Middle Devonian brachiopods, the most characteristic fauna in this region is the *Athyris–Athyrisina* fauna, the important elements being *Bornhardtina*, *Desquamatia*, *Eoreticularia*, *Hypothyridina*, *Productellana*, *Productella*, *Pugnax*, *Stringocephalus*, *Undispirifer*, and *Zdimir*. It appears not only in Hunan–Guizhou–Guangxi but also in Longmenshan and western Qinling, and the most northerly part of southern Tianshan. In the world, it has a close relationship with western Europe and Australia.

Middle Devonian rugose corals were very abundant, with a general decrease of endemic elements and an increase of cosmopolitan ones. The early Middle Devonian fauna was predominated by Ptenophyllidae, Disphyllidae, Stringophyllidae, and Cystiphyllidae. Twenty per cent of genera in Guizhou–Guangxi were endemic, including *Cystitrapezophyllum*, *Dushanophrentis*, *Houershanophyllum*, *Sinospongophyllum*, and *Parasociophyllum*. In addition to the endemic genera, the Australian *Xystriphyllum*, *Trapezophyllum*, and *Sociophyllum*, and the European *Fasciphyllum*, *Dohmophyllum*, and *Digonophyllum* are also found in this region. These close relations with the Tethyan realm suggest that Central China was one part of the realm. Northwards, the endemic elements diminished. Only one of them, *Dushanophrentis*, has been found in Longmenshan, and none in Qinling and southern Tianshan. However, the cosmopolitan elements became more abundant, European elements are more common and even some North American ones are seen. This shows that these areas were located in an open sea with easy biotic exchange, rather than in a restricted epicontinental sea as in Guizhou–Guangxi.

Some endemic genera, such as *Hunanaxia*, *Parasociophyllum*, *Parasunophyllum*, *Pseudodisphyllum*, *Neosyringoxon*, *Dendrostellodes*, *Stringophylloides*, and *Neogrypophyllum*, still appeared in Guizhou–Guangxi in the late Middle Devonian. In addition to the fauna's close relationship with Australian and Europe, it shares great similarity to the coral assemblage in western Canada. Also found in this region is the typical Appalachian element *Heliophyllum*, which indicates a relationship with eastern North America. The mixed nature of elements from all realms does indicate that biotic provincialization was becoming more and more obscure. Northwards, the endemic elements become fewer and fewer, which is just the same as in the early Middle Devonian.

The provincialization of tabulate corals inherited its Early Devonian pattern. South China is characterized by *Oculipora*, *Oculiporella*, *Crassialveolitella*, and other genera. However, the components are not the same between the early and late Middle Devonian. During the earlier period Favositidae were abundant but Alveolitidae and Thamnoporidae appeared in large quantities in the later period. One of the important genera, *Neoroemeria*, occurs in western Qinling as well as in western Sichuan.

Middle Devonian bivalves were much more abundant in the later time than in the earlier; most belong to cosmopolitan genera, such as *Nuculoidea*, *Palaeoneilo*, *Linoptera*, *Leptodesma*, and *Ptychopteria*. Only *Lunania* in eastern Yunnan is endemic.

South China Province

Hunan–Guizhou–Guangxi biome

This is a typical representative of the Central China region, yielding the most numerous endemic elements. Lithofacies was clearly differentiated in this region in the Middle Devonian because of the development of two NE–SW and NW–SE microaulacogens (Xiang and Wei 1978; Ruan 1981; Bai *et al* 1982; Chen *et al.* 1983). The two types of lithofacies are called, respectively, 'Nandan type', of a stagnant, starved environment, and 'Xiangzhou type', of a normal neritic environment. The former contains ammonoids, conodonts, and tentaculites, forming a pelagic community association, while the latter is composed of brachiopods and corals, forming a benthonic community association.

Western Sichuan–Qinling biome

Biotic patterns were similar to those of South China, but there is a lack of endemic elements, with an increasing number of cosmopolitan elements.

Longmenshan biome

This has similar brachiopods to those in Hunan–Guizhou. Rugose corals are rare in endemic elements, except the middle Middle Devonian *Sinospongophyllum*, and are also poor in diversity.

North China–Tarim province

This contains many western European elements.

Southern Tianshan biome

The representative brachiopod genus *Stringocephalus* in South China was widespread in this biome, for example, *Stringocephalus burtini* in the Tuogemaiti Formation and *S. obesus* in the Tuogemaiti and Shaeraming Formations. There are many western European elements of rugose corals and tabulate corals, bearing many similarities to those in the Xingan biome.

Badan Jaran biome

Western European elements of brachiopods increased in great numbers, in addition to the common elements of South China, but the North American elements disappeared. The same applies to rugose corals. during the early Middle Devonian this biome had little connection with South China (Zhang Yan *et al.* 1983; Zhang Yan 1985).

Tibet–Western Yunnan region

Middle Devonian fossil corals are few and fragmental in Tibet, but there are many in western Yunnan. Song Xueliang (1982) has shown that the main genera are *Cylindrophyllum*, *Barrandeophyllum*, *Amplexiphyllum*, *Keriophyllum*, *Temnophyllum*, *Macgeea*, *Hexagonaria*, *Pexiphyllum*, and *Phillipsastraea*, among which *Cylindrophyllum*, *Amplexiphyllum*, *Barrandeophyllum*, and *Pexiphyllum* rarely appear in South China, thus showing some differentiation. However, other genera such as *Temnophyllum*, *Keriophyllum*, *Hexagonaria* and *Phillipsastraea* are not only widely distributed in the region but are also seen in south China. Some species, such as *Temnophyllum posiense* and *T. walteri*, are even the same as those in South China. This suggests a close relationship between western Yunnan and South China. Nevertheless, it is considered by Yu Changmin that western Yunnan belongs to the 'West China fauna', which is similar to the faunas of Australia and Europe, and has no elements in common with the 'East China fauna' in South China.

The brachiopods of western Yunnan are similar to those in Burma, perhaps because both these areas belonged to the southern belt of Proto-Tethys. However, *Eoreticularia*, *Atrypa*, *Levanea*, *Productellana* and *Pugnax* are still more common in South China.

No further provincialization will be attempted because of the lack of data.

Central Asia–Mongolia region

In the Middle Devonian of this region the flora included the arboreal lepidodendrales *Barsassia sibirica* and *Lepidodendropsis*, common in Tunguss and Kazakhstan, although *Protolepidodendron scharyanum* of South China is also found.

The main feature of brachiopods in this region is the lack of the common elements found in the Central China region: *Ambocoelia*, *Athyris*, *Athyrisina*, *Eoreticularia*, *Bornhardtina*, *Productella*, *Stringocephalus* and *Undispirifer*.

Fossil corals occur mainly in eastern and western Janggar, which lack the common Central China elements, such as *Atelophyllum*, *Cystiphyllum*, *Digonophyllum*, *Neospongophyllum*, *Phacellophyllum*, *Phillipsastraea*, *Stringophyllum*, *Spongophyllum*, *Tabulophyllum*, and *Utaratuia*, whereas western European and North American elements, such as *Barrandeophyllum* and *Enteroplasma*, are more common. This situation is the same as in eastern Inner Mongolia, which possesses western European *Glossophyllum* and North American *Cylindrophyllum* and *Heliophyllum*, although this area still has many genera in common with the Central China region (Ding *et al.* 1983; Wei and Li 1983).

The provincialization of tabulate corals is still striking. Elements common in areas to the south of the Ebinur–Gaxunnar–Xar Moron suture, such as *Oculipora*, *Oculiporella*, and *Crassialveolitella* disappear, whereas *Tyrgantites*, rare in Central China, appears in many places in this region, for example in eastern and western Junggar and Mishan, Heilongjiang.

6.3.4 LATE DEVONIAN PALAEOBIOGEOGRAPHICAL PROVINCIALIZATION OF CHINA

The biosphere changed obviously at the transition from the early to the late Late Devonian. In the earlier period brachiopods, rugose, corals and stromatoporoids were very abundant, but they declined abruptly in the latter period.

Central China region

The flora of the Late Devonian is identified as the *Lepidophloeum rhombicum* assemblage. *Lepidophloeum* is restricted to China, Japan, Kazakhstan, eastern North America, and Australia. It is assumed that the Late Devonian flora in South China had a close affinity with the synchronous flora in Kazakhstan.

The brachiopods in this region are characterized by the distribution of the *Yunnanella–Yunnanellina* assemblage, which can be traced almost everywhere, beginning from southern Tianshan in the north, southeastward to Qamdo and southward to most areas in South China. In addition, some genera, such as *Ambocoelia*, *Cyrtiopsis*, and *Hunanospirifer*, are seen only in this region. However, *Cyrtospirifer*, *Productella*, *Schizophoria*, *Schuchertella*, and *Tenticospirifer* are common genera in other regions.

Rugose corals, both solitary and colonial, were abundant in the early Late Devonian, and include *Billingsastraea*, *Disphyllum*, *Hexagonaria*, *Hunanophrentis*, *Macgeea*, *Peneckiella*, *Phillipsastraea*, *Pseudozaphrentis*, *Sinodisphyllum*, and *Temnophyllum*. Sometimes they were combined with stromatoporoids and bryozoans to form reefs. A statistical analysis undertaken by Wang Xunlian (1984) indicated that the coral assemblage of South China had the closest relationship with that of Australia, followed by the relationships with coral assemblages of the Urals and Central Asia. It is interesting that the coral assemblage was more associated with western Canada than with western Europe and the Appalachians of the same realm, which signifies closer communication and more obvious faunal mixing of the Boreal realm and

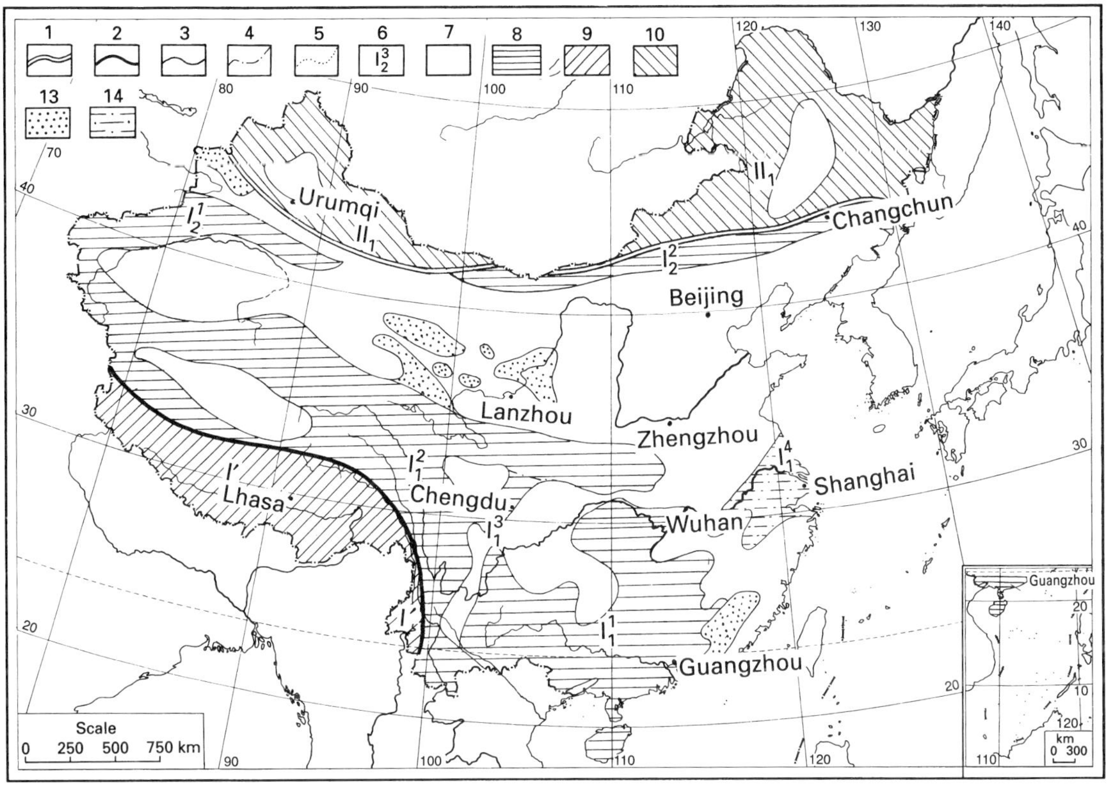

Fig. 6.7 Map of the Late Devonian palaeobiogeographical provincialization of China. 1, Boundary of realms; 2, boundary of regions; 3, boundary of provinces; 4, boundary of biomes; 5, boundary of communities; 6, roman digit = realm mark, subscript = province mark, superscript = biome mark. 7, Oldland. 8–10, Marine and paralic regions: 8, South China; 9, Gondwana; 10, Junggar–Xingan; 11–12, Terrestrial regions: 11, intermontane basins; 12, fluvio–lacustrine basins. I, Proto-Tethyan realm, Central China Region; I_1, South China province, I_1^1, Hunan–Guizhou–Guangxi biome; I_1^2, Qinling–Western Sichuan biome; I_1^3, Longmenshan biome; I_1^4, Lower Yangtze biome; I_2, North China–Tarim province; I_2^1, Southern Tianshan biome; I_2^2, Badan Jaran biome; I′, Proto-Tethyan realm, Tibet–Western Yunnan region, II, Boreal realm, Central Asia–Mongolia region; II_1, Junggar–Xingan province.

Tethyan realm than in earlier epochs. Northward to western Qinling, the situation is the same.

The late Late Devonian (Famennian) was the low ebb of the development of rugose corals. The fossils at this time are rare throughout the world. Some rugose corals have recently been found near Daoxian County, Hunan, including *Pseudozaphrentis*, *Hunanophrentis*, *Disphyllum*, *Temnophyllum*, *Hexagonaria*, *Billingsastraea*, *Phillipsastraea*, *Smithiphyllum*, and *Lophophyllum* (Wang and Zuo 1983); of these, *Pseudozaphrentis* also occurs in Yunnan, *Temnophyllum* in Guangxi, and *Pseudozaphrentis*, *Billingsastraea*, and *Phillipsastraea* in Hubei.

Tabulate corals remained roughly the same as in the Middle Devonian, but their provincialism was not as noticeable as that in the Early and Middle Devonian. *Alveolites*, *Aulopora*, *Crassialveolites*, *Coenites*, *Kueichowpora*, and *Thamnopora* appear in this region and in Junggar–Xingan, but other elements seen this region, such as *Fuchungopora*, *Syringoporella*, and *Thecostegites*, are not found in the Junggar–Xingan region, reflecting the existence of provincialization.

Ammonoids occur mainly in southern Guizhou and northern Guangxi, although they are not abundant. Becuase of remarkable ecological replacement between different regions, they serve as good indices for provincialization. The ammonoids of Guangxi and of Onor, Daxingan Mountains, northeastern China are found in the same lithofacies, but they differ greatly in the diversities of genera and species and the taxonomical content (Table 6.2). There are 18 genera and 34 species in Guizhou–Guangxi but only nine genera and 13 species in Onor. Thus the ratio of generic diversity is 2 : 1 and that of species is about 3 : 1. Evidently, the diversity of Onor ammonoids is low.

In this region, many Late Devonian fish were cosmopolitan (such as *Bothriolepis*) and only *Sinolepis* in the Wutong Formation, Longtan (in Nanjing) is endemic. Pan *et al.* (1987) reported a typical South China ichthyo-assemblage in the Late Devonian Zhongning Formation of Ningxia, North China, which contains *Bothriolepis*, *Remigolepis*, *Sinolepis*, and a polybranchiaspid agnathan. This confirms free communication between South China and the margin of North China. Outside of South and North China *Sinolepis* and *Remigolepis* are known only from eastern Australia (Burrett *et al.* 1990).

South China province

Hunan–Guizhou–Guangxi biome

This still retained strong provincialism. In the late Late Devonian rugose corals were moderately abundant (Wang and Zuo 1984). Lithofacies and biofacies differentiations of the Late Devonian became more striking (Fig. 6.8). In addition to pelagic and benthic biofacies inherited from the Middle Devonian, assemblages in sabkha environments are also displayed in Yunnan–Guizhou, containing very few fossils, with the lithology being composed chiefly of dolomitic limestone and dolomite. Moreover, marine and continental interbedded facies occur in the Hunan–Hubei area. Subdivision of the biome into community associations can be established according to these criteria.

Qinling–Western Sichuan biome

Benthic elements in common with Hunan–Guizhou–Guangxi are the brachiopods *Tenticospirifer*, *Cyrtospirifer*, and *Yunnanella* and the corals *Pseudozaphrentis* and *Sinodisphyllum*.

Longmenshan biome

There is a lack of endemic elements including corals and brachiopods. Only *Pseudoyunnanella* in the middle part of Longmenshan may be an endemic genus. *Yunnanella* and *Yunnanellina* occurred.

Lower Yangtze biome

This consisted of non-marine organisms, for example the plant *Lepidophloeum rhombicum* and the fish *Asterolepis* and *Sinolepis*, the latter being an endemic genus.

Table 6.2 Age and geographical occurrence of some ammonite genera

Genus	Age	S. Guizhou–N. Guangxi	Onor
Chiloceras	D_3^2		+
Cyrtoclymenia	D_3^2	+	
Dicoclymenia	D_3^2	+	
Imitoceras	$D_3^2 - C_1^1$	+	
Lobotornoceras	D_3^2	+	
Piriclymenia	D_3^2	+	+
Prionoceras	$D_3^2 - C_1^1$	+	+
Sporadoceras	D_3^2	+	+

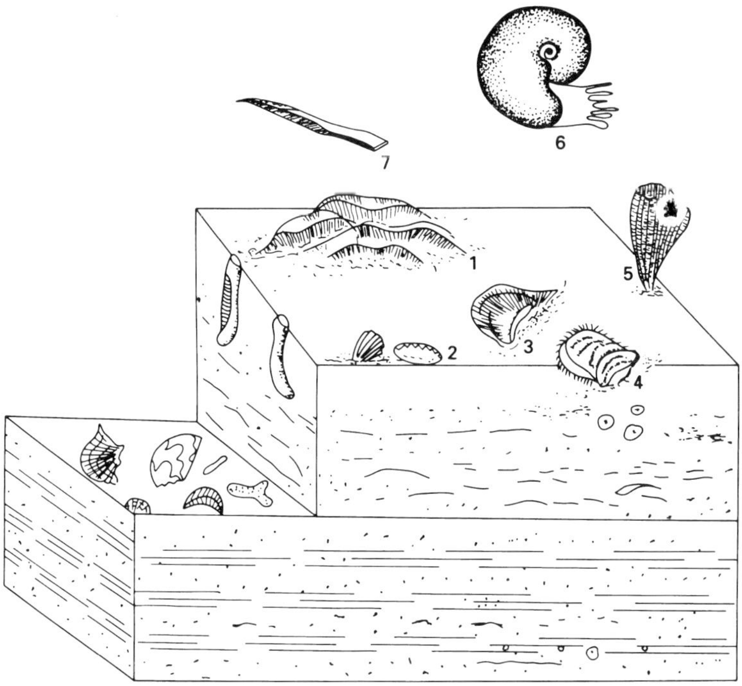

Fig. 6.8 Reconstruction of the Late Devonian communities of the shelly facies and pelagic facies in South China province. 1, *Cyrtospirifer*; 2, *Camerotoechia*; 3, *Schellwienella*; 4, *Productella*; 5, *Fenestella*; 6, *Manticoceras*; 7, conodonts.

North China–Tarim province

This contains some elements in common with Kazakhstan, the Urals and Timan, such as *Lamellispirifer*, *Cryptonella*, and *Desquamatia*, in addition to the common elements of South China (Zhang Yan 1982). The solitary coral *Nalivkinella profunda* of the Boreal realm also appears in the Xipingshan Formation in Badan Jaran of this province.

Tibet–western Yunnan region

There are very few studies of Late Devonian palaeontology in this region. The Yali Formation in Qomolangma, Tibet, yields the coral *Metriophyllum*, brachiopods *Pseudosyrinx* and *Tylothyris*, and the ammonoid *Imitoceras*. The coral *Pseudozaphrentis* occurs in the Chakuoluoma Group in Xainza (Lin Baoyu 1983). The fossil records are insufficient for adequate biogeographical analysis.

The Upper Devonian in western Yunnan is called the Dujiazai Group, and contains *Cyrtospirifer*, *Atrypa*, *Tenticospirifer*, and *Disphyllum*, similar to South China. This probably obscured the provincialization.

Due to lack of material it is better not to suggest further provincialization in this region.

Central Asia–Mongolia region

There is only one province of this region in China, the Junggar–Xingan province (Wang Chenyuan 1983).

The floral content in this region is poorer than in the Central China region, although it is still referred to as the *Lepidophloeum rhombicum* flora. The discovery of *Archaeopteris* in Aihui, Heilongjiang, indicates that this region may belong to the Tunguss flora. Until now *Archaeopteris* has not been found in Central China or Kazakhstan.

As for brachiopods, there is no *Yunnanella–Yunnanellina* assemblage comparable to that of Central China, and strophomenids are relatively predominant in this region. *Mucrospirifer*, *Eleutrokomma*, and *Cariniferella* are found in this region but not in Central China; among these, *Mucrospirifer* is widely distributed in eastern and western Junggar, Xilin Gol Meng of Inner Mongolia and Salair and Altay in Central Asia. In addition, there are many cosmopolitan elements, occurring also in South China, such as *Cyrtospirifer*, *Camerotoechia*, and *Tenticospirifer*.

Rugose corals in Junggar are not many, consisting only of *Barrandeophyllum*, *Nalivkinella*, *Syringaxon*, and *Tabulophyllum*. Late Devonian rugose corals are more developed in Xingan, some of which are not found in other areas, such as *Fistulophyllum*, *Nolaphyllum*, and *Neoheliophyllum*. This region is characterized by the absence of the tabulate corals *Fuchungopora*, *Syringoporelia*, and *Thecostegites*.

The provincialization of ammonoids has been discussed earlier (Table 6.2) and will not be repeated here.

REFERENCES

Bai Shunliang, Jin Shanyu, and Ning Zhongshan (ed.) (1982). *The Devonian biostratigraphy of Guangxi and adjacent area*. Beijing University Press, Beijing (in Chinese).

Burrett, C., Long, J., and Stait, B. (1990). Early–Middle Palaeozoic biogeography of Asian terranes derived from Gondwana. *Geological Society of London, Memoir* **12**, 163–74.

Chen Zhiming, Yang Peiji, Pan Zhengpu, Hou Kui, and Liu Guoliang (1983). Sedimentary features of the 'intraplatform trough' facies in the Middle and Late Devonian, Hunan, Guangxi Province. *Acta Sedimentologica Sinica* **1** (3), 88–98 (in Chinese)

Cocks, L. R. M. and McKerrow, W. S. (1973). Brachiopod distribution and faunal province in the Silurian and Lower Devonian. In *Organisms and continents through time* (ed. N. F. Hughes), pp. 291–304. The Palaeontology Association of London.

Ding Yunjie, Duan Chenhua, and Zhang Jinlin (1983). Devonian stratigraphy from Xilin Gol Meng, Inner Monggol. *Bulletin of the Tianjin Institute of Geology and Mineral Resources, the Chinese Academy of Geological Science* **8**, 162–89 (in Chinese).

Fu Lipu (1983). Early Devonian Rhynchonellida (Brachiopoda) from West Qinling. *Acta Palaeontologica Sinica* **22** (5), 542–50 (in Chinese).

Gierlowski, E. H. (1978). On the geographic distribution of vascular plants in the Devonian. Unpublished senior thesis, University of Chicago.

Hou Hongfei and Boucot, A. J. (1990). The Balkhash–Mongolia–Okhotsk region of the Old World Realm (Devonian). *Geological Society of London, Memoir* **12**, 297–306.

Hou Hongfei and Wang Shitao (1985). Devonian palaeogeography of China. *Acta Palaeotologica Sinica*, **24** (2), 168–97 (in Chinese).

House, M. R. (1971). Devonian faunal distribution. *In Faunal Province in space and time*, (ed. F. A. Middlemiss). Seel House Press, Liverpool.

House, M. R. (1979). Devonian in the east hemisphere. In *Treatise on invertebrate palaeontology* (ed. R. C. Moore), pp. 183–217 Geological Society of America and Kansas University Press.

Johnson, J. G. and Boucot, A. J. (1973). Devonian brachiopods. In *Atlas of paleobiogeography*, (ed. A. Hallam), pp. 89–96. Elsevier, Amsterdam.

Kimura, T. (1984). Paleozoic and Mesozoic flora and flora province of East and Southeast Asia. *Earth* **6**, (2), 18–22 (in Japanese).

Liao Weihua and Ruan Yiping (1988). Devonian of East Asia. In *Devonian of the world*, (ed. N. J. McMillan), Vol. 1, pp. 597–606. Canadian Society of Petroleum Geologists, Calgary, Canada.

Lin Baoyu (1983). Phanerozoic stratigraphy in Xainza County, Xizang (*Tibet*) *Plateau*. *Contribution to the Geology of the Qinghai–Xizang* (*Tibet*) *Plateau*, No. 8, pp. 1–13.

Liu Lu (1979). Distribution of Devonian bivalves in China. *Acta Statigraphica Sinica*, **3**, (2), 127–30 (in Chinese).

Liu Shifan (1983). Biogeography of Silurian and Devonian vertebrates in China. *Vertebrata Palasiatica* **21**, (4), 292–300 (in Chinese).

Liu Shifan (1984). The sequence of Early Devonian vertebrate assemblage in China. *Vertebrata Palasiatica* **22**, (2), 102–7 (in Chinese).

Ni Yunan and Jiao Shiding (1983). Lower Devonian graptolites from Yunnan. *Acta Palaeontologica Sinica* **22**, (3), 295–307 (in Chinese).

Norris, A. W. (1979). Devonian in the western hemisphere. In *Treatise on invertebrate palaeontology*, (ed. R. C. Moore), pp. 218–53. Geological Society of America and Kansas University Press.

Oliver, W. A. and Pedder, A. E. H. (1976). Biogeography of Late Silurian and Devonian rugose in North America. In *Historical biogeography, plate tectonics and the changing environment*, (ed. J. R. Gray and A. J. Boucot), pp. 131–45. Oregon State University Press.

Pan Jiang and Dineley, D. (1988). A review of Early Silurian and Devonian vertebrate biogeography and biostratigraphy of China. *Proceedings of the Royal Society of London*, Series *B* **235**, 26–65.

Pan Jiang, Xiong Jiayong, and Fu Lifen (1982). The first discovery of Middle Devonian *Microbrachium* from Asia. *Earth* (4), 23 (in Chinese)

Pan Jiang, *et al.* (1987) *Continental Devonian of Ningxia and its biotas*, pp. 1–237. Geological Publishing House, Beijing (in Chinese).

Pedder, A. E. H and Oliver, W. (1990). Rugose coral distribution as a test of Devonian palaeogeographic models. *Geological Society of London, Memoir* **12**, 267–76.

Pojeta, J., Jr (ed.) (1986). *Devonian rocks and Lower and Middle Devonian pelecypods of Guanxi, China and the Traverse Group of Michigan*. US Geological Survey Professional Paper, 1394, pp. 1–108.

Ruan Yiping (1981). Devonian and earliest Carboniferous from Guangxi and Guizhou. *Memoirs of Nanjing Institute of Geology and Paleontology, Academia Sinica* **15**, 1–152 (in Chinese).

RGTSNIGP (Regional Geological Team, Shichuan Province and Nanjing Institute of Geology and Palaeontology, Academia Sinica) (1982). *Stratigraphy and paleontology in west Sichuan and east Xizang, China*. People's Publishing house, Sichuan (in Chinese).

Savage, N. M., Perry, D. G., and Boucot, A. J. (1976). A quantitative analysis of Lower Devonian brachiopod distribution. In *Historical biogeography, plate tectonics and the changing environment*, (ed. J. R. Gray and A. J. Boucot), pp. 169–200. Oregon State University Press.

Scotese, C. R. and McKerrow, W. S. (1990). Revised world maps and introduction. *Geological Society of London, Memoir* **12**, 1–24.

Smith, A. G., Hurley, A. M., and Briden, J. C. (1981). *Phanerozoic paleocontinental world maps*, Cambridge University Press.

Song Xueliang (1982). The Carboniferous and Devonian tetracorals from the Baoshan–Shidian region of Western Yunnan. *Contribution to the Geology of the Qinghai–Xizang (Tibet) Plateau* **10**, 18–37 (in Chinese).

Talent, J. A., Gratsianova, R. T., Shishkina, G. R., and Yolkin, E. A. (1987). Devonian faunas in relation to crustal blocks: Kazakhstan, Mongolia, Northern China. *Courier Forschungsinstitut Senckenberg* **92**, 225–33.

Wang Baoyu (1983). New material of tabulate corals from the Kunlun–Altun Mountains with their stratigraphical significance. *Contribution to the Geology of the Qinghai–Xizang (Tibet) Plateau* **2**, 170–86 (in Chinese).

Wang Chenyuan (1983). On the nature of Devonian biogeographical province of the geocynclinal region in Northern China. *Journal of Stratigraphy* (later *Acta Stratigraphica Sinica*) **5** (4), 231–4.

Wang Chenyuan, Ran Yiping, Yu Changmin and Wang Yu (1983). Subdivision of Devonian System in China. In *Studies on stratigraphical boundaries in China*, pp. 36–42. Science Press, Beijing (in Chinese).

Wang Genxian and Zuo Zibi (1984). The Famenian rugose corals from Hunan. *Geology of Hunan* **2**, (1), 54–63 (in Chinese).

Wang Hongzhen (1981). Geotectonic units of China from the viewpoint of mobilism. *Earth Science — Journal of the Wuhan College of Geology* **14**, (1), 42–66 (in Chinese).

Wang Junqing (1984). Geological and paleogeographical distribution of Devonian fishes in China. *Vertebrata Palasiatica* **22**, (3), 219–29 (in Chinese).

Wang Nianzhong (1984). Thelodont, acanthodian and chondrichthyan fossils from the Lower Devonian of Southeast China. *Proceedings of the Linnean Society of New South Wales* **107**, 419–41.

Wang, S. T., Dong, Z. Z., and Turner, S. (1986). Discovery of Middle Devonian Turiniidae (Thelodonti: Agnatha) from western Yunnan, China. *Alcheringa* **10**, 315–25.

Wang Xunlian (1984). Middle–Upper Devonian and its rugose coral assemblage in Zhusilenghaierhan, Ejinaqi, Inner Mongolia. Unpublished Master Degree thesis. China University of Geosciences, Beijing.

Wang Yu and Rong Jiayu (1983). Yukiang (early Emsian) brachiopod zoogeography. In *Paleobiogeographic province of China*, pp. 53–63. Science press, Beijing, (in Chinese).

Wang Zhiping (1985). Early and Middle Devonian rugose coral biogeography of China. *Earth Science—Journal of the Wuhan College of Geology* **10**, 55–70 (in Chinese).

Wei Shuzhang and Li Huankang (1983). The Devonian System of the Dong Ujimqin Qi (Banner), Nei Monggol. *Regional Geology of China*, **6**, 67–80 (in Chinese).

Xiang Liwen and Wei Renyan (1978). The marine 'Nandan type' Devonian of Southern China. In *Symposium of Devonian of Southern China*, pp. 231–9 (in Chinese).

Xu Hankui, Ni Yunan, and Chen Tingen (1981). Silurian and Devonian of Xainza area, Xizang. *Journal of Stratigraphy* (later *Acta Stratigraphica Sinica*) **5** (4), 316–20 (in Chinese).

Young, G. C. (1981). Biogeography of Devonian vertebrates. *Alcheringa* **5**, 102–31.

Young, G. C. (1990). Devonian vertebrate distribution patterns and cladistic analysis of palaeogeographic hypotheses. *Geological Society of London, Memoir* **12**, 243–56.

Yu Changmin and Liao Weihua (1982). Discovery of Early Devonian tetracoral from Xainza, Northern Xizang (Tibet). *Acta Palaeotologica Sinica* **21**, (1), 96–107 (in Chinese).

Yu Changmin, Liao Weihua, and Deng Zhanqiu (1979). The sequence and distribution of Devonian coral assemblages. *Journal of Stratigraphy* (later *Acta Stratigraphica Sinica*) **3**, (2), 96–102 (in Chinese)

Yu Changmin, Liao Weihua, and Deng Zhanqiu (1982). Biostratigraphy, biogeography and paleoecology of Devonian rugose corals from the Beiliu Formation in Beiliu, Guangxi. *Bulletin of Nanjing Institute of Geology and Palaeontology, Academia Sinica* **5**, 41–82 (in Chinese)

Yuan, J. (1988). Proetiden aus dem Jungeren Oberdevon von Sud-China. *Palaeontographica* **A210**, 1–102.

Zhang Yan (1982). The Upper Devonian and brachiopod characters of Zhusilenghaierhan, Western Inner Mongolia. *Journal of Stratigraphy* **6**, (1), 79–81 (in Chinese).

Zhang Yan (1985). Early Devonian (Eifelian) brachiopods from Zhusilenghaierhan Region, Western Nei Monggol, *Acta Palaeontologica Sinica* **24**, (2), 243–58 (in Chinese).

Zhang Yan, Zheng Zhaochang, Zhou Zhiqiang, and Zhang Gang (1983). The Devonian of Zhusilenghaierhan Region, Inner Monggol. *Journal of Stratigraphy* **7**, (2), 92–103 (in Chinese).

Zhao Xiwen (1985). Principles of paleobiogeography and Upper Devonian provincialism in China. *Earth Science — Journal of the Wuhan College of Geology* **10**, 81–8 (in Chinese).

Ziegler, A. M. *et al.* (1981). Paleozoic biogeography and climatology. In *Paleobotany, paleoecology and evolution*, (ed. K. J. Niklas), vol. 2, pp. 231–66 Praeger Publishers.

7 CARBONIFEROUS

Yang Fengqing

7.1 CARBONIFEROUS PALAEO-BIOGEOGRAPHICAL PROVINCIALIZATION OF THE WORLD

Despite of the decline of trilobites, graptolites, and orthocone nautiloids, the Carboniferous Period enjoyed a great diversity of biotas, dominated by corals, brachiopods, ammonoids, and fusulinids, with many other animal groups, such as bivalves, gastropods, crinoids, non-fusulinid forams, ostracods, insects, and vertebrates. Corals, stromatoporoids, and bryozoans became the main reef-building organisms. In addition, there are also a large number of plant fossils. Hill (1973), Ross (1979), and Ross and Ross (1985) have described and summarized in detail the geographical provincialization of Early Carboniferous corals and Carboniferous forams in the world, based on the geographical distribution of the two groups. Read and Mamay (1964), Chaloner and Meyen (1973), Meyen (1982) and Rowley *et al.* (1985) have discussed the phytogeographical provincialization in detail. In China, Wang Zengji (1980*b*), Li and Yao (1983), Lin Yingtang *et al.* (1983), Yang Shipu (1983), Wang Naiwen (1984*a*, *b*), Fan (1985), and Lin Baoyu (1985) have contributed to the study of the biogeographical provincialization of Early Carboniferous brachiopods and corals, Carboniferous plants, and tabulate corals of the country. Based on the work of previous authors, this chapter attempts to approach the Carboniferous biogeographical provincialization by means of statistics and comparisons at the generic level, incorporating palaeoclimatic, palaeogeographical, and plate tectonic data.

Fig 7.1 Map of the Carboniferous palaeobiogeographical provincialization of the world. I, Palaeo-Tethyan realm; II, Boreal realm; III, Gondwana realm; A, Australia; G, Gangdise; I, India; Ic, Indochina; K, Kazakhstan; M, Mongolia; NC, North China; S, Siberia; SC, South China; T, Tarim.

The map of the Carboniferous world, on which the palaeobiogeographical provincialization is based (Fig. 7.1), is basically after Ziegler *et al.* (1977, 1979) and Scotese *et al.* (1979) with some modifications of the Chinese section:

1. In the original map the North and South China blocks lie between 30 and 58°N, but now they are located in the tropical and subtropical areas on the northern side of the palaeoequator in accordance with palaeomagnetic and biotic data. This is supported by the fact that the North China block yields a large amount of Cathaysian floras of humid and warm climate, and South China exhibits well-developed Tethyan biotas. Furthermore, according to their palaeomagnetic data, Bai Yunhong *et al.* (1985) believe that the Tarim block should be at 26°N in the Late Carboniferous. Correspondingly, the palaeomagnetic data obtained by McElhinny *et al.* (1981) suggest that the Emei area of the South China block should be at 1.7°N on the northern side of the equator in the Permian. All these data have established the relative position of the North and South China blocks.

2. In the Carboniferous, the Tarim and North China blocks were separate. The former was connected with the Kazakhstan plate in the late Early Permian and was not connected with North China until the end of the Permian.

3. The Qiangtang and Baoshan microplates, placed in high latitudes of the southern hemisphere on the original reconstruction map, have now been located between the equator and 30°S. because they yield tropical–subtropical Tethyan biotas.

The Carboniferous is traditionally tripartite in China. Its Early Carboniferous corresponds to the Dinantian in western Europe and the Mississippian in North America, while its Middle and Late Carboniferous correspond to the Silesian (western Europe) and Pennsylvanian (North America). This discordance has caused some confusion in the correlation of world and Chinese data. Recently, based on their study of the entire Carboniferous biotas and lithological differentiations in China, Yang Jingzhi *et al.* (1979, 1982) and others (Yang Shipu *et al.* 1980; Rui and Zhang 1986; Wu Wangshi *et al.* 1987) have changed the Chinese Carboniferous subdivision into an early Fengningian and a late Hutian epoch, the latter including what were formerly the Middle and Late Carboniferous of China. This is supported by the author, so this chapter will discuss Carboniferous provincialization in terms of Early and Late Carboniferous according to Yang's subdivision.

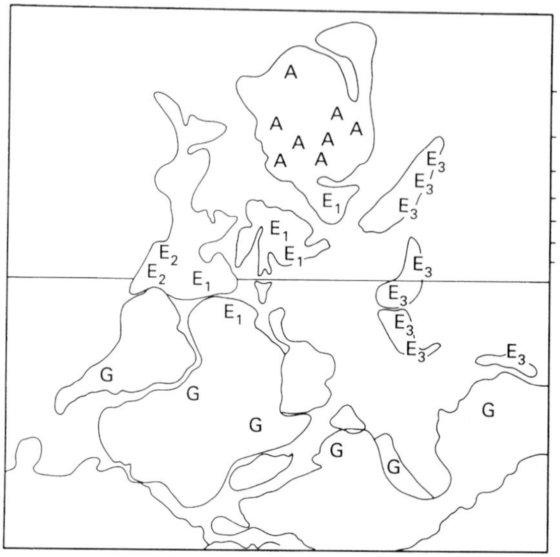

Fig. 7.2 Map showing the Late Carboniferous phytogeographical provincialization of the world (after Chaloner and Meyen 1973). A, Angara realm; G, Gondwana realm. E, Euroamerica realm including E_1, Euroamerica region (*sensu stricto*); E_2, Cordillera region; and E_3, Cathaysia region.

The Carboniferous biosphere is divided into three realms, namely Palaeo-Tethys (or Euramerica in phytogeography), Boreal (or Angara in phytogeography), and Gondwana (Figs 7.1, 7.2).

7.1.1 THE PALAEO-TETHYAN REALM

This includes eastern and western Europe, north-western Africa, central South Asia, eastern and north-western Australia, central and northern North America, and the Tibetan area to the north of the Jieshan Daban–Chabu, Gerze County–Dongqiao line and to the east of the Nujiang River. This realm is located within lower and middle latitudes, with high diversity and endemism. During the Early Carboniferous, corals were characterized by the widespread distribution of *Dibunophyllum*, *Kueichouphyllum*, *Gangamophyllum*, and *Lithostrotionella*. Forams in the early Early Carboniferous were characterized by the relative abundance of some rapidly evolving genera of the Tournayllidae, such as *Quasiendothyra* and *Palaeospiroplectammina*, while the middle and late stages were characterized by abundant *Eostaffella*, *Mediocris*, *Pseudoendothyra*, *Endostaffella*, *Palaeotextularia*, *Cribrostomum*, *Climacammina*, *Howchinia* and *Valvulinella*. Brachiopods were also flourishing, examples being *Gigantoproductus* (Gigantoproductidae), *Delepinea* (Davisiellidae), and *Striatifera*, *Megachonetes*, *Davidsonina*, and *Levitusia* (Linoproductidae). At the end of the Early Carboniferous, plants began to differentiate, showing a certain geographical provincialization. This realm was characterized by the predominance of arboreal lycophytes, arthrophytes, and pteridospermae, and other trees with no indication of growth rings, characterizing the tropical Euramerican (*sensu lato*) flora.

Late Carboniferous biotas were not so diverse as in the Early Carboniferous. Corals presented fewer types, common representatives being *Caninia*, *Bothrophyllum*, and *Arachnastraea*. However, fusulinellacean forams were abundant, widespread, and rapidly evolving, the development of Ozawainellidae, Fusulinidae, and Schwagerinidae with many endemic elements being especially noteworthy. In the brachiopods there existed many genera of Spiriferacea and Productacea. Pectinid bivalves were well developed, with *Dunbarella* and *Pseudomonotis* restricted to this realm. Terrestrial plants flourished, colonizing every ecological zone with distinctive geographical provincialization: Euramerican (*sensu stricto*), Cordilleran, and Cathaysian floras.

During the Carboniferous this realm also contained abundant coral reefs, red beds, evaporites, and desert sandstones (Seyfert and Sirkin 1979), so both organic and inorganic spheres of this realm reflect a tropical–subtropical climate. It should be stated that most areas of the three animal provinces classified by Hill (1973) belong to this realm. *Amygdalophyllum*, the coral said to be an index in eastern Australia, also occurred in China, Japan, and other places, and many coral genera and species of the North American province are scattered in Eurasia as well. Nothwithstanding the coral differentiation, it is more important that all these areas had tropical–subtropical floras, and therefore it is better to unite them in one realm. Here, the North America and eastern Australia provinces can be taken as the second-order subdivision because of their geographical isolation and relatively high endemism (Dutro *et al.* 1979).

7.1.2 THE BOREAL REALM

This is bordered by the Tethyan realm in the south and extends to relatively high latitudes of the northern hemisphere, including northern Asia, which consists of Siberia, Mongolia, and Junggar–Xingan of China. Biotic diversity was low, corals were dominated by small, solitary ones, such as *Zaphrentites*, *Hapsiphyllum*, *Cyathaxonia*, *Sugiyamaella*, *Canadiphyllum*, and *Faberophyllum*. Brachiopods were evidently different from those in the Tethyan realm, with relative abundance of Tolmatchoffiidae, Syringothyridae and Tetracameridae. It is interesting that forams were similar to those in the Tethyan realm. Plants were mainly herbaceous pteridophytes and pteridospermae, forming the Angaran flora (characterized by *Koretrophyllites speranskii*, *Paracalamites tomiensis*, *Angarapteridium grandifoliolatum*, *Angaridium finale*, *Chacassopteris consinna* and *Noeggerathiopsis* spp.), which indicates a temperate climate. There was no Carboniferous glacial activity in the Boreal realm, so a cold climatic zone cannot be recognized. Tropical–subtropical marine elements, such as fusulinids, existed in this realm, probably because there was no geographical barrier to hinder the entrance of tropical, warm surface water into high-latitude areas. Thus the continents were in a temperate and cool environment while the ocean was warmer, resulting in the difference between phyto- and zoogeographical provincializations.

7.1.3 THE GONDWANA REALM

This was bordered by the Palaeo-Tethyan realm in the north. As regards the palaeobiogeography of China, it involved the shallow-water depositional area on the northern margin of eastern Gondwana, namely most of Australia, southern Tibet, central Afghanistan, Nepal, Kashmir, and India. Located in the high latitudes in the southern hemisphere, its most essential characteristic was the development of glaciation in the Late Carboniferous. Glacial deposits and evidence of drifting ice-sheets are found in Bolivia, Argentina, South Africa, and central India. The abrupt cooling down of the climate led to low biotic diversity. In the Early Carboniferous, brachiopods of the families Syringothyridae, Tolmatchoffiidae, and Linoproductidae were quite well developed in Qomolangma (Himalaya), northern Pakistan, Nepal, and adjacent areas. The late Early Carboniferous of Xainza County, Tibet yields abundant fossil corals, such as *Humboldtia*, *Siphonophyllia*, *Parahumboldtia*, *Ekvassophylloides*, and *Endothecium*, but without genera and species in common with the Palaeo-Tethyan realm. In the Late Carboniferous, the diversity of this realm was much lower than in the Early Carboniferous. Representatives include the corals *Cyathaxonia* and *Amplexus*, the presumed cool-water brachiopods *Stepanoviella* and *Trigonotreta*, the cool-water bivalve *Eurydesma*, and the well-known cool *Glossopteris* flora, including *Glossopteris*, *Sphenophyllum*, *Gangamopteris*, *Phyllotheca*, and *Brachyphyllum*. Both psychrophilic biota and glaciation represent the cool–cold climate of these areas, unifying them into a distinctive Gondwana realm.

7.2 CARBONIFEROUS PALAEOBIOGEOGRAPHICAL PROVINCIALIZATION OF CHINA

Chinese Carboniferous marine biotas are abundant, diversified, and widespread, dominated by benthic corals, fusulinids, and brachiopods. Their distribution

Table 7.1 Correlation chart of the Carboniferous of China

			Junggar–Xingan province							S. China province		
			Junggar			Eastern Inner Mongolia	Daxingan Mountains		Central and eastern Heilongjiang	Southeast Guizhou	Northern Sichuan	Northwest Guizhou
Carboniferous	Upper (Hutian) (roughly Silesian)	Mapingian	Pinglinang			Baoligmiao	Haduohe		Yangmugang	Maping	Chuanshan	Maping
		Weiningian	Saer-bulake	Baitam-aeneishan	Shiqiantan / Calagen / Harjiawu		Handahan		Tangjiatun	Huanglong	Huanglong	Weining
	Lower (Fengning) Dinantian	Datangian (Visean)	Nalinkala / Jiangba-sitao	Nanming-shui	Dishuiquan	Halatumiao Group				Shangssu / Chiussu	Upper Tsung-changgou Group	Zhaojiashan / Caohai
		Yanguanian (Aikuanian) (Tournaisian)	Heishantou				Hongshuiquan	Anqing-taihe / Morgenhe	Beixing	Tangpakou / Kolaohe	Lower Tsung-changgou Group	Bojiwan

indicates that China spanned over all three realms. Areas south of the Jieshan Daban–Duoma–Chabu, Gerze County–Dongqiao line and west of Nujiang belonged to the Gondwana realm; the Junggar–Xingan area in the north belonged to the Boreal realm; the rest were constituents of the Palaeo-Tethyan realm. The realm boundaries in the Early and Late Carboniferous were slightly different from each other (Table 7.1).

7.2.1 EARLY CARBONIFEROUS PALAEOBIOGEOGRAPHICAL PROVINCIALIZATION OF CHINA

Early Carboniferous strata can be seen almost everywhere in China except for several oldlands (Cathaysia, Upper Yangtze, Qilian, Tarim, North China, Kam-Yunnan, and Songhuajiang) (Fig. 7.3). Early Carboniferous sediments can be divided into three types. The Tianshan–Xingan type in the far north consists of very thick sediments and is dominated by geosynclinal marine clastics intercalated with a few carbonates and with a large quantity of volcanics and volcaniclastics in the Middle and late Early Carboniferous. The well-researched South–North-western China type is the most widely distributed, dominated by carbonate deposits with intercalations of clastics as well as gypsum and halites (Wu Wangshi *et al.* 1974; He and Chao 1982; Shi 1983; Diaz 1983; Wang Zengji 1983*a*, 1985). The Southern Tibet type in south-western China consists chiefly of thick-bedded littoral clastics and neritic carbonates.

Early Carboniferous fossils are exceedingly abundant in China. Our statistics (Table 7.2) show that there are 622 genera of 14 groups, including 173 genera of corals, 183 of brachiopods, 33 of bivalves, 61 of cephalopods, 30 of gastropods, 34 of forams, 12 of stromatoporoids, 16 of bryozoans, 9 of ostracods, 12 of trilobites, a few genera of conodonts, echinoderms, and palinomorphs, plus 48 genera and 97 species of plant fossils. Among them, benthos predominates, especially corals and brachiopods (taking up about 57 per cent of the total), exhibiting striking provincialism. The coral assemblage

	Tianshan–Julin province			Tarim–Qilian province		N. Tibet–W. Yunnan province		Himalaya–Gangdise province Nianqingkanggula province			
Southern Jiangsu	Northern Tianshan	Central and western Nei Mongolia	Central Jilin	Southwest Tarim	Southern Qilian	Eastern Xizang	Western Yunnan	Himalaya		Northwest Xizang	Gangdise
Chuanshan	Urumqi	Amushan	Shizuizi	Keziliqiman Tahaqi	Taiyuan	Licha	Dingjiazhai		Jilong	Zhanjin Cameng	Langmaria
Huanglong	Qijiagou	Benbatu	Mopanshan	Azigan	Keluke	Aoqu	Shidongsi	Naxing Group	Quzhong		Angjie
	Liushugou			Kalawuyi							Sisuo
	Julideleng	?									
Hochow	Yamansu	Aomugenhedug	Lujuantun	Heshilafu	Huaitoutala	Machala	Pingzhang				Yongzhu Member
Kaolishan			Yufutun								
Jinling	Xiaorequanzi		Beitongqigou	Kelitake	Cheng-chiangou	Wuqingna					Barialangzhai Member
								Yali			Pengga Member
											Duonageli Member

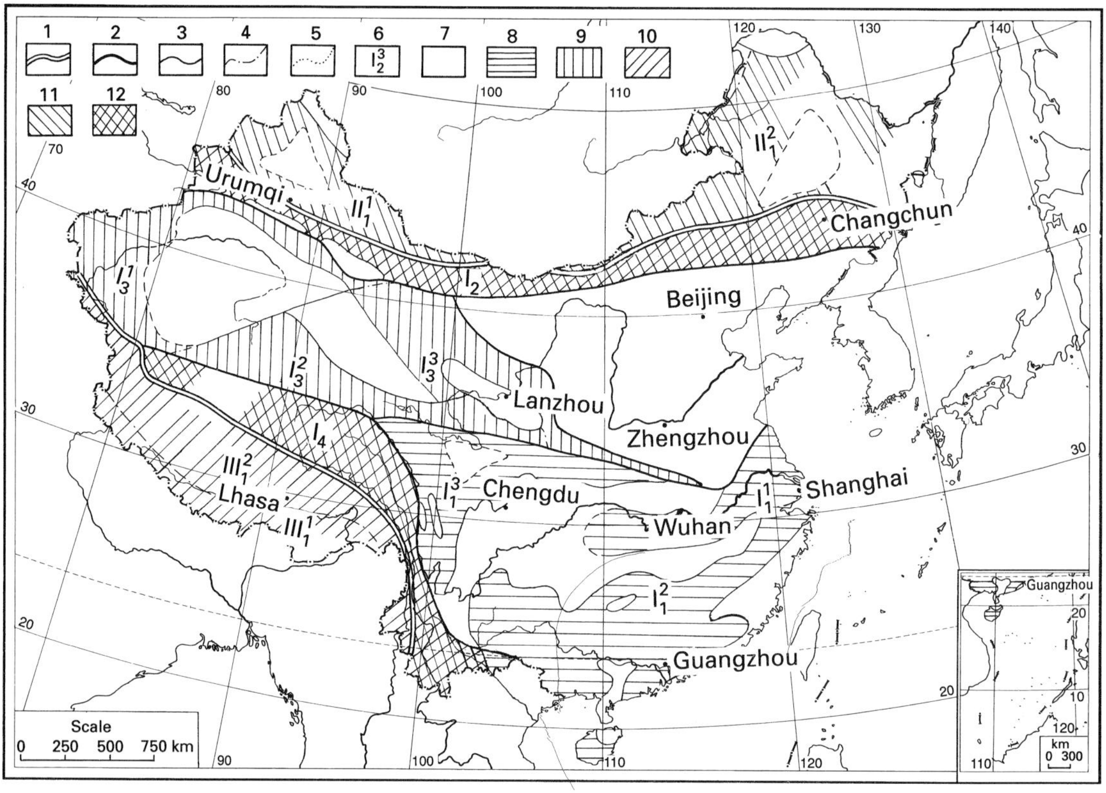

Fig 7.3 Map of the Early Carboniferous palaeobiogeographical provincialization of China. 1, Boundary of realms; 2, boundary of regions; 3, boundary of provinces; 4, boundary of biomes; 5, boundary of communities; 6, roman digit = realm mark, subscript = province mark, superscript = biome mark; 7, Oldland. 8–12, Marine and paralic regions: 8, South China; 9, North China; 10, Gondwana; 11, Junggar–Xingan; 12, transitional. I, Palaeo-Tethyan realm, Cathaysian region, I_1, South China province; I_1^1, Jiangsu–Zhejiang–Anhui biome; I_1^2, Hunan–Guizhou–Guangxi biome; I_1^3, Qinling–Longmenshan biome; I_2, Tianshan–Jilin province; I_3, Tarim–Qilian province; I_3^1, Tarim Basin Margin biome; I_3^2, Qaidam biome; I_3^3, Qilian biome; I_4, Northern Tibet–Western Yunnan province; II, Boreal realm, Central Asia–Mongolia region; II_1, Junggar–Xingan province; II_1^1, Junggar–Altay biome; II_1^2, Xingan biome; III, Gondwana realm; III_1, Himalaya–Gangdise–Nianqingtanggula province; III_1^1, Himalaya biome; III_1^2, Gangdise–Nianqingtanggula biome.

in the Yanguanian (Aikuanian) Period is characterized mainly by the development of ahermatypic corals, mostly without columns, and dominated by taxa endemic to China. Brachiopods are mainly composed of the families Martiniidae, Spiriferidae, Syringothyridae, and Tolmatchoffiidae which are small in individual size but widespread (Yang Shipu 1980). There are a few small forams, many of which belong to the Endothyriidae. Yanguanian (Aikuanian) plants are only found in a few areas of Jiangsu, Zhejiang, and Jiangxi of the Lower Yangtze Oldland where lycophytes and arthrophytes are relatively abundant. In the Datangian the transgression resulted in widening of the area of the sea and connection of basinal and shallow waters, thus enabling various kinds of organisms to flourish. Besides large, solitary ones, the hermatypic corals are also abundant, with the important development of

Table 7.2 Statistics of the Early Carboniferous fossil genera in the provinces of China

Taxa	Total of China (genera)	Junggar–Xingan province	Tianshan–Julin province	South China province	N. Tibet–W. Yunnan province	Tarim–Qilian province	Himalaya–Gangdise–Nianqingtanggula province
Brachiopods	183	60	84	114	61	99	58
Anthozoans	173	22	60	113	68	90	40
Foraminifers	34	0	1	28	12	9	0
Gastropods	30	16	9	10	3	12	1
Bivalves	33	15	7	8	6	11	6
Cephalopods	61	15	5	32	1	33	9
Trilobites	12	5	2	6	0	3	4
Bryozoans	16	4	8	10	1	4	5
Stromatoporids	12	0	0	12	0	0	0
Ostracods	9	0	0	0	3	6	0
Echinoderms	4	0	1	2	0	2	1
Conodonts	7	0	0	5	0	3	1
Fossil plants	48	13	5	29	4	29	0
Total	622	150	150	369	159	301	125

columella and complex columella in the internal structure that allows them to be distinguished from those in the Yanguanian. In the Datangian heterocorals were also developed. In addition to their endemic Chinese features, the coral assemblages also contain western European–Palaeo-Tethyan as well as North American, Siberian, and Australian elements. Brachiopods are large in size and Gigantoproductidae, Linoproductidae, and Daviesiellidae are present. Among forams, the appearance of fusulinids is an important feature. These are composed chiefly of miniature members of the Ozawainellidae, characterized by short, coiling axis and absence of diaphanotheca; non-fusulinid forams are dominated by Endothyriidae with calcareous walls. Phytogeographically the Angaran flora occupied the Junggar–Xingan area, while the Euramerican flora occurred in the rest of terrestrial China, except for southern Tibet.

To sum up, Early Carboniferous biotas in China are both highly diversified and abundant, and different groups of organisms occupied various ecological habitats (Fig. 7.4). It is significant that in the early stage (Yanguanian) corals and brachiopods show distinctive provincialism, while in the late stage (Datangian) the biotic differentiation was obscured except for southern Tibet.

Early Carboniferous biotas of China are attributed to three realms, including six provinces and 11 biomes (Fig. 7.3) on the basis of the differentiation of benthic biotas, and associated palaeoclimatic zonation and palaeotectonic control. Tables 7.2 and 7.3 show the diversity and index fossils of the biotas in each province.

The Palaeo-Tethyan realm

The Chinese section of this realm is called the Cathaysian region, which is subdivided into four provinces.

The South China province

This is situated to the south of the Qinling–Dabie line and the east of Jinshajiang, including most of the administrative provinces in south-western and

Fig. 7.4 Reconstruction of the Late Datangian brachiopod–anthozoan community of the Early Carboniferous of South China. 1, *Gigantoproductus*; 2, *Echinoconchus*; 3, *Buxtonia*; 4, *Megachonetes*; 5, *Kueichouphyllum*; 6, *Lithostrotion*; 7, *Syringopora*; 8, *Chaetetes*; 9, *Yuanophyllum*; 10, *Eostaffella*; 11, *Hexaphyllia*; 12, algae; 13, *Fenestella*; 14, crinoids.

south-eastern China. Its main characteristics are as follows.

1. In this province, brachiopods and corals are highly diversified and abundant, and the number of genera and species of cephalopods and plants exceeds that of other provinces; the total number of genera is about 2.5 times as much as that of the Junggar–Xingan province and three times as much as that of the Himalaya–Nianqingtanggula province (Table 7.2). There are far more fossil-bearing horizons and localities than in other provinces of China (Fan 1963, 1980; Wu Wangshi 1964; Deng Bao 1978; Yang Fengqing 1978; Yang Shipu 1978; Yu *et al.* 1980, 1983; Feng 1981; Ruan 1981a, 1984; Wu *et al.* 1981; Feng *et al.* 1982; Wang Keliang 1984; Wang Zhiping and Ye Gan 1984; Chen Gongxin 1984; Chen *et al.* 1986; Zuo 1986).

2. The Yanguanian faunas originated mainly in the Indo-Pacific and have a strongly regional aspect. Typically, Cathaysian *Cystophrentis* and *Pseudouralinia* are extensively distributed in association with *Zaphrentites* and *Caninia*. Among tabulate corals, *Gorskyites*, *Chia*, *Neomultithecopora*, *Fuchungoporella*, and *Fuchungopora* are characteristic genera of South China (Lin Baoyu 1963). Among brachiopods, *Yanguania*, *Eochoristites*, and *Martiniella* are found widely in Yanguanian strata in South China. All these corals and brachiopods have not yet been reported in the northern geosynclines of the Tianshan, Qilian, and Junggar–Xingan areas (usually abbreviated as Northern Geosynclines). The floras in

Table 7.3 Early Carboniferous palaeobiogeographical provincialization of China

	Biota			
	Anthozoans	Brachiopods	Fossil plants	Others
Tethyan realm				
S. China province	*Cystophrentis* *Pseudouralinia* *Kueichouphyllum* *Thysanophyllum* *Yuanophyllum*	*Eochoristites* *Martiniella* *Gigantoproductus* *Striatifera* *Lochengia*	*Lepidodendron* *Sublepidodendron* *Cardiopteridium* *Triphyllopteris*, *Eolepidodendron*	*Eumorphoceras*
Tianshan–Jilin province	*Sugiyamaella* *Cyathaxonia* *Yuanophyllum* *Kueichouphyllum* *Siphonophyllum*	*Syringothyris* *Rotaia* *Gigantoproductus* *Striatifera* *Vitiliproductus*	Fewer *Neuropteris*	*Goniatites* *Platygoniatites*
Tarim–Qilian province	*Parakassinella* *Siphonophyllia* *Humboldtia* *Kakwiphyllum* *Thysanophyllum*	*Eochoristites* *Syringothyris* *Gradispirifer* *Gigantoproductus* *Balakhonia*	*Sublepidodendron* *Lepidostrobophyllum* *Triphyllopteris* *Lepidodendropsis*	*Cravenoceras* *Neoglyphioceras* *Dombarites* *Cluthoceras*
N. Tibet–W. Yunnan province	*Cystophrentis* *Pseudouralinia* *Humboldtia* *Keyserlingophyllum* *Siphonophyllia*	*Eochoristites* *Martiniella* *Marginatia* *Syringothyris* *Tolmatchoffia* *Grandispirifer*	*Lepidodendron* *Cardiopteridium* *Asterocalamites*	*Goniatites*
Boreal realm Junggar–Xingan province	*Hapsiphyllum* *Cyathaxonia* *Zaphrentites* *Rotiphyllum* *Amplexizaphrentis* *Fasciculophyllum*	*Syringothyris* *Pseudosyrinx* *Fusella* *Brachythyris* *Rotaia*	*Caenodendron* *Chacassopteris* *Demetria* *Cardioneura*	*Gattendorfia* *Muenstroceras* *Pericyclus* *Protocanites*
Gondwana realm Himalaya–Gangdise–Niaqingtanggula province	*Ekvasophylloides* *Parahumboldtia* *Rhopalolasma* *Siphonophyllia* *Humboldtia* *Hapsiphyllum* *Zaphrentites*	*Unispirifer* *Pseudosyrinx* *Syringothyris* *Fusella* *Tylothyris*		*Gattendorfia* *Imitoceras* *Goniatites* *Cravenoceras*

this stage belong to the Euramerican phytogeographic realm. Lycophytes and arthrophytes are relatively well developed, the important elements being *Sublepidodendron mirabile*, *Eolepidodendron wusihense*, *E. juronegensi*, *E. nathorstii*, *Lepidodendron gaolishanense*, *Lepidodendropsis hirmeri*, *Archaeocalamites* sp., and *Rhodeopteridium* sp., among which *Lepidodendron gaolishanense*, *Sublepidodendron mirabile*, and *Eolepidodendron* spp. are the representatives of early Early Carboniferous flora in South China.

3. During the Datangian, accompanying the epicontinental transgression, corals increased to about

113 genera, many of which had columellae and complex columellae; *Yuanophyllum, Kueichouphyllum, Thysanophyllum, Arachnolasma, Aulina, Diphyphyllum, Neoclisiophyllum, Siphonodendron, Lithostrotion, Palaeosmilia,* and heterocorals such as *Heterophyllia* and *Hexaphyllia* occurred commonly. Among these, *Yuanophyllum* and *Kueichouphyllum* are typically Cathaysian. In addition to the endemic aspects, corals show a cerain relationship with western Europe–Palaeo–Tethys. Brachiopods present a high diversity of 114 genera. *Vitiliproductus*, *Kansuella*, *Balakhonia*, *Delepinea*, and *Megachonetes* are widely found in North and South China, while *Lochengia* and *Gondolina* only appear in South China. These brachiopods as a whole show a relationship with western Europe increasing in time. Plant fossils, still belonging to the Euramerican flora (Wu Xiuyuan and Zhao Xiugu 1981), are distributed widely in Hunan, Yunnan, central and western Guizhou, Qamdu and Riwoqe (Tibet), Shanyang and Zhenan (Shaanxi), Dabie Mountains (Henan), Jinzai (Anhui), northern Guangdong, Luocheng (northern Guangxi), Yudu (Jiangxi), Fengcheng (central Jiangxi), Kaihua (western Zhejiang), Zhangping (Fujian) and Shilu (Hainan Island). The major elements are *Cardiopteridium spitsbergense*, *Triphyllopteris collombiana*, *Archaeocalamites scrobiculatus*, *A. prolizus*, *Lepidodendron volkmannianum*, *L. shanyangensi*, *L. rhodeanum*, *Adiantites gothanii*, *Rhodeopteridium hsiangsiangense*, *Lepidostrobophyllum* spp., and *Lepidodendropsis* spp., most of which are index or common elements of the Visean in western Europe. Among them, *Cardiopteridium spitsbergense* and *Triphyllopteris collombiana* are the most widespread and serve as indices. Ammonoids are also abundant including typical Namurian *Eumorphoceras*.

To sum up, in the South China province thermophilic benthos were very abundant, hermatypic corals were extensively distributed, and the Euramerican flora (reflecting a hot and humid climate) was widespread. These facts indicate that the province belonged to the tropical–subtropical climate zone, and was located in low latitudes. South China province consists of three biomes.

The Jiangsu–Zhejiang–Anhui biome

This includes the areas of Jiangsu, northern Zhejiang, southern Anhui, and south-western Hubei in south-east China. At the beginning of the Yanguanian there was no sea in the region, and afterwards very thin marine clastics (Jinling Formation) were deposited. However, in the late Yanguanian, because of the most extensive transgression, several metres of carbonate intercalations were deposited, which bear many brachiopods with the representative genera being *Eochoristites*, *Ptychomarotoechia*, *Martiniella*, and *Productus*, and the corals *Pseudouralinia* and *Syringopora*. The late Yanguanian consists of coal-bearing clastics in Hangzhou (Zhejiang), Nanjing (Jiangsu), and Wuwei, Caoxian, Hexian, Guangde, and Xuancheng (Anhui), which are the areas in China containing the Euramerican flora in the early Early Carboniferous.

The Datangian is dominated by argillaceous limestones and dolomites with a typical Yangtze fauna. The clastics bearing coal seams in Kaihua (western Zhejiang) and Changyang (southwestern Hubei) contain a large quantity of plant fossils, including the index plant *Cardinopteridium spitsbergense* (Zhao and Wu 1982, 1986).

The Hunan–Guizhou–Guangxi biome

This is the main part of the South China province, containing the shallow-sea areas of central Hunan, northern Guangxi, southern Guizhou, northern Guangdong, and eastern Yunnan. It is characterized by the complete sequence of Early Carboniferous strata made of neritic carbonates and clastics, and full development of marine organisms, especially the dominance of benthic corals and brachiopods with strong provincialism and diversity, and the absence of plant fossils in the Yanguanian. Biota of this biome reflect the general pattern of the province at large: the abundance and diversity of biota and the absence of plant fossils in the Yanguanian. Furthermore, the transgression of the Hunan–Guizhou–Guangxi biome moved from south-west to north-east while that of Jiangsu–Zhejiang–Anhui might have moved in the opposite direction, inferred by the shift of coal-bearing horizons. The coal-bearing horizons of the former are all Datangian and get higher and younger eastwards, whereas the latter, lying in the most easterly part of China, already contained coal-bearing layers in the late Yanguanian. This contrast supports the idea of different transgression directions for the two biomes.

The Qinling–Longmenshan biome

This is located to the north of the Hunan–Guizhou–Guangxi biome, south of the Dabie Mountains and Weihe River, and east of Jinshajiang. Zhenan (Shaanxi), the Longmenshan Mountains and Songpan area (north-western Sichuan), and Diebu (Gansu) belong to this biome. The biotas are similar to those of the Hunan–Guizhou–Guangxi biome. It is designated an independent biome because of the following facts. First, it yields some peculiar types of coral, such as *Stelechophyllum*, *Beichuanophyllum*, and *Melanophyllum*, which, being widespread and confined in stratigraphic position, can be regarded as diagnostic elements of the Lower Carboniferous *Cystophrentis*, *Pseudouralinia*, and *Yuanophyllum* zones, respectively. In addition, in the Datangian it yields *Majiaobaphyllum*, *Parathysanophyllum*, *Protocarcinophyllum*, and abundant *Carcinophyllum*, together with *Kizilia* and *Kusbassophyllum*, which also occur in Novaya Zemlya, Kuzbas and Fergana, as well as *Aphrophylloides*, which also occurs in Australia. Secondly, during the acme of transgression, corals and brachiopods show a mixture with elements from the Tarim–Qilian province. For instance, the typical brachiopod of the latter province, *Syringothyris*, has been found in northern Sichuan. *Siphonophyllia* and *Humboldtia*, rich in Delingha–Amnilk (Qinghai), have been discovered in this biome as well as in the Tarim– Qilian province. *Humboldtia* occurs in especially large quantities in the Songpan area. The Qinling–Longmenshan biome also contains *Gangamophyllum*, which is abundant in the Junggar–Xingan and Tarim–Qilian provinces.

The Tianshan–Jilin province

This province includes the northern geosynclinal area to the north of Tianshan–Yinshan and to the south of Junggar–Xingan, consisting of central and northern Tianshan in the west, through Beishan (Jin Songqiao 1974; Yu *et al.* 1978) and Inner Mongolia, to central Jilin in the east, making an approximately latitudinal belt.

The fauna of this province is dominated by Boreal types in the early stage and by Tethyan types in the late stage. Yanguanian corals discovered in northern and central Tianshan and Beishan chiefly consist of small solitary corals, such as *Sugiyamaella*, *Zaphrentites*, *Cyathaxonia*, *Hapsiphyllum*, and *Siphonophyllia*, but lack typical South China elements such as *Cystophrentis* and *Pseudouralinia*. It is the same with the brachiopods, which are dominated by Boreal elements, such as *Syringothyris*, *Rotaia*, *Marginatia*, *Tolmatchoffia*, *Tylothyris*, and *Pseudosyrinx*, but without common South China elements, such as *Yanguania*, *Eochoristites*, and *Martiniella*. In the Datangian, the increasing transgression resulted in the entrance of South China sea-water into the northern geosynclines through the Qilian geosyncline, thus giving rise to the dominant southern-type corals and brachiopods in central Jilin, Aohan Qi in eastern Inner Mongolia, Beishan (Gansu), and southern and northern Tianshan (Xinjiang). The common Datangian corals here are *Arachnolasma*, *Thysanophyllum*, *Yuanophyllum*, *Kueichouphyllum*, *Aulina*, *Neoclisiophyllum*, *Diphyphyllum*, *Siphonodendron*, *Siphonophyllum*, and *Lithostrotion*, the brachiopods include *Gigantoproductus*, *Striatifera*, *Vitiliproductus*, *Antiquatonia*, *Balakhonia*, and *Eomarginifera*. Such a fauna is very similar to the Tethyan type of the South China province. Late Early Carboniferous corals are very abundant in central Jilin. The study of Lin Yingdang *et al.* (1984) shows that besides the types similar to South China there are plenty of late Visean elements of western Europe, the Russian Platform, Kazakhstan, the Urals, and Novaya Zemlya in the Arctic. In addition there are some endemic elements, such as *Palaeosmilia (Parapalaeosmilia)*, *Paradiphyphyllum*, *Fomitchevella*, *Aspidiophyllum*, and *Rhodophyllum*.

There are a few plant fossils in this province, but none of them are typical of the Angaran or Euramerican floras. Thus it may be inferred that this marine area was possibly the main barrier for the intercommunication of the two floras.

The above biogeographical characteristics indicate that this roughly latitudinal province could be a transitional belt from the South China province to the Junggar–Xingan province of the Boreal realm. It is assigned as a province at the northern margin of the Cathaysian Tethys region because of its biota being more related to the southern type.

The Tarim–Qilian province

This lies to the south of the Tianshan–Jilin province, bordering the North China Platform in the east and the northern Tibet–western Yunnan province in the west, including Tarim, the Kunlun Mountains, Qaidam, and the Qilian Mountains.

Its early Early Carboniferous biota was obviously different from the two previous provinces (Wang Zengji 1980*a*, 1985; Li Zhangrong *et al.* 1986). There are no occurrences of the important Cathaysian corals *Cystophrentis* and *Pseudouralinia* typical of the South China Province. Endemic coral assemblages discovered in the Amunike Mountains and Oulongbuluke Mountains in Qinghai, and the Nalinggele River along the south-western margin of the Qaidam Basin, are mainly *Parakassinella* and *Amunikephyllum*. In the Delingha–Huaitoutala area (Qinghai), the assemblage yields a number of *Siphonophyllia* and *Humboldtia*, plus *Siphonodendron*, *Sugiyamaella*, *Dibunophyllum*, *Amygdalophyllym*; it also contains *Kakwiphyllum* and *Zaphriphyllum*, which also occur in North America, and *Kassinella*, *Calminssiphyllum*, *Enygmophyllum*, and *Fasciculophyllum*, comparable to corals found in the Urals and other localities in Russia. *Kakwiphyllum* amounts to about 80 per cent of the total individuals of the coral assemblage (Wang Zengji 1981*a*, *b*). Among the brachiopods, *Eochoristites* of Tethyan type is often seen mixed with *Syringothryris* and *Gradispirifer* of Boreal type.

Due to transgression during the Visean, the biotas generally share a similarity with the South China and Tianshan–Jilin provinces. The coral assemblage, represented by *Yuanophyllum*, is widely distributed on the southern and northern flanks of the Qilian Mountains, east of the Kunlun Mountains and northern Tarim margin. The main elements are *Thysanophyllum*, *Arachnolasma*, *Dibunophyllum*, *Palaeosmilia*, *Aulina*, *Siphonodendron*, *Siphonophyllia*, and *Lithostrotion*, plus abundant heterocorals. *Gangamophyllum*, well developed in the Junggar–Xingan province, appears in large numbers in this province. As regards brachiopods, *Gigantoproductus*, *Antiquatonia*, *Balakhonia*, and *Megachonetes* extended as widely as those in the previous two provinces. However, *Kueichouphyllum*, *Striatifera*, and *Delepinea* are scarce here, although abundant and widespread in the South China province. Furthermore, the Zhongwei and Zhongning areas in Ningxia of this province yield a rich ammonoid assemblage consisting of 23 genera and 43 species. It shares nine genera but no species with a synchronous ammonoid assemblage in the Qiyu (Nandan) area of Guangxi, of the South China province. The Qiyu assemblage is dominated by *Trizonoceras*, *Syngastrioceras*, and *Delepinoceras*, the last two of which are absent in the Zhongwei–Zhongning assemblage, which is dominated by *Cravenoceras*, *Neoglyphioceras*, *Dombarites*, and *Cluthoceras* (Yang Fengqing *et al.* 1983; Yang Fengqing 1984). The ammonoids in the Tianshan–Jilin province are much less diversified and are dominated by *Goniatites* and *Platygoniatites*. Similar to the South China province, this province also contains the Euramerican and lacks the Angaran flora.

The above analysis suggests that this province is dominated by Tethyan but contains a few Boreal elements. It can be divided into three biomes.

The Tarim Basin margin biome

This includes southern Tianshan, Tielikeli, Altun, and the Kunlun Mountains. The Yanguanian biota is poorly exhibited; the only elements reported are the brachiopods as *Eochoristites*, *Schuchertella*, *'Spirifer'*, and *Schizophoria*. In the Datangian organisms began to flourish. Its brachiopods and corals contain no Boreal elements and are closely similar to those of the South China Province. The impoverishment of fossils suggests that Tarim Basin was at that time a relatively restricted epicontinental sea.

The Qaidam biome

This occupies a triangle, elongated in approximately latitudinal direction. It is located to the south-east of the Tarim biome and to the west of the Qinling–Longmenshan biome, containing the eastern Kunlun Mountains, Qimantag Mountains, and Qaidam Basin. It is most strikingly characterized by the mixed nature of a large number of southern-type elements and a few northern-type elements, which distinguishes it from the other biomes.

The Qilian Mountains biome

This is bordered by the Qaidam biome to the south, the Tianshan–Jilin province to the north, the North China Oldland to the east, and the eastern margin of the Tarim Basin in the west. Its Yanguanian biota is very similar to that of the Tarim Basin Margin biome, i.e. no corals are found and brachiopods are represented by *Eochoristites*, *Cleiothyridina*, and *Avonia*. The fossil plants are mainly lycopods, with important genera and species like *Sublepidodendron wusihense*, *Lepidostrobophyllum xiphidium*, and *Lepidodendropsis* sp. In the Datangian the alternation of transgressions and regressions resulted in widespread marine sequences interbedded by terrestrial deposits with intercalations of coal seams. The Datangian biota is rather complicated, containing not only western European *Aulina*, but also South China type *Yuanophyllum*, *Kueichouphyllum*, and *Gigantoproductus*, plus endemic *Kansuella*, and, moreover, the typical Visean element of the Euramerican flora *Triphyllopteris collombiana*. Another noticeable characteristic is that at the top of the Datangian an ammonoid assemblage equivalent to zone E of the Namurian of western Europe appears in Jingyuan (Gansu) and the Zhongwei–Zhongning area of Ningxia, which is similar to the Urals. The alternating marine and terrestrial deposit, the presence of a zone E ammonoid assemblage, and the Euramerican flora in this biome are the basic characteristics to distinguish it from the two previous biomes.

The Northern Tibet–Western Yunnan province

This is situated to the west of the South China province and to the south of the Tarim–Qilian province, including Karakorum, Qiangtang, Tanggula, Qamdu, the Three Rivers area in western Yunnan, and areas in south-eastern Yunnan (east of the Honghe River and south of the Nanpanjiang River). Data in the Qiangtang and Tanggula areas are not sufficient, but the biotas in Qamdu and western and south-eastern Yunnan are fairly well established (Deng 1978; Yang and Fan 1982; Jin and Fang 1983; Chen Genbao 1984; Jin *et al.* 1985; Jin 1985). The Datangian fauna still involves typical South China elements, generally similar to the previous provinces. However, the Yanguanian biota is characterized by the mixed nature of 'northern' and 'southern type' elements. Baoshan and Shidian (western Yunnan), as well as Jianshui and Yanshan (south-eastern Yunnan), yield not only *Cystophrentis* and *Pseudouralinia* typical of the South China Province, but also *Humboldtia*, *Keyserlingophyllum*, and *Siphonophyllia* of 'northern type' (Song Xueliang 1982). The 'northern type' genera are dominant in Baoshan and Shidian, which reflects the close relationship of corals in this province with western Europe and Russia. Similarly, brachiopods display both the common South China elements *Eochoristites* and *Martiniella* and common elements in the Junggar–Xingan province, such as *Marginatia*, *Syringothyris*, *Tolmatchoffia*, and *Grandispirifer* (Yang Shipu 1983). It is to be noted that here the term 'northern type', traditionally used in Chinese literature, actually means Gondwana type, because the above-mentioned corals and brachiopods also occur in the Himalaya–Gangdise–Nianqingtanggula province and other Gondwana localities. It signifies brachiopod and solitary coral bipolarity. The mixed nature of a few 'northern type' and more 'southern type' fossils suggests that this province was a transitional area on the southern margin of the Cathaysian Tethys region.

The Boreal realm

This only has one province in China, belonging to the Central Asia–Mongolia region.

The Junggar–Xingan province

This is located to the north the Tianshan–Jilin province, including eastern and western Junggar (or Xinjiang) as well as the Daxingan Mountains) and Ergun areas (Heilongjiang) (Zheng *et al.* 1982). The main characteristics are as follows.

1. Yanguanian corals are dominated by small, solitary forms, with many genera, such as *Hapsiphyllum*, *Kinkaidia*, *Zaphrentites*, *Zaphrentoides*, *Zaphriphyllum*, *Rotiphyllum*, *Cyathaxonia*, *Fasciculophyllum*, *Amplexi*, *Zaphrentis*, and *Neozaphrentis*, which are similar to those of North America, with the exception of a few western European–Tethyan elements. Similarly, brachiopods exhibit a close relationship to North America and Siberia, all belonging to the Boreal realm. The Yanguanian spiriferids are dominated by *Syringothyris* and *Pseudosyrinx* (Syringothyridae), accompanied by *Fusella* and *Brachthyris*, with degenerate dental plates, and the rhynchonellids are typified by *Rotaia* (Yang Shipu 1983; Yang Shipu and Fan 1983). The ammonoids *Gattendorfia*, *Muenstroceras*, *Protocanites* and *Pericyclus*, and gastropods, are also relatively well developed in this stage.

2. The Datangian faunas, generally similar to the Yanguanian, consist mostly of northern elements. Moreover, its floras show Angaran aspects. In the Nalinkala Formation of Junggar, the Dishuiquan Formation of Kelamaili Mountains and the Shuangjingzi area, as well as in Yumin Co. (western Junggar), Angaran elements have been found, such as *Caenodendron primaevum*, *C. karagandense*, *Chacassopteris consinna, Demetria asiatica*, *Cardioneura microphylla*, and *Koretrophyllites* sp., in which *Chacassopteris consinna* is the index fossil of the Early Carboniferous Angaran flora.

3. In terms of plate tectonics, this province was then a part of the Siberia plate and had not yet been connected with the North China platform.

To sum up, the Early Carboniferous transgression in this province came from the Boreal region and then moved southwards, so that the biotas were closely related to Siberia–North America. The existence of small solitary corals and the Angaran flora reflects a temperate to cool climate in the province.

The province consists of two biomes.

The Junggar–Altay biome

This includes the areas in Xinjiang to the north of Hami and Turpan and to the east of the Jinghe River, and probably also the areas along the Ergun River in northern Heilongjiang. The Yanguanian shares the typical biotic feature of the province, containing brachiopods as the most abundant group (23 identified and 35 unidentified species), followed by corals (12 identified and nine unidentified species), a few ammonoids (two identified and four unidentified species), and plants (one identified and seven unidentified species). In the Datangian, uplift of the continent and northward retreat of the sea resulted in the formation of littoral and non-marine clastics bearing plant fossils. Within the biota, brachiopods still dominate (34 identified and 38 unidentified species), corals are more abundant than in the Yanguanian (eight identified and 20 unidentified species), and ammonoids are represented by six genera. At the same time, the Angaran flora appeared in Junggar (17 identified and 8 unidentified species), represented by Angaropteris. In the eastern part of this biome, the upper part of the Lower Carboniferous contains coal measures of industrial importance.

The Xingan biome

This is bordered by the Tianshan–Jilin province in the south and by Mongolia and Russia in the north, including areas surrounding the Daxingan Mountains of Heilongjiang and Ulanqab Meng of Inner Mongolia. Its faunal characteristics are, in general, similar to the previous biome, and it is differentiated from that by lower diversity and lack of Angaran floral elements.

The Gondwana realm

This has only one province in China, which belongs to the eastern Gondwana region.

The Himalaya–Gangdise–Nianqingtanggula province

This lies to the south of the Northern Tibet–Western Yunnan province and is bordered by the Nujiang fault in the east, containing Lhasa, Xainza, and Gerze in the Gangdise–Nianqingtanggula block, as well as Xigaze, Zhongba, and Qomolangma in the Himalaya. The biotic characteristics are as follows.

1. Early Carboniferous corals are dominated by small solitary ahermatypic corals with low diversity. According to Fan Yingnian (1985), the Yanguanian coral assemblage in Xainza is composed mainly of *Ekvasophylloides* and *Parahumboldtia*, reflecting regional endemism, while containing some North American and Russian elements. This is quite different from the Yanguanian *Cystophrentis* and *Pseudouralinia* coral assemblages in the South China province. Similarly, in the Datangian there were no massive hermatypic corals or even *Lithostrotion*, which was then spread widely over the world. There were only small, solitary corals without dissepiments, dominated by *Rhopalolasma*, and the *Thysanophyllum–Kueichouphyllum–Yuanophyllum* assemblage, which was widespread in other provinces, did not exist here. Although the solitary coral assemblage has some relationship with western Europe and North America, common elements in the Early Carboniferous of western Europe and North America have not been found in this province. This special coral assemblage provides evidence that this province had little exchange with other areas of China, either in the Yanguanian or Datangian.

2. Brachiopods have been discovered mainly in the Himalayas, Xainza Co., and Duoma, Rutog Co. The principal Yanguanian elements are *Unispirifer*, *'Spirifer'*, *Fusella*, *Tylothyris*, *Pseudosyrinx*, and *Syringothyris*, with *Eochoristites* and *Martiniella*, which are common in the South China province, being absent. Datangian gigantoproductids, linoproductids, and daviesiellids, which flourished in the South China province, are replaced here by syringothryids, lichareviids, tolmatchoffiids, and linoproductids.

3. This province lacks those organisms generally assumed to be thermophilic fusulinids.

4. Some corals and brachiopods of this province, such as *Siphonophyllia*, *Humboldtia*, *Hapsiphyllum*, *Zaphrentites*, *Pseudosyrinx*, and *Syringothyris*, are also seen in large quantities in the Junggar–Xingan province although the two provinces are far from each other. This is an example of bipolarity.

5. On the whole, the biota of this province lived in a relatively low-temperature environment owing to the cooling of Gondwana and successive glaciations. The animal assemblage of this province belongs to the Gondwana type and its palaeotectonic position is located on the northern margin of the Indian plate.

It consists of two biomes.

The Himalaya biome

Brachiopods are abundant in this biome. They are divided into the *Fusella–Ovatia* assemblage (Tournaisian), *Marginirugus–Syringothyris* assemblage, and *Gigantoproductus–Balakhonia* assemblage (Visean) (Zhang and Jin 1976). Other important elements are *Tylothyris*, *Pseudosyrinx*, *Composita*, and *Schuchertella*. The assemblages are very similar to the synchronous brachiopod faunas in Kashmir, Nepal, and Australia, and also close to those in the Borohoro Mountains in Xinjiang. The ammonoids *Gattendorfia* and *Imitoceras* occur in the upper member of the Yali Formation in Qomolangma. No corals and fusulinids of warm-water type are found. As a whole, the diversity and abundance are very low in this biome.

The Gangdise–Nianqingtanggula biome

This is located to the north of the Yarlung Zangbo River, to the south of the Northern Tibet–Western Yunnan province, and to the west of the Nujiang fracture zone. At present, detailed study of this biome has been made only in Xainza County. The biota is characterized by strong endemism. During the Yanguanian no index elements of South China have been found, although it has a number of corals, such as *Humboldtia*, *Siphonophyllia*, *Sychnoelasma*, *Ekvasophylloides*, and *Parahumboldtia*. It is the same case with brachiopods, which are dominated by *Unispirifer* and *'Spirifer'* only. In the Datangian, corals consist of a group of solitary corals without dissepiments, the main members being *Rhopalolasma*, *Homalophyllites*, *Hapsiphyllum*, *Zaphrentites*, *Stereolasma*, *Plerophyllum*, and *Mirusophyllum*. This specific coral assemblage is found only in this biome in Chinese territory. It is suggested by Yang and Fan (1982) that the brachiopods on the Datangian stage can be divided into an earlier *Krotovia–Fusella* assemblage and a later *Chonetipustula–Balakhonia* assemblage, with the main fossils being *Torquasitia*, *Overtonia*, *'Productus'*, *Ectochoristites*, *Rhipidomella*, *Eomarginifera*, *Tolmatchoffia*, *Fluctuaria*, *Chonetipustula*, and *Balakhonia*. Some of these brachiopods are found in the Himalaya biome, some are common in western Europe and Fergana, and a few are widespread in other parts of Asia. No fusulinids are seen in the biome, but ammonoids are relatively abundant. It has been pointed out by Sheng (1983) that the ammonoid fauna in Yongzhu (Xainza Co.) contains some typical Visean elements of western Europe and Russia, such as *Goniatites* and *Cravenoceras*. The biotic diversity and abundance of this biome are greater than that of the Himalaya biome, and abundant psychrophilous corals began to occur during the Visean.

Cluster analysis of Early Carboniferous plants

Figure 7.5 shows a computerized binary cluster analysis of Early Carboniferous plants based on published data up to 1984, using Otsuka coefficients (Otc). The samples are stratigraphic regions or subregions according to the *National Chart of Stratigraphic Regionalization*; the variables are 46 genera (genera which are cosmopolitan, generalistic, or only existing in one province having been deleted). The diagram shows that when Otc = 0.2, four groups can be distinguished clearly. The different stratigraphic regions in South China comprise one group and belong to one province. The remaining three are Altay–Junggar, Eastern Tibet–Western Yunnan, and Qilian, which correspond to the Junggar–Xingan, Northern Tibet–Western Yunnan and Tarim–Qilian provinces, respectively. The only remarkable difference between this diagram and the text is that the Tianshan–Jilin province in the text is separated into a Central Tianshan–Northern Tianshan sample and a Songliao–Jilin sample in the diagram. This can be explained as follows: because this province is located between Junggar–Xingan province and Tarim–Qilian province, the biota signifies a transitional nature and the fossil plants contain neither the typical Cathaysian flora elements found in Qilian nor the standard Angaran flora elements of Junggar, but only the intermediate elements of the two floras. Therefore, it can be assumed that this province has a relatively close relationship with Junggar–Xingan and Tarim–Qilian provinces, as shown in the diagram. Generally, the cluster analysis result is coincident with the previous Early Carboniferous provincialization, with the Himalaya–Gangdise–Nianqingtanggula province as the exception since it contains no plant fossils in the Early Carboniferous.

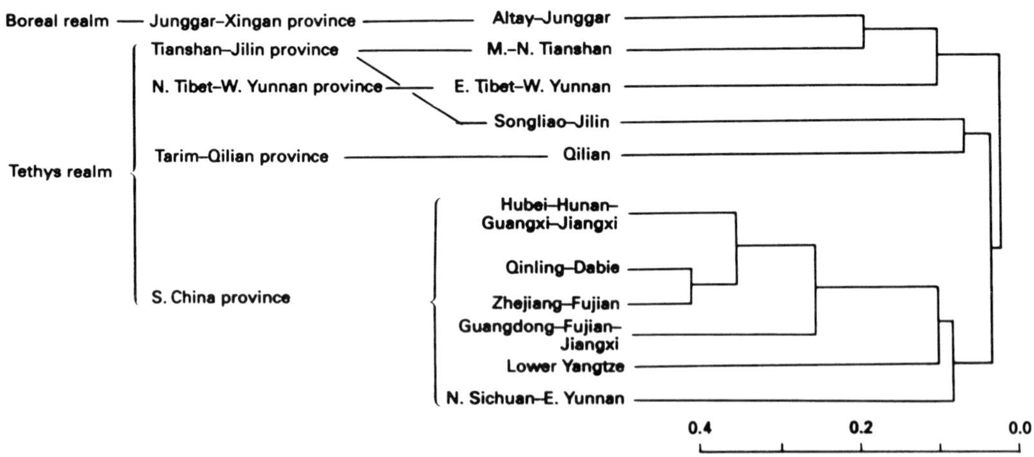

Fig. 7.5 Cluster analysis of the Early Carboniferous plants of China.

7.2.2 LATE CARBONIFEROUS PALAEOBIOGEOGRAPHICAL PROVINCIALIZATION OF CHINA

The Late Carboniferous transgression was more extensive than that in the Early Carboniferous, and Upper Carboniferous strata occur almost everywhere in China. Lithologically, they can be divided into three types: the Northern Geosynclines, or the Tianshan–Xingan area, composed of carbonates with intercalations of volcanics and volcano-clastics; the North China Platform, consisting of alternating marine–terrestrial clastics with intercalations of limestone lenses; and the Yangtze region, made up of neritic carbonates.

Late Carboniferous biotas in China are also exceedingly rich, including 13 groups and 697 genera, among which there are 143 genera of brachiopods, 141 corals, 64 cephalopods, 61 fusulinids, 49 gastropods, 57 bivalves, 30 non-fusulinid forams, 13 bryozoans, 27 ostracods, 28 conodonts, 8 trilobites, 9 echinoderms, and 67 genera of plant fossils. The benthos still dominated, especially corals and brachiopods, which comprise about 41 per cent of the total genera. Late Carboniferous corals are not as abundant as those in the Early Carboniferous and the types are also morphologically simpler, and are characterized by the existence of 'trizoners' with columns and inclined tabulae. North China is dominated by the development of small, solitary corals while South China yields more hermatypic corals, most of which are more-or-less cosmopolitan. Similarly, brachiopods are not as abundant as those in the Early Carboniferous (the generic diversity being reduced to about 22 per cent). They chiefly belong to the Spiriferidae, Productidae, Dictyoclostidae, and Chonetidae, most noticeably *Choristites* and *Dictyoclostus*, Fusulinids began to appear at the end of the Early Carboniferous and flourished in the Late Carboniferous. Early Late Carboniferous fusulinids are small individuals with the development of chomata and three- or four-layered spirothecae, and are dominated by the Fusulinidae. In the later epoch their size increased, their spirothecae becoming keriothec, and members of the schwagerinid family became the main representatives. In this period, goniatites, the most important zone fossil of the Upper Carboniferous in western Europe, are found both in North and South China; Reticuloceratidae and Gastrioceratidae are dominant. Plant fossils were more developed than in the Early Carboniferous, both in their diversity and in their ecological differentiation, and phytogeographical provincialization came into existence. The Boreal temperate Angaran flora appeared in Junggar–Xingan or the Northern Geosynclines, while in Qomolangma, Himalaya, the *Glossopteris* flora first appeared at the end of the Late Carboniferous; this belongs to the cool–cold Gondwanan phytogeographical realm. Apart from these two, in most areas of China a slightly deviated Euramerican flora existed in early Late Carboniferous, then at a late stage the early Cathaysian flora appeared (Fig. 7.6) Both of these represent the tropical–subtropical

Fig. 7.6 Reconstruction of the Late Carboniferous *Lepidodendron–Calamites* community of North China. 1, *Stigmaria*; 2, *Lepidodendron*; 3, *Calamites*; 4, fern tree; 5, *Emplectopteris*.

phytogeographical realm. It should be noted that at the end of the Carboniferous, bivalves of cold-water type, *Eurydesma*, appeared in Rutog, south Qiangtang.

Except for phytogeography, Late Carboniferous provincialization in China has not been studied in detail. In general, the Chinese territory spans all three realms of the world, and falls into five provinces and seven biomes (Fig. 7.7). Diversity and index organisms of each province are shown in Tables 7.4 and 7.5.

The Tethyan realm

This is subdivided into three provinces in China, all belonging to the Cathaysian region of this realm.

The South China–Tarim province

This embraces a vast area of China, south to Dabie–Qinling–Tianshan, north to Jieshan Daban–Duoma–Cabu of Gêrzê County–Dongqiao, and east to Nujiang, including the administrative provinces along the Yangtze and Pearl Rivers, most of western

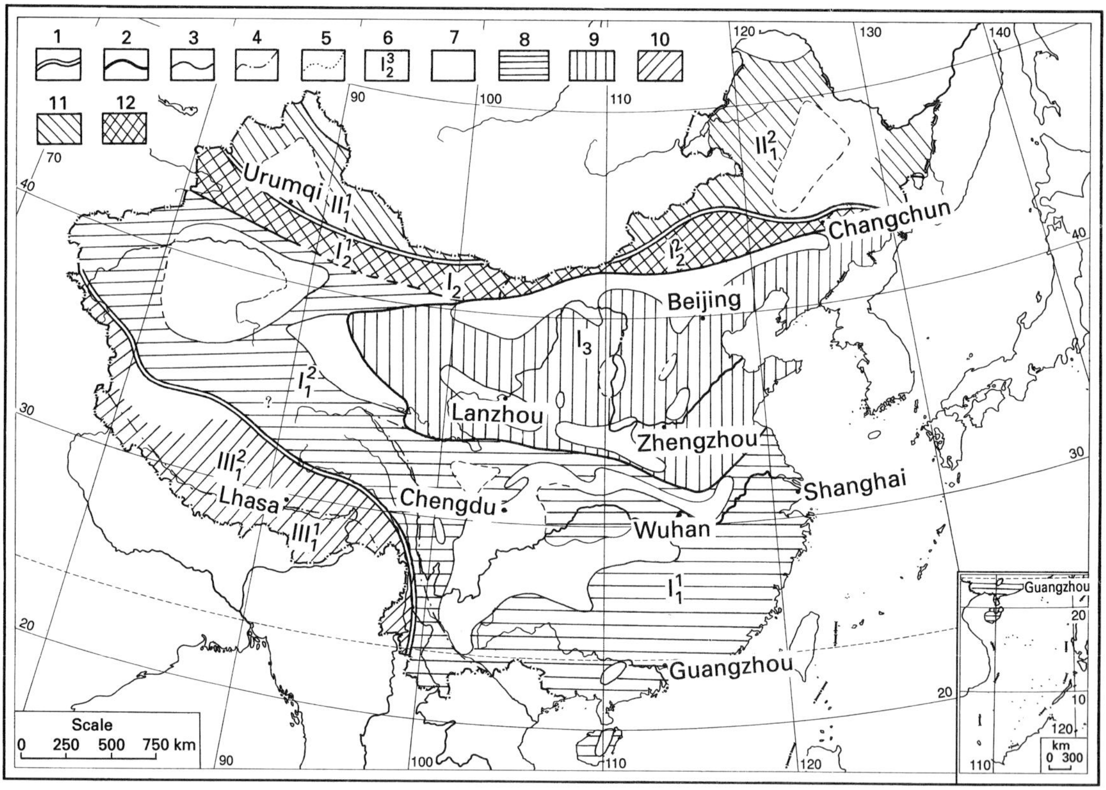

Fig. 7.7 Map of the Late Carboniferous palaeobiogeographical provincialization of China. 1, Boundary of realms; 2, boundary of regions; 3, boundary of provinces; 4, boundary of biomes; 5, boundary of communities; 6, roman digit = realm mark, subscript = province mark, superscript = biome mark; 7, Oldland; 8–12, Marine and paralic regions: 8, South China; 9, North China; 10, Gondwana; 11, Junggar–Xingan; 12, transitional. I, Palaeo-Tethyan realm, Cathaysian region; I_1, South China–Tarim province; I_1^1, South China biome; I_1^2, Tarim–Western Yunnan biome; I_2, Tianshan–Jilin province; I_2^1, Tianshan–Beishan biome; I_2^2, Inner Mongolia–Central Jilin biome; I_3, North China–Qilian province; II, Boreal realm, Central Asia–Mongolia region; II_1, Junggar–Xingan province; II_1^1, Junggar biome; II_1^2, Xingan biome; III, Gondwana realm; III_1, Himalaya–Gangdise–Nianqingtanggula province; III_1^1, Himalaya biome; III_1^2, Gangdise–Nianqingtanggula biome.

Table 7.4 Statistics of the Late Carboniferous fossil genera in the provinces of China

Taxa	Total genera of China	Junggar–Xingan province	Tianshan–Jilin province	N. China–Qilian province	S. China–Tarim province	Himalaya–Gangdise–Nianqingtanggul province
Brachiopods	143	38	64	65	95	45
Anthozoans	141	25	72	26	108	19
Cephalopods	64	11	22	30	35	0
Fusulinids	61	6	32	27	53	0
Gastropods	49	9	15	29	18	1
Bivalves	57	37	12	28	29	16
Non-fusulinid forams	30	0	1	2	29	4
Bryozoans	13	7	3	5	5	3
Ostracods	27	8	1	0	22	0
Conodonts	28	0	0	11	23	0
Trilobites	8	1	2	3	3	0
Echinoderms	9	0	0	4	3	2
Fossil plants	67	7	15	56	20	0
Total	697	149	239	286	443	90

Yunnan, and the marginal areas of the Tarim block. The biota shows following characteristics.

1. The marine fauna is very abundant (Wu 1962; Liao 1975; Lin *et al.* 1977; Liu *et al.* 1978; Zhang Renjie 1980; Hao and Liu 1982; Jiang 1982; Chen and Wang 1983; Tong 1983, 1986; Yang Deli 1984; Wu and Zhao 1984; Wang Zhiping 1987), but the plants are poor, only a few of them being found in Mojiang (south-east Yunnan) and Yushu (Qinghai) (Gu 1977).

2. Non-fusulinid forams tended to decline in the Late Carboniferous while the fusulinids flourished. Fifty-three genera of fusulinids have been found in this province, making up 87 per cent of the total Late Carboniferous genera discovered in China. The most common and important genera are *Profusulina*, *Fusulina*, *Fusulinella*, *Pseudostaffella*, *Triticites*, *Sphaeroschwagerina*, *Robustoschwagerina*, and *Pseudoschwagerina*. This fusulinid fauna is very similar to that of the equivalent horizons in the Russian Platform and south Urals, and is close to some North American fusulinid faunas.

3. Rugose corals in this province are very endemic. Yu Xueguang (1980) reported that among the 40 genera and 109 species, subspecies, and unidentified species, only six species are seen in Europe and two species are found in North America; the rest are endemic species. The genera peculiar to the province are *Chuanshanophyllum*, *Qinglongshanophyllum*, *Pseudocarniaphyllum*, *Jintingophyllum*, *Nephelophyllum*, *Parawentzellophyllum*, *Eokepingophyllum*, *Huanglongophyllum*, *Eokoninckocarinia*, *Eowentzelellites*, *Heritschioides*, *Carinokoninckophyllum*, *Skolekophyllum*, and *Peiraphyllum*. Most of them are Eurasian type and no small, solitary corals of the Junggar–Xingan type are found. In South China tabulate corals include the cosmopolitan genera *Syringopora* and *Multithecopora* as well as the stratigraphical index *Donetzites*, which occurs in the Donetz Basin, the Urals, Central Asia, and this province of China.

4. During this epoch, brachiopods are also abundant and comprise 66 per cent of the total number of genera in the whole of China. Most of these ranged

Table 7.5 Late Carboniferous palaeobiogeographical provincialization of China

	Biota			
	Anthozoans	Brachiopods	Fossil plants	Others
Tethyan realm				
S. China–Tarim province	*Chuanshanophyllum* *Parawentzellophyllum* *Nephelophyllum* *Kepingophyllum* *Pseudocarniaphyllum*	*Buxtonia* *Choristites* *Semicostella* *Dictyoclostus* *Meekella*	Fewer *Pecopteris*	*Profusulina* *Fusulina* *Triticites* *Pseudoschwagerina* *Sphaeroschwagerina* *Retites* *Branneroceras* *Ramosites* *Myalinella* *Selenimyalina*
Tianshan–Jilin province	*Amplexocarinia* *Calophyllum* *Cyathocarinia* *Hillia* *Antheria* *Lomaphyllum* *Akagophyllum* *Caninia*	*Choristites* *Productus* *Dictyoclostus* *Brachythyris*	*Alethopteris* *Lepidodendron* *Linopteris* *Neuropteris* *Pecopteris*	*Fusulina* *Triticites* *Pseudoschwagerina* *Branneroceras* *Gastrioceras* *Aclistoceras* *Stenopronorites*
N. China–Qilian province	*Cyathaxonia* *Cyathocarinia* *Lophacarinophyllum* *Neokoninckophyllum* *Gshelia* *Caninophyllum*	*Choristites* *Dictyoclostus* *Avonia* *Marginifera* *Martinia* *Punctospirifer*	*Neuropteris* *Linopteris* *Lepidodendron* *Callipteridium* *Alethopteris* *Taeniopteris* *Emplectopteridium*	*Fusulina* *Beedeina* *Pseudoschwagerina* *Reticuloceras* *Gastrioceras*
Boreal realm S. Junggar–Xingan province	*Bothrophyllum* *Timania* *Gshelia* *Caninia* *Amplexus* *Cyathocarinia* *Lophocarinophyllum* *Metriophyllum*	*Choristites* *Dictyoclostus* *Sowerbina* *Kozlowskia* *Neospirifer* *Marginifera*	*Angaropteridium* *Angaridium* *Noeggerathiopsis* *Cardioneura* *Neuropteris* *Crassinervia*	*Xinjiangopecten* *Crenipecten* *Girtypecten* *Pinna* *Pterinopectinella* *Somoholites*, *Neopronorites* *Fusulina* *Triticites*
Gondwana realm Himalaya–Gangdise – Jianqingtanggula–province	*Cyathaxonia* *Bradyphyllum* *Stereolasma* *Amplexus*	*Stepanoviella* *Trigonotreta* *Phricodothyris* *Rhipidomella* *Anidanthus* *Marginifera* *Cancrinella*	*Glossopteris* *Sphenophyllum*	*Eurydesma* *Mourlonia*

from the Early Carboniferous and spread extensively with obscured provincialization. They can be divided into three assemblages, namely the *Buxtonia grandis* assemblage, the *Choristites mansuyi–Semicostella panxianensis* assemblage, and the *Dictyoclostus uralicus–Meekella* assemblage.

5. Among molluscs, pteroid bivalves are not as abundant as those in Junggar and Xingan, the important elements being *Selenimyalina*, *Septimyalina*, *Nuculites*, *Myalinella*, and *Modiolus*. Early Late Carboniferous deposits rich in ammonoids have been discovered in Shuicheng (Guizhou) and Nandan (Guangxi). The important elements are *Retites*, *Branneroceras*, *Ramosites*, and *Pseudoparalegoceras*; in addition there are the Russian and North American genera, *Baschkirites*, *Bisatoceras*. *Eogonioloboceras*, *Megapronorites*, *Owenoceras*, and *Winslowoceras*, as well as *Fiqingoceras*, *Leiogastrioceras*, *Machangoceras*, and *Megatrochoceras*, which are found only in China. The abundance of ammonoids is one of the main features to distinguish this province from the others. This province, and the following North China–Qilian province, being located in the Eurasian plate, both have Euramerican and Cathaysian floras and reflect tropical–subtropical climates. The same climatic implication is supported by the abundance of thermophilous benthos.

This province contains two biomes.

The South China biome

It includes all administrative provinces of south-eastern and south-western China and is the main part of the South China–Tarim province. Characterized by carbonate sedimentation, its marine organisms were very abundant, especially fusulinids, brachiopods, and corals, which were endemic and greatly diversified. There are no plant fossils. This biome shows the basic pattern of the province.

The Tarim–Western Yunnan biome

This is located in the west of the province, including the margins of the Tarim block, Kunlun Mountains, northern Tibet, western Sichuan, and western Yunnan. The fauna of this biome is generally similar to that of the previous biome (Wu and Zhang 1979; Li Zhonghai *et al.* 1982; Ruan 1981*b*; Lu 1982; Wang Keliang 1982; Deng and Zhang 1984; Zhu 1985). Fusulinids, brachiopods, and corals are well developed on the western margin of Tarim and in Qamdu–western Yunnan (Zhao Zhixin *et al.* 1984). It is distinguished from the South China Biome by the following characteristics.

1. Besides the common genera and species of rugose corals, some regional elements occur on the western Tarim margin, such as *Eokoninckocarinia*, *Eowentzelellites*, *Heritschioides*, *Carinokoninckophyllum*, and *Skolekophyllum*, most of which are of Eurasian type.
2. A few plant fossils are found in the alternating marine and terrestrial clastics along the margins of Tarim and Mojiang (Yunnan).
3. Besides carbonates, there are also clastics, volcanics, and volcano-clastics.
4. The abundance and diversity of biota in this biome are far less than those in the South China biome. The characteristics of the Carboniferous biota are not well known in Qiangtang and Tanggula (northern Tibet) because of insufficient study there.

The Tianshan–Jilin province

Its northern boundary begins around the area from Mudanjiang to Baicheng, Heilongjiang, westwards through Erenhot along the Chinese border, continuing westwards to Gaxunnur (Inner Mongolia) and Ebinur (Xinjiang), while its southern boundary runs roughly along the Xar Moron River westward and eastward. It includes central and north Tianshan, Beishan, Dong Ujimqin Qi (Inner Mongolia), and central Jilin. This province is dominated by marine deposits with abundant faunas and poor plants (Sheng 1981; Wang Baoyu 1983; Wang Keliang 1985). Fusulinids are extraordinarily abundant, forming fusulinid limestones, and the genera and species are cosmopolitan. Rugose corals are diversified and widely distributed (Guo 1983, 1987*b*), with both small, solitary corals, such as *Amplexocarinia*, *Calophyllum*, and *Cyathocarinia*, and colonial or solitary 'trizoners', such as *Hillia*, *Antheria*, *Lomaphyllum*, and *Akagophyllum*. Here the southern-type corals are mixed with the northern type. Brachiopods are dominated by cosmopolitan genera, the main elements being *Choristites*, *'Productus'*, *Dictyoclostus*, and *Brachythyris*. A small number of plant fossils bear no indicative members of either the Angaran or Cathaysian flora. Considering that the former are found in Junggar–Xingan on its northern side and the latter are seen in the North China–Qilian province on its southern side, the present situation may reflect the fact that the northern geosynclines were the barrier to free communication of the two floras. Because the fauna of this province has only a few Boreal elements, while most of the faunal elements are common in the south, this narrow, latitudinally elongated, biotically transitional belt is taken as a province of the Cathaysian region of the Palaeo-tethyan realm. It is subdivided into two biomes, based on the difference of organisms in the eastern and western sections.

The Tianshan–Beishan biome

This includes the central and northern Tianshan Mountains and Beishan Mountains. The biota is basically similar to that of the province, but the rugose corals display fewer types, and fusulinaceans did not appear in the late Late Carboniferous; thus there is a complete absence of the elements of *Pseudoschwagerina* zone. The tabulate coral, *Cystodendropora*, is an abundant endemic genus most remarkable in northern Tianshan. There are also relatively abundant ammonoids (Wang Mingqian 1981), most genera of which are cosmopolitan and are closely related to those in the Urals and North America, but some of them are similar to those of South China.

The Inner Mongolia–Central Jilin biome

This is different from the Tianshan–Beishan biome in that, fusulinaceans existed up to the late Late Carboniferous, with a complete fusulinid zonation, and corals are exceeding rich. Among tabulate corals, *Sinopora* is especially abundant; *Pseudofavosites* and *Multithecopora* also occur. Rugose corals are abundant and diversified. According to Guo Shengzhe (1987*a*), in the early Late Carboniferous there were both northern-type, small solitary corals, such as *Amplexocarinia*, *Cyathocarinia*, *Metriophyllum*, *Calophyllum*, *Rotiphyllum*, and *Stereostylus*, and polygonal, massive 'trizoners', like *Lithostrotionella* and *Hillia*, which spread widely in Russia, western Europe, North America, and South China. In the late Late Carboniferous, small solitary corals continued to appear in small quantities, but massive and solitary 'trizoners' occurred in large numbers, the main elements being *Hillia*, *Antheria*, *Parastephyllum*, *Lomaphyllum*, *Carinthiaphyllum*, *Akagophyllum*, *Paracarruthersella*, *Empodesma*, and *Amplexizaphrentis*, forming three associations, namely the *Hillia–Antheria* assemblage, *Carinthiaphyllum–Akagophyllum* assemblage, and *Empodesma–Tachylasma* assemblage. Most of them are also found in South China, but the widespread *Kepingophyllum* of South China is not seen in this biome. To sum up, this biome contains not only the solitary corals typical of the Northern Geosynclines but also the massive corals of southern type. At generic level, they are dominated by Chinese elements but mixed with European and North American forms. At species level, they exhibit high endemism, about 40 per cent. The distribution of fusulinids and corals shows that more extensive transgression in the Late Carboniferous resulted in better biotic exchange between South China and the Northern Geosynclines.

The North China–Qilian province

This is bordered by the Tianshan–Jilin province in the north and by the South China–Tarim province in the south, including the southern and northern flanks of the Qilian Mountains, and administrative provinces of Huanghe (Yellow River) drainage. In the Early Carboniferous most parts of this province belonged to the North China Oldland and no Early Carboniferous strata have been reported. The extensive Late Carboniferous transgression covered the platform and resulted in the deposition of widespread alternating marine and terrestrial clastic deposits, with intercalated limestone and coal, making it one of the major coal-mining areas in China. Biotas are characterized by the best development of plants (Li Xingxue *et al.* 1974; Wang Rennong 1979; Wu and Deng 1983; Yang *et al.*

1983). In the early Late Carboniferous, Euramerican elements dominated, mainly *Neuropteris* and *Linopteris* (pteridosperm), secondarily *Lepidodendron* and *Sphenophyllum*, with a few elements of the Cathaysian flora. The important elements are *Neuropteris gigantea*, *N. kaipingiana*, *N. otozamioides*, *Linopteris newropteroides*, *L. brongniartii*, *Eleutherophyllum mirabile*, *Pecopteris aspera*, *Mesocalamites cistiformis*, *Asterophyllites tenuifolius*, *Sphenopteris obtusiloba*, and *Lepidodendron aolungpylukense*. Generally this feature is very similar to the Westphalian flora in the western hemisphere. In the late Late Carboniferous the flora became remarkably differentiated. The early Cathaysian flora appeared in this province and *Linopteris* almost disappeared. Lycophyta began to flourish and *Neuropteris* continued to flourish. *Neuropteris ovata–Lepidodendron* is the representative assemblage of this time (Li and Yao 1983; Wu Xiuyuan and Zhao Xiugu 1983). It is characterized, firstly, by the abundance of oriental lepidodendrales, such as *Lepidodendron oculusfelis*, *L. postumii*, *L. szeiana*, *Cathaysiodendron incertum*, *C. nanpiaoense*, and *Bothrodendron kuianum*, and secondly by the appearance of some distinctive pteridosperms, such as *Alethopteris hallei*, *A. huiana*, *Callipteridium koraiensi*, *C. tachingshanense*, and *Tingia*. Thirdly, there are rich Taeniopteriaceae, such as *Taeniopteris multinervis*, *T. mucronata*, and *T. lingchuanensis*, and pteridophytes and pteridosperms such as *Emplectopteridium alatum*, *Emplectopteris triangularis*, and *Mariopteris hallei*.

Fusulinids, corals, and brachiopods are generally similar to those of the South China–Tarim province (Zhang Yuxiu 1981; Wang Deyou 1982; Wan *et al.* 1983; Yang Shipu 1983; Xia and Zhang 1984, 1985; Guo 1987*a*). The three widespread Late Carboniferous fusulinacean zonations are wholly represented in this province, and in addition there are some genera that are rare in other provinces, such as *Occidentoschwagerina*, *Orientoschwagerina*, and *Oketaella*. Brachiopods are relatively abundant in this province, common elements being *Choristites*, *Dictyoclostus*, *Avonia*, *Marginifera*, *Martinia*, *Punctospirifer*, and *Derbyia*. Among the rugose corals, solitary forms predominated, with small-sized *Cyathaxonia*, *Cyathocarinia*, *Lophocarinophyllum*, and *Sestrophyllum*, as well as large forms including *Neokoninckophyllum*, *Gshelia*, *Caninophyllum*, and *Bothrophyllum*. Goniatitines are relatively abundant in the Zhongwei–Zhongning area of Ningxia (Yang Fengqing *et al.* 1983) and Jingyuan (Gansu), representing the E, R, and G goniatitc zones of the Namurian, with the important genera *Eumorphoceras*, *Reticuloceras*, *Gastrioceras*, and *Cancelloceras*, a fauna which is closely related to that of the Urals.

The Boreal realm

The Junggar–Xingan province is the only Boreal province in China, belonging to the Central Asia–Mongolia Region of this realm. It is located in the north of China, to the north of the Tianshan–Jilin province. A typical Angaran flora developed in this province, distributed in the Junggar Basin and a few areas of Inner Mongolia and Heilongjiang. The representative elements are *Angaropteridium cardiopteroides*, *Angaridium potaninii*, and some species of *Noeggerathiopsis*. The appearance of the Angaran flora is the main feature to distinguish this province from the others. Rugose corals are dominated by Caninidae and Bothrophyllidae, especially by plentiful *Bothrophyllum*, *Timania*, *Gshelia*, *Caninia*, and *Caninophyllym*, low in diversity but abundant in individuals. Small, solitary corals, such as *Amplexocarinia*, *Amplexus*, *Lophyllidium*, *Cyathocarinia*, *Lophocarinophyllum*, and *Metriophyllum*, are relatively common while colonial rugose corals are very rare. The tabulate coral *Cystodendropora* is enriched at specific horizons in the Junggar (Lin and Wang 1985). The general features of the coral assemblage differ from the South China province and the Tianshan–Jilin province, while being very similar to those of the Moscow Basin, Urals, and Novaya Zemlya, and the northern islands of Canada. Brachiopods are characterized by cosmopolitan genera, the most common being *Choristites* and *Dictyoclostus*. There are a few endemic elements, such as *Paramuirwoodia*, *Sowerbina*, and *Kozlowskia*. Bivalves are very abundant. As many as 37 genera have been reported, reaching about 65 per cent of the total reported in the Late Carboniferous of China. The pectinaceans are most abundant, among which *Xinjiangopecten*, *Crenipecten*, *Girtypecten*, *Fasiculiconcha*, *Junggarochondria*, *Pterinopectinella*, *Spathella*, *Goniophorina*, and *Pinna*, appear only in this province of China. Fusulinids are relatively rich and are cosmopolitan. However, they are not reported in the upper Upper Carboniferous in this province. The occurrence of the Angaran flora and small solitary corals show that this province belonged to the temperate belt. Two biomes are distinguished.

The Junggar biome

This is located in north-western China, containing the areas of Junggar and Altay. The most striking character is its Angaran flora. The Tamayineishan and Kalagang Formations in Junggar contain many Angaran elements. In addition to the typical elements, there are *Cardioneura sibirica*, *Mesocalamites cistiformis*, and *Neuropteris* cf. *pulchra*. In addition, the marine strata in southern Junggar display relatively abundant brachiopods, corals, bivalves, goniatitines, and gastropods, each of which present many endemic genera and species. Another noticeable character is the disappearance of fusulinids in the late Late Carboniferous. No elements of the *Pseudoschwagerina* zone appear, but the synchronous goniatitine *Somoholites* and *Neopronorites* occur in beds intercalated in the paralic strata. Generally speaking, during the early Late Carboniferous most areas of this biome were non-marine, and marine deposits accumulated only in its southern part, where the benthos shows clear endemism. Regression occurred in the late Late Carboniferous and the biotas are dominated by gastropods, bivalves, plants, and goniatitines, which clearly distinguished them from the fusulinid-bearing strata of open-sea carbonate facies deposited in other areas.

The Xingan biome

This is situated in north-eastern China, including northern Inner Mongolia and the Daxingan Mountains in Heilongjiang. It is the only biome in China containing Angaran elements of late Late Carboniferous age. According to Huang (1982a,b, 1983), the Yangmugang Formation of Shangzhi Co., Heilongjiang yields *Neuropteris orientalis*, *N. mrassiensis*, *N. dichotoma*, *N.* cf. *paimbaensis*, *Noeggerathiopsis latifolia*, *N. subangusta*, *Angaridium potaninii*, *A. submongolicum*, *Zemiopteris* cf. *glossopterioides*, *Crassinervia mishanensis*, and *Angaropteridium* sp., all being important elements of the Late Carboniferous Angaran flora, among which *Neuropteris* and *Noeggerathiopsis* were the most abundant. The floras in the Haduohe Formation and Baolige Formation in the areas around Zalantan, Arxan, and Dong Ujimqin Qi are similar to those in the Yangmugang Formation. Besides the plant fossils, this biome yields a few brachiopods, such as *Athyris*, *Neospirifer*, *Marginifera*, and *Hustedia*, as well as the small, solitary coral *Amplexus* and fusulinids (Han 1982; Guo 1987*b*).

The Gondwana realm

This exhibits only one province in China, belonging to the eastern Gondwana region.

The Himalaya–Gangdise–Nianqingtanggula province

This is located to the west of Nujiang fracture zone, bordered by the South China–Tarim province in the north and by Bhutan and Nepal in the south. It contains neritic sandstones, shales, slates, and diamictites. We agree with Wang Yizhao (1983) that the diamictites are glacio-marine deposits. In this area they display an accumulation of unsorted boulders with glacial scratches. These glacio-marine deposits contain the cold-water biotas similar to those from the Indian subcontinent of Gondwana, reflecting a marginal sea position and much higher palaeolatitude in the south during the Gondwana ice age in the Late Carboniferous (Chen Bingwei 1982; Hao and Wang 1983; Liang and Wang 1983; Liang *et al.* 1983; Yang Zengrong 1983; Yang Zunyi *et al.* 1984; Cao Renguan 1986). It is subdivided into two biomes according to the biotic differentiation (Liang 1976; Xu 1976; Zhang and Jin 1976; Jin and Sun 1981; Wang Yujing 1981; Yang *et al.* 1981; Lu 1982; Wu *et al.* 1982; Zhang 1981, 1982; Hu 1983; Wang Zengji 1983*a*, *b*; Ruan 1984).

The Himalaya biome

This was located to the south of the Yarlung Zangbo River. The Late Carboniferous strata contain abundant brachiopods and a few bivalves, gastropods, and plant fossils, with an absence of warm-water fusulinids and corals. Brachiopods belong to the cold-water fauna represented by *Stepanoviella gracilis*, *Attenuatella convexa*, and *Trigonotreta* cf. *narsahensis*. In addition, the Gondwanan *Glossopteris* flora, characterized by *Glossopteris communis*, *G. angustifolia*, *G. indica*, and *Sphenophyllum speciosum*, occurs in Tingri and Dinggye on the northern flank of Qomolangma.

The Gangdise–Nianqingtanggula biome

This is composed of the areas between the Yarlung Zangbo River and Tarim–Western Yunnan biome. Brachiopods and bivalves are relatively common but corals are very few. There are no plants nor fusulinids (which are very important elsewhere in the Late Carboniferous of China). The study of the biota in Xainza is more detailed (Yang and Fan 1983). Brachiopods of the early Late Carboniferous include *Phricodothyris asiatica*, *Eomarginifera xizangensis*, and *Pugnax xizangensis*. There are only small, solitary corals, such as *Cyathaxonia*, *Bradyphyllum*, and *Stereolasma*, but *Kionophyllum* and *Lithostrotionella* are absent, although very common in this period in China. In the late Late Carboniferous, the climate further deteriorated and every group of organisms became reduced in numbers. The brachiopods associated with glacial boulders include *Marginifera* cf. *orientalis*, *Stepanoviella xizangensis*, *Cancrinella canciniformis*, *Schizophoria resupinata*, *Dielasma itatubense*, *Leiorhynchoides xizangensis*, *Trigonatreta* sp., and *Cyrtella nagmargensis*, while the only coral is *Amplexus romonovskyi*. Apart from this, the research team of the China University of Geosciences in Tibet have found, in the Zhanjin Formation of Late Carboniferous age around Rutog–Duoma north of the Bangong Lake, abundant cold-water bivalves, including *Eurydesma*, 22 species (Liu and Cui 1983) are associated with the brachiopods *Anidanthus fusiformis*, *Martinia camongensis*, and *Ambikella* sp., and the gastropod *Mourlonia* cf. *freneyensis*. This list shows that this biome contains some Palaeo-Tethyan elements in addition to the elements of Gondwana type.

Cluster analysis of Late Carboniferous plants

Figure 7.8 is the result of cluster analysis of Late Carboniferous plants, which was worked out using

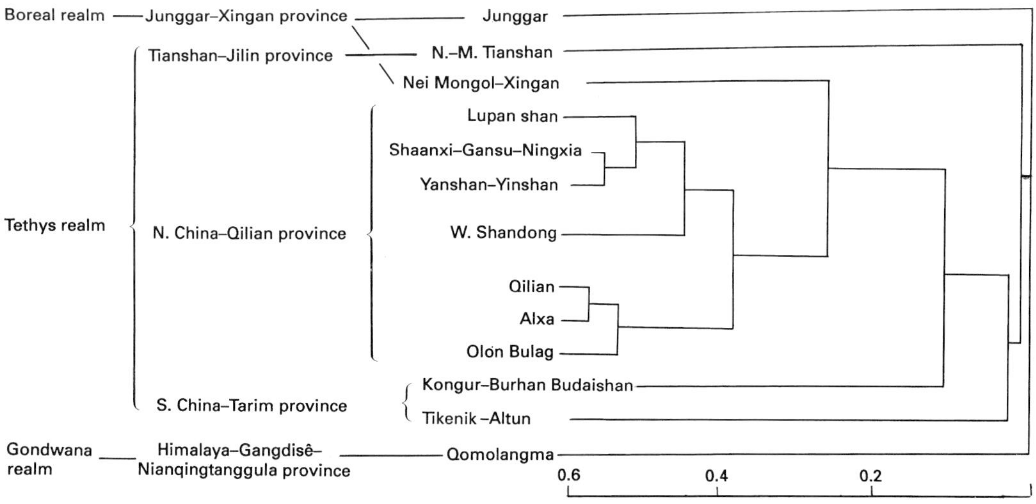

Fig. 7.8 Cluster analysis of the Late Carboniferous plants of China.

the same method and information from publications as that of the Early Carboniferous (Fig. 7.5). Sixty-four genera were dealt with in the computation, with deletion of the genera common to every biome and the genera only occurring in one biome. It can be seen from the diagram that when the Otc value is 0.3, seven clusters are defined. Northern Tianshan–Central Tianshan and Qomolangma, constitute independent clusters, each representing one province. Areas within North China–Qilian come together as one to form an independent province. Kongurshan–Burhan Budai and Tikenik–Altun form one province. As discussed earlier, both Inner Mongolia–Xingan and Junggar contain typical elements of the Angaran flora and should be put into one province. However, Junggar contains fewer plant genera and has few common elements with Inner Mongolia–Xingan, while Inner Mongolia– Xingan includes intermediate elements of the Angaran and Cathaysian floras, many of which are same as those of the North China–Qilian province. Thus, by cluster analysis, the latter is grouped into North China–Qilian. In general, the cluster analysis of Late Carboniferous plants presents results consistent with the provincialization given earlier.

7.3 SOME CHARACTERISTICS OF CARBONIFEROUS PALAEOBIOGEOGRAPHICAL PROVINCIALIZATION IN CHINA

7.3.1 THE RELATIONSHIP BETWEEN PROVINCIALIZATION AND CLIMATIC ZONES

In China the Carboniferous biogeographical units generally extended in an east–west direction, approximately coincident with the climatic zones. In the north the Junggar–Xingan province has small, non-hermatypic corals and the Angaran flora, belonging to the temperate zone. The Himalaya–Gangdise–Nianqingtanggula province in Tibet displays a Gondwana-type cold-water fauna and *Glossopteris* flora, dominated by shrubs and herbs, together with tillites, which reflect a cool–cold climate. In these two non-tropical provinces in northern and south-western China, brachiopods and corals show certain bipolar distributions at the generic level, and the biotas were relatively monotonous with low diversity; most of them were eurythermal. The Cathaysian region lying between the non-tropical provinces belonged to the Tethyan realm and had biotas with very high diversity (Tables 7.2, 7.4). In this region hermatypic corals and thermophilous fusulinids and brachiopods were abundant. Huge arboraceous lycophytes, arthrophytes, and *Cordaites* grew on the continent in association with pteridophytes, and pteridosperms, the trees showing no growth rings. These features indicate clearly the characteristics of a tropical and subtropical climate. Transitional provinces existed between the south-western and northern non-tropical provinces and the Cathaysian Tethys, for example the Tianshan–Jilin province and Northern Tibet–Western Yunnan province. They were characterized by a mixed fauna, with elements from the adjacent provinces, and their biotic diversities were intermediate, probably reflecting a warm, temperate climate. Thus, it can be seen that clear climatic zonation appeared in the Carboniferous in China, and the provincialization reflects the palaeoclimatic and palaeoecological conditions. Hence, the palaeobiogeographical pattern began to reflect climatic zonation clearly in the Carboniferous.

7.3.2 RELATIONSHIP BETWEEN PROVINCIALIZATION AND PLATE TECTONICS

The arrangement of Carboniferous biogeographical units in China was also subjected to the control of the large-scale latitudinal tectonic belts. The Junggar–Xingan area has geosynclinal deposits and its biotic aspect was close to that of Siberia, while clearly different from that of South China. This indicates that the Junggar–Xingan area belonged to the southern slope of the Siberian plate, which explains why it was part of the Boreal realm. The Himalaya–Gangdise–Nianqingtanggula province in Tibet was part of the northern continental margin and slope of Gondwana in Carboniferous times. It was located on the northern margin of the Indian plate and thus was a part of the Gondwana biogeographical realm. The South China and North China blocks in between were drifting microplates in the eastern past of Palaeo-Tethys, i.e. the Cathaysian Tethys. They are now divided by the Kunlun–Qilian–Qinling tectonic belt into two sections, each making a separate province. It is thus clear that the Carboniferous provincialization is closely related to plate tectonics, and the provincial boundaries are usually coincident with plate boundaries.

7.3.3 RELATIONSHIP BETWEEN THE DISTINCTIVENESS OF PROVINCIALIZATION AND TRANSGRESSION

In China there were two transgressions in the Early Carboniferous. The first took place in the Yanguanian (Aikuanian), when the neritic transgression in South China came from south to north-east and north-west. Owing to the barrier of the oldland, highly endemic faunas evolved. However, in North China the transgression of the Tianshan–Xingan geosynclines moved southwards from the Boreal region, and the biotas are closely related to North America and Siberia. The second transgression, in the late Datangian, was more extensive than the first one, when most areas were covered by sea water, but several oldlands remained emergent. The mixed nature of the biotas in each province obscures the provincialization. Because of this, the classification of Early Carboniferous biogeographical provincialization in China is mainly based on the characteristics and distribution of Yanguanian biota. The Himalaya–Gangdise–Nianqingtanggula province in southern Tibet was exceptional, yielding quite different biotas in the Early Carboniferous as a consequence of having little communication with the other areas in China. Even within the same province, the pattern differed greatly between Early and Late Carboniferous. In the Late Carboniferous there was even more extensive transgression, when many massifs were submerged and the whole of North China was covered by water, thus enlarging the marine area. Hence the provincialization is not so evident as that of Early Carboniferous time. This time, however, continental plants began to show clear provincialism, with the temperate-climate Angaran flora in the far north of China, the tropical and subtropical Euramerican and Cathaysian floras in North and South China, and the Gondwana *Glossopteris* flora of cool–cold type in the Qomolangma area. This serves as the main basis for Late Carboniferous biogeographical provincialization. It is evident from the above analysis that the distinctiveness of Carboniferous provincialization in China is directly related to the extensiveness of transgression and isolation of land and seas.

REFERENCES

Bai Yunhong, Chen Guoliang, Sun Qingge, and Sun Yuhang (1985). Late Paleozoic polar wander path for the Tarim Block and tectonic significance. *Seismology and Geology* **7** (**1**), 71–80 (in Chinese).

Cao Renguan (1986). Discovery of Late Carboniferous glaciomarine deposits in western Yunnan. *Geological Review* **23**, (3), 236–42 (in Chinese).

Chaloner, W. G. and Meyen, S. V. (1973). Carboniferous and Permian floras of the Northern Continents. In *Atlas of palaeobiogeography*, (ed. A. Hallam), pp. 169–84. Elsevier, Amsterdam.

Chen Bingwei (1982). The discovery of Middle and Upper Carboniferous tilloids in Baxoi, Xizang (Tibet) and their significance. *Geological Review* **28**, (2), 148–51 (in Chinese).

Chen Genbao (1984). The Carboniferous of the Baoshan region, west Yunnan. *Journal of Stratigraphy* **8**, (12), 129–35 (in Chinese).

Chen Gongxin (1984). Fusulinids, ammonites and fossil plants. In *Paleontological Atlas of Hubei Province*, pp. 12–73, 322–8, 548–915. Hubei Science and Technology Publishing House, Wuhan (in Chinese).

Chen Jiahuai, Li Hanming, Ling Jiaxing, Feng Shaonan, and Shi Guorun (1986). New contribution to the knowledge of Early Carboniferous strata in western Hubei. *Journal of Stratigraphy*, **10**, (1), 49–53 (in Chinese).

Chen Xu and Wang Jianhua (1983). The fusulinids of the Maping Limestone of the Upper Carboniferous from Yishan, Guangxi. *Palaeontologica Sinica, New series, B* **19**, 1–139 (in Chinese).

Deng Bao (1978). Discovery of Early Carboniferous flora from Shanyang, south Shanxi, and its stratigraphic significance. *Acta Geologica Sinica* **52**, (1), 15–21 (in Chinese).

Deng Zhanqiu and Zhang Yansheng (1984). Paleozoic corals from Hengduan Mountains, Southwestern China. *In Stratigraphy and paleontology in west Sichuan and east Xizang, China*, Part 4, pp. 1–102. Sichuan Sciences and Technology Publishing House, Chengdu (in Chinese).

Diaz, D. M. (ed.) (1983). *The Carboniferous of the world. I, China, Korea, Japan and S.E. Asia*, pp. 16–238. IUGS Publication.

Dutro, J. T., Jr, Gordon, M., Jr, and Huddle, J. M. (1979). Palaeontological zonation of the Mississippian System. *In Paleotectonic investigations of the Mississippian System in the United States*, Part A, *Interpretive summary and special features of the Mississippian System*, pp. 407–26.

Fan Yingnian (1963). Lower Carboniferous corals from Kueichow and Hunan. *Acta Palaeontologica Sinica* **11**, (2), 270–91 (in Chinese).

Fan Yingnian (1980). Early Carboniferous strata and coral of northwestern Szechuan. *Professional Papers on Stratigraphy and Palaeontology* **9**, 1–47 (in Chinese).

Fan Yingnian (1985). A division of zoogeographical province by Permo-Carboniferous corals in Xizang (Tibet). *Contribution to the Geology of the Qinghai-Xizang (Tibet) Plateau* **16**, 87–106 (in Chinese).

Feng Rulin (1981). Late Lower Carboniferous brachiopods from central–southern Guizhou with their stratigraphic sequence. *Acta Palaeontologica Sinica* **20**, (3), 248–54 (in Chinese).

Feng Shaonan, Hu Yufan, and Zhu Jianan (1982). Fossil plants and their assemblages from the Early Carboniferous in Guangdong. *Acta Botanica Sinica* **24**, (4), 374–82 (in Chinese).

Gu Yingqu (1977). Discovery of Upper Carboniferous coal bed from Yunnan. *Journal of Stratigraphy* **3**, (3), 191 (in Chinese).

Guo Shengzhe (1983). Middle and Upper Carboniferous rugose corals from southern Dahinganling (Mts. Great Khingan). *Acta Palaeontologica Sinica*, **22**, (2), 220–30 (in Chinese).

Guo Shengzhe (1987*a*). Carboniferous corals from eastern and southern Liaoning, China. *Bulletin of the Shenyang Institute of Geology and Mineral Resources, Chinese Academy of Geological Sciences* **15**, 99–113 (in Chinese).

Guo Shengzhe (1987*b*). The marine Carboniferous System of northern China geosynclinal region. *Bulletin of the Shenyang Institute of Geology and Mineral Resources, Chinese Academy of Geological Sciences* **15**, 121–44 (in Chinese).

Han Jianxiu (1982). Middle and Upper Carboniferous fusulinids from the Nadanhada Range, Heilungjiang province. *Acta Palaeontologica Sinica* **21**, (3), 315–21 (in Chinese).

Hao Tonglin and Wang Naiwen (1983). Carboniferous glaciamarine deposits in northern Xizang. *Bulletin of the Chinese Academy of Geological Sciences* **7**, 41–50 (in Chinese).

Hao Yichun and Lin Jiaxin (1982). Foraminiferal assemblages of Upper Carboniferous Huanglung Formation in Yangchun of Guangchun of Guangdong. *Earth Sciences–Journal of the Wuhan College of Geology* **3**, 19–33 (in Chinese).

He Baozhen and Chao Jixiang (1982). The Carboniferous paleogeographic outline in Northwestern China and its sedimentary characteristics. *Coal Geology and Exploration* **2**, 14–20 (in Chinese).

Hill, D. (1973). Lower Carboniferous corals. In *Atlas of palaeobiogeography*, (ed. A. Hallam), pp. 133–45. Elsevier, Amsterdam.

Hu Changming (1983). New genera and species of spiriferacean brachiopods in the Late Carboniferous to Early Permian from Duoma district, Rutog, Xizang (Tibet), China. *Earth Sciences – Journal of Wuhan College of Geology* **1**, 105–18 (in Chinese).

Huang Benhong (1982*a*). Permo-Carboniferous flora of Tianshan–Xingan foldbelt. *Kexue Tongbao* **25**, (15), 703–5 (in Chinese).

Huang Benhong (1982*b*). Permo-Carboniferous terrestrial deposits and palaeogeographic features in the northern part of northeast China. *Geological Review* **28**, (5), 395–402 (in Chinese).

Huang Benhong (1983). On late Late Paleozoic paleophytogeographic regions of eastern Tianshan–Xingao foldbelt and its geological significance. In *Contribution for the project of 'Plate Tectonics in Northern China'* Vol. 1, pp. 138–55. Geological Publishing House, Beijing (in Chinese).

Jiang Shuigen (1982). Rugose corals of Hunan. In *Palaeontological Atlas of Hunan Province*, pp. 81–161. Geological Publishing House, Beijing (in Chinese).

Jin Songqiao (Chin Sungchiao) (1974). Subdivision and correlation of the Lower Carboniferous of the Beishan region, Kansu. *Acta Geological Sinica* **2**, 159–74 (in Chinese).

Jin Suhua (1985). Early Carboniferous (Tournaisian) fossil brachiopods from Qingshuigou of Baoshan County, Yunnan. *Contribution to the Geology of the Qinghai-Xizang (Tibet) Plateau* **16**, 75–86 (in Chinese).

Jin Yugan and Sun Dongli (1981). Palaeozoic brachiopods from Xizang. In *Palaeontology of Xizang*, Part 4, pp. 27–176. Science Press, Beijing (in Chinese).

Jin Yugan and Fang Runsen (1983). Early Carboniferous brachiopods from Shidian, Yunnan. *Acta Palaeontologica Sinica* **22**, (2), 139–52 (in Chinese).

Jin Yugan, Wang Yu, Sun Dongli, and Shi Quan (1985). Late Palaeozoic and Triassic brachiopod from the east of the Qinghai-Xizang plateau. In *Stratigraphy and*

Palaeontology in west Sichuan and east Xizang, China Part 3, pp. 182–249. Sichuan Science and Technology Publishing House, Chengdu (in Chinese).

Li Xingxue and Yao Zhaoqi (1983). Carboniferous and Permian floral provinces in East Asia. In *Palaeobiogeographic province of China*, pp. 74–82. Science Press, Beijing (in Chinese).

Li Xingxue, Yao Zhaoqi, Cai Chongyang, and Wu Xiuyuan (1974). Carboniferous biostratigraphy of Tsingyuan district, east Kansu, China. *Memoirs of Nanjing Institute of Geology and Palaeontology, Academy Sinica* **6**, 99–118 (in Chinese).

Li Zhangrong, Lu Yiju, Chen Guolong, and He Guoxiong (1986). Carboniferous and Triassic biostratigraphy of Mt. Burhan Budai, central Qinghai. In *Carboniferous and Triassic strata and fossils from the southern slope of Mt. Burhan Budai, Qinghai, China* pp. 1–72. Anhui Science and technology Publishing House, Hefei (in Chinese).

Li Zhonghai, Gu Xueda, Yang Shengkai, Zhu Zhanxiang, Huang Shengbi, and Wu Tingyi (1982). Carboniferous stratigraphy of eastern Xizang and western Sichuan. In *Stratigraphy and palaeontology in west Sichuan and east Xizang,China* Part 1, pp. 93–130. Sichuan People's Publishing House, Chengdu (in Chinese).

Liang Dingyi and Wang Weipin (1983). Preliminary discussion on the Carboniferous and Permian Systems and its fauna at Kangmar district and Qushia of Lhaza, Xizang (Tibet) plateau. *Contribution of the Geology of the Qinghai-Xizang (Tibet) Plateau* **2**, 226–36 (in Chinese).

Liang Dingyi, Nie Zetong, Guo Tieying, Xu Baowen, Zhang Yizhi, and Wang Weipin (1983). Permo-Carboniferous Gondwana–Tethys facies in Southern Karakorum, Ali, Xizang (Tibet). *Earth Sciences – Journal of Wuhan College of Geology* **11**, 9–28 (in Chinese).

Liang Xiluo (1976). Carboniferous and Permian ammonoids from the Mount Jolmo Lungma region. In *A report of scientific expedition in the Mount Jolmo Lungma Region (1966–1968), Palaeontology*, fasc. 3, pp. 215–22. Science Press, Beijing (in Chinese).

Liao Zhuoting (1975). Uppermost Carboniferous brachiopods from western Guizhou. *Acta Palaeontologica Sinica* **18**, (6), 527–46 (in Chinese).

Lin Baoyu (1963). Tabulates of Early Carboniferous from Nanlin region. *Memoirs of Institute of Geosciences, Ministry of Geology, P.R. China* **4**, (1), 1–75 (in Chinese).

Lin Baoyu (1985). A preliminary study on the stratigraphical distribution and zoogeographical province of the Carboniferous tabulate corals of China. *Professional Papers of Stratigraphy and Palaeontology* **12**, 27–46 (in Chinese).

Lin Baoyu and Wang Baoyu (1985). Middle Carboniferous tabulate corals from Jinghe district of Xinjiang. *Geological Review* **31**, (6), 512–17 (in Chinese).

Lin Jiaxing, Chen Gongxin, Zhou Zuren, and Zhang Bufei (1977). Fusulinids. In *Palaeontological atlas of Central–South China*, Vol. 2, pp. 4–95. Geological Publishing House, Beijing (in Chinese).

Lin Yingdang, Huang Zhuxi, and Wu Shizhong (1983). The classification of the zoogeographical regions of Lower Carboniferous corals in China. *Journal of Changchun College of Geology* **3**, 1–7 (in Chinese).

Lin Yingdang, Wu Shizhong, and Xu Shouyong (1984). The Datangian corals of the Lower Carboniferous in central Jilin. *Journal of Changchun College of Geology* **1**, 3–11 (in Chinese).

Liu Benpei and Cui Xinsheng (1983). Discovery of *Eurydesma* fauna from Rutog, northwest Xizang (Tibet), and its biogeographic significance. *Earth Sciences – Journal of Wuhan College of Geology* **1**, 79–92 (in Chinese).

Liu Zhouan, Xiao Ximing, and Dong Wenlan (1978). Fusulinid. In *Palaeontological atlas of Southwest China–Guizhou*, Vol. 2, pp.12–98. Geological Publishing House, Beijing (in Chinese).

Lu Linhuang (1982). Carboniferous and Early Permian bryozoans in the Qamdo area, eastern Xizang. In *Stratigraphy and palaeontology in west Sichuan and east Xizang*, Part 2, pp. 267–81. Sichuan People's Publishing House, Chengdu (in Chinese).

McElhinny, M. W., Embleton, B. J. J., Ma, X. H., and Zhang, Z. K. (1981). Fragmentation of Asia in the Permian. *Nature* **293**, 212–16.

Meyen, S. V. (1982). The Carboniferous and Permian floras of Angaraland (A synthesis). *Biological Memoir* **7 (1)**, 1–109.

Read, C. B. and Mamay, S. H. (1964). *Upper Paleozoic floral zones and floral provinces of the United States.* US Geological Survey, Professional Papers, 454k, pp. 1–35.

Ross, C. A. (1979). Carboniferous. In *Treatise on invertebrate palaeontology*, Part A, pp. 254–89. Geological Society of the University of Kansas Press.

Ross, C. A. and Ross, J. R. P. (1985). Carboniferous and Early Permian biogeography. *Geology* **13**, (1), 27–30.

Rowley, D. B., Raymond, A., Parrish, J. T., Lottes, A. L., Scotese, C. R., and Ziegler, A. M. (1985). Carboniferous paleogeographic, phytogeographic, and paleoclimatic reconstructions. *International Journal of Coal Geology* **5**, 7–42.

Ruan Yiping (1981*a*). Devonian and earliest Carboniferous ammonoids from Guangxi and Guizhou. *Memoirs of Nanjing Institute of Geology and Palaeontology, Academia Sinica* **15**, 1–152 (in Chinese).

Ruan Yiping (1981*b*). Carboniferous ammonoid faunas from Qixu in Nandan, Guangxi. *Memoirs of Nanjing*

Institute of Geology and Palaeontology, Academia Sinica **15**, 153–232 (in Chinese).

Ruan Yiping (1984). Some Devonian and Carboniferous ammonoids from Xizang. *Acta Palaeontologica Sinica* **23**, (5), 597–604 (in Chinese).

Rui Lin and Zhang Linxin (1986). On the Carboniferous–Permain boundary. *Journal of Stratigraphy* **10**, (4), 249–61 (in Chinese).

Scotese, C. R., Bambach, R. K., Barton, C., Van der Voo, R. and Zieler, A. M. (1979). Paleozoic Base Maps. *Journal of Geology* **87**,(3), 217–34.

Seyfert, C. K. and Sirkin, L. A. (1979). *Earth history and plate tectonics*. Harper and Row, New York.

Sheng Huaibin (1981). Late Carboniferous ammonites from northern Tianshan, Xinjiang. *Bulletin of Institute of Geology, Chinese Academy of Geological Sciences* **3**, 83–96 (in Chinese).

Sheng Huaibin (1983). The ammonoids of late Lower Carboniferous from Yongzhu village, Xainza district in North Tibet. *Contribution to the Geology of the Qinghai-Xizang (Tibet) Plateau* **8**, 41–68 (in Chinese).

Shi Xide (1983). The discovery of Tournaisian in the Mt. Amunike of the northeastern Qaidam Basin, with a discussion on the division of the Lower Carboniferous in the Mt. Oulongbuluke. *Contribution to the Geology of Qinghai-Xizang (Tibet) Plateau* **2**, 187–205 (in Chinese).

Song Xueliang (1982). The Carboniferous and Devonian tetracorals from the Baoshan–Shidian region of western Yunnan. *Contribution to the Geology of the Qinghai-Xizang (Tibet) Plateau* **10**, 18–37 (in Chinese).

Sun Qiaoli and Zhang Linxin (1985). Upper Carboniferous fusulinids from the Altun Mountains, Xinjiang. *Acta Micropalaeontologica Sinica* **2**, (1), 49–60 (in Chinese).

Tong Zhengxiang (1983). The Carboniferous brachiopod assemblage zone of Yanbian, Sichuan. *Journal of Stratigraphy* **7**, (4), 299–304 (in Chinese).

Tong Zhengxiang (1986). Early Early Carboniferous brachiopod fauna in northwest Sichuan. *Acta Palaeontologica Sinica* **25**, (6), 672–85 (in Chinese).

Wan Shilu, Ding Hui, and Zhao Songyin (1983). Middle and Late Carboniferous conodont biostratigraphy of North China. *Journal of the China Coal Society* **2**, 62–72 (in Chinese).

Wang Baoyu (1983). New material of tabulate corals from the Kunlun-Altun Mountains with their stratigraphical significance. *Contribution to the Geology of the Qinghai-Xizang (Tibet) Plateau* **2**, 170–86 (in Chinese).

Wang Deyou (1982). Some marine pelecypods from the Middle Carboniferous in Gushi of Henan. *Acta Palaeontologica Sinica* **21**, (4), 456–68 (in Chinese).

Wang Keliang (1982). Carboniferous and Permian Foraminifera of Xizang. In *Palaeontology of Xizang*, fasc. 4, pp. 1–32. Science Press, Beijing (in Chinese).

Wang Keliang (1984). Early Carboniferous Foraminifera from Shaoyang area of Hunan province and their stratigraphic significance. *Bulletin of Nanjing Institute of Geology and Palaeontology Academia Sinica* **6**, 209–24 (in Chinese).

Wang Keliang (1985). Fossils in the south of Tuomuer region. 2. Foraminifera of the Late Carboniferous epoch. In *Geology and Palaeontology of Tuomuer Region, Tianshan*, pp. 137–40. Xinjiang People's Publishing House, Urumqi (in Chinese).

Wang Mingqian (1981), Carboniferous ammonoids from eastern Xinjiang. *Acta Palaeontologica Sinica* **20**, (5), 468–81 (in Chinese).

Wang Naiwen (1984*a*). On the palaeobiogeography and plate tectonics of Qinghai–Tibet Plateau. *Bulletin of the Institute of Geology, Chinese Academy of Geological Sciences* **9**, 1–28 (in Chinese).

Wang Naiwen (1984*b*). The rise of Gondwana research in China and some key problems on paleobiogeography related to plate tectonics. *Bulletin of the Chinese Academy of Geological Science* **10**, 103–11 (in Chinese).

Wang Rennong (1979). Carboniferous stratigraphy of the Hulusitai Basin, Helanshan. *Journal of Stratigraphy* **3**, (1), 57–64 (in Chinese).

Wang Yizhao (1983). The characteristics and significance of Carboniferous gravel bed in Tengchong and Baoshan area, western Yunnan. *Contribution to the Geology of the Qinghai-Xizang (Tibet) Plateau* **2**, 71–7 (in Chinese).

Wang Yujing (1981). Fusulinids from Xizang of China. In *Paleontology of Xizang*, fasc. 3, pp. 1–80. Science Press, Beijing (in Chinese).

Wang Zengji (1980*a*). Early Carboniferous rugose corals from the Dulan and the Naling Guole River areas of Qinghai, northwest China. *Acta Palaeontological Sinica* **19**, (6), 493–9 (in Chinese).

Wang Zengji (1980*b*). Geographic distribution and stratigraphic significance of the heterocoral *Hexaphyllia* in China. *Bulletin of Nanjing Institute of Geology and Mineral Resources, Chinese Academy of Geological Sciences* **1**, (1), 42–8 (in Chinese).

Wang Zengji (1981*a*). Rugose corals from the early Lower Carboniferous Chuanshangou Formation in Amunike Mountain, Qinghai province. *Acta Geologica Sinica* **55**, (3), 170–8 (in Chinese).

Wang Zengji (1981*b*). Early Lower Carboniferous stratigraphy of northwest and north China. *Geological Review* **27**, (6), 533–8 (in Chinese).

Wang Zengji (1983*a*). On the Carboniferous stratigraphy from the western section of eastern Kunlunshan. *Contribution to the Geology of the Qinghai–Xizang (Tibet) Plateau* **2**, 207–25 (in Chinese)

Wang Zengji (1983*b*). Upper Carboniferous and Lower Permian compound rugose corals in the western Kunlun

Mountains, Xinjiang, *Geological Review* **29**, (6), 499–505 (in Chinese).

Wang Zengji (1985). The Lower Carboniferous in the western section of the eastern Kunlun Mountains. *Geological Review* **31** (2), 111–17 (in Chinese).

Wang Zhiping (1987). Upper Carboniferous rugose corals from central Hunan. *Acta Palaeontologica Sinica* **26**, (4), 471–85 (in Chinese).

Wang Zhiping and Ye Gan (1984). Early Carboniferous heterophyllids from Majiaoba, Jiangyou, Sichuan province. *Earth Sciences – Journal of Wuhan College of Geology* **3**, 33–7 (in Chinese).

Wu Wangshi (1962). Upper Carboniferous corals from Yishan, Kwangsi. *Acta Palaeontologica Sinica* **10**, (3), 326–42 (in Chinese).

Wu Wangshi (1964). Lower Carboniferous corals in central Hunan. *Memoirs of Nanjing Institute of Geology and Palaeontology, Academia Sinica* **3**, 1–100 (in Chinese).

Wu Wangshi, Liao Weihua, and Zhao Jiamin (1982). Palaeozoic rugose corals from Xizang. In *Palaeontology of Xizang*, fasc. 4, pp. 107–51. Science Press, Beijing.

Wu Wangshi and Zhang Yansheng (1979). Late Palaeozoic rugose corals from Batang and Yidun, western Szechuan. *Acta Palaeontologica Sinica* **18**, (1), 25–40 (in Chinese).

Wu Wangshi and Zhao Jiaming (1984). On the biological characteristics and the stratigraphical distribution of the family Kepingophyllidae. *Acta Palaeontologica Sinica* **23**, (4), 411–19 (in Chinese).

Wu Wangshi, Zhang Linxin, and Ching Yuhan (1974). The Carboniferous of western Kueichow. *Memoirs of Nanjing Institute of Geology and Palaeontology, Academia Sinica* **6**, 72–98 (in Chinese).

Wu Wangshi, Zhang Linxin, Zhao Xiuhu, Jin Yugan, and Liao Zhuoting (1987). *Carboniferous stratigraphy in China*, pp. 1–160. Science Press, Beijing.

Wu Wangshi, Zhao Jiamin, and Jiang Shuigen (1981). Corals from the Shaodong Formation (Etroeung) of South China. *Acta Palaeontologica Sinica* **2**, (1), 49–60 (in Chinese).

Wu Xiuyuan and Deng Bao (1983). Fossil plants from the Caoliangyi Formation of Fengxian, Shaanxi. *Acta Palaeontologica Sinica* **22**, (2), 183–92 (in Chinese).

Wu Xiuyuan and Zhao Xiugu (1981). Fossil plants from the Kaolishan Formation (Lower Carboniferous) in Jurong, southern Jiangsu. *Acta Palaeontologica Sinica* **20**, (1), 50–9 (in Chinese).

Wu Xiuyuan and Zhao Xiugu (1982). Subdivision and correlation of the Carboniferous continental strata in China. In *Stratigraphic Correlation Chart in China with Explanatory Text*, pp. 137–52. Science Press, Beijing (in Chinese).

Xia Guoying and Zhang Zhicun (1984). Foraminifera. In *Palaeontological atlas of North China*, Vol. 3, pp. 5–59. Geological Publishing House, Beijing (in Chinese).

Xia Guoying and Zhang Zhicum (1985). Foraminifera: Fusulinacea. In *Palaeontological atlas of North China*, Vol. 1, pp. 39–149. Geological Publishing House, Beijing (in Chinese).

Xu Ren (1976). On the discovery of a *Glossopteris* flora in southern Xizang and its significance in geology and palaeogeography. *Scientia Geologica Sinica* **4**, 324–31 (in Chinese).

Yang Deli (1984). Brachiopoda of the Yangtze Gorges area. In *Biostratigraphy of the Yangtze Gorges area*, Vol. 3, pp. 203–39. Geological Publishing House, Beijing (in Chinese).

Yang Fengqing (1978). On the Lower and Middle Carboniferous subdivisions and ammonites of Western Guizhou. *Professional Papers of Stratigraphy and Palaeontology* **5**, 143–200 (in Chinese).

Yang Fengqing (1984). Late Early Carboniferous ammonoids from Chenmazijing of Ningxia. *Acta Palaeontologica Sinica* **25**, (3), 260–71 (in Chinese).

Yang Fengqing,, Zheng Zhaochang, and Liu Zhicai (1983). Carboniferous ammonoid zones from Zhongning and Zhongwei, Ningxia region, China. *Earth Sciences – Journal of Wuhan College of Geology* **2**, 9–23 (in Chinese).

Yang Guanxiu, Chen Chonghui, and Zhang Shuliang (1983). Floral character of Taiyuan Formation and Shanxi Formation in eastern Zhungerqi, Nei Monggol. *Earth Sciences – Journal of Wuhan College of Geology* **4**, 69–83 (in Chinese).

Yang Jingzhi, Wu Wangshi, Zhang Linxin, Liao Zhuoting, and Ruan Yiping (1979). New classification of the Carboniferous System of China. *Journal of Stratigraphy* **3**, (3), 188–92 (in Chinese).

Yang Jingzhi, Lu Linhuang, and Xia Fengsheng (1981). Late Palaeozoic bryozoans from Xizang. In *Palaeontology of Xizang*, fasc. 3, pp. 81–100. Science Press, Beijing (in Chinese).

Yang Jingzhi, *et al.* (1982). Classification and correlation of the Carboniferous in China. In *Stratigraphic correlation chart in China with explanatory text*, pp. 124–36. Science Press, Beijing (in Chinese).

Yang Shipu (1978). Lower Carboniferous brachiopods of Guizhou province and their stratigraphic significance. *Professional Papers of Stratigraphy and Palaeontology* **5**, 78–142 (in Chinese).

Yang Shipu (1980). The stratigraphical and geographical distribution of Fengninian brachiopods of China. *Geological Review* **26**, (6), 471–8 (in Chinese).

Yang Shipu (1983). Palaeozoogeographic provinces of the Lower Carboniferous brachiopods of China. In

Palaeobiogeographic provinces of China, pp. 64–73. Science Press, Beijing (in Chinese).

Yang Shipu and Fan Yingnian (1982). Carboniferous strata and fauna in Shenzha district, northern Xizang (Tibet). *Contribution to the Geology of the Qinghai-Xizang (Tibet) Plateau* **10**, 46–9. Geological Publishing House, Beijing (in Chinese).

Yang Shipu and Fan Yingnian (1983). Carboniferous brachiopods from Xizang (Tibet) and their faunal provinces. *Contribution to the Geology of the Qinghai-Xizang (Tibet) Plateau* **11**, 265–89 (in Chinese).

Yang Shipu, Hou Hongfei, Gao Lianda, Wang Zengji, and Wu Xianghe (1980). The Carboniferous System of China. *Acta Geologica Sinica* **54**, (3), 167–75 (in Chinese).

Yang Zengrong (1983). Subdivision of Carboniferous System in Baoshan area, Yunnan. *Contribution to the Geology of the Qinghai-Xizang (Tibet) Plateau* **11**, 61–70 (in Chinese).

Yang Zungyi, Liang Dingyi, Guo Tieying, and Nie Zetong (1984). The Lower Gondwana Series of Xizang (Tibet) and adjoining region with emphasis on its palaeogeographic and palaeotectonic significance. In *Scientific Papers on Geology for International Exchange–Prepared for the 27th International Geological Congress*, Vol.1, pp. 88–110. Geological Publishing House, Beijing (in Chinese).

Yu Jiangzhang (Yu, C. C.), Lin Yingdang, Huang Zhuxi, and Cai Tuei (1978). Early Carboniferous stratigraphy and corals of eastern Xinjiang. *Professional Papers of Stratigraphy and Palaeontology* **5**, 1–77 (in Chinese).

Yu Jiangzhang, Lin Yingdang, and Huang Zhuxi (1980). Lower Carboniferous Heterocorallia in China. In *Scientific Papers on Geology for International Exchange – Prepared for the 26th International Geological Congress*, Vol. 4, pp. 108–12. Geological Publishing House, Beijing (in Chinese).

Yu Jiangzhang, Lin Yingdang, Shi Yen, Huang Zhuxi, and Yu Xueguang (1983). *Carboniferous and Permian corals*, pp. 1–354. Jilin People's Publishing House, Jilin (in Chinese).

Yu Xueguang (1980). Upper Carboniferous Chuanshanian tetracorals of southern Jiangsu. *Professional Papers of Stratigraphy and Palaeontology* **9**, 492–7 (in Chinese).

Zhang Linxin (1981). The Carboniferous and Permian fusulinid successions in eastern Qinghai-Xizang Plateau. In *Geology and Ecology Studies of Qinghai-Xizang Plateau*, Vol. 1, pp. 187–92. Science Press, Beijing (in Chinese).

Zhang Linsin (1982). Fusulinids of eastern Qinghai-Xizang plateau. In *Stratigraphy and palaeontology in west Sichuan and east Xizang, China*, Part 2, pp. 119–244. Sichuan People's Publishing House, Chengdu (in Chinese).

Zhang Renjie (1980). Bivalves from Middle Carboniferous, Dongfang county, Hainan island, Guangdong province. *Geological Review* **26**, (4), 281–6 (in Chinese).

Zhang Shixing and Jin Yugan (1976). Upper Paleozoic brachiopods from the Mount Jolmo Lungma region. In *A report of the Scientific Expedition in the Mount Jolmo Lungma Region (1966–1978), Palaeonotology*, fasc. 2, pp. 159–271. Science Press, Beijing (in Chinese).

Zhang Yuxiu (1981). Discovery of Late Carboniferous bivalves from Shanxi Formation in central and south Shanxi. *Geological Review* **27**, (6), 483–90 (in Chinese).

Zhao Xiugu and Wu Xiuyuan (1982). Fossil plants from the Tzushan Series in Zudu of southern Jiangxi. *Acta Palaeontologica Sinica* **21**, (6), 699–708 (in Chinese).

Zhao Xiugu and Wu Xiuyuan (1986). Carboniferous flora in western Zhejiang. *Memoirs of Nanjing Institute of Geology and Palaeontology, Academia Sinica* **22**, 1–70 (in Chinese).

Zhao Zhixin, Han Jianqiu, and Wang Zenji (1984). *The Carboniferous strata and its fauna from southwestern margin of Tarim Basin in Xinjiang*. Geological Publishing House, Beijing (in Chinese).

Zheng Zhaochang, Zhang Gang, Li Yuzheng, Liu Zhicai, and Liu Ruiqi (1982). An outline of the strata in the Badan Jaran area, Inner Mongolia. *Journal of Stratigraphy* **6**, (3), 225–33 (in Chinese).

Zhu Xiufang (1985). Early Upper Carboniferous fusulinids form Ulan county, Qinghai province. *Bulletin of Institute of Geology, Chinese Academy of Geological Sciences* **11**, 1–20 (in Chinese).

Ziegler, A. M., Scotese, C. R., McKerrow, W. S., Johnson, M. E. and Bambach, R. K. (1977). Paleozoic biogeography of continents bordering the Iapetus (pre-Caledonian) and Rheic (pre-Hercynian) Oceans. In *Palaeontology and plate tectonics*, Special Publications in Biology and Geology, Milwaukee Public Museum, No. 2, pp. 1–34.

Ziegler, A. M., Scotese, C. R., McKerrow, W. S., Johnson, M. E. and Bambach, R. K. (1979). Paleozoic paleogeography. *Annual Review Earth Planetary Sciences* **7**, 437–502.

Zuo Zibi (1986). Discovery of *Kueichouphyllum* from Yanguan Stage of Lower Carboniferous. *Acta Palaeontologica Sinica* **25**, (3), 296–300 (in Chinese).

8 PERMIAN

Xu Guirong and Yang Weiping

8.1 PERMIAN PALAEO-BIOGEOGRAPHICAL PROVINCIALIZATION OF THE WORLD

The basic map of Permian palaeobiogeographical provincialization of the world used in this chapter is by Scotese *et al.* (1979) with a modification of the Chinese part according to Chinese palaeomagnetic and palaeoclimatic information, discussed in detail in this chapter and Chapter 15. Palaeolatitudinal locations for South and North China have been fixed to a high degree of reliability, and for Tarim, Siberia, and Tibet to a lesser degree (Khramov *et al.* 1981; Chen *et al.* 1983; Bai *et al.* 1985; Lin *et al.* 1985; Zhao and Coe 1989; Nie *et al.* 1990). South China was near the equator while North China was about 10°–20°N, Tarim was about 30°N, and Himalaya about 20°S.

The distribution of Permian faunas and floras in the world has been discussed extensively. For more recent work, reference should be made to Asama (1985), Ishii *et al.* (1985), Nakamura *et al.* (1985), Ross and Ross (1985, 1990), Chaloner and Creber (1988), and Ziegler (1990). The provincialization is shown in Fig. 8.1 and Table 8.1.

We have still followed the traditional Chinese demarcation of the Carboniferous–Permian boundary at the base of the Sakmarian instead of the Asselian, pending a reconsideration from the National Committee of Stratigraphy (Table 8.2).

8.1.1 THE BOREAL REALM

This includes the area to the north of Junggar–Xingan in China, Siberia, Greenland, Novaya Zemlya (Russia), Spitsbergen, and northern Canada. In China the distribution area of the Angaran flora lies to the north of the Ebinur–Gaxunnur–Xar Moron River region. In the Early Permian the Angaran flora of temperate–cool climate mainly occurred in this area. The Early Permian Angaran flora within China's northern territory consists of a north-eastern *Nephropsis–Crassinervia–Zamiopteris* assemblage (Huang Benhong 1982, 1983) found in Cuiluan (Yichun County), Shenshu (Tieli County), Mishan County, and areas in the Daxingan Mountains; and a north western *Zamiopteris–Walchia* assemblage (Wu Shaozu 1983), found in the southern and eastern margins of the Junggar Basin. There are about 20 common genera of marine animals (Wang

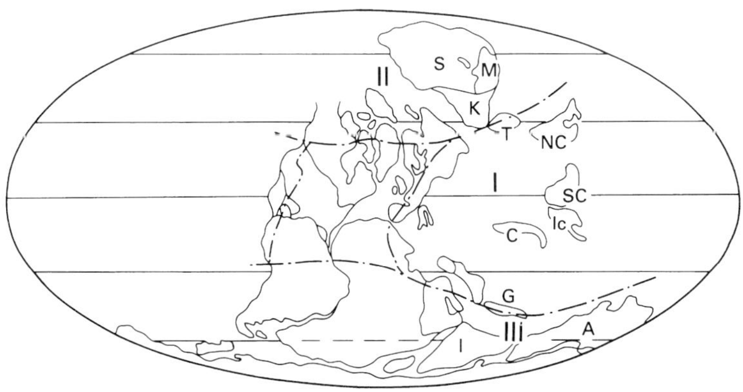

Fig. 8.1 Map of the Permian palaeobiogeographical provincialization of the world. I, Palaeo-Tethyan realm; II, Boreal realm; III Gondwana realm; A, Australia; C, Central Tibet and Western Yunnan; G, Gangdise; I, India; Ic, Indochina; K, Kazakhstan; M, Mongolia; NC, North China; S, Siberia; SC, South China; T, Tarim.

Table 8.1 Permian palaeobiogeographical provincialization of the world

			Early Permian	Late Permian
North cold temperate zone	Boreal realm	C. Asia–Mongol region	*Yakovlevia–Horridonia* assemblage *Lytvolasma* assemblage *Uraloceras* assemblage *Nephropsis–Crassinervia–Zamiopteris* flora	*Callipteris–Supaia–Viatscheslavia* flora *Callipteris–Comia–Iniopteris* flora
North warm temperate zone			*Lytvolasma* assemblage *Monodiexodina* assemblage *Neoschwagerina–Yabeina* assemblage *Horridonia–Liosotella–Spiriferella* *Licharewia–Yakovlevia* and *Leptodus, Meekella* assemblages	*Comia–Iniopteris, Lepeophyllum* flora *Tychtopteris, Nephropsis,* *Zamiopteris, Callipteris* flora
Tropics–subtropics	Tethyan realm	Cathaysia–Tethyan region	*Misellina–Cancellina* and *Verbeekina–Neoschwagerina* assemblages *Ipciphyllum–Iranophyllum* assemblage *Tyloplecta yangtzeensis–Monticulifera–Cathaysia* assemblage *Emplectopteris–Taeniopteris* spp.–*Emplectopteridium alatum–Cathaysiopteris whitei* flora	*Codonofusiella–Palaeofusulina–Gallowayinella* assemblage *Gigantonoclea guizhouensis,* *Lobatannularia multifolia,* *Ullamannia bronni* flora and *Gigantopteris nicotianaefolia,* *Otofolium* flora
South warm temperate zone			*Lytvolasma–Verbeckiella* assemblage *Wannerophyllum,* *Iranophyllum, Wentzelella* assemblage *Stenoscisma, Costiferina* assemblage	*Palaeofusulina, Gallowayinella* assemblage *Peltichia, Oldhamina,* *Cathaysia* assemblage *Gigantopteris nicotianaefolia,* *Lobatannularia multifolia* flora
South cold temperate zone	Gondwanan realm	Gondwana–Tethyan region	*Lytvolasma* assemblage *Eurydesma* assemblage *Uraloceras* assemblage *Steponoviella, Costiferina* assemblage *Glossopteris* flora	*Glossopteris* flora

Table 8.2 Correlation chart of the Permian of China

Permian					
Lower		Upper			
Qixia'ian	Maokouan	Wujiapingian	Changxingian		
Selong Group		Selong Baga Luoshiza		Qomolangma	Himalayan province
Qubu	Qubuerg Xiukang	?		Dingri	
?	Lasarla Gaqoi		?	Zhongba	
Chitichun	Limestone			Pulan	
Wululong	?			Central Tibet	Central Tibet province
Luobadui		Mengza Group		Lhasa	South China province
Luobadui		? Chaka coal measures		Shuanghu	
Luobadui		Tuoba coal measures		Tuoba	
Qixia	Maokou	Wujiaping Longtan	Changxing Dalong	South China	
Shanxi	Lower Shihezi	Upper Shihezi	Shiqianfeng	North China	N. China province
Bayinhe		Nuoyinhe		Southern Qilian	Tarim-Qilian province
Shoushangou Taheshen Fanjiatun Yilaxi		Yangjiagou Madatun		Huadin	Inner Mongo–Jilin province
Daisuanbian Miaoling Kedao		Kaishantun		Yanji	
Tumenling Yuquan		Hongshan		Acheng	Junggar–Xingan province
Gaojiaoupang Sijiashan Liutiaogou ?		Shunjiafen	Laolongtou	Zalaiteqi	
Shuangbaotang Jinta Jushitan ?		Hongliuxia		Beishan	

Naiwen 1984), dominated by solitary corals of cool-water type, such as *Lytvolasma*, and thick-shelled brachiopods of cool-water type, such as *Yakovlevia* and *Horridonia*. Ammonoids are characterized by the abundance of *Uraloceras* and *Paragastrioceras* (Liang Xiluo 1981). *Uraloceras* occurs in the Urals, north-eastern Siberia, and arctic Canada, and is thus typically Boreal. Other genera, such as *Metalegoceras* and *Neouddenites* are also typical elements of the Boreal realm. Bivalves are represented by the *Atomodesma* fauna, chiefly distributed in eastern and northern Siberia, Novaya Zemlya, Greenland, Canada, Alaska, and northern Nevada (United States).

The Late Permian Angaran flora is found, in the north-east, in Shenshu (Yichun Co.) and Mishan Co. (Heilongjiang), as well as in Yanbian County and Dunhua County (Jilin) (Huang 1982, 1983), and consists of a *Callipteris–Supaia–Viatscheslavia* assemblage in early Late Permian and a *Callipteris–Comia–Iniopteris* assemblage in the late epoch. In the north-west this flora is found along the southern and eastern margins of the Junggar Basin, containing the *Walchia–Ullmannia–Cordaianthus–Noeggerathiopsis–Psygmophyllum* assemblage in the early epoch and *Pecopteris ontriscifolia*, *Callipteris zeilleri*, *Iniopteris sibirica*, and *Comia dentata* in late Late Permian. The northward movement of the Tarim Platform and western North China (north Qilian) into this floral area in the Late Permian resulted in the mixture of elements from the Boreal realm with those from the Tethyan realm. In the Late Permian there were only a few fusulinids, corals, and other marine invertebrates, but terrestrial animals and plants were quite abundant in the Chinese part of the Boreal realm.

8.1.2 THE PALAEO-TETHYAN REALM

The Permian middle–low latitude area between the Boreal realm and the Gondwanan realm is divided into two oceans by the continent of Pangaea II (southern part of North America, northern part of South America and West Africa); the eastern section, or the main ocean, was the wedge-shaped Palaeo-Tethys opening eastwards, and the western section was a smaller tropical ocean occupying the south-western part of North America and north-western part of South America. During the Early Permian in the Delaware Basin in western Texas (North America) a famous bioherm contained a large diversity of calcareous algae, sponges, bryozoans, bivalves, brachiopods, and fusulinids. Bioherm limestones, bearing abundant *Pseudoschwagerina* and *Parafusulina*, are seen widely in the Carnic Alps in north-eastern Italy, and bioherms also occur in the Sakmarian Stage in the lower part of the Lower Permian of the Russian Platform. In South China, Permian bioherms consist of colonial corals, sponges, hydrozoans, and stromatoporoids. This east–west trending belt of bioherms and accompanying evaporites roughly represents the area located in the tropical and subtropical climatic zones. (Drewrey *et al.* 1974; Flakes 1979; Cui and Zhen 1984). The relevant area in China with tropical biotas belonged to the Palaeo-Tethyan realm.

The Palaeo-Tethyan realm includes the Chinese territory of Pamir, Northern Tibet, and South China; for its whole range, reference should be made to Sengör (1984) and Sengör *et al.* (1988). There are more than a dozen marine invertebrate groups, among which fusulinids, corals, brachiopods, ammonoids, and bivalves are very abundant. The number of fossil genera and species in this realm is several times as many as that of the Boreal realm; thus, in China, the common genera of the common phyla were about three times as numerous as those of the Boreal realm (Wang Naiwen 1984). The fusulinids in southern China, for instance, were at the peak of their diversity, i.e. over 100 species and more than 20 genera. Bioherms, gypsiferous and red beds are found widely. From all these facts and the palaeomagnetic data put together (for details refer to Table 2 of Nie *et al.* 1990), it can be deduced that this realm was located near the equator, in a zone of tropical and subtropical climate.

In western North America, the Tethyan fusulinids *Neoschwagerina*, *Pseudodoliolina*, *Yabeina*, and Verbeekinidae, representing the tropical type, occurred in alternation with the North American type (or endemic type) *Pseudoschwagerina* and *Pseudofusulinella*. It is seen in western North America that verbeekinids are not distributed along latitude but is a north–south trend. Other organisms, such as corals, also show this kind of phenomenon. This has been explained by terrane accretion theory (Yancey 1976). Because no part of China is concerned, North and South America will not be discussed in this chapter.

The Early Permian fusulinids of Palaeo-Tethys, and especially in Cathaysian Tethys are composed mainly of Verbeekinidae and Neoschwagerinidae (the common elements are *Neoschwagerina*, *Neomisellina*, *Yabeina*, *Lepidolina*, *Verbeekina*, *Sumatrina*, *Afghanella*) and also some genera of the Schwagerinidae, (such as *Schwagerina*, *Parafusulina*, *Pseudofusulina*, *Chusenella*, *Rugososchwagerina*, and *Polydiexodina*). Corals are characterized by Waagenophyllidae and Wentzellophyllidae, the common genera being *Ipciphyllum*, *Iranophyllum*, *Wentzelella*, *Wentzelloides*, *Polythecalis*, *Hayasakaia*, and *Wentzellophyllum*. The important brachiopod taxa are *Monticulifera*, *Urushtenia*, *Neoplicatifera*, *Cathaysia*, *Tyloplecta yangtzeensis*, *Vediproductus*, *Asioproductus gratiosus*, *Edriosteges*, *Spinomarginifera*, *Tschernyschewia*, *Orthothetina*, and *Haydenella*. Ammonoids include *Daubichites*, *Waagenoceras*, *Stacheoceras*, *Strigogoniatites*, and *Tauroceras*. Plants are dominated by lepidodendrales and gigantopterids. The common Early Permian associations are the *Emplectopteris–Taeniopteris* spp.—*Emplectopteridium alatum* and *Emplectopteris triangularis–Taeniopteris* spp.—*Cathaysiopteris whitei* assemblages (Li Xingxue and Yao Zhaoqi 1983).

In the Late Permian, the distribution area of fusulinids became limited to southern China and Tibet and the number of genera and species also decreased greatly, (the main elements are *Palaeofusulina*, *Codonofusiella* and *Gallowayinella*). The same is true for corals, with only *Liangshanophyllum*, *Lophophyllidium*, and *waagenophyllum* being widespread. In addition, there were brachiopods such as *Oldhamina* and *Leptodus*, constituting the so-called *Oldhamina* assemblage. The Late Permian plant assemblages are divided into two areas, namely South China and North China, with a border along the Kunlun–Qinling–Dabie line. The earlier assemblage of the South China area included *Gigantopteris nicotianaefolia–Otofolium* spp. and *Lobatannularia multifolia*, and the later one included *Ullmannia* cf. *bronni* and *Gigantonoclea guizhouensis*. The earlier assemblage of the North China area included *Gigantonoclea hallei*, *Fasciopteris* spp., and *Loba tannularia ensifolia*, while in the later epoch there were a few elements such as *Lobatannularia* cf. *heianensis*, *Fascipterias* cf. *hallei*, *Taeniopteris* spp., and *Oligocarpis* sp. (Li Xingxue and Yao Zhaoqi 1983).

8.1.3 THE GONDWANAN REALM

In China this includes the areas to the south of the Yarlung Zangbo River. Gondwana was the region characterized by the wide distribution of *Glossopteris* flora and glacial deposits. In the early Early Permian glacial deposition covered a large part of Gondwana. Tillites are found widely in Africa, Australia, India, and South America, with the Palaeo-South Pole inferred to be in South Africa. Correspondingly, there were temperate–cool water faunas around the outer side of this realm, surrounding the tillite-bearing zone. To sum up, the main body of Gondwana in the Early Permian was situated in the polar or subpolar region of the southern hemisphere, belonging to a cool to cold climate zone, but its marginal parts may have reached temperate or even warm climate zones, especially by the late Early Permian (Nie *et al.* 1990). In the Early Permian the biotas in the Qomolangma area in China were also typical of the Gondwanan type, with corals, brachiopods, and bryozoans including specific cold-water elements such as *Costiferina*, *Fusispirifer*, *Taeniothaerus*, *Callomarginatia*, and *Uraloceras*, clearly pointing to the fact that Qomolangma was then located in a relatively high-latitude area of the southern hemisphere and was part of the northern margin of Gondwana. It is interesting that Qomolangma also yields *Lytvolasma*, *Strophalosia*, *Wyndhamia*, *Aulosteges*, *Neospirifer*, and *Spiriferella*, similar to the association of the Tianshan–Xingan province (Boreal realm). This verifies the bipolar distribution of faunas in the Early Permian.

The Late Permian strata of the Gondwanan realm outcrop in remote and scattered areas in China, so they have not been studied in detail. However, the *Glossopteris* flora is also found in India and adjacent countries.

It should be noted that palaeoclimatic conditions were not fixed for each realm. In the earliest age of the Permian (Sakmarian and Artinskian) the Gondwanan and Boreal realms were located, respectively, in southern and northern cool–warm climate zones and even cold climate zones, while in the late age of the Early Permian (Maokouan), the margins of the two realms might have entered regions of the southern and northern warm temperate zone, with a change to warm–hot climate and the entrance of many Tethyan organisms. During the Late Permian, Tarim, Beishan, and Qilian moved northward to amalgamate with Siberia or Mongolia. (According to Pruner 1987, the palaeomagnetic APW (apparent polar wandering) path of Mongolia and North China became compatible after the Carboniferous, but the Mongolia–Okhotzk suture did not close until the Jurassic, so there is the possibility that North China did not amalgamate with Siberia but with Mongolia.) Thus they moved from the tropical and subtropical zone to a warm climate zone and hosted mixed or transitional floras. The relationship of the realms to climatic zones shifts in time, and is therefore shown to be oblique in Table 8.1. Based on a comparison with present-day vegetation (Walter 1985), Ziegler (1990) suggested a tropical rainforest for the Cathaysian flora; cool-climate temperate Angaran and *Glossopteris* floras; and a warm, temperate transitional Late Permian flora for Qilian and Beishan, due to northward drift of North China into a higher latitude. This agrees fully with our conclusions.

8.2 PERMIAN PALAEO-BIOGEOGRAPHICAL PROVINCIALIZATION OF CHINA

Permian strata in China (Table 8.2) are widely distributed and well exposed with abundant fossils. Biogeographical study has been pursued recently by many scientists (including Wu Wangshi 1975; Zhang Shouxin and Ching Yukan 1976; Sheng Jinzhang and Wang Yujing 1981; Zhan Lipei *et al.* 1982; Huang Benhong 1983; Li Xingxue and Yao Zhaoqi 1983; Wu Shaozu 1983; Yang Jingzi and Lu Linhuang 1983; Lin Baoyu 1984; Zhang and He 1985).

8.2.1 EARLY PERMIAN PALAEOBIOGEOGRAPHICAL PROVINCIALIZATION OF CHINA

The Early Permian biotas in China (Fig. 8.2) are rich in both marine and terrestrial organisms. According to our incomplete statistics, in the Early Permian there were over 60 genera and more than 330 species of fusulinids, about 64 genera and over 210 species of corals, and about 55 genera and 395 species of bryozoans. In addition, there were tabulate corals, brachiopods, cephalopods, bivalves, gastropods, forams, trilobites, conodonts, sponges, worms, crinoids, belemnites, echinoids, ophiuroids, fish, marine reptiles, and algae; in total more than 20 marine phyla. Terrestrial life included plants, vertebrates, conchostracans, ostracods, and insects.

The Early Permian biogeography of China can be considered in terms of three realms, three regions, and six provinces, as shown in Fig. 8.2, and Tables 8.3 and 8.4.

The Cathaysian region of the Palaeo-Tethyan realm

This region includes the areas to the south of the Ebinur–Gaxunnur line and to the north of the Longmuco–Jinshajiang line. The Cathaysian Tethys region belongs to the eastern part of tropical Palaeo-Tethys, or Laurasian Tethys. South China is the centre of this region, which also contains Japan, Sikhote-Alin, northern Tibet, and parts of South-East Asia. It is characterized by the development of Cathaysian fauna and flora.

The Cathaysian fauna in the Early Permian includes fusulinids, corals and brachiopods. The Tethyan fusulinids include *Misellina*, *Neoschwagerina*, *Yabeina*, *Verbeekina*, and *Parafusulina*, which are widely distributed in South-East Asia, New Zealand, Japan, Pamir, Yugoslavia, Sicily, Greece, Turkey, the Crimea, the northern Caucasus, and China; the typical fusulinids of Cathaysian

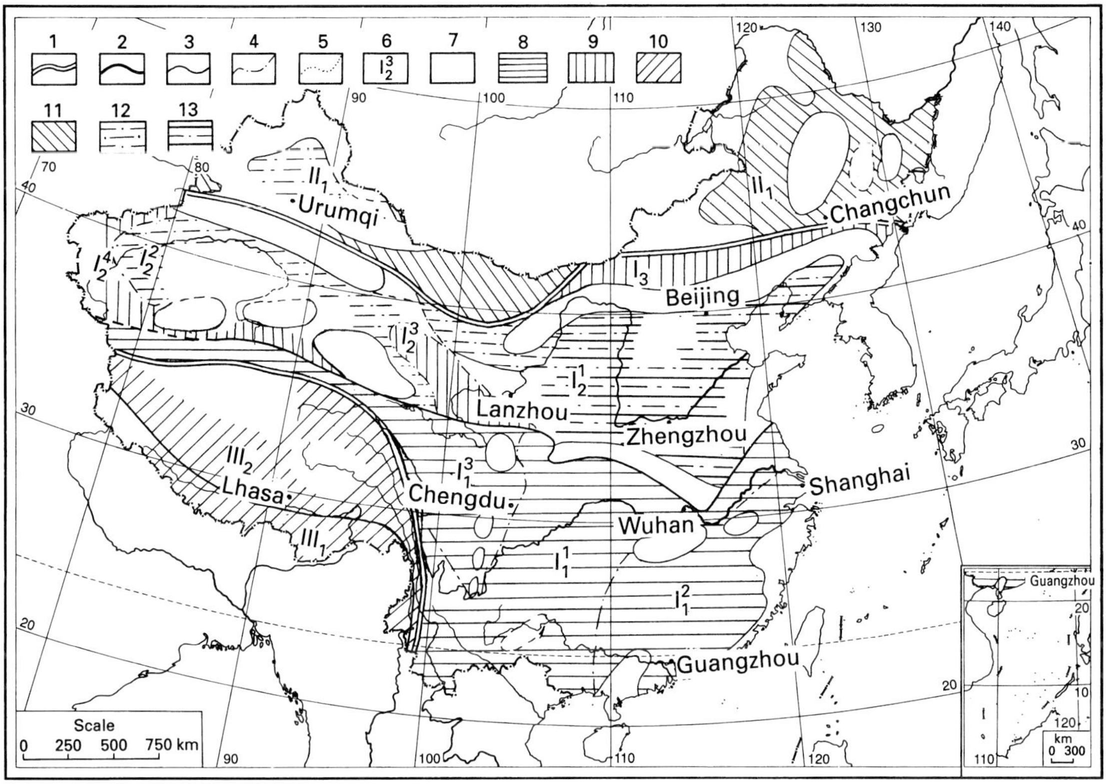

Fig. 8.2 Map of the Early Permian palaeobiogeographical provincialization of China. 1, Boundary of realms; 2, boundary of regions; 3, boundary of provinces; 4, boundary of biomes; 5, boundary of communities; 6, roman digit = realm mark, subscript = province mark, superscript = biome mark; 7, Oldland. 8–11, Marine and paralic regions: 8, South China; 9, North China; 10, Gondwana; 11, Junggar–Xingan. 12–13, Terrestrial regions (Late Permian and Mesozoic–Cenozoic): 12, fluvio–lacustrine basins; 13, coal-bearing basins. I, Palaeo-Tethyan realm, Cathaysian region; I_1, South China province; I_1^1, Upper Yangtze biome; I_1^2, South-eastern China biome; I_1^3, Songpan biome; I_2, North China–Tarim province; I_2^1, North China biome; I_2^2, Tarim biome; I_2^3, Qinling–Qilian biome; I_2^4, South-western Xinjiang biome; I_3, Inner Mongolia–Jilin province; II, Boreal realm, Central Asia–Mongolia region; II_1, Junggar–Xingan province; III, Gondwanan realm, Gondwanan Tethyan region; III_1, Himalayan province; III_2, Central Tibet–Western Yunnan province.

Tethys are *Cancellina*, *Afghanella*, *Eoverbeekina*, *Neofusulinella*, and *Yangchenia*. The Early Permian Cathaysian coral assemblage is called the *Iranophyllum–Ipciphyllum* assemblage, previously the Yangtze coral fauna of the Tethyan realm (Wu and Zhang 1979). *Iranophyllum* is also seen in Sicily, Turkey, Tunisia, Iraq, Iran, Afghanistan, Pakistan, Kampuchea, Laos, and Japan. In China it occurs in Kunlun and Qinling as well as in Tibet to the north of the Yarlung Zangbo River, e.g. in Zhongba, Lhunzhub (found in the lower part of the Luobadui Formation), and Baxoi (Wu Wangshi 1975). *Ipciphyllum* is chiefly distributed in the Maokou Formation in south-western and southern China. The typical Cathaysian genera of widely distributed brachiopods in South China are *Urushtenia*, *Neoplicatifera*, *Monticulifera*, *Cathaysia*, *Squamularia* and *Orthothetina*; they form an independent brachiopod assemblage occupying roughly the same areas as the coral assemblage. Moreover, the areas

Table 8.3 Permian palaeobiogeographical provincialization of China

Early Permian		Late Permian	
Boreal realm		C. Asia–Mongol region	
C. Asia–Mongolian region Junggar–Xingan Province	*Nephropsis–Crassinervis–Zamiopteris–Walchia* *Yakovleria–Horridonia* *Monodiexodina, Lytvolasma* *Uraloceras*	Junggar–Xingan province	*Zamiopteris–Noeggerathiopsis* *Callipteris–Comia* *Iniopteris* *Hamiapollenites–Striatobietites*
Tethyan realm		Cathaysian Tethys Region	
Cathaysian Tethys region		Tarim–Qilian province	*Callipteris–Iniopterissibirica* *Zamiopteris glossopteris* *Lobatannularia heianinsis*
Inner Mongolia–Jilin province	*Spiriferella–Yakovlevia* *Misellina–Parafusulina* *Waagenophyllum–Yatsengia* *Lobatannularia–Pecopteris–Gigantonoclea–Tingia–Taeniopteris*	N. China province	*Gigantonoclea hallei–Lobatannularia ensifolia* *Protohaploxypinus* *Shanxisaurus xuecunensis*
South China province		S. China province	
SE China biome	*Schouchangoceras–Daubichites* *Misellina* *Urushtenia–Neoplicatifera* *Polythecalis, Waagenophyllum* *Gigatopteris nicotianaefolia* *Compsopteris wongi*	SE China biome	*Gigantopteris nicotianaefolia* *Lobatannularia* *Altudoceras–Sanyangites* *Spinomarginifera–Neochonetes*
Songpan–Upper Yangtze biome	*Schwagerina tschernyschewi* *Misellina claudiae* *Parafusulina–Cancellina* *Ipciphyllum–Wentzellophyllum* *Hayasakaia, Iranophyllum* *Monticulifera–Crytospirifer*	Upper Yangtze biome	*Edriosteges–Tyloplecta* *Palaeolima minima* *Lobatannularia cathaysiana* *Palaeofusulina–Rotodiscoceras–Pseudotirolites–Pleuronodoceras*
N. China–Tarim province	*Emplectopteris triangularis* *Emplectopteridium alatum* *Taeniopteris–Lobatannularia*	Songpan biome	*Colaniella, Palaeofusulina* *Gallowayinella* *Squamularia grandis*
		Qiangtang–Qamdo biome	*Gigantopteris nicotianaefolia* *Lobatannularia miltifolia* *Palaeofusulina–Gallowayinella* *Peltichia–Oldhamina*
Gondwana realm		Gondwana–Tethyan region	
Gondwana–Tethyan region			
C. Xizang–W. Yunnan province	*Polythecalis–Wentzellophyllum* *Parafusulina–Misellina* *Polydiexodina* *Lytvolasma–Monodiexodina* *Stenoscisma–Costiferina* *Eurydesma*	Himalaya province	*Neogondolella subcarinata* *Glossopteris?*
Himalaya province	*Glossopteris* *Taeniothaerus–Calliomarginatia* *Lytvolasma–Uraloceras*		

Table 8.4 Occurrence of the main Permian fossils in China

	Gondwanan realm – Gondwana–Tethys region: Himalayan province	Tethyan realm – Cathaysia–Tethyan region: C. Tibet W. Yunnan province	Tethyan realm – S. China province: Songpan biome	Tethyan realm – S. China province: Yangtze biome	Tethyan realm – S. China province: SE China biome	Tethyan realm: Tarim province	Tethyan realm: Inner Mongolia–Jilin province	Boreal realm – C. Asia–Mongolian region – Junggar–Xingan province: Junggar biome	Boreal realm – C. Asia–Mongolian region – Junggar–Xingan province: Xingan biome	Age
Monodiexodina	×	×	×				×	×	×	Early Permian
Parafusulina		×		×	×	×	×			
Misellina		×		×	×		×			
Schwagerina				×	×					
Neoschwagerina	×	×	×	×	×		×			
Verbeekina		×	×	×			×			
Rugososchwagerina	×			×	×					
Cancellina		×		×	×					
Neomisellina		×		×	×		×			
Polydiexodina			×	×	×		×			
Chusenella		×		×			×			
Polythecalis				×	×		×			
Lytvolasma	×	×					×	×	×	
Iranophyllum		×	×							
Ipciphyllum			×	×	×					
Wentzelella		×	×				×			
Allotropiophyllum				×	×		×			
Waagenophyllum				×	×		×			
Orthotichia				×			×	×		
Yakovlenia							×	×	×	
Eurydesma	×	×								
Urushtenia	×?	×	×		×					
Cryptospirifer				×						
Horridonia						×	×	×		
Callifera								×		
Stepanoviella	×									
Waagenoceras					×		×			
Uraloceras	×						×			
Schouchangoceras					×					
Altudoceras		×			×		×			Late Permian
Daubichites					×		×			
Cyclolobus	×	×			×					
Paragastrioceras					×		×			

Table 8.4 (*cont.*)

Glossopteris	×							
Emplectopteris triangularis					×			
Taeniopteris multinervis					×			
Gigantonoclea fukienensis					×			
Tingia carbonica					×			
Zamiopteris							×	×
Noeggerathiopsis lutifolia							×	×
Comia–Iniopteris					×		×	×
Callipteris						×		
Oldhamina			×	×	×			
Edriosteges				×	×			
Squamularia grandis			×	×	×			
Peltichia zigzag				×	×			
Spinomaginifera				×	×			
Liangshanophyllum				×	×			
Waagenophyllum				×	×			
Pseudotirolites				×	×			
Rotodiscoceras				×	×			
Reichlina		×		×	×			
Gallowayinella		×		×	×			
Palaeofusulina		×		×	×			
Codonofusiella		×		×	×			

occupied by these brachiopods and corals coincide with those occupied by the Cathaysian bivalve assemblage (Fang Zongjie 1985), ammonoid assemblage (Zhao Jinke *et al.* 1965, 1978), and gastropods (Wang and Xie 1980; 1982). All these assemblages, put together, represent a typical marine biota—the Cathaysian fauna—developed in Cathaysian Tethys during the Permian. The corresponding flora is the typical Cathaysian flora, consisting mainly of gigantopterids; in the Early Permian it constituted the *Emplectopteris triangularis–Emplectopteridium–Cathaysiopteris* assemblage (Li and Yao 1983).

Three provinces can be distinguished in this region.

The South China province

This is characterized by abundant phyla with high diversity (Fig. 8.3), and has been subjected to detailed study (Rui 1981; Zhan *et al.* 1982; Yang Zunyi *et al.* 1987). In this province benthonic organisms were exceedingly abundant. Fusulinids, corals, brachiopods, bryozoans, cephalopods, conodonts, forams, sponges, gastropods, algae, and ophiuroids are commonly seen.

Early Permian brachiopods are best characterized by *Monticulifera* and *Cryptospirifer*. Other widely distributed invertebrates in this province include the fusulinids *Schwagerina tschernyschewi*, *Misellina claudiae*, *Parafusulina*, *Cancellina*, *Neoschwagerina*, *Neomisellina*, and *Yabeina*, and the corals *Wentzellophyllum volzi*, *Hayasakaia elegantula*, *Polythecalis yangtzeensis*, *Szechuanophyllum* and *Ipciphyllum*.

The South China province can be subdivided into three biomes according to its lithofacies and biotic types, i.e. Upper Yangtze biome, South-eastern China biome, and Songpan biome. The Upper Yangtze biome had environments ranging from coastal plain to epicontinental shallow sea. It was dominated by benthos, and contained abundant corals, fusulinids, conodonts, and brachiopods, but fewer ammonoids. The important fossils are the brachiopods *Crypto-*

spirifer and *Monticulifera*, the corals *Cystomichelinia* and *Ipciphyllum*, and fusulinids of the Neoschwagerinidae and Verbeekinidae.

The South-eastern China biome represents a restricted sea environment, where deep water facies often occurred owing to block faulting, usually in a north-easterly trend. It is characterized by the relatively high development of siliceous rocks. The biota are composed chiefly of ammonoids, brachiopods, and bivalves, whereas there are only a small number of fusulinids and corals. The ammonoids are characterized by Shouchangoceratidae, including *Altudoceras* and *Schouchangoceras*. The brachiopods include *Urushtenia* and *Neoplicatifera*, and the bivalves *Gujocardita* and *Paradoxipecten*. Clastic and carbonate deposits of different ages are found on the margin of the biome, containing the biotas similar to those of the previous biome.

The Songpan biome is located west of the Longmen–Xikang–Yunnan massif, covering the Songpan–Lixian, Yushu–Zhongdian areas, and mountains of Burhan Budai, Qimantag, Kunlun, and Anyemaqen. Geotectonically it was characterized by a mixture of small plates and small interplate ocean basins. The deposits are shallow marine carbonates and clastics (Songpan and Lixian) and flysch formations (Kunlun, Anyemaqen, and Qimantag Mountains). The corals include not only *Ipciphyllum* but also *Iranophyllum*, which typifies this biome. The distribution of *Iranophyllum* and *Wentzelella* begins in the east from the fold-belt of the Longmen Mountains–Xikang–Yunnan Oldland to the areas, further west, of the Karakorum Mountains and to the north of Yarlung Zangbo, extending further also to Iran, Turkey, Tunisia, and Italy. Its northern boundary is inferred to be approximately the southern slope of the Kunlun Mountains (Wu and Zhang 1979) and its southern boundary, the Yarlung Zangbo River. In this biome there are not only shallow-water corals, fusulinids (*Misellina* and *Neoschwagerina*), and brachiopods (*Monticulifera*, *Callifera*, *Urushtenia*), but also deep-water ammonoids (*Artinskia*, *Popanoceras*, *Perrinites*, *Waagenoceras*).

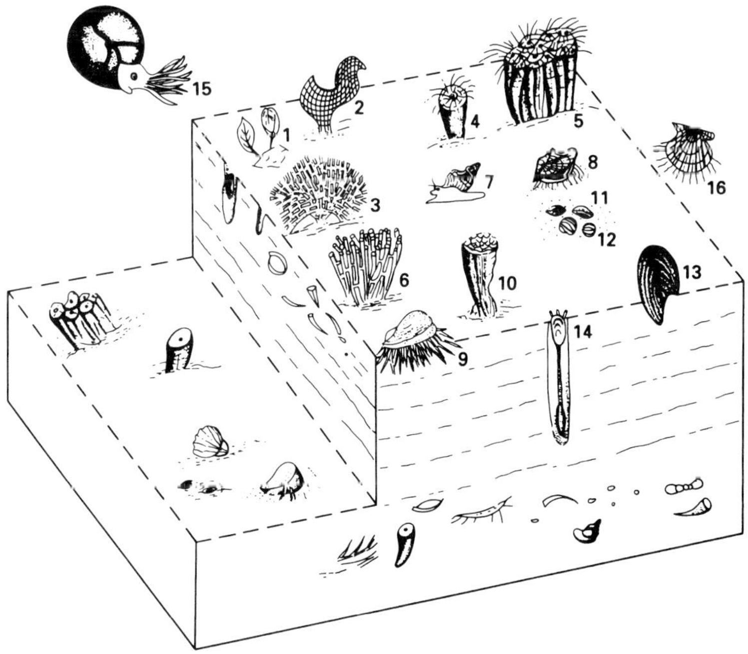

Fig. 8.3 Reconstruction of the Early Permian brachiopod, anthozoan, and bryozoan community of the Upper Yangtze biome. 1, *Athyris*; 2, *Fenestella*; 3, *Fistuliramus*; 4, *Lophophyllidium*; 5, *Polythecalis*; 6, *Hayasakaia* 7, *Worthenia*; 8, *Hayasakapecten*; 9, *Waagenoconcha*; 10, *Richthofenia*; 11, *Polydiexodina*; 12, *Verbeekina*; 13, *Cryptospirifer*; 14, *Lingula*; 15, *Neoaganides*; 16, *Acanthopecten*.

The North China–Tarim province

This province is mainly a terrestrial basin. Its major characteristic is the development of the North China type Cathaysian flora. The common terrestrial and freshwater animals are bivalves, ostracods, conchostracans, and vertebrates. The province can be subdivided into three biomes according to biotas and depositional facies: the North China biome, the Tarim biome, and the Qinling–Qilian biome.

The North China biome consists of lacustrine coal-bearing clastic deposits in the early Early Permian. There are abundant plant remains, such as *Emplectopteris triangularis*, *Emplectopteridium alatum*, *Taeniopteris*, *Lobatannularia sinensis*, and *Cathaysiopteris whitei*, and coal beds that can be mined. In the basal beds of the clastic deposits marine intercalations contain the brachiopod *Lingula*. This evidence shows that the North China biome had a relatively humid climate and was close to the sea in the Early Permian.

The Tarim biome was an intracontinental basin, so its climate was comparatively dry, with a series of mainly red variegated clastics, occasionally intercalated with basalts. Its flora is composed mainly of Cathaysian elements, such as *Sphenophyllum scopulatum*, but lacks the most characteristic ones, *Gigantopteris* and *Gigantonoclea*. The Kunlun area to the south-west of the Tarim Basin and the southern Tianshan area to its north-west yield marine fossils, mostly of Tethyan type, such as *Polydiexodina*, *Neoschwagerina*, *Iranophyllum* and *Wentzelella* (*Ipciphyllum*), which indicate that the sea came from the south. However the lower diversity and occurrence of *Polydiexodina* imply a more northerly location than North China, thus in accordance with that of Tarim.

The Qinling–Qilian biome occupied a bay trending north-westward, thus it was an area with marine strata, different from the two neighbouring terrestrial biomes to its east and west. The biota are still of Tethyan type. The Bayinhe Formation in southern Qilian, for example, yields *Verbeekina*, *Wentzelella*, and *Polydiexodina*, but with low diversity. On either side of the bay the strata change gradually into terrestrial facies, connected with the ancient Qilian Mountains and Qaidam Oldland.

The Inner Mongolia–Jilin province

The palaeotectonic position of this province was the continental slope of the northern margin of the North China–Tarim platform. This province is characterized by the mixed occurrence of both Tethyan and Boreal marine biotas, and by the domination of the North China Cathaysian flora in terrestrial facies, both indicating the continuation of the Carboniferous Tianshan–Jilin province. The orogeny at the end of the Carboniferous led to the enclosure of the western part of the Tianshan–Xingan geosynclines and change of the northern Tianshan area of that province to land. Therefore, in the Early Permian there remained only the central and eastern parts of the previous Carboniferous province, i.e. the Inner Mongolia–Jilin province, which includes the south-eastern part of Inner Mongolia, the north-western part of Hebei, and the central and eastern parts of Jilin.

This province is characterized chiefly by the abundance of fusulinids. *Pseudodoliolina* and *Schwagerina* occur in Yujiabeigou of the Guangxing commune on the southern bank of the Xar Moron River. Westwards, to the area of Sanmianjing (Baokang) on the southern margin of the ancient marine basin, the fusulinids *Misellina claudiae* and *Parafusulina*, and brachiopods of the *Orthotichia–Dictyoclostus* assemblage can be found (Li Li *et al.* 1984). Huadian county of central Jilin (Shoushangou Formation) yields the brachiopods *Echinoconchus*, *Athyris*, and *Compressoproductus*, the fusulinids *Parafusulina*, *Pseudofusulina kraffti*, *Schwagerina chernyschewi*, and *Monodiexodina** sp., and the coral *Yatsengia*. The Fanjiatun Formation in Zhoujiayao of Shuangyang County and Shuanhe town of Yongji County, and the Miaoling Formation in Yanbian, Yanji, and Wangqing Counties yield the brachiopods *Orthotetes*, *Waagenoconcha*, *Anidanthus*, *Leptodus*, *Stenoscisma*, *Spiriferella**, and *Yakovlevia**, the fusulinids *Neoschwagerina*, *Chusenella*, *Yabeina*, *Codonofusiella*, and *Dunbarula*; the corals *Waagenophyllum*, *Lophophyllidium*, and *Yatsengia*; the bryozoan *Hayasakapora* (Yu *et al.* 1981, 1983); and the ammonoids *Waagenoceras* and *Stacheoceras*. These faunas are dominated by Tethyan types but mixed with Boreal elements (marked by *). As far as plants are concerned, Fusong and Hunjiang Counties, as well as other areas, yield North China-type Cathaysian floral elements including '*Pecopteris feminaeformis*', *Taeniopteris taiyuanensis*, *Sphenophyllum kawasakii*, and *Gigantonoclea*; and Yujiabeigou of the Guangxing commune on the southern bank of the Xar Moron River yields *Gigantonoclea yujiaensis*, *Taeniopteris integra*, *Pecopteris wongii*, *Annularia gracilescens*, and '*Nilssonia*' *huabeiensis* (Huang Benhong 1983).

The Boreal realm and its Central Asia–Mongolia region

This realm includes the Chinese territory to the north of the Ebinur–Gaxunnur–Xar Moron River line, and is determined mainly by the distribution of the Angaran flora, while the Inner Mongolia–Jilin province of the Palaeo-Tethyan realm, to the south of the Xar Moron River, contains a relatively mixed fauna. In the past these areas were roughly called the 'Xingan–Mongolia geosynclines', 'Northern geosynclines' or 'Beishan–Xingan sea area', the major part of which is the Inner Mongolia–Jilin–Heilongjiang geosynclines. Based on the available data, this realm may extend to Kuruktag (Xinjiang) at its westernmost end. Further westwards the deposits turn into terrestrial facies. The biota of this geosyncline area comprise the Junggar–Xingan province.

The Junggar–Xingan province

This embraces chiefly northern Xinjiang, northern and north-eastern Inner Mongolia, northern Liaoning and Jilin, and most parts of Heilongjiang. In Inner Mongolia and north-eastern China, the dominant fossils are brachiopods (Li Wenguo 1983), corals (Guo 1980), bryozoans (Liu 1982), and ammonoids; fusulinids (Han 1980, 1981) are rare. Fossils of *Jakutoproductus* of the early Early Permian, found in Beishan and Yanchi Counties of Dong Ujimqin Qi, Inner Mongolia, can be correlated with the occurrence of the same genus in Verkhojansk and Yakutsk (Siberia), and are thus Boreal. In the early Early Permian the cool-water brachiopod assemblage *Spiriferella–Kochiproductus–Yakovlevia* (Li Wenguo 1983) developed in this province. The upper part of the Ulanbanxing Formation in western Inner Mongolia yields the ammonoids *Popanoceras* and *Uraloceras*, which are usually considered to be cool-water types. The Suolon area in north-eastern China yields the Boreal-type brachiopod assemblage *Waagenites deplanata–Horridonia–Liosotella*. Similar brachiopod assemblages are found in the Tumenling Formation and Yuquan Formation of Yichun County and Yuquan County, and the Songjiang of Acheng County in Heilongjiang. Boreal characteristics are also seen in the associated small solitary corals *Tachylasma* and *Calophyllum* and the fusulinids *Pseudodoliolina* and *Monodiexodina* (Han 1981) of this province. The Liutiaogou Formation of late Early Permian age in Jalaid Qi, northern Inner Mongolia, yields a *Richthofenia–Leptodus–Enteletes* assemblage (Hou *et al.* 1979), which is a warm-water brachiopod fauna. The assemblage occurs together with Tethyan corals and fusulinids, such as *Waagenophyllum*, *Wentzelella*, and *Codonofusiella*. The mixture of these warm-water organisms may be associated with the fact that the climate became warmer and the warm-water zone expanded in the late Early Permian, as was the case in the coeval Himalayan region. However, the possibility of a southward and clockwise shift of Siberia to close the Central Asia–Mongolian sea cannot be excluded. As regards the plants, Early Permian plant components are elements of the Angaran flora, represented in the east (Inner Mongolia and Xingan) by *Zamiopteris*, *Crassinervia*, *Nephropsis*, and *Neuropteris daheshonensis*, and in the west (the southern and eastern margins of Junggar in northern Xinjiang) by a *Zamiopteris–Walchia–Noeggerathiopsis–Latifolia* assemblage ranging in age from the late Early Permian to early Late Permian (Wu Shaozu 1983).

In Beishan, the Early Permian was characterized by ammonoids and a specialized brachiopod fauna, with fewer fusulinids. The ammonoid *Uraloceras* assemblage, with abundant species and individuals, was discovered in the early Early Permian Shuangbaotang Formation. The common elements are *Uraloceras*, *Tumaroceras*, *Demarizites*, *Neocrimites*, *Propinacoceras*, and *Medlicottia*. *Ulraloceras* is found in the Urals and north-eastern Siberia, Canada, and Australia, as well as in Gansu, western Inner Mongolia, and the Qomolangma area of China; it is thus a typical non-tropical genus with a bipolar distribution. *Tumaroceras* was first seen in western Verkhojansk (Russia), and has also been found in arctic Canada. This fauna possesses characteristics of a cool-water type. The late Early Permian Jushitan Formation in Beishan yields *Richthofenia–Leptodus–Enteletes*, a specialized warm-water brachiopod assemblage identical with that of Inner Mongolia; some Tethyan corals and fusulinids also coexist with this assemblage. Also found in this formation are the ammonoids *Waagenoceras*, *Strigogoniatites*, and *Stacheoceras*. *Waagenoceras* is distributed in Sicily, Timor, the Crimea, Sikhote-Alin, British Columbia in western Canada, and Japan, as well as in Gansu, Jilin, Fujian, and Hunan in China, and is thus tropical–subtropical. *Stacheoceras* is usually associated with *Waagenoceras*. There are still Boreal elements in the Jushitan Formation, such as *Spiriferella*, *Yakovlevia*, and *Uraloceras*. As a whole it is better to assign Beishan to the Boreal realm. The situation of faunal mixing and warming in the late Early Permian is identical with that of Inner Mongolia and north-eastern China, and was evidently a consequence of the same causes.

The Gondwanan realm and its Gondwana–Tethyan region

The Himalayan province

The area to the south of Yarlung Zangbo is regarded as the Himalayan province. Recent investigations have unearthed a batch of new fossil discoveries in this previously little known area (Wang and Mu 1980; Wang *et al.* 1981; Liang *et al.* 1982, 1983; Wu *et al.* 1982; Liu and Cui 1983; Lin 1984; Sheng 1984; Yu and Dai 1985). The overall characteristics of this province are the same as those of the other areas of the Gondwanan realm such as Kashmir, Nepal, Timor, and Australia. The biotas are typified by small, solitary corals, large thick-shelled cool-water brachiopods, and *Glossopteris* flora. The solitary corals *Lytvolasma*, *Wannerophylllum*, *Verbeekiella*, the ammonoid *Uraloceras* (Fan 1985), and the brachiopods *Taeniothaerus*, *Streptorhynchus tibetana* and *Spiriferella* have been found in the Early Permian strata of this province. The biota of the Selong Group in the Qomolangma area consists of a *Glossopteris* flora with *Glossopteris communis*, *G. angustifolia*, and *G. indica* (Xu Ren 1976); a large number of brachiopods, commonly *Taeniothaerus*, *Calliomarginatia*, and *Chonetella* (Zhang and Ching 1976); the coral *Lytvolasma* assemblage with more than 20 species (Wu Wangshi 1975); the ammonoid *Uraloceras xizangensi*; and over 20 kinds of bryozoans, among which *Streblotrypa* and *Streblascopora* occur commonly (Yang and Lu 1983). All the above elements are similar to those of the Sakmarian and Artinskian in the Urals (fauna) and Sydney Basin of Australia (flora). The fauna of the early Early Permian was chiefly distributed either in the Boreal Ocean (*Streptorhynchus kempet* from the Arctic Bear Islands has affinity with *S. tibetana*; Zhang and Ching 1976) or within Gondwanan Tethys (Australia, Timor). The faunal distribution shows clear bipolarity, representing biota characteristic of a non-tropical region. Although the biotas of different facies are developed in the Early Permian strata in this province, they contain elements showing affinities to neighbouring areas of the Tethyan realm. As a whole they should be put into one province, the Himalayan province, belonging to Gondwana.

However, there was some change in the late Early Permian (Maokouan). In addition to the continued occurrence of the cool-water corals *Lytvolasma*, *Wannerophyllum*, *Verbeekiella*, *Timorphyllum*, *Pseudofavosites*, and *Trachypsammia*, and the brachiopods *Neospirifer*, *Spiriferella*, and *Stenoscisma*, newcomers appeared in the areas to the south of Yarlung Zangbo. It should be noted that the ammonoids *Neocrimitites* and *Adrianites* (common elements both in the Arctic and in Tethys) are abundant (Sheng Huaibin 1984) in the Xiukang Formation in Lhaze. The fusulinids *Reichelina cribroseptata*, *Dunbarula pusilla*, and *Chusenella extensa*, found in the Lasaila limestone in Zhongba and the Chitichun limestone in Burang (southern Tibet), can be correlated with the *Neoschwagerina* zone of the Tethyan region. The coral *Wentzelella zhongbaensis* found therein is a component of the *Iranophyllum* coral assemblage in the same region (Wang and Mu 1980). A concept of 'exotic blocks' was once suggested to explain the occurrence of these tropical elements in Maokouan (Guadalupian) strata in southern Tibet (Yin and Guo 1976), but now it seems that such assemblages are not only widely distributed *in situ* in the Himalayas but are also typical of much of the northern margin of Gondwana (Nakazawa and Dickins 1985). During the Early Permian there was a world-wide transgression and climatic warming. It is therefore accepted here that the mixture of Tethyan elements of Maokouan age in the areas to the south of Yarlung Zangbo resulted from the rise of temperature and the widening of the tropical–subtropical climate belt, which coevally influenced not only the northern Gondwanan margin but also the southern Siberian margin, as has been indicated in the previous paragraphs. Of course, the possibility exists that the northward drift of the plate into the widened tropical area also caused the invasion of tropical biotas.

To conclude, it can be seen that the tropical–subtropical area and the corresponding biotas extended southwards into this realm during the late Early Permian. In the early Early Permian the area south of Yarlung Zangbo was dominated by cool biota with their northern boundary extending to Karakorum. By the late Early Permian the warm-water biota not only occupied the areas to the north of the Yarlung Zangbo River but also extended to where they mixed with Gondwanan biotas.

The central Tibet–Western Yunnan province

This province is called by some the South China transition or mixed region. It includes areas to the north of Yarlung Zangbo (such as Gangdise–Niainqingtanggula, Karakorum, Tanggula, and Hohxil) to the southern slope of Kunlun, and the areas to the west of the Hengduan Mountains in eastern Tibet and western Yunnan. The deposits are mainly composed of marine carbonates, argillites, and other clastics interbedded with medium–basic and medium–acidic volcanics. Fusulinids and brachiopods are the major elements. The early Early Permian *Polythecalis* and *Wentzellophyllum* are found in Rutog of the Ngari region of north-western Tibet (He and Weng 1983). *Polythecalis* is also found in Angyi to the north of Gerze, and *Misellina* and *Parafusulina* are found in Xiaobangda (Markam, eastern Tibet). All of them are the important elements of Qixia'an (Sakmarian to Kubergandian) age in South China. This province is characterized by the alternation and mixture, in time and space, of tropical and temperate–cool biota. So far as temperate–cool biota are concerned, the *Lytvolasma* assemblage is found in the early Early Permian Wululong Formation in Zasong of Lhunzhub County, and in Majula of the Wululong area to the north of Yarlung Zangbo, where the Tethyan element *Tachylasma minor* is also found. The *Eurydesma* assemblage is also found in early Early Permian strata in Rutog of the Ngari region (Liu and Cui 1983). The Gondwana fauna existed in stratigraphic alternation with the Tethyan fauna, the genera commonly seen including *Monodiexodina*, *Lytvolasma*, *Neospirifer*, and *Parafusulina* (Liang *et al.* 1983). A mixed Gondwana–Tethyan flora has been discovered in Xiagangjiang (Tibet) (Li Xingxue *et al.* 1985) and a *Lytvolasma–Wannerophyllum* assemblage is found in Xiala of Xainze County in central Tibet. In the Karakorum in Xinjiang a *Monodiexodina* assemblage is found in Konggashankou, and *Monodiexodina*, *Pseudofusulina*, *Schwagerina*, and *Parafusulina* occur in an area about 10 km to the south-east of Jiawendapo of Yichengli Mountain (Sun and Zhang 1985). Amenggangri of Rutog yields the corals *Lytvolasma*, *Paracaninia*, *Tachylasma*, and *Lophophyllum*, and fusulinids of the *Monodiexodina–Parafusulina* assemblage (Liang *et al.* 1983). All these data tell us that a temperate–cool-water biota had already crossed the Bangong Lake–Nujiang River suture and reached the Karakorum and Qiangtang regions. As for tropical organisms, a large number of corals with dissepiments, the most abundant being colonial hermatypic corals, occur in central Tibet, such as in Xainze, Lugu of Gerze, Duoma of Rutog, Shiquanhe, Zhongba, Burang, Lagu of Baxoi, Lhunzhub, Markam, and Haitong. The typical area, Xainze, yields *Waagenophyllum*, *Iranophyllum*, *Ipciphyllum*, *Wentzelloides*, *Wentzellites*, *Chusenophyllum*, *Yatsengia*, and *Yokoyamaella*, belonging to the warm-water *Iranophyllum–Wentzelella* assemblage. Most of the brachiopods are of warm-water type, but also mixed with some cool-water genera, such as *Stenoscisma* and *Costiferina*.

Nakamura *et al.* (1985) subdivided a western Tethyan subprovince for the Middle Permian (Japanese usage) brachiopod provincialization, which includes roughly the corresponding part of China. They also emphasized its mixed nature by indicating the presence of Gondwana-type brachiopods plus *Strophalosiina* and *Orthothetina*.

This transitional fauna is attributed to the Gondwana–Tethyan region according to Huang Jiqing, Wang Naiwen and others (Wang Naiwen 1984). Wang Naiwen (1984) has suggested five reasons for such an attribution. He has emphasized that, in view of the whole world, tropical Tethyan faunas and Cathaysian floras with low diversity may have appeared on the northern margin of Gondwana and in the shallow seas of Gondwanan Tethys. It should be noticed here that the two microplates of Qiangtang–Tanggula and Gangdise–Nianqingtanggula successively split off from Gondwana and shifted northwards to join with Eurasia during the Permian (Nie *et al.* 1990). This process is reflected by the progressive increment of tropical biotas starting from the late Early Permian. However, attribution of this region to the Gondwanan realm did not change until the Late Permian.

Cluster analysis results and other parameters of Early Permian biotas

In this chapter rugose corals are selected for cluster analysis. The Otsuka coefficient is used and the results are shown in Fig. 8.4. The similarity coefficient of northern Xinjiang and Inner Mongolia with the remaining regions is very low, less than 0.1, and can be clustered first as parts of the Boreal realm. Secondly, at a similarity coefficient of 0.22 or less, the areas of the northern China–Tarim Province and of the South China province can be separated. The division of the three biomes in the South China province is reflected in the clustering. Therefore, it can be said that the result of the cluster analysis is

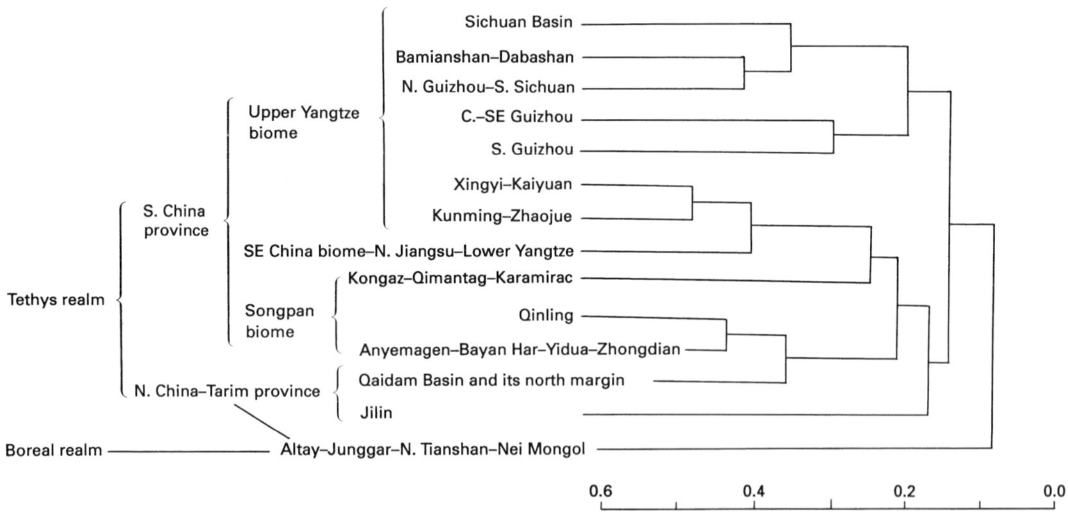

Fig. 8.4 Cluster analysis of the Early Permian rugose corals of China.

approximately consistent with the provincialization scheme adopted in this chapter. The Gondwanan realm is not included in this figure because rugose corals are too few. However, provincialization in some areas (Qinling for example) are confused by several similarity coefficients. The positions of these areas appear to be ambiguous and the provincialization boundaries appear to some extent unclear.

The diversity statistics of fusulinids and corals for each realm make it clear that there are three peak areas of diversity in southern China, i.e. the Upper Yangtze (92–115 species), Yunnan–Guizhou–Guangxi (107 species), and Jiangxi–Guangdong (*c.* 120 species), and the peak areas of corals are roughly the same; the data are not presented here.

8.2.2 LATE PERMIAN BIOGEOGRAPHICAL PROVINCIALIZATION OF CHINA

The fusulinids, rugose and tabulate corals that had flourished in the Early Permian declined sharply in the Late Permian, while brachiopods remained abundant, and some endemic types, such as *Oldhamina*, started to appear. Ammonoids mostly belong to the Ceratitida, no longer the goniatitids of the Early Permian. In the Late Permian gymnosperms dominated, so that the flora had a mesophytic aspect. Other Early Permian biotas continued into the Late Permian.

According to our incomplete statistics, in the Late Permian of China there were about 23 genera of fusulinids; over 20 genera and more than 60 species of corals; 17 genera and 45 species of bryozoans; over 80 genera and more than 190 species of brachiopods, and 70 genera and about 230 species of bivalves.

The Late Permian palaeobiogeographical provincialization in China almost follows the pattern of the Early Permian. However, the removal of the biogeographical unit boundary, started in the Maokouan, became rather obvious in the Late Permian. The Yarlung Zangbo fault belt evidently separated the two realms of Gondwana and Tethys. In the north, the Angaran flora and Cathaysian flora exchanged and alternated in occurrence, leading to a change of the realm boundary. Within each realm the marine area became relatively reduced and oldlands increased in extent, so that the late Late Permian biogeographical provincialization exhibited the initial form of Mesozoic palaeobiogeographical provincialization.

The Late Permian palaeobiogeographical provincialization in China can be expressed in terms of three realms, three regions, and five provinces (Fig. 8.5).

The Cathaysian Tethyan region of the Palaeo-Tethyan realm

In the Late Permian the southern border of this region moved southwards to the areas along Yarlung Zangbo, and its northern boundary remained unchanged. It is a typical tropical biotic region with the following aspects.

1. The *Gigantonoclea–Lobatannularia* flora of gigantopterids was widely distributed in this region, as based on discoveries in South China, North China, Indochina, and adjacent areas. In South China elements of wet-climate type dominate, while North China was characterized by a less wet tropical type, thus forming different subtypes (Li Xingxue *et al.* 1982*a*, *b*).

2. The marine invertebrates were still characterized by a high abundance and diversity of biotas. Bioherms were developed, e.g. in Jiantianba and Huangnitang of Lichuan County, western Hubei (Fan Jiasong *et al.* 1982), and calcareous algae were also abundant, e.g. in western Guizhou (Mu 1981). Fusulinids, and other forams, corals, conodonts, brachiopods, bryozoans, bivalves, cephalopods, spores, and pollen are all present in this region. According to our incomplete data, there are over 40 genera and more than 200 species of non-fusulinid forams; 78 genera and over 191 species of brachiopods; 23 genera and 97 species of ammonoids; and 69 genera and 288 (including 61 unidentified) species of bivalves.

The Upper Permian stratigraphy in South China can be widely correlated with that of other major areas of Tethys, such as the Dzhulfa region of Transcaucasia, the Kuh-e-Ali Bashi region in

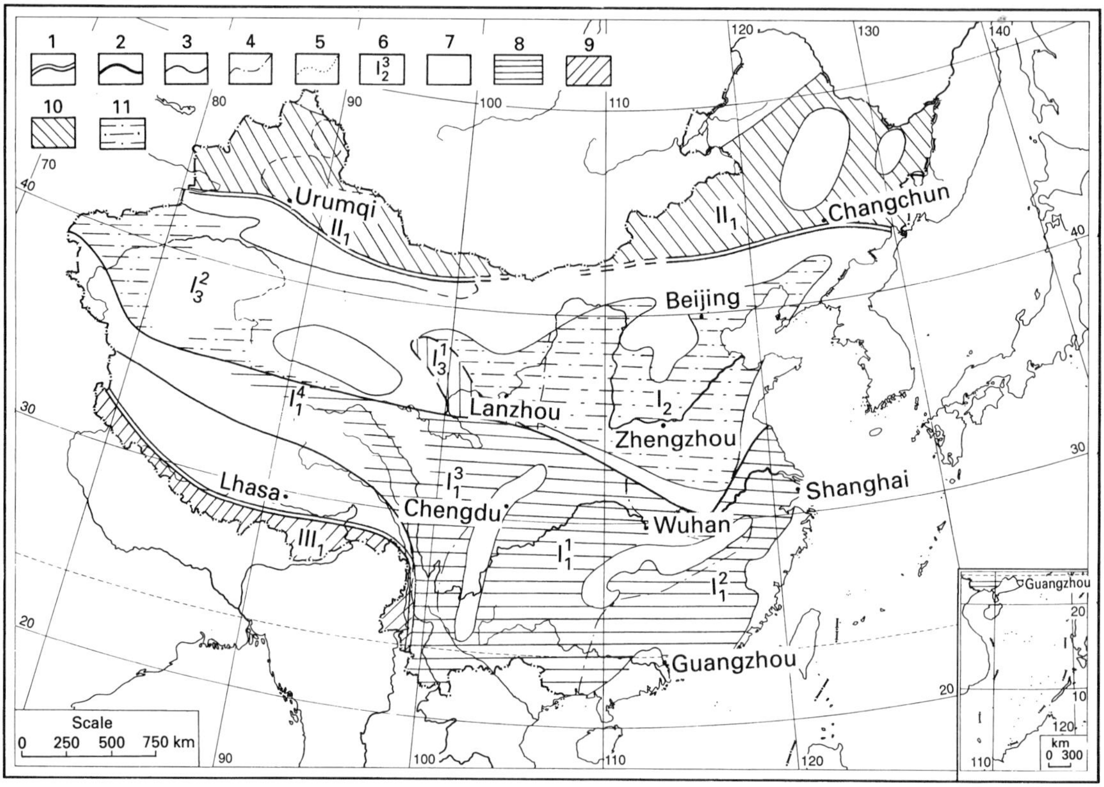

Fig. 8.5 Map of the Late Permian palaeobiogeographical provincialization of China. 1, Boundary of realms; 2, boundary of regions; 3, boundary of provinces; 4, boundary of biomes; 5, boundary of communities; 6, roman digit = realm mark, subscript = province mark, superscript = biome mark; 7, Oldland. 8–10, Marine and paralic regions: 8, South China; 9, Gondwana; 10, Junggar–Xingan. 11, Terrestrial region (Late Permian and Mesozoic–Cenozoic), fluvio–lacustrine basins. I, Palaeo-Tethyan realm, Cathaysian region; I_1, South China province; I_1^1, Upper Yangtze biome; I_1^2 South-eastern China biome; I_1^3, Songpan biome; I_1^4, Qiangtang–Qamdo biome; I_2, North China province; I_3, Tarim–Qilian province; I_3^1, South Qilian biome; I_3^2, Tarim–Beishan biome; II, Boreal realm; Central Asia–Mongolia region; II_1, Tianshan–Xingan province; III, Gondwanan realm, Gondwanan Tethyan region; III_1 Himalayan province.

northern Iran and Abadeh in central Iran, and the Pamir and Kashmir areas in Central Asia. Just as described for the Early Permian, this region in the Late Permian contained the same unique Cathaysian fauna. It was suggested by Zhao Jinke *et al.* in 1965 that the ammonoid *Pseudotirolites–Rotodiscoceras–Pleuronodoceras* assemblage, found in the Changxing and Dalong Formations in southern China, be called the Cathaysian ammonoid fauna. After his study of the gastropods in Late Permian strata of western Guizhou, Wang Huiji (1982) argued that the snail assemblage with *Guizhouspira sequens*, *Euphemites indicus*, *Bellerophon* cf. *triangularis*, *Ptychobellerophon*, *Turbinilopsis*, *Paleostylus*, and *Cylindritopsis* is of Tethyan affinity, being especially closely related to the fauna of South-East Asia; he thus named it the Cathaysian gastropod fauna. Similarly, Fang Zongjie (1985), in his study of the bivalve fauna of southern China, concluded that there existed a Cathaysian bivalve assemblage including *Tambanella* and *Hunanopecten*. Working on bryozoans, Ross and Ross (1990) also grouped a Cathaysian bryozoan assemblage in the central Tethyan regions (corresponding to our realms), which, in this chapter, roughly corresponds to the Cathaysian region.

Three provinces can be recognized in this region.

The South China province

This province inherited the characteristics of the Early Permian and continued to be characterized by the abundant and diverse biotas which have been subjected to detailed study. The biotas are still dominated by fusulinids and other forams, brachiopods, ammonoids, corals, conodonts, bivalves, gastropods, ostracods, and plants, and also contain some vertebrate fossils. Fusulinids are abundant in the whole South China province and can be widely correlated in the province with the lower *Codonofusiella* zone and upper *Palaeofusulina* zone. Corals are represented by *Liangshangnophyllum* and *Lophophyllidium*, and brachiopods by *Oldhamina* and *Tyloplecta*. The ammonoid assemblage of the Changxingian stage is also well developed, with typical elements of Cathaysian fauna such as *Rotodiscoceras*, *Pseudotirolites*, and *Pleuronodoceras*, which are also distributed throughout the province.

The South China province can be subdivided into four biomes based on different depositional and biotic facies.

The Upper Yangtze biome

The lower stage (Wujiapingian) of the Upper Permian is made up of coal-bearing clastic facies (the Longtan Formation) or limestone facies (the Wujiaping Formation). Basic volcanic rocks (Emeishan basalt) are widely distributed in its south-western part, i.e. in western Guizhou, western Guangxi, and eastern Yunnan. Its upper stage (Changxingian) is dominated by limestone facies (the Changxing Formation). However, its south-western margin is always dominated by clastic facies. This biome chiefly contains brachiopods, fusulinids and other forams, and conodonts. The ammonoid *Anderssonoceras* assemblage of the lower stage of the province does not occur in the biome.

The South-eastern China biome

The lower stage of the Upper Permian is made of coal-bearing clastic facies (the Longtan Formation). In the upper stage a sequence of siliceous rocks (the Dalong Formation) is found in many places, and its south-eastern margin is always dominated by clastic facies. In this biome, the main groups are brachiopods, ammonoids, and fusulinids. In the lower stage the ammonoid *Anderssonoceras* assemblage occurs, but conodonts are very few.

The two above biomes are only roughly divided. Their characteristics cannot be distinguished clearly because changes in the Longtan–Wujiaping and Dalong–Changxing facies often occur within short distances.

The Songpan biome

A large area of medium–basic eruptions is found in the Late Permian epoch in the north-eastern, southern, and western parts of this biome. Other areas are dominated by clastic deposits, but limestone predominates in the bordering parts of Qinghai and Gansu in the north. The rocks are generally metamorphosed and very thick, and the lithofacies are rather complex. However, the biotas in this biome are basically coincident with those of the Upper Yangtze biome, being rich in brachiopods, fusulinids, and plants, with rare ammonoids. There is little information about the western or Hohxil section of this biome.

The Qiangtang–Qamdo biome

There is only a little information for this biome. It consists mainly of coal-bearing, terrestrial, littoral, and shelf clastic facies intercalated with volcanics (according to the information obtained from the Shuanghu area of northern Tibet and Tuoba and Leiwuqi near the Qamdo area, eastern Tibet). All the plants found belong to the typical Cathaysian flora of South China type, commonly called the *Gigantopteris nicotianaefolia–Lobatannularia multifolia* assemblage (Li Xingxue *et al.* 1982*a*, *b*). Also, the fusulinids (*Codonofusiella*, *Palaeofusulina*, and *Gallowayinella*), brachiopods, and corals are the same as those found in South China. For example, the plants *Gigantonoclea lagrelii*, *Pecopteris* *echinata*, *Annularia pingloensis*, and *Compsoperis contracta* have been collected from the non-marine strata in Shuanghu and Lanju (northern Tibet). The coal measures of Tuoba in Qamdo and Rijuechaka in Shuanghu yield diverse ferns, reflecting a wet tropical climate, with accompanying faunas dominated by warm-water elements such as the fusulinids *Palaeofusulina* and *Galowayinella*, and the brachiopods *Peltichia*, *Oldhamina*, and *Cathaysia*. A typical element of Palaeo-Tethyan type, *Waagenophyllum*, is found in the Upper Permian in Duoma of Rutog County, north-western Tibet.

There has been no report about the Upper Permian in most of the areas in the Gangdise–Nianqingtanggula region. Only microfossils and poorly preserved bivalves of the genus *Guizhoupecten* are reported to have been found in sandstones and shales interbedded with limestones in the Mengla Group near Lhasa. *Guizhoupecten* is a common Wujiapingian bivalve in South China, and is also reported from central Iran and south-western North America.

In the Late Permian of either Qiangtang–Qamdo or Gangdise–Nianqingtanggula a temperate–coal biota is no longer found. It is certain that the Qiangtang– Qamdo region was already on the southern margin of Eurasia, judging either from the depositional facies, floras, or palaeomagnetic data. Also, it is very probable that Gangdise–Nianqingtanggula had already separated from Gondwana and had became a component of the tropical Tethys realm. Recently, Nie *et al.* (1990) suggested a Permian rifting away from Gondwana of Cimmeria, including Gangdise– Nianqingtanggula.

The North China province

During the Late Permian, non-marine deposition occurred in almost all of the province. The strata contain Cathaysian flora, terrestrial bivalves, ostracods, conchostracans, spores and pollen, and vertebrates. Strata of the North China province are represented by the upper Shihezi Formation and the Shiqianfeng Formation in Shanxi and Shaanxi, including mainly variegated sandstones interbedded with siltstones and mudstones, and a *Gigantonoclea hallei–Lobantannularia ensifolia* assemblage. Most of the pollen is of finely carinated bisaccates, such as *Protohaploxypinus*, *Striatopodocarpites*, and *Lueckisporites*. The bivalves include *Abiella* and *Microdontella*, and the vertebrates include *Shanxisaurus xuecunensis*. This province differs from the Tarim–Qilian province in that there is no mixture with Angaran flora here.

The Tarim–Qilian province

This province embraces the Tarim, Qilian and Beishan areas, and characterized mainly by a mixture of Cathaysian flora with varying amounts of Angaran flora. *Compsopteris wongi*, *Taeniopteris multinervis*, *Cordaites wongi*, and *Cordaites principalis*, discovered at Celeaqike along the southern margin of Tarim, can be correlated with the upper Shihezi Formation of North China, but there are few index Cathaysian elements, such as *Gigantopteris* and *Gigantonoclea*. At a higher horizon *Callipteris* and *Comia* are found in Duwa and Qiemo on the southern margin of the Tarim basin, and *Callipteris zeilleri* and *C*. cf. *altaica* are found at Biyouleguzigangou and the Kelasuhe River to the north of Baicheng on the northern margin. These are the common elements of the Angaran flora. In the northern marginal area, *Schizoneura manchuriensis*, a common member of Cathaysian flora, has been found, thus indicating of mixture (Wu Shaozu 1983). In the Qilian region, Gansu, the Cathaysian flora *Lobatannularia heianensis* assemblage is mixed with an Angaran flora *Iniopteris sibirica–Zamiopteris glossopteris* assemblage (Wang Dexu *et al.* 1984). It is significant that a representative of the Cathaysian flora, *Lobatannularia heianensis*, appears in the same sample with a representative of the Angaran flora, *Pursongia*, in a gully on the banks of the Yangluhe River in the northern Qilian Mountains. Similarly, elements of the Angaran flora are also found in the Upper Permian of the Beishan area.

This province is subdivided into two biomes. The Tarim–Beishan biome consists totally of non-marine facies. The southern Qilian biome is of marine facies, with the Nuoyinhe Formation containing the brachiopods *Urushtenia* and *Megaderbya*, but a typical Tethyan coral–fusulinid–brachiopod assemblage is absent.

The Boreal realm

During Late Permian times the boundary of the Angaran floral distribution seems to have undergone some changes; in the eastern section, along the Xar Moron River line, the Angaran flora was distributed to its north and Cathaysian flora distributed to its south during the Late Permian, but in the western section along the Ebinur–Gaxunnur line the boundary shifted at different times during the Late Permian. In the early Late Permian the demarcation line was along the Ebinur line, north Xinjiang was a region of Angaran flora and south Xinjiang was an area of Cathaysian flora. However, in the middle–late Late Permian, north Xinjiang still belonged to the Angaran floral region, but south Xinjiang (Wu Shaozu 1983) and Qilian and Beishan (Wang Dexu *et al.* 1984) became an area of mixed flora. Accordingly, the Tarim–Qilian province of the Palaeo-Tethyan realm can be attributed to the Boreal realm if the southern boundary of Angaran flora is taken as a standard. In this chapter we have taken this mixed region to be a province of the Palaeo-Tethyan realm, considering the historical inheritance of provincialization. There is no mixture of Cathaysian flora in the Boreal realm, but some elements of Cathaysian flora are found in the Hongshan Formation in the south-eastern Xiaoxingan Mountains in Heilongjiang.

The organisms in this realm are all non-marine, those of especial abundance being plants, ostracods, conchostracans, bivalves, palynomorphs, and vertebrates.

Late Permian deposition in the western section can be represented by the lower Canfanggou group in Xinjiang. Terrestrial organisms are represented by microspores, megaspores, plants, bivalves, and vertebrates (Yang Jiduan *et al.* 1987). Early Late Permian strata are most developed along the eastern and southern margins of the Junggar and Yining Basins, yielding the important elements of the Angaran flora—*Zamiopteris* and *Noeggerathiopsis latifolia*,—which may be correlated with the Kuznetz Basin in Central Asia that belongs to the same realm. Gymnosperm pollen is dominant, represented by *Hamiapollenites*, *Striatobietites*, and *Cordatina*. The bivalve *Palaeoanodonta solonensis* and the vertebrate *Kunpania scopulusa* are important elements. In middle–late Late Permian the flora of northern Xinjiang is represented by the *Comia–Iniopteris* assemblage. *Callipteris*, *Zamiopteris*, and *Pargongia* extend widely. They can be found around the Junggar Basin and can be correlated with the Late Permian flora of the Kuznetz Basin.

In the eastern section, areas north of the Xar Moron River contain a typical Angaran flora, including *Comia*, *Leptophyllum*, *Iniopteris*, *Nephropsis*, *Zamiopteris*, and *Tychtopteris*, which have also been discovered at localities to the east of the Songliao Plain, such as Yichun, Shenshu, Mishan Counties, and the Yanbian region of Jilin. The bivalves *Palaeomutula keyserlingi* and *Palaeoanodonta* cf. *longissima* occur locally.

The Himalayan province of the Gondwanan Tethyan region, the Gondwanan realm

It was generally believed in the past that there were few Upper Permian strata in southern Tibet because, as was widely held in China, in the whole Himalayas the highest Permian deposits, including the Malung shale of Ladakh, the *Cyclolobus* bed and Kuling shale of Spiti and Kumaon, and Quburika Formation of the Qomolangma area, are at most basal Upper Permian. However, some reports about the Himalayan Upper Permian have been published recently. Yu and Dai (1985) have identified the Changxingian conodonts *Neogondolella subcarinata changxingensis* and *N. deflecta* in the Baga Formation (Selong Group) in Shengmi village of Pazuoqu, Tingri Co., southern Tibet. Rao and Zhang (1985) reported that the Nimaloshiza Formation in Tulong village of Nyalam County, southern Tibet contains 22 genera of palynomorphs, which are close to the Late Permian association in India and Australia. Liang (personal communication) has collected *Palaeofusulina* and Late Permian corals in Mayang of Zanda County. These discoveries, together with the Late Permian fauna found by Nakazawa *et al.* (1975, 1981) in the upper Zewan Formation of Kashmir, indicate that higher horizons of Upper

Permian are likely to exist in Himalayan areas, including southern Tibet. The biota of this part of the sequence are still unclear in detail, but they have a close relationship to Gondwana according to data on ammonoids (*Cyclolobus* in Spiti and Kumaon), palynomorphs (Nimaloshiza Formation in Tulong), and some individual bivalves (*Atomodesma*? in Tulong). On the other hand, however, there are reports about the presence of South China-type brachiopods (a *Chonetella nasuta* assemblage in the Baidingpu limestone of Kangma, for example) and warm-water fusulinids and forams (for instance *Reichelina* and *Colaniella* in the Chitichun limestone). Hence, there is the possibility that a faunal mixture in distribution and horizon may have existed, similar to that of the late Early Permian. At present, this realm is tentatively divided with the Yarlung Zangbo River as its northern boundary.

8.3 THE RELATIONSHIP BETWEEN PERMIAN GEOTECTONICS, PALAEOGEOGRAPHY, PALAEOCLIMATE, AND PALAEOBIOGEOGRAPHICAL PROVINCIALIZATION

The framework of Early Permian geotectonics and palaeogeography in China can be summarized as 'three blocks and two belts'. The northern block is the Junggar residual marine basin and its eastern equivalents, the southern block is the shallow continental shelf of Tibet, and the middle block is the China Oldland and its marginal sea, all being more or less stable. Between them existed two geosynclinal belts, the Three Rivers (Sanjiang)–Karakorum Geosynclines and the Tianshan–Xingan Geosynclines, or Northern Geosynclines, both being submarine and tectonically active. This framework controlled Early Permian palaeobiogeographical provincialization: the northern block becoming the Boreal realm, the southern one the Gondwanan realm, and the central one the Cathaysian Tethyan region. The southern boundary of the Cathaysian Tethyan region moved over Three Rivers– Karakorum and reached Yarlung Zangbo in the late Early Permian.

This kind of provincialization is also consistent with palaeoclimate zonation. The northern block, with sediments consisting mainly of clastics interbedded with marine beds and containing the Angaran flora, shows the characters of the Boreal realm, which indicates that it existed in a temperate–cool climatic zone in a high northern latitude. The sediments of the middle Cathaysian Tethyan region are composed mainly of carbonate deposits with bioherms and colonial corals, containing the Cathaysian flora, coal or gypsum, and halite, which indicate that this region existed in tropical and subtropical areas of low latitude near the equator. The southern block, especially the Himalayan areas, contains tillites and deposits bearing cool–cold-water organisms from the Late Carboniferous to the Early Permian, which reflect the fact that this region still remained in a high latitude region of the southern hemisphere in Early Permian times. All of these coincide with the Permian palaeogeographical positions of different parts of China provided by palaeomagnetic data. The mixed or transitional areas between the three realms are believed to have resulted mainly from the widening of the warm-climate belts, with the rise of temperature in the Late Permian and equatorward drift of Siberia(?) and formerly Gondwanan microplates, not excluding, of course, the effect of ocean currents on these mixed or transitional areas.

It is worth noting that in an area around Yanji, north-eastern China, there are many South China-type fossils, among which *Misellina claudiae* is the most characteristic, mixed with cool-water brachiopods of the Boreal realm. A similar mixture is also seen in the Kitakami area of Japan and in Sikhote-Alin of the Russian Far East. Some authors, on the basis of palaeomagnetic data, suggest that these areas were located within a dozen degrees of northern latitude or several degrees of southern latitude in the Permian, or even belonged to a Pacifica continent (Yancey 1976; Chang Zhengkun 1984). Permian fusulinids of Tethyan type are found in North American terrains. The existence of terrain movements around Yanji must be established. Of course, the effects of warm currents moving north-eastwards from Palaeo-Tethys cannot be excluded.

Both cases imply the existence of a Pacific continent or Palaeo-Pacific Ocean. Another phenomenon worth noting is that the Early Permian ammonoid *Timorites* seems to have been distributed on both sides of the Pacific, since it is found in Timor, Himalaya, south-western China, Honshu (Japan), Sikhote-Alin, and south-western North America (Sheng Huaibin 1984). Early Permian Tethyan fusulinids also spread to western North America (Bostwick and Nestell 1967). All these examples appear to illustrate the fact that the circum-Pacific province that characterized the palaeobiogeographical provincialization in Mesozoic and Cenozoic times had already begun to exist in the Permian.

REFERENCES

Asama, K. (1985). Permian to Triassic floral change and some problems on the paleobiogeography, parallelism, mixed floras and origin of the angiosperms. In *The Tethys, her paleogeography and paleobiogeography from Paleozoic to Mesozoic* (ed. K. Nakazawa and J. M. Dickins), pp. 199–218. Tokai University Press.

Bai Yunhong, Chen Guoliang, Sun Qingge, and Sun Yuhang (1985). Late Paleozoic polar wander path for the Tarim Block and tectonic significance. *Seismology and Geology* **7**, (1), 71–80 (in Chinese).

Bostwick, D. A. and Nestell, M. K. (1967). Permian Tethyan fusulinid faunas of the Northwestern United States. In *Aspects of Tethyan Biogeography* (ed. C. Adams and D. V. Ager), Systematics Association Publication No. 7, pp. 93–102.

Chaloner, W. G. and Creber, G. T. (1988). Fossil plants as indicators of late Palaeozoic plate positions. In *Gondwana and Tethys*, (ed. M. G. Audley-Charles and A. Hallam), Geological Society Special Publications, No. 37, pp. 201–10.

Chang Zhengkun (1984). Sino-Korean Block and Yangtze Block as part of the Pacifica Continent in the late Paleozoic. *Bulletin of the Chinese Academy of Geological Sciences* **9**, 45–54 (in Chinese).

Chen Guoliang, Bai Yunhong, and Li Yongan (1983). Palaeomagnetism of Lower Permian in the Wushi-Aksu area of Xinjiang. *Seismology and Geology* **5**, (4), 12 (in Chinese).

Cui Kexin and Zhen Yongyi (1984). On the Palaeoclimates from Sinian to Permian in China. *Scientia Geologica Sinica* **1**, 1–11 (in Chinese).

Drewrey, G. E., Ramsay, A. T., and Smith, A. G. (1974). Climatically controlled sediments, the geomagnetic field, and trade wind belts in Phanerozoic time. *Journal of Geology* **82**, (5), 539–45.

Fan Jiasong, Ma Xing, Zhang Yinben, and Zhang Wei, (1982). The Upper Permian reefs in West Hubei, China. *Facies* **6**, 1–14 (in Chinese).

Fan Yingnian (1985). A division of zoogeographical provinces by Permo-Carboniferous corals in Xizang (Tibet), China. *Contribution to the Geology of the Qinghai-Xizang (Tibet) Plateau* **16**, 87–106, Geological Publishing House (in Chinese).

Fang Zongjie (1985). A preliminary study of Cathaysia faunal province. *Acta Palaeontologica Sinica* **24**, (3), 344–49 (in Chinese).

Frakes, L. A. (1979). *Climate throughout geologic time*. Elsevier, Amsterdam.

Guo Shengzhe (1980). A general account of early Permian rugose coral fauna from the geosynclinal region of northeastern China. *Bulletin of the Chinese Academy of Geological Sciences*, 5, **1**, (1), 103–14 (in Chinese).

Han Jianxiu (1980). The morphology, evolution and distribution of the genus *Monodiexodina* and its allied genera. *Bulletin of The Chinese Academy of Geological Science*, 5, **1**, (1), 90–102 (in Chinese).

Han Jianxiu (1981). General accounts on the fusulinid fossil zones of Early Permian Age in the northern part of Northeast China. *Geological Review* **27**(6), 539–42 (in Chinese).

He Xinyi and Weng Fa (1983). New material of Early Permian corals from Ali (Ngari), Northern Xizang (Tibet). *Earth Science – Journal of Wuhan College of Geology* **19**, 69–78 (in Chinese).

Hou Hongfei, Xiang Liwen, Lai Caigen, and Lin Baoyu (1979). Advances in the Palaeozoic stratigraphy of Tianshan-Xingan region. *Acta Stratigraphica Sinica* **3**, (3), 175–87 (in Chinese).

Huang Benhong (1982). Permo-Carboniferous terrestrial deposits and paleogeographic features in the northern part of Northeast China. *Geological Review* **28**, (5), 395–401 (in Chinese).

Huang Benhong (1983). On late Late Paleozoic palaeophytogeographic regions of eastern Tianshan-Hingan foldbelt and its geological significance. *Contributions for the Project of Plate Tectonics in Northern China* **1**, 138–55 (in Chinese).

Ishii Ken-ichi, Okimura, Y., and Ichikawa, K. (1985). Notes on Tethys biogeography with reference to Middle Permian fusulinaceans. In *The Tethys, her paleogeography and paleobiogeography from*

Paleozoic to Mesozoic, (ed. K. Nakazawa and J. M. Dickins), pp. 139–56. Tokai University Press.

Khramov, A. N., Petrova, G. N., and Pechersky, D. M. (1981). Palaeomagnetism of the Soviet Union. In *Paleoreconstruction of the continents*, (ed. M. W. McElhinny and D. A. Valencio), pp. 177–94. American Geophysical Union and the Geological Society of America.

Li Li, Gu Feng, and Li Wenguo (1984). Early Permian productids from Xi Ujimqin Qi, Nei Monggol Autonomous Region. *Professional Papers of Stratigraphy and Paleontology* **11**, 71–128. Geological Publishing House (in Chinese).

Li Wenguo (1983). A study of Early Permian brachiopods from Zhesi (Jisu Honguer), Inner Mongolia. *Regional Geology of China* **4**, 52–61 (in Chinese).

Li Xingxue and Yao Zhaoqi (1983). Carboniferous and Permian floral provinces in East Asia. In *Palaeobiogeographic provinces of China*, pp. 74–82. Science Press, Beijing (in Chinese).

Li Xingxue, Yao Zhaoqi, and Deng Longhua (1982a). An early Late Permian flora from Toba, Qamdo District, Eastern Xizang. *Palaeontology of Xizang* **5**, 17–44. Science Press, Beijing (in Chinese).

Li Xingxue, Yao Zhaoqi, Zhu Jianan, Duan Shuying, and Hu Yufan (1982b). Late Permian plants from northern Xizang. *Palaeontology of Xizhang* **5**, 1–16. Science Press, Beijing (in Chinese).

Li Xingxue, Wu Yiming, and Fu Zaibin (1985). Preliminary study on a mixed Permian flora from Xiagangjiang of Gerze District, Xizang and its paleobiogeographic significance. *Acta Palaeontologica Sinica* **24**, (2), 150–70 (in Chinese).

Liang Dingyi, Nie Zatong, Guo Tieying, Zhang Yizhi, and Wang Weiping (1982). Some advances in research on Permian and Triassic stratigraphy and paleontology of northern Ali, Xizang (Tibet) Autonomous Region. *Geological Review* **28**, (3), 245–6 (in Chinese).

Liang Dingyi, Nie Zatong, Guo Tieying, Xu Baowen, Zhang Yizhi, and Wang Weipin (1983). Permo-Carboniferous Gondwana-Tethys facies in southern Karakoran Ali, Xizang (Tibet). *Earth Sciences – Journal of Wuhan College of Geology* **19**, 9–28 (in Chinese).

Liang Xiluo (1981). Early Permian cephalopods from northwestern Gansu and western Nei Monggol. *Acta Palaeontological Sinica* **20**, (6), 485–99 (in Chinese).

Lin Baoyu (1984). Lower Permian stratigraphy and coral faunas from both flanks of Yarlung Zangbo River in central-southern Xizang (Tibet). In *Sino-French Co-operative Investigation in Himalayas (1980)* (ed. Li Guangcen and J. L. Mercier), pp. 63–84. Geological Publishing House, Beijing (in Chinese).

Lin Jinlu, Fuller, M., and Zhang Wenyou (1985). Preliminary Phanerozoic polar wander path for the North and South China blocks. *Nature* **313**, 444–9.

Liu Benpei and Cui Xinsheng (1983). Discovery of Eurydesma fauna from Rutog, northwest Xizang (Tibet), and its biogeographic significance. *Earth Science – Journal of Wuhan College of Geology* **19**, 1–8 (in Chinese).

Liu Xiaoliang (1982). The Permian bryozoa and their biogeographic region in the northern part of northeast China. *Geological Review* **28**, (3), 193–200 (in Chinese).

Mu Xinan (1981). Upper Permian calcareous algae from western Guizhou. *Acta Palaeontologica Sinica* **20**, (1), 33–47 (in Chinese).

Nakamura, K., Shimizu, D., and Liao Zhuoting (1985). Permian palaeobiogeography of brachiopods based on the faunal provinces. In *The Tethys, her paleogeography and paleobiogeography from Paleozoic to Mesozoic*, (ed. K. Nakazawa and J. M. Dickins), pp. 185–98. Tokai University Press.

Nakazawa, K. and Kapoor, H. M. (ed.) (1981). The Upper Permian and Lower Triassic faunas of Kashmir. *Palaeontologica Indica* **46**, 204.

Nakazawa, K., Kapoor, H. M. Ishii, L., Bando, Y., Okimura, Y., and Tokuoka, T. (1975). The Upper Permian and the Lower Triassic in Kashmir, India. *Memoirs of the Faculty of Science, Kyoto University, Series of Geology and Mineralogy* **47**, (2), 106.

Nakazawa, K. and Dickins, J. M. (ed.) (1985). *The Tethys, her paleogeography and paleobiogeography from Paleozoic to Mesozoic*. Tokai University Press.

Nie Shangyou, Rowley, D. B., and Ziegler, A. M. (1990). Constraints on the location of the Asian microcontinents in Palaeo-Tethys during the Late Palaeozoic. *Geological Society of London Memoir*, **12**, 397–410.

Pruner, P. (1987). Palaeomagnetism and palaeogeography of Mongolia in the Cretaceous, Permian and Carboniferous – preliminary data. *Tectonophysics* **139**, 155–67.

Rao Rongbiao and Zhang Zhenggui (1986). A discovery of Permo-Triassic transitional fauna in the Qomolangma Feng Area: its implications for the Permo-Triassic boundary. *Xizang (Tibet) Geology* **1**, 19–31 (in Chinese).

Ross, J. R. P. and Ross, C. A. (1985). Carboniferous and Early Permian biogeography. *Geology* **13**, 27–30.

Ross, J. R. P. and Ross, C. A. (1990). Late Palaeozoic bryozoan biogeography. *Geological Society of London Memoir* **12**, 353–62.

Rui Lin (1981). The Permian system of Yangzi stratigraphic province. *Journal of Stratigraphy* **5**, (4), 262–75 (in Chinese).

Scotese, C. R., Bambach, R. L., Barton, C., Van der Voo, R., and Zielger, A. M. (1979). Paleozoic base maps. *Journal of Geology* **87**, 217–27.

Sengör, A. M. C. (1984). *The Cimmeride orogenic system and the tectonic of Eurasia*. Geological Society of America, Special Paper No. 195, pp. 1–82.

Sengör, A. M. C., Altiner, S., Cin, A., Ustaomer, T., and Hsu, K. J. (1988). Origin and assembly of the Tethyside orogenic collage at the expense of Gondwanaland. In *Gondwana and Tethys* (ed. M. G. Audley-Charles and A. Hallam), Geological Society of London, Special Publication, No. 37, pp. 119–82.

Sheng Huaibin (1984). Late Lower Permian ammonoids from Xiukang Formation, Lhaze District, Xizang (Tibet). In *Himalayan Geology*, Vol. 2, pp. 219–48. Geological Publishing House, Beijing, China.

Sheng Jinzhang and Wang Yujing (1981). Permian fusulinids from Xizang with reference to their geographical provincialism. *Acta Palaeontologica Sinica* **20**, (6), 546–51 (in Chinese).

Sun Qiaoli and Zhang Linxin (1985). Early Permian fusulinids from Alge Mountain of Xinjiang. *Acta Palaeontological Sinica* **24**, (5), 503–10 (in Chinese).

Walter, H. (1985). *Vegetation of the Earth* (3rd edn). Springer-Verlag, Berlin.

Wang Dexu, He Bo, and Zhang Shuling (1984). On a mixed Cathaysia and Angara flora from Qilianshan region. In *Scientific Papers on Geology for International Exchange. Prepared for the 27th International Geological Congress* Vol. 1, pp. 13–24. Geological Publishing House, Beijing (in Chinese).

Wang Huiji (1982). Late Permian gastropods from Heshan of Laibin, Guangxi Province. *Acta Palaeontologica Sinica* **21**, (4), 495–501.

Wang Huiji and Xie Yuhua (1980). The Late Permian–Early Triassic gastropods of western Guizhou. In *The Late Permian coal-bearing strata and fauna of western Guizhou and eastern Yunnan*, (ed. Nanjing Institute of Geology and Palaeontology), pp. 195–240, Science Press, Beijing.

Wang Naiwen (1984). The rise of Gondwana research in China and some key problems on paleobiogeography related to plate tectonics. *Bulletin of the Chinese Academy of Geological Sciences* **10**, 46–58 (in Chinese).

Wang Yujing and Mu Xinan (1980). Some new observations on the Permian Biostratigraphy of the Himalayan province in Southern Xizang. *Journal of Stratigraphy* **4**, (2), 145–51 (in Chinese).

Wang Yujing, Sheng Jinzhang, and Zhang Linxin (1981). Fusulinids from Xizang of China. In *Palaeontology of Xizang*, Vol. 3, pp. 1–81. Science Press, Beijing (in Chinese).

Wu Shaozu (1983). Preliminary study on the Permian phytogeography of Xinjiang. In *Palaeobiogeographic provinces of China*, pp. 91–9. Science Press, Beijing (in Chinese).

Wu Wangshi (1975). The coral fossils from the Qomolangma Feng region. In *A report of a scientific expedition in the Mount Jolmo Lungma Region (1966–1968), Palaeontology*, Fasc. 1, pp. 83–128. Science Press, Beijing (in Chinese).

Wu Wangshi and Zhang Yansheng (1979). Late Palaeozoic rugose corals from Batang and Yidun, Western Szechuan. *Acta Palaeontologica Sinica* **18**, (1), 25–38 (in Chinese).

Wu Wangshi, Liao Weihua, and Zhao Jiaming (1982). Palaeozoic rugose corals from Xizang. In *Palaeontology of Xizang*, Vol. 4, pp. 107–151. Science Press, Beijing (in Chinese).

Xu Ren (1976). On the discovery of a *Glossopteris* flora in southern Xizang and its significance in geology and palaeogeography. *Scientia Geologica Sinica* **4**, 324–31 (in Chinese).

Yancey, T. E. (1976). Permian positions of the northern hemisphere continents as determined from marine biotic provinces. In *Historical biogeography, plate tectonics, and the changing environment*, (ed. J. Gray and A. Boucot), pp. 239–49. Oregon State University Press.

Yang Jiduan, *et al.* (1984). Late Permian and Early Triassic continental strata and fossil assemblages in Northern China. In *Scientific Papers on Geology for International Exchange. Prepared for the 27th International Geological Congress*, Vol. 1, pp. 88–96. Geological Publishing House (in Chinese).

Yang Jingzhi and Lu Linhuang (1983). Geographic distributions of the Permian bryozoa and its fossil assemblages in China. In *Palaeobiogeographic provinces of China*, pp. 83–90. Science Press, Beijing (in Chinese).

Yang Zunyi, Yin Hongfu, Wu Shunbao, Yang Fengqing, Ding Meihua, and Xu Guirong (1987). *Permian–Triassic boundary stratigraphy and fauna of South China*. Geological Publishing House, Beijing, (in Chinese).

Yin Jixiang and Guo Shizeng (1976). On the discovery of the stratigraphy of Gondwana facies in northern slope of the Qomolangma Feng in southern Xizang, China. *Scientia Geologica Sinica 4*, 291–322 (in Chinese).

Yu Hongjin and Dai Jinye (1985). Late Permian and Early Triassic conodont faunas and their significance in stratigraphy in south Xizang (Tibet), China. *Contribution to the Geology of the Qinghai–Xizang (Tibet) Plateau* **16**, 139–51. Geological Publishing House, Beijing (in Chinese).

Yu Jianzhang, Lin Yintang, and Huang Zhuxi (1981). Early Permian corals from Central Jilin. *Acta Palaeontologica Sinica* **20**, (6), 273–84 (in Chinese).

Yu Jianzhang (Yu C. C.), Liu Yingdang, Shi Yen, Huang Zhuxi, and Yu Xueguang (1983). *Carboniferous and Permian corals*. Jilin People's Publishing House, Jilin (in Chinese).

Zhan Lipei, Chen Yuling, Li Li, and Chen Binguei (1982). The Permian system of China. In *Stratigraphy of China*, Vol. 1 *An outline of the stratigraphy in China*, pp. 219–62. Geological Publishing House, Beijing (in Chinese).

Zhang Shanzhen and He Yuanliang (1985). Late Palaeozoic palaeo-phytogeographic provinces in China and their relationships with plate tectonics. *Palaeontologia Cathayana* **2**, 77–86.

Zhang Shouxin and Ching Yukan (1976). Late Paleozoic brachiopods from the Mount Jolmo Lungma Region. In *A report of a scientific expedition in the Mount Jolmo Lungma Region (1966–1968), Palaeontology* Fasc. 2, 159–271 (in Chinese).

Zhao Jinke, Liang Xiluo, Zou Xiping, Lai Caigen, and Zhang Ridong (1965). *Fossil Cephalopoda in China*. Science Press, Beijing, (in Chinese).

Zhao Jinke, Liang Xiluo, and Zheng Zhuoguan (1978). Late Permian cephalopods of South China. *Palaeontologia Sinica* (*New Series B*) **12** (in Chinese).

Zhao, X. X. and Coe, R. S. (1989). Tectonic implications of Permo-Triassic palaeomagnetic results from North and South China. In *Deep structure and past kinematics of accreted terranes*, (ed. J. W. Hillhouse). American Geophysical Union Monograph 5, pp. 267–83.

Ziegler, A. M. (1990). Phytogeographic patterns and continental configurations during the Permian Period. *Geological Society of London Memoir* **12**, 363–79.

9 TRIASSIC

Yin Hongfu and Lin Qiuxian

9.1 TRIASSIC PALAEOBIOGEOGRAPHICAL PROVINCIALIZATION OF THE WORLD

The Triassic Pangaea was indented by a wedge-shaped Tethyan Ocean spanning tropical and subtropical areas, with the Antarctic and Siberian Platform located in southern and northern high latitudes respectively. Configuration of the eastern Tethys is modified in this chapter on the basis of new data from palaeomagnetic, palaeogeographical, and geotectonic researches. This region displayed the interspersion of microcontinents and micro-ocean basins, reminiscent of modern South-East Asia. North China (centred at Taiyuan, Shanxi) and Yangtze (centred at Emei, Sichuan) were at the palaeolatitudes of about 15°N and 2°N, respectively, during the Late Permian. However, North China had a rapid northward migration during the Permian and collided with Siberia–Mongolia in the Early Triassic. Later, Yangtze started to move northwards and merged with North China by the late Middle Triassic, which implies either the merging of Yangtze with North China or intraplate crust shortening and final welding of the two. Their Late Triassic palaeolatitudes being about 22°N and 32°N respectively. The Himalaya area was in a southern warm–temperate zone and the palaeolatitude of Tulong was 24.2°S (Zhu Zhiwen *et al.* 1981), but this region migrated into the subtropical belt during the Late Triassic. Gangdise was also located at 20–30°S, but its faunas were quite different from those of the Himalayan area (McElhinny *et al.* 1981; Lin Jinlu 1987). The Yarlung Zangbo and Nujiang belts experienced remarkable tension and expansion during the Triassic, and both Gangdise and Qiangtang shifted northwards considerably during the Triassic–Jurassic interval (Table 14.1). A majority of microcontinents, i.e. Qiangtang–Qamdo (Q), Gangdise (G), Indochina (Ic), and South China (SC), were located near the palaeoequator (Fig. 9.1). The southern margins of North China (NC) and

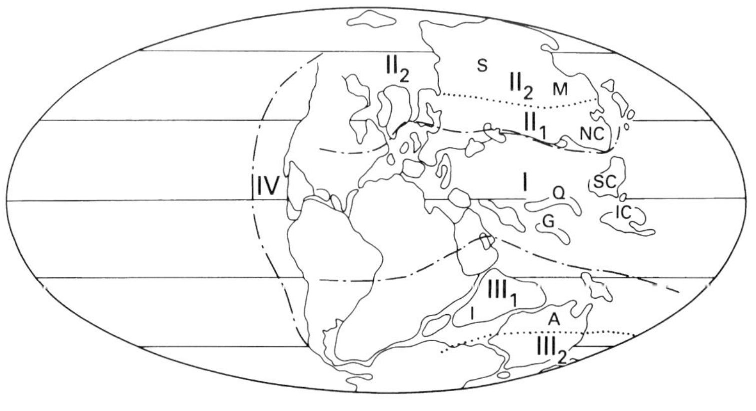

Fig. 9.1 Map of the Triassic palaeobiogeographical provincialization of the world. I, Tethyan realm; II_1, Boreal realm, central Laurasian region; II_2, Boreal realm, north Laurasian and Boreal sea region; III_1, Gondwanan realm, Gondwanan Tethys, and central–south Gondwanan region; III_2, Gondwanan realm, Austral Sea region; IV, East Pacific realm; A, Australia; G, Gangdise; I, India; Ic, Indochina; Q, Qiangtang; M, Mongolia; NC, North China; S, Siberia; SC, South China.

Tarim, as well as the northern margin of the Indian subcontinent, entered low latitudes.

The Early and Middle Triassic biogeographical provincialization of the world can be expressed in terms of three realms, namely the Tethyan realm, the Gondwanan realm (which is subdivided into a Gondwanan Tethyan region and a Southern Gondwanan region), and the Boreal realm (which is subdivided into a Central Laurasian region and a Northern Laurasian and Boreal Sea region). In the Late Triassic the Gondwanan Tethyan region became a part of the Tethyan realm. The eastern Pacific is established as an independent realm because of its peculiar fauna (Table 9.1).

In the Tethyan realm there flourished tropical marine biotas and tropical plants such as the *Voltzia–Pleuromeia* and *Dictyophyllum–Clathropteris* (abbreviated as *D.–C.* henceforth) floras. Oxygen isotopic analysis of ammonoid shells give an average seawater temperature of 23–27 °C (Zakharov 1973), confirming warm conditions. *Voltzia* has also been reported from the Urals and Siberia; *Pleuromeia*, from northern Siberia (Olenek, Taimyr), India (Rewa), and Australia (Gosfold Bed). *Dictyophyllum* and *Clathropteris* have both been reported from middle palaeolatitudes in the Triassic, so only when they are found abundantly can they be regarded as tropical–subtropical.

During the Early and Middle Triassic the Gondwanan Tethyan region lacked tropical–subtropical biotas, and was characterized by ammonoids of middle to high latitudes (such as *Otoceras*, *Svalbardiceras*, and *Gurleyites*) and bivalves of low diversity (Yin 1981), thus reflecting a warm temperate climate. Marine faunas of the Central Laurasian region show a mixture of temperate (Boreal) and tropical (Tethyan) elements; meanwhile, its floras contain considerable temperate elements, typified by the Late Triassic *Danaeopsis-Bernoullia* (*Symopteris*) (abbreviated as *D.–B.* henceforth) flora. Different authors regard the nature of the *D.–B.* flora as temperate, temperate–subtropical, or subtropical. We hold this flora to be temperate because it is distributed north of the Early–Middle Triassic evaporite belt, and mostly in the less humid continental areas, and because it contains more ginkgoales, fewer cycadales, no dipteridaceans, and a few Gondwanan elements with doubts, such as *Glossopteris?* (Tongchuan, Shaanxi Province) and *Gondwanidium*? (Yushe, Shaanxi Province). Moreover, the coexistence of this flora with any tropical Tethyan fauna has not yet been reported. The northern distribution area of this flora extends into northern Eurasia, being cool temperate, and is named the Siberian flora (Dobruskina 1982). Its southern distribution extends over central Eurasia, from Greenland, southern Sweden, Kazakhstan, South Xinjiang, and the Qilian Mountains to North China and western Liaoning. This belt corresponds roughly to the European–Chinese flora of Dobruskina (1982), and is mixed with tropical *D.–C.* elements; therefore it is probably warm temperate. This is in accordance with the flourishing of warm temperature vertebrates: *Lystrosaurus*, *Chasmatosaurus*, cynognathids, and kannemeyerians along the Dvina River (Russia), Xinjiang, and Shanxi (see Chapter 1).

The Southern Gondwanan and the Northern Laurasian regions are similar in that both consist of peculiar marine biotas and temperate floras: for the former the *Dicroidium* flora, and for the latter the Early–Middle Triassic Angaran (Korvonchanian) flora and Late Triassic Ural (Siberian) flora. According to palaeomagnetic data, both already extended into polar regions but were not growing under cold weather conditions because the Triassic poles were not capped by ice. Oxygen isotopic data of ammonoid shells from the Boreal Sea give a temperature of 14 °C (Zakharov 1973). The biota and other criteria show that both regions were, at most, in a cool temperate belt. The Southern Gondwanan region had vertebrates similar to those in Central Eurasia, such as *Lystrosaurus*, *Chasmatosaurus*, *Cynognathus*, and *Kannemeyeria* (South Africa, southern South America, India). This reptile fauna seems to have had a bipolar distribution in the red beds along the middle latitudes of two hemispheres (Cox 1973) and to be representative of a temperate, especially a warm temperate, climate.

Late Triassic palaeobiogeography underwent alterations due to the equatorward drift of former Gondwanan fragments, expansion of the humid climatic zone, and effects from the Palaeo-Pacific

Table 9.1 Triassic palaeobiogeographical provincialization of the world

Climatic zone	*Realm*	*Region*	*Early Triassic*	*Middle Triassic*	*Late Triassic*
N. temperate zone	N. Laurasian realm	N. Laurasia	Korvonchan flora Maltzev flora	Ural flora	(*Cladophlebis* assemblage)
		Boreal Sea	*Otoceras* *Keyserlingites* *Cloraia stachei*	*Amphipopanoceras* *Nathorstites*	*Entomonotis* *Sirenites* Spiriferinidae
N. warm temperate zone	C. Laurasian realm	C. Laurasia	*Lystrosaurus*	*Kannemeyeria* *Shanxiconcha–Sibireconcha*	
		Marginal Sea	*Voltzia–Pleuromeia*	*Danaeopsis–Bernoullia*	
				mixed by tropic–subtropic forms	
Tropical–subtropical	Laurasia–Tethyan realm	S. Laurasia	*Voltzia–Pleuromeia*	*Annalepis*	*Dictyophyllum–Clathropteris*
		Europe–Asia–Tethys Sea	*Owenites* *Tirolites* *Claraia aurita* *Eumorphotis inaequicostata*	madreporarians *Paraceratites* *Protrachyceras* *Daonella lommeli* *Diplopora*	madreporarians *Megalodontidae* Halorellinae Heteromorphic ammonites
S. warm temperate zone	Gondwanan Tethyan realm	Gondwana–Tethys Sea	*Otoceras* *Svalbardiceras* *Claraia decidens*	*Keyserlingites* *Tulongospirifer strachei* *Daonella lommeli*	attr. Laur. Tethys realm *Serania* *Nuculana misolensis* *Hagabirhynchia*
S. cool temperate zone	S. Gondwanan realm	C.–S. Gondwana	*Glossopteris* *Lystrosaurus*	*Dicroidium* *Kannemeyeria*	
		Maorian Sea	*Claraia–Eumorphotis* assemblage rare	*Amphipopanoceras* *Praegonia*	*Hokonuia* *Maoritrigonia* Spiriferinidae
Tropical temperate zone	E. Pacific realm				*Entomonotis subcircularis* *Gnomohalorites–Neohimavavites*

Plate. Concurrent with the disappearance of evaporitic belts, the humid *D.–C.* flora expanded from low palaeolatitudes northward to Bogeslovsk in the Urals and Scoresby Sound in East Greenland, and southward to St. Juan Province (Argentina). La Tenera (Chile), Lesotho (South Africa), and Ipswich (Australia). The tropical–subtropical marine biota, symbolized by the *Burmesia–Indopecten* fauna, spanned the whole of Tethys, not only along Eurasian Tethys but also along Gondwanan Tethys, so that in the Late Triassic the southern margin of Tethys—the Gondwanan Tethyan region—was classified within the tropical–subtropical Tethyan realm. Because the climates changed from latitudinal zonation towards a more uniform distribution, and the Palaeo-Pacific began to act as a biogeographical unit (Section 14.1), the Palaeo-Pacific faunas brought over the continental margins by the circum-Pacific transgression were usually longitudinally distributed along coasts. Subsequent terrain movement or strike-slip shift strengthened the longitudinal distribution of originally tropical faunas along the western coast of North America. Consequently, the eastern Pacific areas have similar biotas from south to north, and constitute an independent realm consisting of peculiar ammonite heteromorphs (Wiedmann 1973), hexacorals (Stanley 1979), and gastropods (Haas 1953), as well as monotids and trigoniids (Westerman 1973; Kobayashi and Tamura 1983*b*).

9.2 TRIASSIC PALAEO-BIOGEOGRAPHICAL PROVINCIALIZATION OF CHINA

9.2.1 EARLY AND MIDDLE TRIASSIC PALAEOBIOGEOGRAPHICAL PROVINCIALIZATION

Early and Middle Triassic biotas of China are among the most diversified in the world. As determined up to 1984, the marine organisms consisted of some 20 groups, including: bivalves, Early Triassic *c.* 182 species (28 genera), Middle Triassic *c.* 280 species (57 genera); ammonites, Early Triassic *c.* 350 (97), Middle Triassic *c.* 280 (80); brachiopods, Early Triassic *c.* 35 (14), Middle Triassic *c.* 180 (57); conodonts, forams, sponges, hexacorals, worms, snails, nautiloids, ostracods, bryozoans, crinoids, ophiuroids, fish, aquatic reptiles, and various kinds of algae. Terrestrial organisms included plants, vertebrates, estheriids, ostracods, charophytes, and insects.

Biogeographically, the Lower and Middle Triassic of China is attributed to ten biomes, four provinces, four regions, and three realms (Fig. 9.2, Tables 9.2, 9.3, 9.4).

The Gondwanan Tethyan region of the Gondwanan realm

Aspects of this region have been discussed in Section 9.1. Areas south of the Yarlung Zangbo River belong to the Himalayan province of this region, which has the following two characteristics.

1. Its biota differs from that of the tropical Tethyan realm, but it is similar to other provinces of the Gondwanan Tethyan region and even to the Boreal Sea region of the Boreal realm.

(a) Many Early and Middle Triassic ammonites, such as *Otoceras*, *Svalbardiceras*, *Gurleyites*, *Budhaites*, and *Anacrochordiceras*, are elements either mainly distributed in the Boreal Sea or restricted to the Gondwanan Tethyan region, thus displaying bipolarity. Many typical tropical Tethyan ammonites, such as *Tirolites*, *Dinarites*, *Columbites*, *Subcolumbites*, *Paraceratites*, *Prohungarites*, and *Xenoceltites* are lacking here. This shows an extra-tropical aspect.

(b) Its Early Triassic *Claraia* assemblage resembles that of Gondwanan Tethyan areas, such as Kashmir and Spiti, while differing from that of Eurasian Tethys, such as South China and even North Tibet (Yin 1981). Late Early Triassic bivalves, including *Claraia*, *decidens*, *Leptochondria minima*, and *Eumorphotis* aff. *huancangensis*, recently discovered in the Ngari area of western Tibet, resemble those from south Qilian (Yang *et al.* 1983) and the Primoria area of central Laurasia, thus again displaying bipolarity.

(c) As in other areas of Gondwanan Tethys, this province lacks hermatypic hexacorals, reef-building red algae, and reef-associated forams.

2. The diversity of organisms in this province is significantly lower than that of the tropical Tethyan realm. For example, the number of Early Triassic ammonite species is about one-quarter, and the number of bivalve species less than one-tenth, that of Youjiang province, tropical Tethyan realm. Allowing for other factors, including a smaller amount of

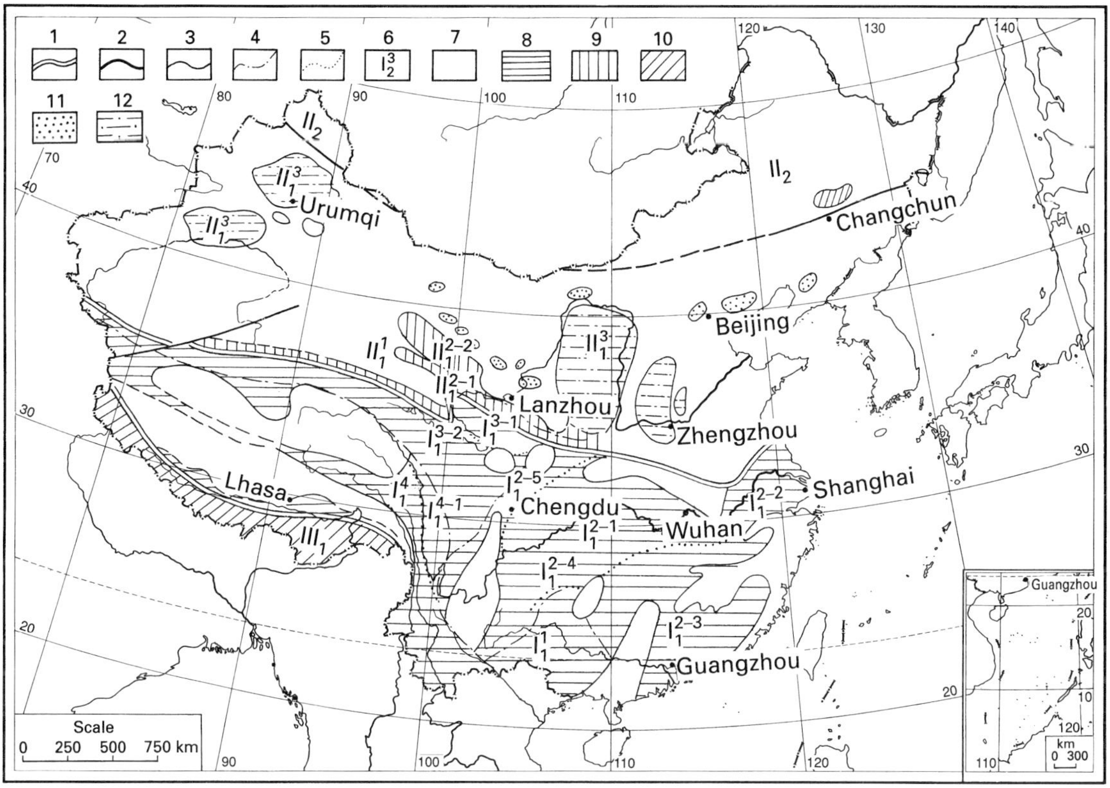

Fig. 9.2 Map of the Early and Middle Triassic palaeobiogeographical provincialization of China. 1, Boundary of realms, 2, boundary of regions; 3, boundary of provinces; 4, boundary of biomes; 5, boundary of communities; 6, roman digit = realm mark, subscript = province mark, superscript = biome mark; 7, Oldland. 8–10 Marine and paralic regions: 8, South China; 9, North China; 10, Gondwana. 11–12, Terrestrial regions (Late Permian and Mesozoic–Cenozoic); 11, intermontane basins; 12, fluvio–lacustrine basins. I, Tethyan realm, Cathaysian region; I_1, Southern China province; I_1^1, Youjiang biome; I_1^2, Yangtze biome; I_1^{2-1}, Upper Yangtze; I_1^{2-2}, Lower Yangtze; I_1^{2-3}, South-eastern China; I_1^{2-4}, Qingyan–Ceheng; I_1^{2-5}, Longmen Mountains; I_1^3, Bayan Har–Southern Qinling biome; I_1^{3-1}, Southern Qinling communities; I_1^{3-2}, Bayan Har communities; I_1^4, Tibet–West Yunnan biome; I_1^{4-1}, Yidun; II, Laurasian realm; II_1, Central Laurasian region, North China–North-western China Province; II_1^1, Burhan Budai biome; II_1^{2-1}, Northern and Central Qinling Communities; II_1^{2-2}, Southern Qilian–Southern Ordos communities; II_1^3, North China–Xinjiang biome; II_2, Northern Laurasian region, Altay–Xingan province; III, Gondwanan realm; Gondwanan Tethyan region; III_1, Himalayan province.

research, the temperate climate of this area, due to the relatively high latitudes, must be largely responsible for this phenomenon.

The South China province of the Cathaysian Tethyan region, Tethyan realm

The Tethyan realm is subdivided into a Mediterranean Tethyan region to the west and a Cathaysian Tethyan region to the east of the Caucasus. The Cathaysian Tethys is distinct from the Mediterranean Tethys in its Early and Middle Triassic bivalves (Yin 1981; Kobayashi and Tamura 1983*a*) and ammonites (Tozer 1980), Middle Triassic bryozoans (Hu 1984), and Late Triassic *Burmesia–Indopecten* fauna (see Table 9.6).

During the Early and Middle Triassic, areas of the Cathaysian Tethyan region within China, i.e.

Table 9.2 Correlation chart of the Triassic of China

			Gondwanan Tethyan region	Cathaysian Tethyan region				Pacific Tethyan region		Central Eurasian region			
			Himalaya	Northern Tibet	Bayan Har	Yangtze	Youjiang	SE China	Jiling and Heilongjiang	Western Qinling	Southern Qilian	Northern China	Northern Xinjiang (Junggar)
Triassic	Upper	Norian and Rhaetian	Derirong	Quehala Gr. or Mailonggang Gr.	Yidung Gr. or Jeza Gr.	Xujiahe Xiaotangzi	Fulong'au	Anyuan Gr. or Xiaoping Gr.	Nanshuangyashan Gr.	?	Mole Gr.	Yanchang	Xiaochuangou Gr.
			Quelonggonba			Ma'antang	Pingdong			Dengdengqiao			
		Carnian											
	Middle	Ladinian			Zagashan Gr.	Tianjingshan	Hekou			Qinyu		Tongchuan	Karamay
		Anisian	Tulong Gr.	Caquepu Gr. or Xihu Gr.		Leikoupuo	Baifong	Yangjia Gr.		Guojiashan	Junzihe	Ermaying	
	Lower	Olenekian			Bocigou	Jialingjiang	Luoluo Gr.	Daye Gr.		Maresongduo	Jianghe	Heshanggou	Shaofanggou
		Induan				Feixianguan				Zalishan	Xiahuancang	Liujiagou	Jiucaiyan

Table 9.3 Triassic palaeobiogeographical provincialization of China

Realm	*Region*	*Province*	*Biome*	*Early Triassic*	*Middle Triassic*	*Late Triassic*
Boreal	N. Laurasia	Altay–Xingan		*Cornia* *Dawinula*	*Shanxicocha Sibirceconcha* *Danaeopsis–Bernoullia* flora	
Boreal	C. Laurasia	N. China–NW China	N. China–Xinjiang	*Pleuromeia* *Ginkgoites*	*Shanxiconcha* *Bernoullia*	*Danaeopsis–Bernoullia* Flora mixed with *Dictyophyllum–* *Clathropteris* elements
Boreal	C. Laurasia	N. China–NW China	S. Qilian –Ordos	*Eumorphortis iwamowi* *Glyptophiceras polare*	*Lenotropites* *Leptochondria illyrica*	Same as above
Boreal	C. Laurasia	N. China–NW China	W. Qinling	*Svalbardiceras* *Gurleyites*	*Lenotropites* *Arctohungarites*	Incorporated with Bayan Har province of the Cathaysian Tethys region
Boreal (T_{1-2}) / (T_3)	C. Laurasia	N. China–NW China	Burhan Budai	*Eumorphotis mulitformis* *Claraia stachei*	*Lenotropites* *Leiopyllites*	Incorporated with Bayan Har province of the Cathaysian Tethys region
Boreal (T_{1-2})	C. Laurasia (T_{1-2}) / Pacific Tethys (T_3)	Jilin–Heilong-jiang		Included in C. Laurasia realm		*Entomonntis ochotica* *Dictyophyllam–* *Clathroperis*
Tethys	Cathaysian Tethys (T_{1-2}) / Pacific Tethys (T_3)	SE China		Included in the Yangtze province of the Cathaysian Tethys region		*Bakevelloides hekiensis* *Guangdongella*
Tethys	Cathaysian Tethys	S. China	Yangtze	*Pseudoclaraia wangi* *Clara aurita* *Tirolites*	*Progonoceratites* *Leptochondria illyrica*	*Costatoria napengensis* *Yunnanophorus* *Dictyophyllum*
Tethys	Cathaysian Tethys	S. China	Youjiang	*Subcolumbites* *Columbites*	*Daonella indica* *Daonella ignorbilis* *Paraceratites*	Included in Yangtze province
Tethys	Cathaysian Tethys	S. China	Bayan Har	*Subvishnuites* *Cordillerites*	*Daonella pichleri* *Daonella grabens*	*Halobia rugosoides* *Halobia comata*
Tethys	Cathaysian Tethys	S. China	W. Yunnan–Xizang	*Claraia clarai* *Procarnites kokeni*	*Coststoria gold-fussi mansuyi* *Balatonites*	*Indopecten* *Costatoria napengensis*
Gondwana (T_3)	Gondwanan Tethys	Himalaya	Yarlung Zangbo R.		Undivided	*Monotis salinaria* Halobiidae
Gondwana (T_3)	Gondwanan Tethys	Himalaya	Qomo-langma	*Otoceras*, *Claraia griesbachi* *Claraia decidens*	*Daonella indica* *Nudirostralina griesbachi*	*Tibetites* *Burmesia–* *Indopecten* fauna

Table 9.4 Occurrence of the main Triassic fossils in China

	Gondwanan Tethyan region				Cathaysian Tethyan region			C. Laurasian region					Boreal region	
	Qomolangma	Yarlung Zangbo R.	Tibet–E. Yunnan	Bayan Har	Yangtze	Youjiang R.	SE China	Jilin–Heilongjiang	Burhan Budai	W. Qinling	S. Qilian–S. Ordos	Xinjiang–N. China	Altay–Xingan	Age
Otoceras	+				?									Early–Middle Triassic
Svalbardiceras	+									+				
Gurleyites	+									+				
Claraia decidens	+		+		+	+								
Nudirostralina griesbachi	+													
Daonella indica						+								
Claraia aurita			+	+	+	+								
Pseudoclaraia wangi			+	+	+	+	+							
Eumorphotis inaequicostata			+	+	+	+	+							
Tirolites					+	+								
Annalepis					+									
Costatoria goldfussi mansuyi			+	+	+	+								
Leptochondria illyrica			+		+	+								
Daonella mousoni				+		+								
Daonella grabens				+		+								
Thamnasteria			+		+									
Montlivaultia			+		+									
Tremadictyon					+									
Nudirostralina subtrinodosi					+									
Lenotropites									+	+	+			
Glyptophiceras polare											+			
Eumorphotis multiformis			+	+	+	+	+		+	+	+			
Parantiptychia											+			
Pseudospiriferina tsinghaiensis								+	+	+				
Pleuromeia												+		
Voltzia				+								+		
Ginkgoites												+		
Cornia											+	+		

between the Kunlun Mountains–Qinling Mountains–Dabie Mountains line and the Yarlung Zangbo River, can be attributed to one province—the South China province. It has the following characteristics:

1. Numerous organic groups (>20 groups), a high diversity, and a high endemic percentage. For example, 372 species of 84 genera of Early and Middle Triassic bivalves have been reported here,

Table 9.4 (*cont.*)

	Gondwanan Tethyan region				Cathaysian Tethyan region				C. Laurasian region				Boreal region	
	Qomolangma	Yarlung Zangbo R.	Tibet–E. Yunnan	Bayan Har	Yangtze	Youjiang R.	SE China	Jilin–Heilongjiang	Burhan Budai	W. Qinling	S. Qilian–S. Ordos	Xinjiang–N. China	Altay–Xingan	Age
Dawinula					+							+	+	Late Triassic
Danaeopsis fecunda											+	+	+	
Bernoullia zeilleri					+						+	+	+	
Sibireconcha					+		+					+	+	
Shanxiconcha					+		+					+	+	
Anatibetites	+		+		+									
Halobia yunnanensis	+		+	+	+	+								
Burmesia lirata	+		+		+	+								
Monotis salinaria		+	+											
Halobia himalayaensis		+	+				+							
Thisbites			+		+									
Trachyceras			+		+	+								
Costatoria napengensis			+		+	+								
Dictyophyllum *Clathropteris*			+	+	+	+	+	+	+	+	+	+	+	
Guangdongella						+								
Bakeveloides hekiensis						+								
Oxytoma mojsisovicisi						+								
Jiangxiella						+								
Entomonotis ochotica							+							

including *c*. 40 per cent endemic species; many tropical taxa of the Laurasian Tethyan realm are known here (Tables 9.3, 9.4)

2. Early Triassic *Claraia* and *Eumorphotis* are represented here by several dozens of species each, equipped with varied forms and sculptures. During the Middle Triassic in this province there flourished reef-builders such as red algae (He *et al.* 1980), hexacorals, sphinctozoans, and dasycladaceans (Deng and Kong 1984), accompanied by very rich faunas of brachiopods, bivalves, and gastropods, as in Qingyan, Guizhou. Many biotas dispersed along Eurasian Tethys and can be traced from South China to as far as the Alps and Germany, e.g. the early Early Triassic *Claraia stachei–C. aurita–Eumorphotis multiformis* bivalve assemblage; the middle and late Early Triassic *Tirolites–Eumorphotis inaequicostata* assemblage; the early Middle Triassic *Daonella* (*Longidaonella*) assemblage and the late Middle Triassic *Paraceratites trinodosus* assemblage of the open-sea type; the Middle Triassic *Progonocer-*

atites assemblage and *Myophoria* (*Costatoria*) *goldfussi* assemblage of the restricted-sea type; and crinoids, ostracods, etc. (Kristan-Tollmann 1988). The palaeolatitude (*c.* 20°N), widespread evaporites, and bioherms (Yang *et al.* 1983) of this east–west belt indicate that it belonged to the tropical–subtropical zone.

3. Early and Middle Triassic floras (Ye 1979; Zhou and Li 1979) were similar to contemporaneous European floras, but different from those of the Gondwanan and Boreal realms.

4. There was a narrow belt along the northern margin of this province, i.e. from the Lower Yangtze Valley to Huangshi (Hubei), Guangyuan (Sichuan), Dangchang (Gansu), and Maduo (Qinghai), where non-tropical bivalves, brachiopods, and cephalopods infiltrated. For example, a few ammonites typical of mid- and high latitudes are found here: *Otoceras*? (Changxing; Wang 1984), *Hypophiceras* (Changxing, Guangyuan, Daye), *Stenopopanoceras*, *Pearylandites*, *Arctohungarites* (Dangchang; Yin *et al.* 1992), *Anotoceras*, *Wordieoceras*, and ?*Dunedinites* (Maduo; Wang and Chen 1984).

The Early and Middle Triassic of the South China province can be subdivided into four biomes.

The Yangtze biome

Except for the Youjiang area, this biome covered the whole of South China proper, west to the Kam-Dian Oldland, which was at that time covered by a tropical, epicontinental sea. These marine faunas are diversified, have been considerably investigated, and are dominated by shallow benthos, mainly bivalves (totalling more than 160 species in the Early Triassic) and ammonites, together with brachiopods, conodonts, gastropods, corals, forams, and algae (Fig. 9.3). Palaeogeographically this biome was separated by the Jiangnan Oldland into two parts. In the south-eastern part, or the transgressed area on the Lower Palaeozoic Huanan Orogenic Belt, both carbonates and clastics were deposited. Later this part was gradually uplifted into land from south-east to north-west during the Middle Triassic.

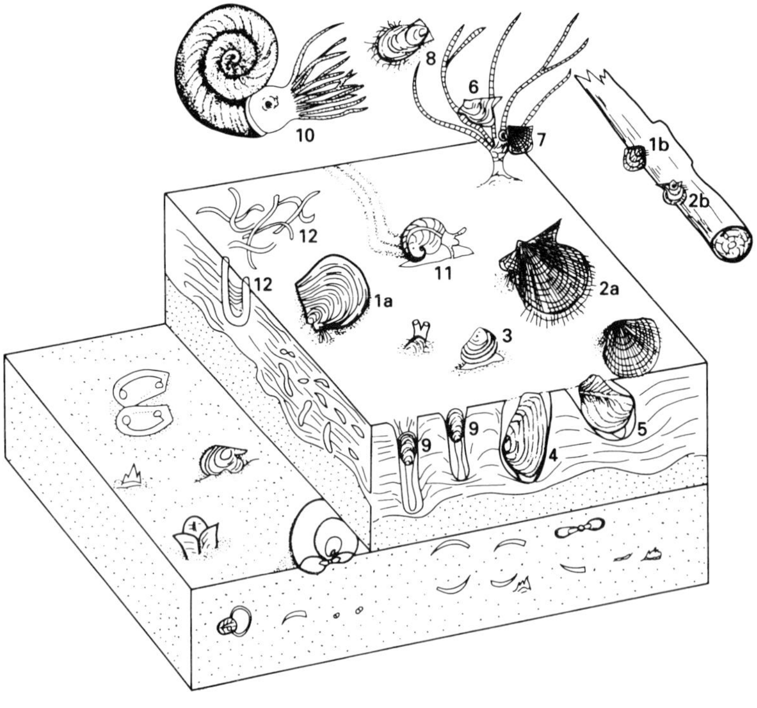

Fig. 9.3 Reconstruction of the *Claraia–Ophiceras* community of the Early Triassic Feixianguan Formation of the Yangtze biome. 1, *Claraia* (a) and its juvenile (b); 2, *Eumorphotis* (a) and its juvenile (b); 3, *Nucula*; 4, *Unionites*; 5, *Neoschizodus*; 6, *Pteria*; 7, *Towapteria*; 8, *Entolium*; 9, *Lingula*; 10, *Ophiceras*; 11, *Bellerophon*; 12, worm traces.

The north-western part, or the Yangtze Platform, was mainly covered by the wide, shallow, level-bottomed Upper Yangtze epicontinental sea. Its southern (Guiyang, Zhenfeng, and Kaiyuan), north-western (Longmeng Mountains), and north-eastern (north Lower Yangtze) sides were relatively depressed. Beyond the southern and north-western depressions, along the demarcating margins with the Youjiang and Songpan Basins, lay shallow-water shelf-margin belts where organism diversity reached a climax. In all these subareas there developed different lithofacies and biofacies, thus constituting the following five community complexes

The Upper Yangtze communities

The typical community sequence of this area is: early Early Triassic normal neritic, benthic *Claraia–Eumorphotis* community; late Early Triassic normal–hypersaline, neritic, benthic *Pteria* cf. *murchisoni–Tirolites* community; and Anisian hypersaline, restricted neritic *Leptochondria illyrica–Progonoceratites* community. The area experienced regression in the Ladinian, and the Ladinian neritic benthic–semipelagic *Protrachyceras–Halobia comatoides* community is only distributed in south-western Guizhou and south-eastern Yunnan. This area is specified by the dominance of benthos, abundance of bivalves and development of restricted hypersaline communities.

The South-eastern China communities

Early Triassic communities of this area resembled the Upper Yangtze ones, and were also dominated by benthic bivalves. However, during the Middle Triassic regression in this area terrestrial or littoral red clastics with scarce fossils were deposited. Even in the Early Triassic, its diversity was much lower, e.g. only 38 species of bivalves are known, which is about one-third that of the Upper Yangtze; moreover, there are a few elements occurring only in one of the two community groups, such as *Claraia annulata*, *C. bijiensis*, *Eumorphotis senta*, *E. orientalis* of the Upper Yangtze and *C. pingxiangensis*, *C. jiangxiensis*, *Pteria longiaurita* of South-east China.

The Lower Yangtze communities

An important characteristic of these communities is that the number of Early Triassic ammonite species is double the number of bivalve and brachiopod species, and pelagic types surpass benthics, although ammonites were mainly concentrated along the northern margin (Caoxian of Anhui). Another characteristic is the considerable endemism of these communities in the Early Triassic, which amounts to 37 per cent in ammonoites and 24 per cent in bivalves, including two new genera *Periclaraia* and *Guichiella*. The third notable characteristic is that, accompanying the regression, the Ladinian community was composed of brackish–freshwater fauna. These characteristics seem nearly enough for an independent biome, but we retain this community in the Yangtze biome on account of the dominance of benthos in a large part of the area and the existence of hypersaline communities in the late Early Triassic and Anisian.

The Qingyan–Ceheng communities

This shelf-margin belt between the Yangtze Platform and Youjiang Basin developed into a barrier due to the build up of carbonate, which extended intermittently for 500 km. Many authors held that the chief reef-builders were red algae, but lately Fan Jiasong (personal communication) firmly denied its organic origin and interpreted the alleged red algae to be caliche structure. Anyhow, there were abundant fore-reef or pre-buildup benthos, among which gastropods (Yin and Yochelson 1983) and brachiopods (Yang and Xu 1964) account for more than half the total species number.

The Longmen Mountains communities

This area includes the shelf-margin belt between the Yangtze Platform and Bayan Har Basin, the elongate depression east of it, and, tentatively, the Yanyuan–Lijiang area, a narrow belt alongside the western margin of the Kam-Dian Oldland. The Early Triassic and Anisian communities were analogous to those of the Upper Yangtze, namely *Claraia–Eumorphotis*, *Pteria* cf. *murchisoni–Tirolites*, and hypersaline *Leptochondria illyrica–*

Costatoria goldfussi mansuyi, successively. However, during the Ladinian it did not undergo regression but formed a foram–*Halobia* cf. *kui* community. It differs from Qingyan–Ceheng in that it lacks a Middle Triassic build-up with its associated benthos; only in Zhen'an, south Shaanxi have stromatolitic limestones up to 50 m thick been discovered.

The Youjiang biome

This biome covers western Guangxi, south-eastern Yunnan, and south-western Guizhou, which comprises the Youjiang drainage. This is a rejuvenated basin, due to rift-and-sag from the Yangtze Platform, existing since Permian times. Biogeographically, ammonites (Yang *et al.* 1982, Table 2) and bivalves belonged to the Cathaysian Tethyan type and were similar to those of the Yangtze biome. However, most fossils are found in Lower Triassic thin-bedded micrites and Middle Triassic distal turbidites or basinal shales, pertaining to slope and open-sea pelagic biofacies. Plankton and nekton far surpass benthos, reaching a species ratio of 5 to 1. In the Early Triassic, ammonites dominated (>210 species of 64 genera), followed by bivalves (38 species of 18 genera). In the Middle Triassic, ammonites constituted 42 species of 28 genera, whereas 73 species of 16 genera of bivalves have been found, among which 58 per cent are supposedly pseudoplanktonic daonellids 50 (species). As to endemic Cathaysian Tethys elements, there are more than a dozen ammonite genera, such as *Eoacrochordiceras*, *Eodanubites*, *Preflorianitoides*, *Pseudowenites*, *Proharpooceras*, *Paradinarites*, *Mesohedenstroemia*, *Lingyunites*, and *Fengshanites*, and 245 bivalve species, including *Daonella guizhouensis*, *D. producta*, *D. ignorbilis*, *Halobia kui*, *H. rugosoides*, and *Posidonia bakevelliaeformis*.

The Bayan Har–Southern Qinling biome

This biome can be subdivided into the Bayan Har communities and southern Qinling communities.

The Southern Qinling communities

These developed on the northern carbonate shelf margin of the Yangtze Platform and Songpan Paraplatform during the Early and early Middle Triassic. They are mainly benthic communities, consisting of bivalves of more than 80 species (40 genera), gastropods (59 species, 33 genera), brachiopods (19 species, 15 genera), ammonoites (17 species, 14 genera), and conodonts. In Dangchang Co. they were aggregated together abundantly with dasycladaceans, but the aggregation did not build up bioherms as in the Qingyan area, probably due to the area's more northerly location. The faunas were basically Tethyan, but mixed with considerable temperate or Boreal elements; the bivalves and brachiopods were very similar to those of southern Qilian, northern and central Qinling, and even Primoria (southern Sikhote-Alin) of the Central Laurasian region. The Anisian ammonites are mostly known in northern and central Qinling, with Boreal types such as *Stenopopanoceras*, *Pearylandites*, and *Arctohungarites*. In these respects it seems reasonable to assign these communities, together with northern and central Qinling, to the Central Laurasian region. However, this area belongs to the Yangtze margin whereas northern and central Qinling belong to the North China margin, and there was a deep-sea basin with nannofossils and abyssal trace fossils separating the two (Yin *et al.* 1988); because of this, we have set this area within the South China province.

The Bayan Har communities

These occupied a triangle delineated by the Hohxil–Ganze–Litang line to the southwest, the Xiugou–Maqen line to the north, and the Longmeng Mountains–Kam-Dian deep fault to the south-east. The biofacies are analogous with those of the Youjiang biome, i.e. dominated by plankton and nekton and with few benthos. Fossils are found mainly in pelagic and distal turbitite facies, deposited on a wide, passive continental margin, formed through riftogenesis and depression of the Songpan Paraplatform from the Yangtze Platform since the late Early Permian. Palaeontological research is at an intial stage. So far there have been reports of Early Triassic ammonoites (16 species of 12 genera, of which most genera plus four species are known in the Youjiang and Yangtze biomes) and bivalves (27 species of 5 genera, of which most are also known from Yangtze). The Middle Triassic fauna was dominated by the supposedly pseudoplanktonic bivalves *Daonella*, *Halobia*, and *Posidonia* (totalling 19 species of 3 genera), together with ammonites, and also a few *Traumatocrinus*, corals, and plant remains. The biofacies and faunal composition are similar to the Youjiang biome; both belonged to the same South China province but were separated by the Yangtze biome into two independent biomes.

The Tibet and Western Yunnan biome

In the past decade a number of Early and Middle Triassic bivalves, ammonites, brachiopods, and other groups have been reported from Shuanghu (Qiangtang area), Lhasa, and Qamdo (Tibet), as well as from western Yunnan and southern Qinghai (Zhang *et al*. 1976; Wen 1979; Gu *et al*. 1980; Yang and Lian 1980; Wang and He 1981; Chen 1982; Sichuan Regional Survey *et al*. 1982). The data show that the ammonites belong to the Tethyan realm and were closely related to those of the Cathaysian Tethyan region, and the bivalves were typically of Cathaysian Tethyan type. For example, an Anisian ammonite fauna near Lhasa contains 16 known species, of which 15 are from the Tethyan realm; ammonites from the Qamdo area (Early Triassic, 14 species of 12 genera; Middle Triassic, 18 species of 7 genera) show the same aspect; all 20 known bivalve species of 10 genera from the Qiangtang area are Cathaysian Tethyan elements, identical with elements of the Yangtze biome. Middle Triassic bivalves from the Jinggu and Simao areas, western Yunnan, show the same features. To sum up, it is now safe to say that this is a neritic, benthic biome pertaining to the South China province of the Cathaysian Tethyan region, and is different from the contemporaneous Himalayan faunas.

Along the eastern segment of the Bangong Lake–Nujiang fault running through the central part of this area, there occurs the widespread Shading slate which has been recently proved to contain geosynclinal Triassic deposits, so this fault may subdivide the area into a northern and a southern biome. However, only a few benthic bivalves and corals have been reported and palaeontological data are not enough to confirm this possibility, so, tentatively, we retain the whole area in one biome.

The Yidun–Zhongdian area used to be allocated to the Bayan Har stratigraphic region. However, aspects of both its Early–Middle and Late Triassic faunas are different from those of the latter, which is dominated by plankton and nekton, both similar to the Tibet–Western Yunnan biome. It is either chiefly benthic or benthic plus pelagic with subequal importance. Furthermore, tectonically it has recently been clarified that a small Triassic ocean existed along the Ganze–Litang fault between Yidun–Zhongdian and Bayan Har. As a result, the Yidun–Zhongdian communities are classified as a part of the Tibet–Western Yunnan biome. Bivalves are most abundant, followed by brachiopods and gastropods. There are 33 bivalve species of 8 genera, most of which are elements of the Cathaysian Tethys. Most Middle Triassic bivalves, brachiopods, and gastropods are identical with those of the Yangtze biome.

The North and North-western China province of the Central Laurasian region, Boreal realm

The characteristics of this region and realm have been discussed in Section 9.1. In China, during the Early and Middle Triassic, it contained plants (137 species, 78 genera), conchostracans (23 species), non-marine bivalves (more than 20 species, 7 genera), ostracods (more than 10 species, 2 genera), charophytes, vertebrates, etc. Its marginal transgressed areas (southern Qilian and Burhan Budai) yielded marine bivalves (120 species, 40 genera), ammonoites (38 species, 28 genera), brachiopods (69 species, 27 genera), gastropods, ophiuroids, etc. In contrast to the mixed biota along the northern margin of Cathaysian Tethys, Central Laurasia as the southern part of the Boreal realm was a region infiltrated by tropical–subtropical Tethyan elements. On this basis, the Boreal realm was subdivided into a Central Laurasian region mixed with tropical elements and an unmixed Northern Laurasian region. There are not enough data to delineate a demarcation line clearly. However, because the biota were no more mixed with tropical elements in the Altay–Mongolia–Xingan region, the boundary is postulated to run along Yinshan–Altay Shan and extend westward throughout central Eurasia (Dobruskina 1982, 1988). The whole areas of this region within China are assigned to the North and North-western China province, which includes four biomes:

The North China–Xinjiang biome

This biome was totally composed of non-marine biotas, including plants (51 species, 28 genera), vertebrates (20 genera), insects (5 species, 5 genera), bivalves (20 species, 5 genera), ostracods (24 species, 5 genera), and conchostracans (32 species, 16 genera). The flora of this biome was dominated by temperate elements, among which were ginkgoales, more than 10 species (6 genera) (*Baiera*, *Sphenobaiera*, *Glossophyllum*, *Ginkgo*, *Ginkgoites*, and *Stenorachis*), conifers (*Araucarites*), and temperate ferns (*Danaeopsis* and *Bernoullia* (*Symopteris*)). There are also a few tropical elements, e.g. *Pleuromeia* and Voltzia, which occur frequently in Shaanxi, Shanxi, and Hebei. The lycopsid *Pleuromeia* is known from France, Germany, Spain, Italy, Austria, Central Russia, Lower Volga, the northern Caucasus, eastern Precaspian, Pamir, Japan, Mongolia, and Primoria. With a few exceptions in Siberia (Olenek River, Taimyr) and Gondwana (Rewa of India and the Gosfold Bed of Australia), this genus is principally distributed along the northern Tethys and southern Eurasia margins, and is thus tropical–subtropical. *Voltzia* is also a tropical element typical of western Europe, although in eastern Asia it also occurs as far south as Hainan Island and as far north as the Urals, and the Kuznetsk and Tongusk basins. The flora as a whole signifies a warm temperate climate.

Abundant vertebrates have been found in Shaanxi, Shanxi, southern Inner Mongolia, and Junggar. Representatives of the Early Triassic are *Lystrosaurus*, *Chasmatosaurus*, and *Fugusuchus*, and of the Middle Triassic, kannemeyerians, parakannemeyerians, and *Shanxisuchus*. Similar assemblages also occur in southern and eastern Africa, South America, India, and the southern Urals. These faunas used to be regarded as tropical, especially because they usually occur in red beds. However, because they are mainly distributed in a bipolar manner (Gondwana and Central Eurasia) with sporadic occurrences in Indochina (Lamprabang), and are accompanied by the above-mentioned flora, it is deemed here that they are warm temperate rather than tropical elements.

Ostracods of this province mainly belong to *Darwinula*. Many species of *Darwinula*, as well as the genus *Tungchuania*, are not known in South China, whereas most of the genera of South China, such as *Londaia* and *Hollinella*, are not known here. As regards the conchostracans, the *Xiangxiella–Protomonocarina* assemblage (23 species, 5 genera) of South China is also not known here. The non-marine bivalves are dominated by *Shaanxiconcha antiqua* and *S. longa* which are quite different from the bivalves of South China.

The Southern Qilian–Northern and Central Qinling biome

This is a marine biome which can be subdivided into the epicontinental Southern Qilian–Southern Ordos communities and the Northern and Central Qinling continental slope communities.

The Southern Qilian–Southern Ordos communities

These occupy the epeiric sea area over the southern margin of the North China Platform and the newly emerged Qilian Paraplatform, including southern Qilian, Xiang'a'dong (Jianzha County), the Tianshui–Qin'an area, and Qishan–Linyou area. Located in palaeolatitudes higher than South China, this area has temperate elements which had access through the eastward seaway to the Palaeo-Pacific. They are mixed with tropical elements from the Tethys to the south. The Early and Middle Triassic biotas are mainly benthic. Chief groups are benthic bivalves (120 species, 40 genera), brachiopods (69 species, 27 genera), ammonites (30 species, 12 genera), and gastropods (15 species, 14 genera). They possess the following characteristics:

1. Among the Early Triassic bivalves, 68 per cent of species are known from Cathaysian Tethys, while 58 per cent of species are known from Primoria, Russia (Yang *et al.* 1983). The Middle Triassic bivalves show a similar pattern. Early and Middle Triassic species endemism reaches 33 per

cent. One of the temperate aspects is that *Claraia* is neither diversified nor abundant here; instead there are such species as *Leptochondria minima*, *L. bittneri*, *L. virgalensis*, *Eumorphotis multiformis*, and *E. ivanowi*, which usually flourished in more northern areas. *Eumorphotis E. ivanowi* is a typical species in Primoria and Japan. The Middle Triassic myophoriids here are mostly smooth or weakly ornamented, while the Tethyan myophoriids include both smooth and radially ornamented forms.

2. Of the 17 Early Triassic ammonite species, 59 per cent (10 species) are identical with those of Siberia, Greenland, and the arctic part of North America, while the same percentage (10 species) is identical with those of Laurasian Tethys. The Middle Triassic ammonites are similar in this respect. *Glyptophiceras*, *Wordeioceras*, and *Lenotropites*, and many species, are common with those of the Boreal Sea.

3. Brachiopods show strong endemism, reaching 80 per cent at species level (Sun 1982).

The Northern and Central Qinling communities

These are developed along the southern slope of the platform and are dominated by pelagic forms, especially ammonites (Early Triassic, 31 species, 26 genera; Middle Triassic, 38 species, 23 genera, next are bivalves and brachiopods.) During the Early Triassic in central Qinling, deep-sea mudstones were deposited together with distal turbidites and contourites, in which have been found trace fossils of the *Nereites* ichnofacies and calcareous nannofossils, one of the earliest occurrencies so far known (Yin *et al.* 1992).

The Burhan Budai biome

In this biome have been discovered ammonites (88 species, 44 genera), bivalves (62 species, 42 genera), brachiopods (24 species, 17 genera), and other marine groups. Like in the Southern Qilian–Northern and Central Qinling biome, it is specified by mixed fauna of both Tethyan and Boreal forms. The bivalve and brachiopod faunas are very similar to the southern Qilian ones; in the Early Triassic there are a few claraiids, with *Claraia stachei* and *Eumorphotis multiformis* dominating, a pattern that is typical of the Boreal region (Yin 1981). The ammonites are more abundant than in southern Qilian, showing more offshore biofacies; especially notable is the remarkable mixed Boreal and Tethyan nature of early Anisian ammonites (Wang and Chen 1984). Boreal-type Triassic ammonites usually occur north of 55°N (present-day latitude). This area marks the extreme southerly distribution (33°N) of Boreal-type Triassic ammonoids (another location being in Nevada, USA).

The Altay–Xingan province of the Northern Laurasian region, Boreal realm

This terrestrial region was a highland with a few basins during Early Triassic times. In the Triassic it was wholly uplifted and left little fossil record. Early and early Middle (?) Triassic plants do not include *Pleuromeia* and *Voltzia*, but temperate ferns, which constitute the Korvonchanian flora, flourished in Tongusk and Kuznetsk. A few important elements, such as *Paracalamites* are also found along the Onon River (Transbaikalia) and in Jiutai of Jilin. Lower Triassic invertebrates are known as the Maltsev fauna, which is composed of the bivalve *Palaeanodonta* assemblage (*Palaeomutela*, *Microdontella*, '*Utschamiella*'), the estheriid *Pseudoestheria tomiensis-Glyptoasmussia wetlugensis* assemblage, and the ostracod *Darwinula* assemblage. The *Palaeanodonta* assemblage has been reported from Russia (Tonguss, Kuznetsk) and north-east China (Taohai'yinzi and Jiutai). *Cornia*, an important element of the estheriid assemblage, is abundant in Junggar, Jiutai, and Pechola of European Russia.

Cluster analysis of Early and Middle Triassic bivalves and brachiopod

Statistics based on records up to 1984 give the following data. Early Triassic bivalves, 182 species (27 genera); Middle Triassic bivalves, 80 species (57 genera), Middle Triassic brachiopods, *c.* 180 species (57 genera). Q-mode binary clusterings have been analysed by computer after the deletion of a few cosmopolitan species, using the stratigraphical region or subregion (according to the *National Chart of Stratigraphic Regionalization of China*) as a sample unit and adopting the Otsuka coefficient (Otc). The results are basically in accord with the provincialization indicated earlier in this chapter.

For Fig. 9.4a, after deletion of cosmopolitan, pan-Cathaysian, and unconfirmed species, 107 species of Early Triassic bivalves were analysed. This figure shows that, at the level of Otc = 0.3, three major groups can be distinguished, i.e. southern Qilian, Qomolangma, and South China, which includes eight subregions. This result corresponds to the tripartite regionalization of Lower Triassic marine bivalves, i.e. the marine belt of Central Eurasia, Gondwanan Tethys, and Cathaysian Tethys.

For Fig. 9.4b, 117 species of Middle Triassic bivalves were analysed. Southern Qilian and Qomolangma remain independently separated, whereas the South China province is evidently bipartite, one group being the pelagic Bayan Har and Youjiang biomes, the other constituting the restricted epicontinental Yangtze biome. Central and southern Guizhou as well as Yanyuan–Lijiang are subregions traditionally attributed to Yangtze, but here they are clustered together with Bayan Har and Youjiang, which reflects the transitional nature of biotas of the marginal belt between platform and slope.

In Fig. 9.4c, 128 species of Middle Triassic brachiopods were analysed. This figure is quadripartite, i.e. divided into South China (Qiangtang–Qamdo–

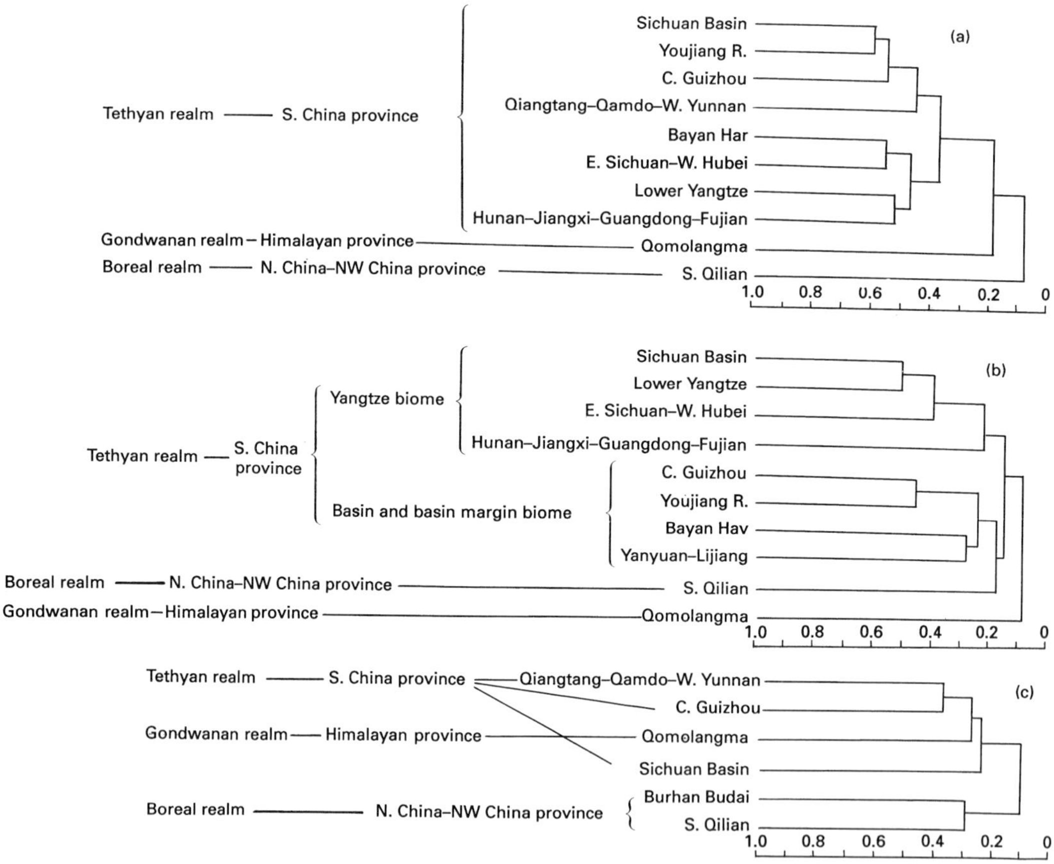

Fig. 9.4 Cluster analysis of the Early–Middle Triassic fossils of China. (a) Early Triassic bivalves; (b) Middle Triassic bivalves; (c) Late Triassic brachiopods.

western Yunnan, central Guizhou), Himalaya, North and north-western China (southern Qilian and Burhan Budai), and the Sichuan Basin. The extraordinary separation of the Sichuan Basin from the rest of South China is due to its restricted marine biofacies, which give rise to 10 new species out of a total of 20 species (50 per cent new species) although most of its genera are identical with those of central and southern Guizhou.

9.2.2 LATE TRIASSIC PALAEOBIOGEOGRAPHICAL PROVINCIALIZATION OF CHINA

Two significant changes took place in the Late Triassic palaeobiogeography of China. The Indosinian Orogeny caused general regression of Tethys from the eastern half of China. Due to the interaction of the Palaeo-Pacific Plate with the Eurasia Plate, an uplift region was formed along the so-called Inner Neocathaysian Zone, i.e. the mountains of Yunkai Dashan, Xuefengshan, Wulingshan, Taihangshan, and Lueliangshan, as well as along Qianshan and Zhangguangcai Ling in north-eastern China. The areas to the east of this region, subjected to the complex interaction of the influence of Palaeo-Pacific, the expansion of the humid belt, and the effect of currents along the north western Palaeo-Pacific (like the Kuro and Oya currents today), formed a Pacific Tethyan biogeographical region, which combines aspects of both the Pacific and the tropical Tethys. The expansion of the humid belt, symbolized by the extensive replacement of Middle Triassic evaporites by Late Triassic coal-bearing deposits, enhanced the invasion of Tethyan biotas into Gondwana; meanwhile extension of Neotethys (*sensu* Sengor 1984) caused northward migration of Tethyan microplates. As a result, biotas of the whole of eastern Tethys tend to become homogeneous and are incorporated into one tropical–subtropical Tethyan realm, while differentiation between the Gondwanan and Eurasian Tethys falls to regional level (Fig. 9.5).

Chinese Late Triassic fossils are noticeably more diverse than those of the Early and Middle Triassic. According to statistics up to 1984, there are bivalves (638 species, 141 genera), among which 36 species (7 genera) are non-marine, plants (425 species, 107 genera), ammonites (205 species, 72 genera), brachiopods (*c*. 90 species, 50 genera), gastropods (80 species, 56 genera), radiolarians, forams, chaetitids, conodonts, porifers, hexacorals, bryozoans, nautiloids, belemnites, crinoids, ophiuroids, ostracods, conchostracans, red and green algae, charophytes, palynomorphs, insects, and vertebrates (including fish); altogether more than 20 major groups.

Late Triassic plants are biogeographically significant because the two realms are mainly differentiated phytogeographically (Table 9.5). The southern, or *Dictyophyllum–Clathropteris* (*D.–C.*), flora and northern, or *Danaeopsis–Bernoullia* (*D.–B.*), flora are distinguished not only through a different composition of genera and species but also by other criteria. The *D.–C.* flora has a higher diversity (twice that of the *D.–B.* flora), with more cycadales (>20 per cent), fewer ginkgoales (<10 per cent), and fewer sphenopsids and conifers; the *D.–B.* flora, on the other hand, has a lower diversity, fewer cycadales (<20 per cent) and more ginkgoales (>10 per cent). In Table 9.5 the Wangqing flora of east Jilin and the flora of western Liaoning and southern Jilin are transitional examples which, albeit allocated to *D.–C.* or *D.–B.* flora, respectively, according to the dominant elements, are mixed with elements from the other flora, thus showing untypical taxonomical percentages.

The Late Triassic biotas of China are divided into six provinces of five regions and two realms (Table 9.3).

The Tethyan realm (eastern part)

This tropical–subtropical unit shows the following features:

1. The terrestrial regime of the whole of South China is signified by the *Dictyophyllum–Clathropteris* flora (Fig. 9.6) which also extends northwards to Japan, Primoria, the Amnok drainage of North Korea (Lower Daido Group), and the Wangqing, Shuangyang, Linjiang, and Yihe areas of north-eastern China.

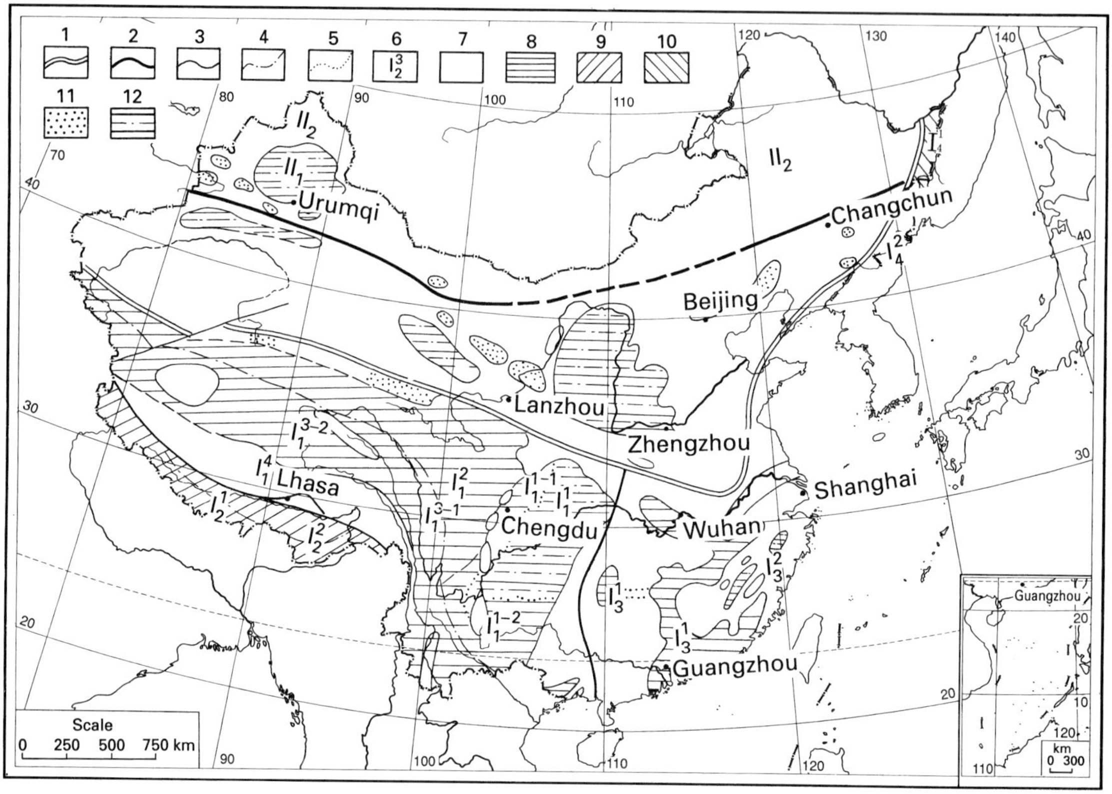

Fig. 9.5 Map of the Late Triassic palaeobiogeographical provincialization of China. 1, Boundary of realms; 2, boundary of regions; 3, boundary of provinces; 4, boundary of biomes; 5, boundary of communities; 6, roman digit = realm mark, subscript = province mark, superscript = biome mark; 7, Oldland. 8–10, Marine and paralic regions: 8, South China; 9, Gondwana; 10, Junggar–Xingan. 11–12, Terrestrial regions (Late Permian and Mesozoic–Cenozoic): 11, intermontane basins; 12, coal-bearing basins. I_1, Tethyan realm, Cathaysian Tethyan region, Southern China province; I_1^1, Yangtze biome; I_1^{1-1}, Longmen Mountains; I_1^{1-2}, Yunnan–Guizhou–Guangxi; I_1^2, Bayan Har–Qinling biome; I_1^{3-1}, Yidun-Yushu; I_1^{3-2}, Tanggula–Western Yunnan; I_1^4, Gangdise–Nianqingtanggula biome; I_2, Tethyan realm, Gondwanan Tethyan region, Himalayan province; I_2^1, Qomolangma biome; I_2^2, Yarlung Zangbo biome; I_3, Tethyan realm, Pacific Tethyan region, South-eastern China province; I_3^1, Hunan–Jiangxi–Fujian–Guangdong Bay biome; I_3^2, Cathaysian Oldland biome; I_4, Tethyan realm, Pacific Tethyan region, Jilin–Heilongjiang province; I_4^1, Ussuri biome; I_4^2, Yalu River biome; II, Laurasian realm, Central Laurasian region, North China–North-west China province; II_4, Laurasian realm, Northern Laurasian region, Altay–Xingan province.

2. The tropical and subtropical biota, symbolized by hermatypic hexacorals and megalodontid and heteromorph ammonites, extends north to Tetiukh of Primoria, south-western Japan, and Ryukyu Island, and south to the Himalayas and Indonesia.

3. The eastern part of Tethys, centred on East Asia and extending westward to Transcaucasia and Jordan, is biogeographically different from its western or Mediterranean part. The difference is displayed in bivalves (Kobayashi and Tamura 1983a), plants (Dobruskina 1982,1988), and bryozoans (Hu 1984). The eastern part can be called the *Burmesia–Indopecten* fauna, characteristic taxa of which are listed in Table 9.6.

Table 9.5 Comparison of the Late Triassic floras of China

	Number of species (genus)	L	A	P	PS	C	G	Co	O	Floral feature
Tethyan realm										
Xujiahe (s.l.) flora of S. China	191	1(1)	11(4)	63(31)	24(0)	60(21)	10(5)	17(5)	10(4)	*Dictyophyllum–Clathropteris* Flora
	(80)	0.5%	6%	33%	13%	13%	5%	6%	5%	
Anyuan (including Fujian and Zhejiang) flora	110	1(1)	7(3)	22(16)	8(5)	48(16)	6(3)	9(4)	10(6)	
	(54)	/	5%	23%	7%	48%	6%	9%	1%	
Flora of Babaoshan Group and Jieza Group, Qinghai	30	/	5(2)	13(8)	3(3)	6(4)	1(1)	/	2(1)	
	(19)		17%	43%	10%	20%	3%		7%	
Wangqing flora E. Jilin	45	/	5(2)	9(5)	1(1)	8(4)	6(6)	14(4)	2(1)	Mainly *D.–C.* flora
	(23)		11%	20%	2%	18%	13%	31%	4%	
Boreal realm										
Yanchang flora, N. China	91	/	15(3)	24(9)	14(6)	11(6)	13(8)	1(1)	13(5)	*Danaeopsis–Bernoullia* flora
	(38)		16.5%	26.3%	15.4%	121%	14.3%	1%	14.3%	
Flora of Muli Group, S. Qilian	50	/	9(3)	15(8)	8(4)	8(6)	6(4)	3(2)	3(2)	
	(29)		18%	30%	16%	16%	11%	4%	6%	
Flora in Baicheng area, Tarim	22	/	4(3)	10(6)	1(1)	/	2(2)	5(2)	/	
	(17)		18%	45%	5%		9%	23%		
Tianshan–Junggar flora	16	/	5(2)	7(4)	1(1)	/	2(2)	2(2)	1(1)	
	(10)		31%	44%	6%		11%	11%	6%	
W. Liaoning–S. Jilin flora	13	/	3(1)	3(2)	1(1)	3(2)	2(2)	1(1)	/	Mainly *D.–B.* flora
	(9)		23%	23%	8%	23%	15%	8%		

L, Lycophytes; A, arthrophytes; P, pteridophytes; PS, pteridospermophytes; C, cycadophytes; G, gingophytes; Co, coniferophytes, O, others; *D.–C.* flora, *Dictyophyllum–Clathropteris* flora; *D.–B.* flora, *Danaeopsis–Bernoullia* flora.

The Southern China province of the Cathaysian Tethyan region, Tethyan realm

Compared with the Early and Middle Triassic, the Late Triassic dimensions of this province shrank because its eastern part shifted into the Pacific Tethyan region. The new boundary is approximately along the Yunkai Mountains, Wuling Mountains, and eastern Yangtze Gorges. Its western boundary remains the same as in the Early and Middle

Fig. 9.6 Reconstruction of the floral community of the Late Triassic Xujiahe Formation of Sichuan in the Yangtze biome (from Xu Ren 1982). 1, bennettitales; 2, cycadales; 3, dipteridaceans; 4, *Lepidopteris stuttgartiensis*; 5, *Todites*; 6, *Equisetum arenaceum*; 7, *Neocalamites*.

Table 9.6 The occurrence of some fossils of *Burmesia–Indopecten* provincialism

	Transcaucasia N. Caucasia	Iran	Oman	Indonesia	Himalaya	Qinghai–Xizang	S. China	Japan	Coastal region
Bivalves									
Burmesia	+			+	+	+	+	+	
	(Jordan)								
Indopecten	+	+	+	+	+	+			+
Prolaria	+			+	(Burma)	+	+		(Balkan)
Pergamidia				+	+	+	+		
Costatoria		+							
verbeeki	+	+		+		+			
Palaeocardita									
buruca	+	+		+		+	+		
Ammonoids									
Tibetites				+	+	+			+
Paratibetites				+	+	+	+		
Bryozoans									
Paraleioclema	+								
Leioclema					+	+		+	

Triassic. This is a typical tropical Tethyan unit with the following features:

1. The *D.–C.* flora was widespread in Yangtze and Bayan Har and reaches the Tanggula Mountains and western Yunnan, but is unknown so far south of the Bangong Lake–Nujiang Line.

2. Tropical biotas, especially benthos, flourished. These include hexacorals (*c.* 63 species, 29 genera), megalodontid bivalves (*c.* 20 species, 5 genera), and heteromorph ammonites, especially *Hannaoceras* which is a tropical genus also occurring in California, Nevada, the Alps, the Balkans, Cyprus, Ailaoshan (western Yunnan), the Longmeng Mountains, and Ryukyu Island (Japan). The following taxa, mainly tropical, are so far unknown from the Himalayas but exist here: oysters and oyster-like bivalves (there are a few exceptions to this group, *Gryphaea* and *Liostrea* have been reported from the South-eastern China province of Pacific Tethyan region; *Lopha* is also found in southern Qilian), halorellid brachiopods, chaetitids, dasycladaceans, cryptonemial algae, and calcareous sponges, which form patch reefs in the pre-Longmeng Mountains area.

The Carnian and early and middle Norian ammonites of this province are closely related to those found in the Alps, Himalayas, and Timor. Contemporary bivalves are homogeneous throughout East and South-East Asia. Characteristic taxa are the ammonites, *Trachyceras*, *Protrachyceras*, *Tropites*, *Tibetites*, *Thisbites*, and *Paratibetites*, bivalves, certain species of *Halobia*, *Burmesia*, *Costatoria* (*kueichowensis*, *napengensis*), and *Angustella angusta*; brachiopods, *Oxocolpella*, *Neoretzia*, and *Halorella*; and the gastropod, *Sisenna*. After the late Norian a large part of this province was converted into a brackish or swamp environment with the bivalve *Yunnanphorus–Indosinion* fauna (including *Unionites*, *Trigonodus*, and *Modiolus)* and the *D.–C.* flora, different from the normal marine Tethyan *Monotis salinaria* fauna in the Himalayas or the Pacific type *Waagenoperna* fauna in the South-eastern China province.

The Norian estheriid *Euestheria minuta* assemblage corresponds to a similar assemblage in North China, but is different from the contemporary *Anyuanestheria* assemblage of south-eastern China.

The South China province is subdivided into four biomes.

The Yangtze biome

In the Late Triassic, south-eastern China was excluded from Yangtze, but the former Youjiang biome, whose Late Triassic deposits are restricted to Shiwandashan and its southern slope, yields Yangtze-type plants, estheriids, and the bivalve *Yunnanophorus* faunule (17 species of 10 genera), and is hence incorporated with the Yangtze biome. This biome shows the following features:

1. As in the Early and Middle Triassic, this biome exhibits the most diversified Late Triassic biotas in China. Initial statistics provide more than 20 major groups, including bivalves (433 species), plants (173 species), ammonites (59 species), brachiopods (25 species), and gastropods (19 species).

2. Plants are typically of the D.–C. flora, dominated by dipteridaceans which are modern index flora of tropical rain forests. Only in sporadic localities have a few D.–B. elements been infiltrated (Yipinglang, Shiwandashan, and Xujiahe). The tropical taxa described in Fig. 9.6 are mostly known here.

3. There are fewer genera and species of halobiid bivalves and ammonites than in the Tibet–western Yunnan and Bayan Har biomes. Other taxa that flourished in the Tethyan open sea are also scarce or unknown here; these include the bivalves *Neomegalodon*, *Indopecten*, *Monotis*, *Protocardia*, *Palaeocardita* (scarce), *Pergamidia* (scarce), *Sichuania*, and *Tulongella*, and the ammonites *Placites*, certain tibetitids, and certain juvavitids. The brackish *Yunnanophorus Indosinion* assemblage was well developed.

4. Many Pacific-type taxa, including trigoniids from the south-eastern China province are unknown here.

Points 1, 2, and 4 signify its tropical Tethyan character; point 3 denotes its restricted epicontinental environment.

This biome can be further subdivided according to deposits and community types. The Longmeng Mountains and the eastern Yunnan–southern Guizhou–Guangxi areas belong to a marine regime that was inhabited by littoral–neritic benthic communities. Most of this biome is non-marine throughout the Late Triassic. By the latest Triassic, the sea had entirely retreated, and the whole area was then inhabited by the D.–C. flora and freshwater taxa such as unioniids (*Shaanxiconcha*, *Sibericoncha*, and *Unio*).

The Bayan Har and Southern Qilian biome

These two areas form the famous and vast triangular turbidite basin in the rejuvenated miogeosyncline series which formed as a result of rift-and-sag tectonics on the continental slopes of Yangtze and Songpan after Ladinian times. The biome is characterized by the dominance of pelagic organisms, especially halobiid bivalves. In the Bayan Har area, there are 30 species of *Halobia* and its relatives (*Parahalobia* and *Posidonia*), in contrast to the 15 remaining bivalve species; among the latter the most important is *Pergamidia*. There are scarcely any other groups except a few ammonites, plant remains, and radiolarians. Important species include *Halobia comata*, *H. superba*, *H. yunnanensis*, *H. fallax*, *Pergamidia timorensis*, and *P. eumenea*. This list is almost identical to the leading pelagic elements in the Yidun–Yushu communities, and are basically of eastern Tethyan type. The southern Qinling area is little known, but basically corresponds with Bayan Har.

The Northern Tibet–Western Yunnan biome

This biome is basically analagous with Yangtze and should have been incorporated with it if it were not geographically separated by Bayan Har. Most taxa that exist here have also been discovered in Yangtze. However, located adjacent to the Himalayan province, this biome contains a few taxa in common with those of the Himalayas, which are unknown or scarce in the Yangtze biome. They include the bivalves *Indopecten* (4 species) and *Neomegalodon* (19 species), and the ammonites *Cladiscites* (5 species) and *Cyrtopleurites* (3 species). Certain Pacific-type trigoniids, such as *Kumatrigonia*, *Minetrigonia*, and *Myophorigonia* have been found here and also occur sporadically in the Himalayas and Bayan Har. These small differences probably denote that the sea was more open here than in Yangtze.

This biome is further subdivided into the Yidun–Yushu and Tanggula–Western Yunnan communities.

The Yidun–Yushu area, bordered by the Ganze–Litang fault to the east and Jinshajiang fault to the west, used to be assigned to the Bayan Har stratigraphic region, because both are dominated by Late Triassic geosynclinal deposits. However, recent research has revealed that a small piece of pre-Carnian oceanic crust, denoted by ophiolites and radiolarian cherts, developed and subducted during the Late Triassic along the Ganze–Litang fault, and the Late Triassic arc–trough system of this area is associated with this fault rather than with the Jinshajiang fault. According to some authors, the Jinshajiang fault had already sutured together by the Late Permian, but other authors hold that it was still active in the Triassic. The biome, controlled by and in harmony with the active-margin-type arc-and-basin environments, has a diversified fauna which changes rapidly up the succession. Fossils of shallow benthic dwellers on the carbonate platform can pass into a pelagic assemblage in turbidite deposits within a few hundred metres. The general ratio of benthic to planktonic plus nektonic species is 1 : 0.25, in contrast to that of Bayan Har. So far, more than 300 species have been reported in this mixed-type biota, including biofacies from inter- and supra-tidal zones to abyssal zones, plus reef facies, as denoted by hexacorals, hydrozoans, and sponges.

The Tanggula–Western Yunnan communities are developed on the continental shelf of the Qiangtang–Tanggula paraplatform, which in Late Triassic times first encountered widespread transgression, followed by regression, leading to the deposition of coal-bearing swamp deposits. The *Halobia–Burmesia–Yunnanophorus* bivalve sequence, *D.–C.* flora, and other biotas found here are identical with those of Yangtze.

The Gangdise–Nianqingtanggula biome

Differentiation of this biome from the Northern Tibet–Western Yunnan biome, with which it was formerly incorporated, is based on the following factors:

1. Late Triassic geosynclinal deposits (Wuga Group) have been discovered along their demarcation line, the Bangong Lake–Nujiang line, which implies that rift-opening had already taken place between them.

2. In contrast to the basically marine Northern Tibet–Western Yunnan region, this region was basically an emergent oldland during the Late Triassic; only along the Naqu–Luolong zone on its north-eastern margin (Quehala Group of the Shading slate series) and in the vicinity of Lhasa on its southern margin (Mailonggang Group) do there exist marine deposits bearing corals, bivalves (mainly halobiids), conodonts, gastropods, and plants. Because the Upper Triassic rocks along the Yarlung Zangbo River mark a much wider and deeper ecological barrier than the Bangong Lake–Nujiang Line, we deem that this area is more closely related to the South China province than to the Himalayan province.

The Himalayan province of the Gondwanan Tethyan region, Tethyan realm

As reasoned above, the Himalayan province and Gondwanan Tethyan region were merged into the Tethyan realm in the Late Triassic. They remained an independent region and province due to the following peculiarities:

1. In its marine regime, biotas are not fully tropical. A series of organisms flourishing in tropical belts, such as heteromorph ammonites, oysters and oyster-like forms, halorellid brachiopods, sponges and chaetitids, and solenopore algae, are not reported here.

2. The plants that flourished in the adjacent Indian Subcontinent belong to the temperate *Dicrodium* flora, not the tropical *D.–C.* flora.

3. The province shares a peculiar assemblage with other Gondwanan Tethyan areas including Aman, southern Iran, and the Outer Banda Island Arc (Timor, Buru, Seran, Misol); this includes the brachiopods *Hagabirhynchia*, *Spiriferina abichi*, and *Misolia*, and the bivalves *Serania* and *Nucula misolensis*.

4. The freshwater *Shaanxiconcha–Sibireconcha* bivalve assemblage and the *Euestheria minuta* estheriid assemblage occur in eastern Tibet and western Yunnan, but so far have not been discovered in the Himalayas.

Points 1, 2, and 3 imply that the Himalayas were still in higher latitudes, thus having a lower temperature than South China. Point 4 probably signifies the existence of a marine barrier to freshwater faunas between the Himalayas and northern Tibet. These differences support the establishment of an independent province.

The Himalayan province is subdivided into two biomes. The southern Qomolangma biota is characterized by high diversities of benthic bivalves, brachiopods, and nectonic ammonites, inhabiting a carbonate–clastic continental shelf. In contrast, the northern Yarlung Zangbo biota consists of pelagic types of ammonites, halobiids, and radiolarians of low diversity, occurring in miogeosynclinal flysch; benthos are almost lacking.

The Pacific Tethyan region of the Tethyan realm

The Pacific Tethyan region is assigned to the Tethyan realm because both share the tropical *D.–C.* flora. It is separated as a region because it has marine biotas of Pacific type. Two provinces are proposed. The South-eastern China province is typified by tropical Pacific biotas, similar to contemporary ones in south-western Japan. In the Jilin–Heilongjiang province, biotas are of mixed type, containing both tropical and temperate elements, and the marine faunas are mainly of Boreal type. In this regard, it is plausible to put this province in the southern mixed belt of the Boreal realm, i.e. the Central Eurasian region. However, because of the widespread *D.–C.* flora in this province, it is provisionally attributed to the Tethyan realm.

The South-eastern China province

This includes elements of both Pacific type and endemic embayment type. The former are mainly bevalves, including *Bakevelloides* (7 species), *Chlamys mojsisovicsi, Falcimytilus* (1 species), *Inovceramus* (1 species), *Isognomon* (6 species), *Tosapecten* (1 species), *Oxytoma* (9 species), *Palaeophalus* (2 species), *Radiastarte* (1 species), *Waagenoperna* (4 species), *Protocardia*, *Retinuculana*, and *Trigonucula*. Endemic elements are also conspicuous in ostracods, estheriids, and insects, e.g. the ostracod *Gomphocythere* faunule consists of 21 endemic species, in contrast to three known species. Important endemic elements are bivalves, *Guangdongella* (2 species), *Isocardioides yini*, *Jiangxiella* (6 species), *Nanglingella* (2 species), and *Trigonodus liuyangensis*; gastropods, *Fascivalvata pericarinata*; ostracods, *Oncocythere* and *Minicythere*; conchostracans, *Anyuanestheria*; and insects, *Triassoblatta* and five other genera. Furthermore, perhaps more significant is the absence of many Cathaysian Tethyan groups, including ammonites and other cephalopods, halobiids, megalodontids, corals, sponges, and most of the important elements of the *Burmesia* assemblage. Such absences may be due to both the Pacific influence and abnormal conditions in embayment circumstances.

So far 105 species of 49 genera of plants are known in this province, dominated by *D.–C.* elements. However, two conditions are different from the Cathaysian Tethys.

1. In many areas there is an admixture of elements of the northern Yanchang flora, e.g. *Danaeopsis fecunda*, *Bernoullia* (*Cymopteris*) *zeilleri*, *Glossophyllum shensiensis*, and *Todites shensiensis*. Examples are the Jiuligang and Xiaoping of Formations Jingmen-Dangyang Basin, western Hubei (all four species mentioned above); Xiaoping Formation, northern Guangdong (first two species); Anyuan Formation, Jiangxi (first and fourth species); Dakeng and Wenbingshan Formations, Fujian (last three species). Such admixtures are also seen in Yangtze, e.g. the Yipinglang Formation, but to a much lesser degree.

2. Cycadales (48 per cent in the Anyuan flora, Table 9.5) are in excess of ferns (pteridophytes and pteridosperms), whereas in Yangtze and Qinghai–Tibet, ferns are relatively more abundant (52 per cent in the Xujiahe flora, S. China and 70 per cent in Babaoshan and Jieza flora, Qinghai, Table 9.5). Such a situation probably reflects a lower temperature or humidity than in Yangtze.

The south-eastern China province is subdivided into a marine and a terrestrial biome. The marine biome is called the Hunan–Jiangxi–Fujian–Guangdong embayment biome. It is similar to south-eastern Japan and Primoria (Russia) but, due to an embayment environment, lacks the halobiids, monotids, and ammonites typical of the latter two areas. This embayment in turn consists of Guangdong Bay and the Huanan–Jiangxi–Fujian estuary, of which the communities are defined differently by various authors. According to Chen J. 1983, the former is composed of an early stage *Guangdongella–Bakevellia* community, a middle stage *Palaeophalus–Oxytoma* community, and a late stage *Waagenoperna* community; the latter is composed of an early stage *Unionites–Myophoriopsis* community, a middle stage *Bakevelloides–Jiangxiella* community, and a late stage *Waagenoperna* community. On the other hand, the Academy of Coal Science (1980–83) named the Hunan–Jiangxi communities as an early *Bakevelloides hekiensis–Palaeophalus oblongatus* community and a Late *Trigonodus*? *liuyangensis* community. The distribution of these embayment faunas through river and lake connections reached westward to the Sangzhi–Hueihua area of western Hunan, north-westward to the Jingmen–Dangyang Basin of western Hubei (*Waagenoperna*, *Bakevelloides*, etc.) and north-eastward to Jinxian, southern Anhui (*B. hekiensis*). Brackish bivalves, such as *Mytilus*, are also found in the Fanjiatang Formation in the vicinity of Nanjing. This embayment system generally extends in a north-easterly direction, evidently in accordance with the Neocathaysian trough formed during the Indosinian Orogeny, a pattern paralleling the embayment along the lower Amur and Ussuri Rivers far away in north-eastern China.

The terrestrial biome is called the Xuefengshan–Cathay Oldland biome. The floral aspects have been discussed above. The main freshwater bivalves, such as *Unio* (4 species), *Sibireconcha* (4 species), and *Shaanxiconcha* (5 species), are identical with central Eurasian bivalves at the generic level, but have different species. Many of them have been reported from the Huangmaqing Formation of south Jiangsu, have been which has been suggested by some authors to be of Middle Triassic age.

The Jilin–Heilongjiang province

Data from this province are barely adequate to allow subdivision into two biomes. The marine Ussuri biome embraces the Chinese part of the Sikhote-Alin geosyncline and yields Boreal Pacific elements such as *Entomonotis ochotica* and *Plagiostoma nuitoensis*, quite different from the contemporary communities in the south-eastern China embayments. This area includes the Nadanhada Mountain, where blocks of Upper Palaeozoic limestones bearing fusulinids and Upper Triassic to Middle Jurassic radiolarian cherts are slumped into Late Jurassic exotic clastics, and hence are regarded as an accreted terrain from the Tethys–Pacific (Kojima 1989; also Fig. 14.6).

The terrestrial Amnok (Yalu) River biome is dominated by *D.–C.* flora, as seen in Tianqiaoling of Wanqing County, Laoheishan of Dongning County, Yihe of Linjiang County, and Dachoupai, North Korea. However, the D.–C. flora here is not typical but transitional (Table 9.5) with fewer cycadales and more ginkgoales. The occurrence of the Boreal genus *Arctobaiera* is also notable. This tropical–subtropical flora occurs near the temperate *D.–B.* flora in Changchun, Hunjiang, and other localities. Thus the realm boundary has to turn northeastward and bend between the two floristic zones. This bending may be due to either the influence of the Palaeo-Pacific and its Ussuri embayment or large-scale post-Indosinian sinistral strike-slip faulting along the famous Tanlu Fault.

The Boreal realm

The North and North-western China province of the Central Eurasian region (eastern part)

The Central Eurasian region is specified by the *Danaeopsis–Bernoullia* (*Symopteris*) flora. Dobruskina (1982) separated an eastern floral section of this region. The North and North-western China province produces 91 species and 38 genera of this flora.

Freshwater bivalves account for more than 30 species of 7 genera the (Yanchang Formation alone yields 22 species of 4 genera) characteristic ones being *Shaanxiconcha* ('*Utshamiella*'), *Sibireconcha*, and *Unio*. This faunule has been reported in Central Asia, Pamir, Xinjiang, the Qilian Mountains, Ordos Basin, western Henan, western Liaoning, southern Jilin, and a few localities in Yangtze and south-eastern China, but is as yet unknown from Gondwana, the Himalayas, and Gangdise.

Ostracods, the *Darwinula–Lutkevichinella–Tongchuania* faunule, are also most abundant in the Yanchang Formation (18 species of 4 genera). The principal elements of the faunule have also been discovered in Great Britain, Germany, Poland, the Dnieper–Donetz Basin of Ukraine, the Precaspian, the Emba area and Tulgai Basin of Kazakhstan, Xinjiang, and the Ordos Basin. The faunule is approximately distributed along the Central Eurasian belt and is different in composition from South China elements.

The conchostracan *Euestheria minuta* faunule (7 species) has been discovered in Qilian (Jingyuan, Lingwu), Ordos (Tongchuan, Hanchen), and western Henan (Jiyuan).

The eastern part of the Central Eurasian region is distinguished by differences in the above groups between Europe and East and Central Asia, which imitates the east–west partition in Eurasian Tethys.

The northern boundary of this province is hard to delineate due to insufficient data. A tentative line along the Altay–Xar Moron River, duplicating that of the Early and Middle Triassic, is adopted here. This province, and the region as well, is a belt, different from but considerably influenced by the tropical Tethys. Support for this is as follows.

1. Its *D.–B.* flora is mixed with *D.–C.* elements. Such localities include Kuqa, Xinjiang (*D. nathorsti, Clathropteris, Hausmannia ussuriensis, and Maratiopsis muensteri*); Jinyuan and Wuwei, Gansu (*D. nathorsti*, and *Clathropteris*); southern Qilian (*Dictyophyllum?* and *Clathropteris*); Beipiao and Caoyang, western Liaoning (*D. nathorsti* and *C. meniscioides*); and Ordos (*Anomozamites Loczyi*, *Lepicopteris ottonis*, *Nilssonia acuminata*, *N. pterophylloides*, and *Ptilozamites*).

2. In the Datong coal mine of the Qilian Mountains area, there have been discovered marine Upper Triassic interbeds yielding the tropical Tethyan *Burmesia–Yunnanophorus* faunule, as well as the oyster-like *Lopha* and the Pacific trigoniid *Kumatrigonia*. This definitely indicates a connection with Tethys through the seaway along the western Qinling–Qaidam gap, where Late Triassic molluscs occur.

3. It shares the same *Euestheria minuta* faunule with Yangtze and Tibet–western Yunnan. The ostracod *Tongchuania*, typical of the Ordos Basin, is reported doubtfully from central Yunnan (Xiangyun County).

The above points seem to justify the inference of the warm temperate or even subtropical character of this province. The mixed flora probably extends through Central Asia into southern Russia, Austria, and Switzerland.

The Altay–Xingan province of the Northern Eurasian region

This province covers Junggar, Tulfan, and other northern Xinjiang basins, as well as the Beishan area in Gansu. It also includes Mongolia, Transbaikalia, parts of Central Asia (the Fergan Basin and Sayan Mountains), and perhaps the Daxingan and Xiaoxingan Mountains. Its *D.–B.* flora is not mixed with *D.–C.* elements. The bivalve faunule resembles that of the North and North-western China province. Data from other groups are too fragmentary to give any general information. The Mohe area of northern Heilongjiang belongs to the Mongolia–Okhotsk geosyncline, but so far no information about this area has been obtained.

Cluster analysis of Late Triassic bivalves and ammonites

The data source and method of operation are the same as for the Early and Middle Triassic. For Fig. 9.7b, 93 Late Triassic ammonites were analysed, after the deletion of cosmopolitan and pan-Chinese species. The result shows little discrimination between Cathaysian and Gondawanan Tethys. The cluster reflects biofacies rather than biogeography. Ammonites of Qomolangma, Qiangtang–Qamdo (northern Tibet)–western Yunnan and Yidun–Zhongdian are clustered in the first group of open-sea biofacies. The restricted-sea biofacies taxa are segregated in the second group. The three areas within the third group are all located at the western margin of the Yangtze Platform which faced the Bayan Har and western Yunnan open seas, and thus belong to the transitional type.

In Fig. 9.7a, after the same deletion, 361 Late Triassic bivalve species are analysed. The results clearly display quadripartite grouping. All the non-Pacific Tethyan marine bivalves are clustered in the largest group, including both Cathaysian and Gondwanan areas, of which the further subdivision is, as in ammonites, ecological rather than biogeographical. The two Pacific Tethyan provinces each form an independent group. This means that the two provinces are faunistically different, and are united under the same region only by the common *D.–C.* flora. The fourth group comprises the freshwater bivalves.

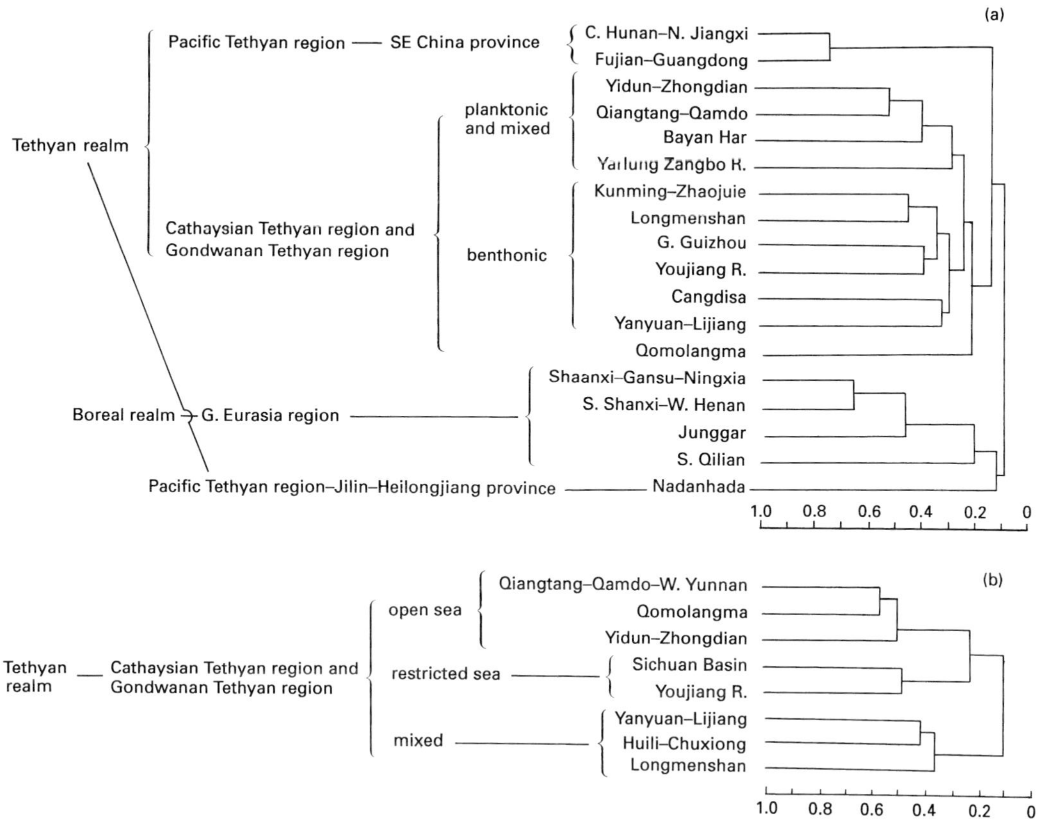

Fig. 9.7 Cluster analysis of the Late Triassic fossils of China. (a) Late Triassic bivalves; (b) Late Triassic ammonoids.

9.3 REMARKS ON TRIASSIC PALAEOBIOGEOGRAPHY

9.3.1 RELATIONS BETWEEN CLIMATE ZONATION, PLATE TECTONICS, AND BIOGEOGRAPHY

The Triassic provincialization of China is basically palaeolatitudinal. This is caused by the latitudinal arrangement of both climate zonation and geographical barriers formed by the Tethys Ocean and on the Eurasian Plate. Only in the Late Triassic did northeastward biogeographical boundaries appear in eastern China. This is because land–sea boundaries, currents, drainages, and monsoon rains of more or less the same direction brought about by the Palaeo-Pacific greatly changed the distribution patterns of biota in this part of China. The simultaneous East Pacific realm was also established under the influence of the Palaeo-Pacific.

The southern boundary of the Boreal realm shifted from the Ebinur–Xar Moron line in the Permian southward to the Kunlun–Qinling line in the Early and Middle Triassic. During this time the northern boundary of Gondwanan Tethys did not shifted northward, so such a shift may not have been caused by expansion of the zone of cool climate. It is more readily interpreted as due to the rapid northward migration of North China and Tarim blocks during the Permo-Triassic interval, which is in accordance with palaeomagnetic data.

On the other hand, in the Late Triassic the Tethyan realm expanded poleward; tropical rain forest (*D.–C.* flora) and Tethyan *Burmesia* fauna invaded both Gondwanan Tethys and marginal areas of the Boreal realm. This was evidently due to expansion of the zone of humid tropical climate.

The absence of a latitudinal transitional or mixed zone implies large-scale subduction. Normally, climatically demarcated biogeographical units should have a latitudinally mixed belt in between, because climates only change gradually relative to latitudes. The widespread mixed belt between Cathaysian Tethys and the Boreal realm indicates that the relative location of the two plates in modern times may be roughly what they were in the Triassic, and no postulation of large-scale subduction is necessary. Conversely, the clear-cut delineation between the Himalayas and Cathaysian Tethys in the Early and Middle Triassic implies the disappearance of a mixed belt which occupied a considerable latitudinal range. This range and its accompanying mixed biota was eliminated by subduction. Climatically similar biomes should have occupied similar latitudinal ranges, even though separated by geographical barriers, as in the case of Yangtze and northern Tibet–western Yunnan.

9.3.2 THE SIGNIFICANCE AND DISTRIBUTION PATTERN OF BIOFACIES

Biomes, communities, and other biofacies play an important role in the third to fifth level (province and under-province levels) of provincialization. This supports our general idea of biogeographical provincialization in Chapter 1. However, biofacies factors also function strongly at the first and second levels, and these must be eliminated when making biogeographical subdivisions. For example, it is common sense that in the Late Triassic, Bayan Har should be in the same Cathaysian region with northern Tibet–western Yunnan and Yangtze. However, because of its pelagic biofacies, its difference from the latter two, based on its fossil list, would be more than 50 per cent, surpassing the value required by a regional level differentiation. Such effects should always be considered and avoided where possible.

Distribution patterns of biofacies clearly reflect tectonic and sedimentary environments. The Triassic of South China shows the common rule that from continents to slope and basin the biofacies change from epicontinental benthos to marginal (transitional) and finally to pelagic. Detailed examples are as follows:

1 Arid tropical zones: hypersaline biofacies → barrier reef biofacies → turbidite pelagic biofacies. Example: late Early to Middle Triassic, the Upper Yangtze hypersaline benthos → Qingyan barrier reef(?) fauna (see arguments in the text) → Youjiang turbidite *Daonella* fauna.

2. Humid tropical zones. Example: Late Triassic, Upper Yangtze coal-bearing swamp plants and molluscs → Longmeng Mountains fauna of marginal biofacies → Bayan Har turbidite *Halobia* fauna. Along such marginal zones there occur corals, sponges, oysters, often in patch reefs, which are usually absent from temperate zones.

3. Temperate zones. Example 1: Anisian, North China and Qilian Oldland with coal basins → southern Qilian marginal zone with abundant benthos → western Qinling turbidite ammonites and ichnofossils. Example 2: Late Triassic, Indian Subcontinent with plants in coal basins → Qomolangma marginal epeiric benthos → Yarlung Zangbo turbidite *Monotis–Halobia*–radiolarians.

In all these cases the marginal zones of the continental shelf yield the most diversified biofacies, combining dominant benthos and a variety of planktons and nekton. These marginal zones are usually elongated, forming barrier reefs in tropical areas. They are good sources and accumulation beds of petroleum.

Areas within these marginal zones are usually epicontinental, restricted, littoral, or swampy, depending on the particular local conditions. Under hot-arid climates, oil and gas reservoirs can form, as well as evaporitic deposits, as in the Sichuan Basin during the Early and Middle Triassic. Under humid climates there would accumulate coal measures, such as the Tumeng'gla, Xujiahe and Houbachong coal series in the Late Triassic.

REFERENCES

Chen Chuzhen (1982). Fossil bivalves of Late Palaeozoic to Triassic in Tibet. *Palaeontology in Tibet* **4**, 211–24 (in Chinese with English abstract).

Chen Jinhua (1983). Late Triassic and Early Jurassic bivalve assemblages and palaeogeography of South-eastern China, with a bivalve-biogeographic division in China. In *Palaeobiogeographic provinces of China,* (ed. Compiling Committee of Treatise of Basic Theories in Palaeontology), pp. 100–20. Science Press (in Chinese).

Cox, C. B. (1973). Triassic Tetrapods, In *Atlas of Palaeobiogeography*, (ed. A. Hallam), pp. 213–24. Elsevier Scientific, Amsterdam.

Deng Zhanqiu and Kong Lei (1984). Middle Triassic corals and sponges from Southern Guizhou and eastern Yunnan. *Acta Palaeontologica Sinica* **23**, (4), 489–504 (in Chinese with English abstract).

Dobruskina, I. A. (1982). Triassic flora of Eurasia. *Trudy Geologicheski Instituta, Akademia Nayka, USSR*. **365**, 1–196 (in Russian).

Dobruskina, I. A. (1988). The history of land plants in the Northern Hemisphere during the Triassic with special reference to the floras of Eurasia. *Geologie und Paläontologie der Mitterungen der Innsbruck* **15**, 1–12.

Gu Qingge, He Guoxiong, and Wang Yigang (1980). Discovery of the Late Anisian *Paraceratites trinodosus* fauna (Ammonoidea) from Duilungdeqen. Tibet and its significance. *Acta Palaeontologica Sinica* **11**, (5), 343–56 (in Chinese with English abstract).

Haas, O. (1953). Mesozoic invertebrate faunas of Peru, parts 1–2. *Bulletin of the American Museum of Natural History* **101**.

He Zhiai, Yang Hong, and Zhou Jingcai (1980). The Middle Triassic reef in Guizhou Province. *Scientia Geologica Sinica* **1980**, (3), 256–64 (in Chinese with English abstract).

Hu Zhaoxun (1984). Triassic Bryozoa from Xizang (Tibet) with reference to their biogeographical provincialism in the world. *Acta Palaeontologica Sinica* **23**, (5), 568–77 (in Chinese with English abstract).

Kobayashi, T. and Tamura, M. (1983a). On the oriental province of the Tethyan realm in the Triassic period. *Proceedings of Japan Academy* **59B,** (7), 203–6.

Kobayashi, T. and Tamura, M. (1983b). The Arcto-Pacific realm and the Trigoniidae in the Triassic period. *Proceedings of the Japan Academy* **59B**, (7), 206–10.

Kojima, S. (1989), Mesozoic terrane accretion in Northeast China, Sikjote-Alin and Japan regions. *Palaeogeography, Palaeoclimatology, Palaeoecology* **69**, 213–32.

Kristan-Tollmann, E. (1988). Unexpected microfaunal communities within the Triassic Tethys. In *Gondwana and Tethys*, (ed. M. G. Audley-Charles and A. Hallam). Geological Society Special Publication No. 37, pp. 213–24.

Lin Jinlu (1987). China Paleomagnetic data list(1). *Scientia Geologica Sinica* **1987**, (2), 183–7.

McElhinny, M. W. *et. al.* (1981). Fragmentation of Asia in the Permian. *Nature* **293**, 221–16.

Sengör, A. M. C. (1984). *The Cimmeride orogenic system and the tectonic* of *Eurasia*. Geological Society of America, Special paper No. 195, pp. 1–82.

Sichuan Regional Geological Survey and Nanjing Institute of Geology and Palaeontology (1982). *The stratigraphy and palaeontology of West Sichuan and East Tibet*, parts 1 and 2. Sichuan People's Publishing House (in Chinese).

Stanley, G. D. (1979). *Paleoecology, structure and distribution of Triassic coral buildups in western America*. University of Kansas. Paleontological Contributions No. 65.

Sun Dongli (1982). Triassic brachiopoda of China. *Rivista Italiana Paleontologia* **85**, (3–4), 1175–88.

Tozer, E. T. (1980). Triassic Ammonoida, geographic and stratigraphic distribution. In *The Ammonoidea*, (ed. M. R. House and J. R. Senior). Systematics Association Special Volume 18, pp. 397–432.

Wang Yigang (1984). Earliest Triassic ammonoid faunas from Jiangsu and Zhejiang and their bearing on the definition of Permo-Triassic boundary. *Acta Palaeontologica Sinica* **23**, (3), 257–70 (in Chinese with English abstract).

Wang Yigang and Chen Guorong (1984). Triassic ammonite beds in Qinghai province. *Contribution to the Geology of Qinghai-Xizang (Tibet) Plateau* **14**, 137–46. Geological Publishing House (in Chinese).

Wang Yigang and He Guoxiong (1981). Some ammonites in Tibet. *Paleontology in Tibet* **2**, 103–31. Science Press (in Chinese).

Wen Shixuan (1979). New materials of biostratigraphy in the northern Qingtang Plateau. *Acta Stratigraphica Sinica* **3**, (2), 150 6 (in Chinese).

Westerman, E. G. (1973). The Late Triassic bivalve *Monotis*. In *Atlas of Palaeobiogeography*, (ed. A. Hallam), pp. 251–8. Elsevier Scientific, Amsterdam.

Wiedmann, J. (1973). Upper Triassic heteromorph ammonites. In *Atlas of Paleobiogeography*, (ed. A. Hallam), pp. 235–50. Elsevier Scientific, Amsterdam.

Xu Ren (1982). *Reconstructions of fossil floral communities*. Science Press, Beijing.

Yang Xibin and Lian Yuchi (1980). New material of Triassic of North Tibet. *Journal of the Changchun College of Geology* **1980**, (4), 39–42 (in Chinese).

Yang Zunyi and Xu Guirong (1964). *Middle and Upper Triassic brachiopods of Central Kueichou*. Publishing House of Industry of China, Beijing (in Chinese).

Yang Zunyi, Li Zishun, and Qu Lifan (1982). The Triassic system of China. *Acta Geologica Sinica* **56**, (1), 1–21 (in Chinese with English abstract).

Yang Zunyi, et. al. (1983). *Triassic of the South Qilian Mountains*. Geological Publishing House (in Chinese).

Ye Meina (1979). On some Middle Triassic plants from Hubei and Sichuan. *Acta Palaeontologica Sinica* **18**, (1), 73–81 (in Chinese with English abstract).

Yin Hongfu (1981). Palaeogeographical and stratigraphical distribution of the Lower Triassic *Claraia* and *Eumorphotis* (Bivalvia). *Acta Geologica Sinica* **55**, (3), 161–9 (in Chinese with English abstract).

Yin Hongfu and Yochelson, E. L. (1983). Middle Triassic Gastropoda from Qingyan, Guizhou Province, China. *Journal of Paleontology* **57**, (1), 162–87 (part 1); (3), 515–38 (part 2); (5), 1098–127 (part 3).

Yin Hongfu, Yang Fengqing, Lai Xulong, and Yang Hengshu (1988). Triassic belts and Indosinian development of the Qinling Mts. *Geoscience* **2**, (3), 355–65 (in Chinese with English abstract).

Yin Hongfu, Yang Fengqing, Huang Qishen, Yang Hengshu, and Lai Xuelong (1992). *Triassic of the Qinling Mts. and adjacent areas*. China University Press, Wuhan in Chinese with English summary).

Zakharov, E. T. (1973). The importance of the palaeobiogeographical data for the solution of the problem on the Lower Triassic division. In *The stratigraphy of the Alpine–Mediterranean Triassic*, (ed. H. Zapfe), pp. 237–44. University of Vienna Press, Vienna.

Zhang Wentang, Chen Peiji, and Shen Yanbin (1976). *The Fossil Conchostraca of China*. Science Press (in Chinese).

Zhou Zhiyan and Li Baoxian (1979). A preliminary study of the Early Triassic plants from the Qionghai District, Hainan Island. *Acta Palaeontologica Sinica* 18, (5), 444–63 (in Chinese with English abstract).

Zhu Zhiwen, Zhu Xiangyuan, and Zhang Yiming (1981). Palaeomagnetic observation in Xizang and continental drift. *Acta Geophysica Sinica* **24(1)**, 35–43 (in Chinese).

10 JURASSIC

Wu Shunbao and Tong Jinnan

10.1 JURASSIC PALAEOBIOGEOGRAPHICAL PROVINCIALIZATION OF THE WORLD

The plate tectonic history of the Jurassic is characterized by the disintegration of Gondwana and the initiation of the three main modern oceans. Another aspect of the Jurassic is its higher and more evenly distributed palaeotemperature than in modern times. The Jurassic palaeobiogeographic provincialization of the world is suggested as follows (Fig. 10.1, Table 10.1).

1. The Laurasian Tethyan realm belonged to the tropical–subtropical zone. According to palaeomagnetic data, the palaeolatitude of this realm ranged from about 30°S to 40°N during the Jurassic period. It included the major part of the Tethys Ocean, the northern parts of South America and Africa, and the southern part of Laurasia. It can be subdivided into two regions, the marine Laurasian Tethyan region and the terrestrial Central Laurasian region (Hallam 1971, 1975, 1977; Westermann 1981, 1988; Imlay 1984).

2. The Boreal realm belonged to the temperate–subtropical zone, the palaeolatitude of which ranged from about 40°N to 70°N at that time. Its land area was vast but intermittently underwent transgression from the surrounding seas. Marine strata are mainly distributed in the circum-Pacific and a major part of Europe. The Boreal realm may be subdivided into three regions, the Boreal Sea region, the Northern Laurasian region, and the Arctic region.

3. The Gondwanan realm was situated in the southern temperate–subtropical zone, with its southern latitude ranging from 25°S to 70°S. However, post-Jurassic plate movements have widened the modern latitudinal ranges from 15°S to the South Pole. The Indian plate had already migrated to about 30°N. The realm may be subdivided into two regions: the Gondwanan Tethyan region (sea) and the Southern and Central Gondwanan region (land) (Stevens 1965; Willeg 1973; Rage 1988).

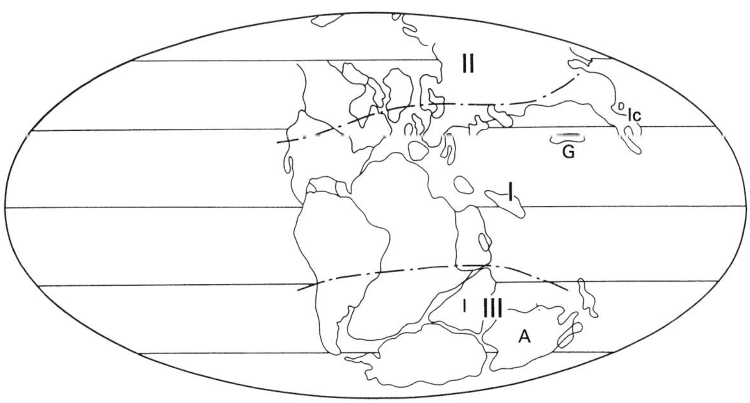

Fig. 10.1 Map of the Early Jurassic palaeobiogeographical provincialization of the world. I, Laurasian Tethyan realm; II, Boreal realm; III, Gondwanan realm; A, Australia; G, Gangdise; I, India; Ic, Indochina.

Table 10.1 Jurassic palaeobiogeographical provincialization of the world

Climatic zone	Realm	Region	Early Jurassic	Middle Jurassic	Late Jurassic
N. temperate–subtropical	Boreal realm	Arctic Sea region	*Arctasteroceras–Lugodoctylites*	*Arkelloceras–Arcticoceras* *Lenella–Arctotis*	*Streblites–Chetaites* *Buchia*
		Boreal Sea region	*Dactylioceras–Asteroceras* *Acrocoelites–Cuspiteuthes* *Oxytoma cygnipes–Hippopodium ponderosum*	*Seymourites–Kosmoceras* *Megateuthis* *Isognomon isognomonoides–Mytiloceramus*	*Amoeboceras–Kochpurites* *Cylindroteuthis–Pachyteuthis* *Buchia rugosa–B. mosquensis–Nipponitrigonia*
		N. Laurasian region	*Coniopteris–Phoenicopsis* *Ferganoconcha–Sibireconcha brevis*	*Pseudograpta* *Pseudocardinia–Yananconcha*	*Cladophlebis aldanensis–Raphaelia diamensis* *Keratestheria* *Sphaerium–Monyinia*
Tropical–subtropical	Laurasia–Tethyan realm	C. Laurasian region	*Ptilophyllum–Coniopteris* *Eosolimnodiopsis* *Lilingella–Qiyangia* *Lufengosaurus*	*Mamenchisaurus* *Eolamprotula–Psilunio*	*Cupressinocladus–Pagiophyllum* *Eoseutheriopsis*
		Laurasian Tethys Sea region	*Arietites–Schlotheimia* *Passaloteuthis* *Pseudopecten–Gryphaea arcuata* *Lithiotis–Cardinia*	*Sonninia–Oppelia–Parkinsonia* *Gryphaea bilobita–G. dilatata*	*Virgatosphinctes–Perisphinctes* *Pachymytilus–Spondylopecten* *Neoburmesia–Nipponitrigonia*
S. temperate–subtropical	Gondwanan realm	Gondwanan Tethys Sea region	*Bouleiceras–Protogrammoceras* *Weyla ambongoensis–Pseudoacella–Kalentera*	*Ermoceras–Thamboceras* *Macrocephalites–Indocephalites* *Kaleatera–Haastina*	Perispinctids–aspidoceratids *Belemnopsis–Hibolithes* *Indotrigonia–Opisthotrigonia* *Malayomaorica*
		C.-S. Gondwanan region			

10.2 JURASSIC PALAEOBIOGEOGRAPHICAL PROVINCIALIZATION OF CHINA

10.2.1 AN OUTLINE

In the Jurassic (Table 10.2) terrestrial environments were predominant in China (Wang Hongzhen 1978), accounting for about 85 per cent of the total area. The mountainous areas included Palaeo–Tianshan, Palaeo–Kunlunshan, Palaeo–Qinling, Palaeo–Qilianshan, Palaeo–Dabieshan, Palaeo–Hengduanshan, mountainous areas in southern Jiangxi, Zhejiang, and Fujian, and the mountain chains of Altay and Yinshan. There were also plateaus, such as north-eastern China, North China, the northern bank of the Huai He, and Yunnan–Guizhou. There were lake basins, such as Qingyang Lake, Bashu Lake, Dian Lake, Yunmong swamp, Tarim Basin, and Junggar Basin. In the eastern circum-Pacific region there were also many faulted basins. Three areas of marine environment were present, in south-western China (Wang Si'en *et al.* 1985), south-eastern China, and north-eastern China; the largest sea area was the region where a great transgression took place in the Middle Jurassic, covering the greater part of Tibet, southern Qinghai, western Yunnan, and a part of Xinjiang. The south-eastern sea only existed during the Early Jurassic, covering parts of Guangdong, Hunan, Jiangxi, Fujian, and Taiwan; and the north-eastern sea occupied the Wandashan area of eastern Heilongjiang during the middle Late Jurassic transgression. Such varied physical environments with tropical–subtropical palaeoclimates over a vast area were favourable for the existence of a wide variety of organisms. Ammonites and bivalves dominated in marine facies, whereas plants, vertebrates, and bivalves flourished in terrestrial environments. In China, the arid zone obviously shifted from south in the Early Jurassic to north in the Late Jurassic, as indicated by changes in the distribution of coal-bearing beds and red beds (Fig. 10.2).

The Jurassic in China embraces three realms, the Laurasian Tethyan realm (tropical semi-arid to arid), the Boreal realm (subtropical–northern temperate humid to semi-humid), and the Gondwanan realm (southern temperate), with the boundary between the first two realms shifting in time. Generally, the boundary shifted from south to north from the Early to the late Middle Jurassic and then shifted southward in the Late Jurassic. This is explained by the clockwise rotation of the Chinese continent, especially during the Early and Middle Jurassic, but some people have argued that the widening of the tropical zone was caused by the world temperature increase during the Jurassic period. The authors believe that both situations were operative in the Jurassic. However, Kosygin and Parfenov (1981), noted that a complete welding of the Palaeo-Tethyan microcontinents to the Siberian margin was not completed until Late Jurassic to Early Cretaceous times, with a counterclockwise rotation. A counterclockwise rotation is possible in the Late Jurassic and Early Cretaceous because there was a shift to cooler conditions over the whole of China, but certainly did

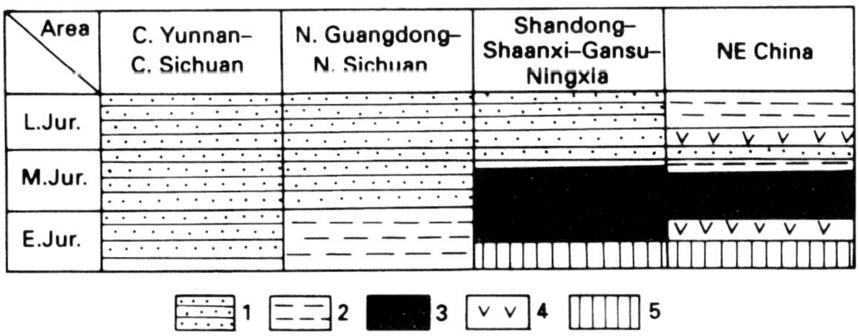

Fig. 10.2 Migration of the Jurassic coal beds and red beds in eastern China. 1, Arid or semi-arid red or variegated clastics; 2, coal-bearing strata; 3, minable coal beds; 4, volcanics; 5, no deposits.

Table 10.2 Correlation chart of the Jurassic of China

<table>
<tr><th colspan="3"></th><th>Qomolongma</th><th colspan="2">Tanggula</th><th>Northern Guangdong</th><th>Northern Sichuan</th><th>Western Yunnan</th></tr>
<tr><td rowspan="10">Jurassic</td><td rowspan="3">Upper</td><td>Tithonian</td><td rowspan="3">Menbu</td><td rowspan="3">Qiang-molequ</td><td rowspan="5">Yanshiping Gr.</td><td rowspan="4">Baizushan Gr.</td><td rowspan="2">Penglaizhen</td><td rowspan="3">Bazhulu</td></tr>
<tr><td>Kimmeridgian</td></tr>
<tr><td>Oxfordian</td><td>Suining</td></tr>
<tr><td rowspan="3">Middle</td><td>Callovian</td><td rowspan="2">Larongla</td><td rowspan="2">Wenquan</td><td>Upper Shaximiao</td><td rowspan="3">Hepingxiang</td></tr>
<tr><td>Bathonian</td><td rowspan="2">?</td><td>Lower Shaximiao</td></tr>
<tr><td>Bajocian</td><td>Niehniehhiungla</td><td>Seywa</td><td></td><td>Qianfoyan</td></tr>
<tr><td rowspan="4">Lower</td><td>Toarcian</td><td rowspan="4">Pupugaqiao</td><td rowspan="4">Zhamnaqu</td><td rowspan="4"></td><td rowspan="3">Qiaoyuan</td><td rowspan="4">Baitianba</td><td rowspan="4">Yangjiang</td></tr>
<tr><td>Pliensbachian</td></tr>
<tr><td>Sinemurian</td></tr>
<tr><td>Hettangian</td><td>Jinji</td></tr>
</table>

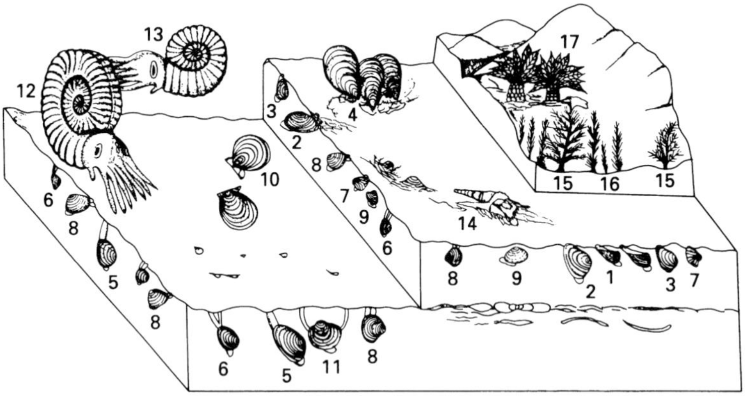

Fig. 10.3 Reconstruction of the Early Jurassic *Arietites–Hongkongites* community of northern Guangdong in the Guangdong–Taiwan–Hunan–Jiangxi–Fujian Bay province. 1, *Lilingella*; 2, *Qiyangia*; 3, *Hunanella*; 4, *Isognomon*; 5, *Pleuromya*; 6, *Nuculana*; 7, *Cardium*; 8, *Protocardia*; 9, *Palaeonucula*; 10, *Camptonectes*; 11, *Luciniola*; 12, *Arietites*; 13, *Hongkongites*; 14, gastropods; 15, pteridophytes; 16, *Neocalamites*; 17, cycadales.

Northern Xinjiang	Qilian	Ordos Basin	Northern Hebei	Western Liaoning	Xingan	Wandashan
Lower Tugulu Gr.	Hekou Gr.	Lower Zhidan Gr.	Xiguayuan Dabei Zhangjiakou Baiqi	Jiufutang Jianchang Yixian	Volcanic rock	Yunshan
	Datonghe				Baoshi	
Kalaza		Fenfanghe			Xingren	Chaoyangten
Qigu	Kushuixia	Anding	Houcheng	Tuchengzi	Fujiawazi	
Toutunhe	Xinghe	Zhiluo	Tiaojishan	Lanqi	Jubao	Peide
			Jiulongshan			
Xishanyao	Yaojie	Yanan	Longmen	Haifanggou	Wanbao	Dongsheng
Sangonghe		Fuxian	Yaopo	Beipiao		
Badaowan	Daxigou		Nandaling	Hinglongou	Hongqi	
			Xingshikou	Kuntoupoluo		

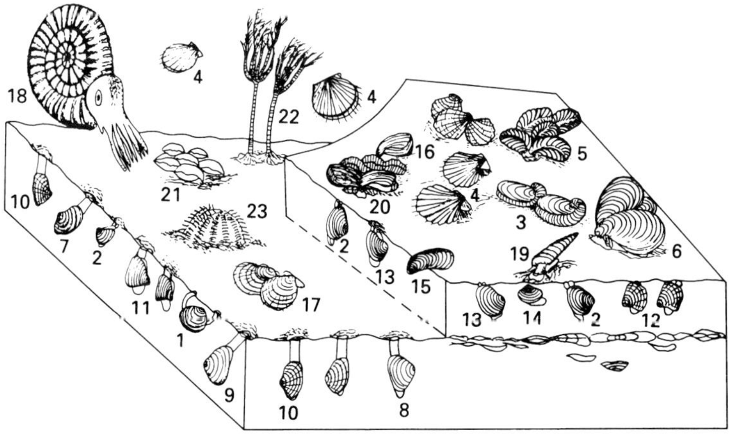

Fig. 10.4 Reconstruction of the Middle Jurassic *Burmirhynchia–Holcothyris* community of eastern Tibet in the Tibet–Western Yunnan province. 1, *Mactromya*; 2, *Protocardia*; 3, *Gryphaea*; 4, *Chlamys*; 5, *Lopha*; 6, *Liostrea*; 7, *Goniomya*; 8, *Pleuromya*; 9, *Homomya*; 10, *Pholadomya*; 11, *Pseudotrapezium*; 12, *Anisocardia*; 13, *Myophorella*; 14, *Palaeonucula*; 15, *Modiolus*; 16, *Meleagrinella*; 17, *Camptonectites*; 18, ammonoides; 19, gastropods; 20, *Burmirhynchia*; 21, *Holcothyris*; 22, crinoids; 23, echinoids.

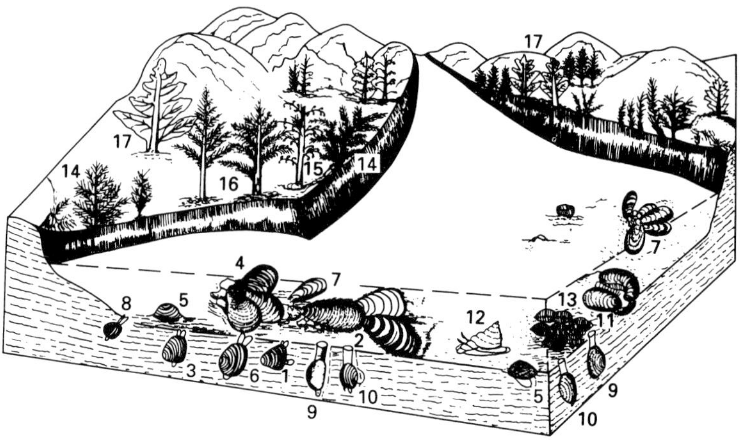

Fig. 10.5 Reconstruction of the Late Jurassic *Buchia–Isognomon* community in Wandashan province. 1, *Astarte*; 2, *Isognomon*; 3, *Corbula*; 4, *Aguilerella*; 5, *Nuculoma*; 6, *Sinopsammobia*; 7, *Mytilus*; 8, *Nuculana*; 9, *Pachymya (Arcomya)*; 10, *Pleuromya*; 11, *Buchia*; 12, gastropods; 13, *Thurmannella*; 14, pteridophytes; 15, ginkgoales; 16, taxodiaceans; 17, conifers.

not occur before the Late Jurassic. Each realm could be subdivided into regions and provinces. The names of each region and province and their representative fossils are given in Table 10.3. Figures 10.3, 10.4, and 10.5 partially demonstrate the fauna and flora in China during the Jurassic period, represented by northern Guangdong (Early Jurassic), eastern Tibet (Middle Jurassic), and the Wandashan area (Late Jurassic).

10.2.2 EARLY JURASSIC PALAEOBIOGEOGRAPHICAL PROVINCIALIZATION OF CHINA

In the Early Jurassic of China there was much land but little sea, and many mountains but few lake basins. There were only two marine areas; one was at the corner of Tibet, southern Qinghai, western Yunnan, and south-western Xinjiang, and the other was a gulf covering parts of Guangdong, Taiwan, Hunan, Jiangxi, and Fujian. The latter clearly experienced two transgressions after the Late Triassic; one of which occurred in the Early Jurassic. All the large mountain systems mentioned above already existed then. Among them were the Palaeo-Kunlunshan, Palaeo-Bayan Har Shan, and Palaeo-Qinling, extending approximately from west to east and dividing China into northern and southern regions. According to the distribution of climatic zones, a more important boundary for dividing biogeographical realms passes through south-western Guangxi, south-western Guizhou, western Sichuan, southern Qinghai, and the south-western corner of Xinjiang (Fig. 10.6). The area to its south belonged to an arid or semi-arid tropical climate zone and is attributed to the Laurasian Tethyan realm; whereas to the north, the area belonged to a humid or semi-humid, subtropical–north temperate zone, and was attributed to the Boreal realm. The Gondwanan Tethyan region of the Gondwanan realm was situated in a subtropical to south temperate zone. Based on organism distributions, depositional patterns, and the results of cluster analysis of bivalves, the Early Jurassic in China can be biogeographically divided into six provinces pertaining to three realms.

Laurasian Tethyan realm

This realm can be divided into the Laurasian Tethyan region and the Central Laurasian region. Each region in China contains one province.

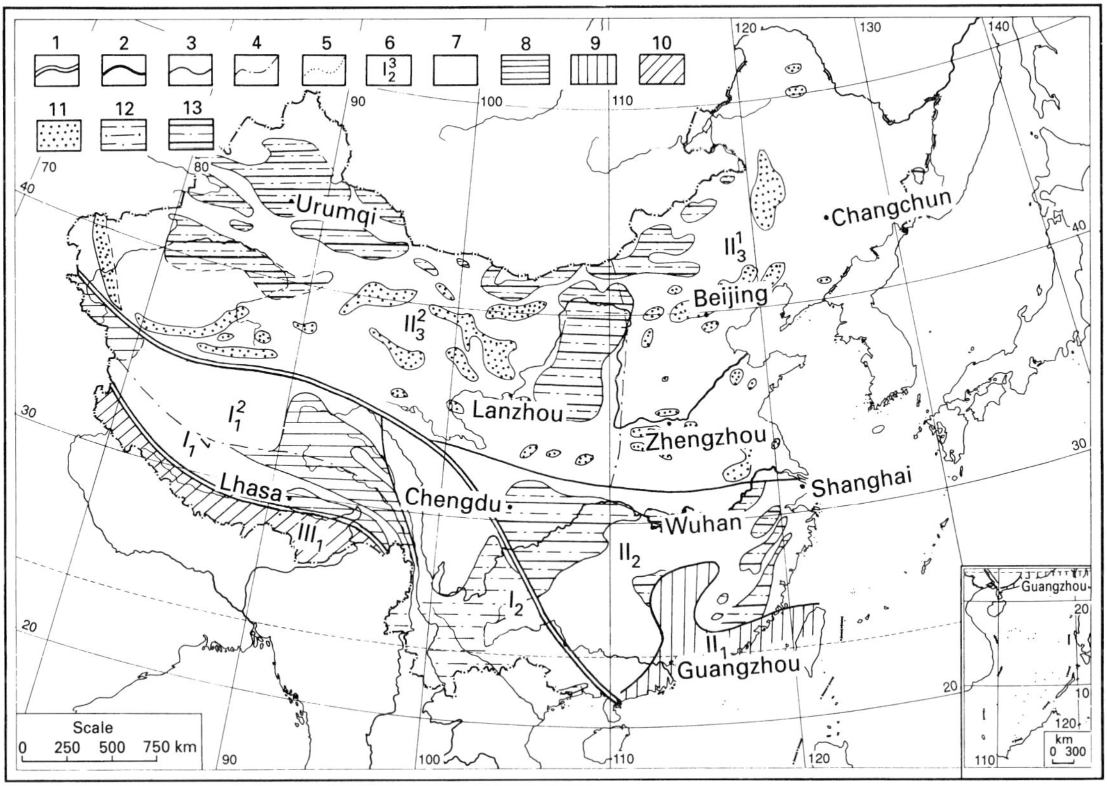

Fig. 10.6 Map of the Early Jurassic palaeobiogeographical provincialization of China. 1, Boundary of realms; 2, boundary of regions; 3, boundary of provinces; 4, boundary of biomes; 5, boundary of communities; 6, roman digit = realm mark, subscript = province mark, superscript = biome mark; 7, Oldland. 8–10, Marine and paralic regions: 8, South China; 9, North China; 10, Gondwana. 11–13, Terrestrial regions (Late Permian and Mesozoic–Cenozoic): 11, intermontane basins; 12, fluvio–lacustrine basins; 13, coal-bearing basins. I, Laurasian–Tethyan realm; I_1, Tibet–Qinghai (marine) province; I_1^1, Gangdise–Nianqingtanggula biome; I_1^2, Tanggula biome; I_2, South-western China (terrestrial) province; II, Boreal realm; II_1, Guangdong–Taiwan–Hunan–Jiangxi– Fujian Bay province; II_2, South China province; II_3, Northern China province; II_3^1, North–North-eastern China biome; II_3^2, North-western China biome; III, Gondwanan realm; III_1, Himalayan province.

Qinghai–Tibet province (marine)

This contained a part of the northern marginal sea of the Tethyan Ocean and the area influenced by its transgression, as well as the Gangdise block, which appeared as an island in the tropical Tethyan Ocean at that time. It consisted of two parts: the Gangdise–Nianqingtanggula and Tanggula–Karakorum. Apparently the Early Jurassic sea did not reach as far as western Yunnan. Due to pronounced environmental differentiation, fossils common to all constituent parts are few.

Gangdise–Nianqingtanggula biome

This biome lay between the Yarlung Zangbo River and the Bangong Lake–Gerze–Nujiang line. The Neo-Tethyan Ocean was to the south and the Tanggula–Karakorum block to the north. This condition persisted until its collision with the Qiangtang block at the end of the Late Jurassic. Therefore, there was a low biotic similarity of this biome to those of Tanggula and western Yunnan. Due to a lack of data, it is difficult to establish this as an independent province. Thus it is regarded as a biome in this chapter.

Table 10.3 Jurassic palaeobiogeographical provincialization of China

<table>
<tr><th>Realm</th><th>Reg.</th><th>Pr.</th><th>Early Jurassic</th><th colspan="2">Early
Middle Jurassic</th><th colspan="2">Late
Middle Jurassic</th><th colspan="2">Late Jurassic</th></tr>
<tr><td rowspan="4">Boreal</td><td>Boreal Sea</td><td>Wandashan</td><td colspan="3">Incorporated with N. China Pr.</td><td colspan="2">Arctocephalites–Stenocodoceras amm. Ass.
Ginkgo–Coniopteris plant Ass.</td><td colspan="2">Isognomon isognomoides–Buchia mosquensis biv. Ass.
Coniopteris burejensis–Gleichenites norden-skioldi plant Ass.</td></tr>
<tr><td rowspan="3">N. Laurasia</td><td>N. China</td><td>Coniopteris–Phoenicopsis plant Ass.
Ferganoconcha–Shaanxiconcha–Sibireconcha biv. Ass.</td><td>N. China–NE China Pr.</td><td>Coniopteris–Phoenicopsis plant Ass.
(no thick-shell freshwater mussels)</td><td>NE China Pr.</td><td>Coniopteris–Phoenicopsis plant Ass.</td><td>N. China–NE China Pr.</td><td>Jehol fauna
(Fuxin Formation excluded in this paper)
Ephemeropsis
–Eosestheria
–Lydoptera</td></tr>
<tr><td>S. China</td><td>Ptilophyllum–Coniopteris plant Ass.
Qiyangia–Apseudocardinia biv. Ass.</td><td>C. China–NW China Pr.</td><td>Coniopteris–Phoenicopsis plant Ass.
Psilunio–Eolamprotula biv. Ass.</td><td rowspan="2">N. China–NW China Pr.
(belonging to C. Laurasian Reg.)</td><td rowspan="2">Coniopteris–Phoenicopsis plant Ass.
(rare)
Eolamprotula–Psilunio biv. Ass.</td><td colspan="2" rowspan="2">Incorporated with S. China–NW China Pr.</td></tr>
<tr><td>Guangdong–Taiwan–Hunan–Jiangxi–Fujian Bay</td><td>Arietites–Hongkongites amm. Ass.
Teinonuculana–Inoceramus
–Lilingella biv. Ass.</td><td colspan="2">Incorporated with Kunlun–S. China Pr.</td></tr>
</table>

Laurasia–Tethys	Laurasian Tethys	Qinghai–Tibet (Sea)	*Gleviceras–Arnioceras* amm. Ass. (only in Tanggula Mts) *Nuculana–Weyla* biv. Ass.	Qinghai Tibet W. Yunnan Pr.	*Burmirhynchia–Holcothyris* bra. Ass. *Lopha–Gryphaea–Trigonia* biv. Ass. *Psilunio–Cuneopsis* biv. Ass	*Eolamprotula* biv. Ass.		*Virgatosphinctes –Indosphinctes –Perisphinctes* amm. Ass. (not in W. Yunnan) *Peregrinoconcha* biv. Ass.
	C. Laurasia	SW China (Land)	*Lufengosaurus* fauna *Podozamites–Otozamites* flora	Kunlun–S. China Pr	*Ptilophyllum–Coniopteris* plant Ass. *Pseudocardinia–Tutuella* biv. Ass. *Psilunio–Cuneopsis–Eolamprotula* thick-shell freshwater mussels Ass. *Mamenchisaurus* fauna		S. China–NW China Pr.	No Jehol fauna *Acanthopteris–Onychiopsis* plant Ass.
Gondwanan	Gondwanan Tethys	Himalaya	*Echioceras–Schlotheimia–Kondiloceras* amm. Ass.	*Dorsetenia–Sonninia* amm. Ass. *Trigonia–Indogrammatodon* biv.Ass.		*Macrocephalites–Indocephalites* amm. Ass. *Belemnopsis lalonglensis–Hibolithes catlinensis* bel. Ass.	*Haplophylloceras–Virgatosphinctes* amm. Ass. *Belemnopsis gerardi–B. uhligi* bel. Ass. *Buchia extensa–B. blanfordiana* biv. Ass.	

Abbreviations:
amm., ammonite; Ass., assemblage; bel., belemnite; biv., bivalve; bra., brachiopod; Mts, mountains; Pr., Province; Reg., Region.
* In the early Jurassic W.Yunnan was incorporated with the SW China province.

An increasing amount of evidence shows that a suture exists along the line from Bangong Lake to Nujiang River, i.e. an ocean basin probably existed between the two blocks in the Jurassic. Relatively extensive ultrabasic rocks have been found along Dengqen, Amdo (northern Tibet), and the northern side of Bangong Lake in north-western Tibet. An ophiolite suite consisting of ultrabasic rocks, pillow lavas, and radiolarites has also been found in the Jiangcuo area, 200 km to the west of Naqu. According to Wang Naiwen (1984), the strata contain such abundant radiolarian fossils as *Sethocyrtis*, *Criptocapsa, Dicolocapsa,* and *Tricolocapsa,* indicating a date not earlier than the Jurassic and not later than the Early Cretaceous. In some localities, Early Cretaceous carbonates overlie Late Jurassic flysch sediments, indicating that geosynclinal conditions had disappeared by the Early Cretaceous. Another piece of relevant evidence is that freshwater bivalves in the northern continent did not reach this biome until the end of the Jurassic period (Wen Shixuan 1981).

The central area of the biome was occupied by the Gangdise lowland, accounting for half the area, with no sedimentation. Sea water mainly covered the faulted gulfs in the eastern part. In the south, in the Lhasa to Nyingchi area, the Mailonggang Group consists of tuffaceous sandstones, sandstones, and shales with a thickness of 1877 m, locally containing coal seams, a part of which probably belongs to the Upper Triassic. The relatively impoverished biota was dominated by bivalves of the *Pseudotrapezium–Ceratomya* assemblage. The gastropod *Nerinea* also existed here. The flora was mainly represented by the *Coniopteris–Equisetum* assemblage. In the Naqu–Luolong area in the northern part of this biome, the lower part of the Jurassic consists of conglomerates, sandstones, marls, and dolomitic limestones, whereas the upper part is dominated by limestones up to 3594 m thick, called the Sangba Formation. A *Plicatostylus–Weyla* assemblage is representative of molluscs, while brachiopods include *Dinoproductus*.

Tanggula–Karakorum biome (terrestrial)

This biome extends from Karakorum in the west through the Tanggula Mountains to the Qamdo area in the east. Its southern boundary extends from Bangong Lake, through Garze to the Nujiang River, while its northern boundary lies along the line of Karakorum, the Hohxil Mountains, Yushu, and the Jinshajiang River. Upland areas predominated in this biome, with partly shallow, marine sediments during the Early Cretaceous. The Zhamnaqu Formation in the Tanggula area consists predominantly of fine clastics, limestones, and marls up to 2114 m thick. Ammonites are represented by the *Gleviceras–Arnioceras* assemblage and bivalves include *Trigonia* and *Pleuromya*. The Bagongbulansha Group of Early–Middle (undifferentiated) Jurassic in the Karakorum area consists of conglomerates, tuffaceous shales, and limestones of paralic facies with a *Cladophlebis–Ptilophyllum* floral assemblage and a *Camptonectes–Chlamys* bivalve assemblage. In the Qamdo area there are purple conglomerates, siltstones, and shales, interbedded with limestones, coal seams, and gypsum, the Early Jurassic fossils of which have not been described in detail.

South-western China province (terrestrial)

This biome covered Guangxi, Guizhou, western Sichuan, the Yushu area in Qinghai, and the whole of Yunnan. It was situated in the tropical arid to semi-arid climatic zone and had a terrestrial environment. The strata are mainly composed of purple and variegated clastics, deposited in inland basins and mountain basins, such as the Yangjiang Formation of western Yunnan, the Yimen Formation in the Xichang district, the Lufeng Formation of central Yunnan, and the Wangmen Formation and Baise Formation in southern Guangxi. The fauna in this biome contains vertebrates, freshwater bivalves, ostracods, and estheriids. In China, Early Jurassic reptiles, belonging to the so-called *Lufengosaurus* fauna, were most abundant in this area. Important members of this fauna were the reptiles *Kunmingosauras* and *Yunnanosauras,* bivalves *Cuneopsis, Unio,* and *Sibireconcha,* and the ostracod *Darwinula*. The flora is represented by the *Podozamites–Otozamites* assemblage, including *Coniopteris, Clathropteris*. Due to the arid climate, plants did not thrive. Chen Peiji and Shen Yanbin (1970) found that the conchostracan *Palaeolimnadia* assemblage represented by *Palaeolimnadia dinnanensis* contains fossils that are large in size but with thin shells. No *Euestheria taniiformis* was found. It is generally believed that the thinning of conchostracan shells within a given genus was caused by a warmer climate.

Boreal realm

In China there were four provinces, all belonging to the Northern Laurasian region of this realm.

Guangdong–Taiwan–Hunan–Jiangxi–Fujian province (gulf)

During the Early Jurassic the boundary between the tropical–subtropical and subtropical–north temperate zones extended from Nanning through Guiyang to Tanggula. Thus this biome should belong to the non-tropical Boreal realm. It was an embayment of the Palaeo-Pacific Ocean that transgressed onto the Laurasian continent, leading to the deposition of sediments of paralic facies, with a large gulf in Guangdong narrowing towards Hunan and Fujian. Two facies can be identified:

1. Shallow sea–offshore facies. This includes a large part of Guangdong, with the normal marine ammonites *Arietites–Hongkongites* assemblage and the bivalves *Teinonuculana*, *Oxytoma*, and *Astarte*. The ammonites reach north-westward to Ruchen County (Hunan).

2. Offshore–gulf facies. This was located at Zixing, Sandu, Lingling, and Liling (Hunan) and southern Fujiang. This region was characterized by bivalves of shallow marine or transitional facies called the *Bakevellia–Modiolus* assemblage, including *Modiolus*, *Mytilus*, *Bakevellia*, and *Chlamys* (Chen Zhonghui *et al.* 1982). It should be noted that *Lilingella* and *Hunanella* are always distributed in coal measures without any association with typical marine fossils. It remains to be further examined whether they are actually marine or freshwater bilvalves (Wu Shunbao 1981). Terrestrial sediments intercalated with the marine strata contain coal seams, with plants characteristic of the *Ptilophyllum–Coniopteris* assemblage. Early Jurassic marine sediments apparently also occur in Taiwan, but data are sparse.

South China province

This covers Guangxi, Guizhou, eastern Sichuan, Hunan, Fujian, Zhejiang, southern Hubei, southern Anhui, and southern Jiangsu. It was referred to the northern region of Laurasia because within China's territory the Early Jurassic tropical climate zone was shifted southwards. It was within the subtropical–north temperate zone, humid to semi-humid, with a sedimentary environment dominated by inland lakes and faulted basins. The sediments are grey to dark grey clastics, intercalated with coal seams (Chen and Cao 1986), e.g. the Baitianba Formation of Sichuan, the Xiangxi Formation (*sensu stricto*) of western Hubei, the Wuchang Formation of eastern Hubei, the lower part of the Xiangshan Group of Jiangsu–Anhui, and the Zaoshang Formation of Hunan–Jiangxi. Freshwater bivalves are abundant and represented by the *Apseudocardinia–Qiyangia* assemblage. Plant fossils are rich near the coal seams and characterized by the *Ptilophyllum–Coniopteris* assemblage, including *Ptilophyllum contiguum*, *Coniopteris hymenophylloides*, *Otozamites hsiangchiensis*, *Phlebopteris polypodioides*, *Sagenopteris*, *Nilssonia*, and *Baiera*. Diversity statistics for the Early Jurassic flora have been undertaken by Zhang Zicheng (Wang Si'en *et al.* 1985) (Table 10.4).

Table 10.4 Diversity statistics for the Early Jurassic plant genera and species from the South China and North China provinces of the Boreal realm (from Wang Si'en *et al.* 1985)

S. China province	Early	39/14	35/12	8/5	7/5	7/2
	Late	39/13	33/9	21/6	12/7	7/3
N. China province		24/9	51/14	33/11	16/9	18/5

In the early part of the Early Jurassic the flora of South China had some properties in common with that of eastern Greenland, Sweden, Germany, and France, with about 29 species of 17 genera similar or identical to those of these regions. Estheriids are represented in the Baitianba Formation of northern Sichuan; altogether *c*. 30 species form the *Palaeolimnadia baitianbaensis* assemblage, which includes *Euestheria* and *Bulbilimnadia* (Chen and Shen 1979). The many species of *Palaeolimnadia* are characterized by small individuals with thick shell having no decoration, thus differing from those of the South-western China province.

Northern province

This covers a vast area, including the whole of North China, north-eastern, and north-western China. In western China the boundary with the Laurasian Tethys passes through Kunlun and Bayan Har; in the east, its boundary with the South China province goes through Dabashan and Dabieshan. The flora is characterized by the *Coniopteris–Phoenicopsis* assemblage, the diversity statistics of which are shown in Table 10.4. Compared with the contemporary flora of the South China province, there were quite a number of common elements among the pteridophytes, cycadophytes, conifers, ginkgoales, and equisetales, which indicated that both provinces can reasonably be referred to the same realm. Yet there is a difference between the two. In South China the thermophilous cycadophytes are rather abundant, accounting for about 40 per cent of species, while the ginkgoales only account for 13 per cent, reflecting the subtropical characteristics. On the other hand, the northern province yields 17 per cent cycadophytes and 24 per cent ginkgoales, reflecting the warm temperate characteristics of this area. The bivalves are mainly characterized by the *Ferganoconcha–Sibireconcha–Shaanxiconcha* assemblage. Estheriids are represented by the *Eosolimnadiopsis* assemblage which, in the Fuxian Formation of Shaanxi and Inner Mongolia, yields many species such as *E. shanbeiensis* and *E. gushanensis*. This genus also occurs in the Menkoushan Formation of Liuyang County, Hunan.

The province may be subdivided into two biomes, i.e. the North China–North-eastern China biome and the North-western China biome.

North China–North-eastern China biome

This covers the whole of North China and North-eastern China. A very important feature of this area was circum-Pacific volcanic activity. There were rather few depositional regions, chiefly scattered fault basins, while the major area was occupied by the northern China and north-eastern China plateaus. As a representative of North China, the western hills of Beijing display the conglomerates, sandstones, shales, and coal seams, as well as medium-basic lavas, tuffs, and agglomerates. The plants are characterized by the *Coniopteris–Phoenicopsis* assemblage including *Coniopteris bureiensis, Phoenicopsis angustifolia, Todites denticulata, Ctenis chinensis,* and *Czekanowckia rigida*. The freshwater bivalves include *Naiadites?, Shaanxiconcha, Sibireconcha,* and *Ferganoconcha*. In north-eastern China, the Early Jurassic volcanic activity occurred frequently, but the late stage of this epoch was an important coal-forming period. The characteristics of the plant assemblage are similar to those of North China, being dominated at an early stage by arthrophytes, but with no *Coniopteris*. The bivalves are represented by *Tutuella* and *Pseudocardinia*. The Monying Basin of Shandong yields the estheriid fossils *Palaeolimnadia chuanbeiensis* and *P. longmenshanensis*, very similar to the fossils of South China.

North-western Biome

This covers Shaanxi, Qinghai, most of Xinjiang, Gansu, the whole of Ningxia, and the western part of Inner Mongolia. Palaeo-Tianshan and Palaeo-Qilianshan cross cut this biome. There were two kinds of sedimentary environments: large-scale inland basins represented by the Ordos Basin in the east and the Tarim Basin and Junggar Basin in the west, and narrow intermontane basins in the central part.

The Ordos and Guyang basins cover Shaanxi and central Inner Mongolia. The deposits are fluvial–lacustrine and swamp sediments laid down in wide inland basins, often in the form of coal-bearing clastics. Plants are represented by the assemblage of

Phoenicopsis–Coniopteris with *Phoenicopsis angustifolia, Coniopteris hymenophylloides, Todites williamsoni, Otozamites benglensis, Ginkgoites digitata*, and *Equisetites serrani*. The bivalves are dominated by *Ferganoconcha*, including many species such as *F. subcentralis, F. curta, F. shaanxiensis*, and *Shaanxiconcha*. The estheriids include *Eosolimnodiopsis gushanensis*. The Qilian–western Qinling intermontane basins cover a large part of Gansu, northern Qinghai, and western Inner Mongolia. The depositional environment was mainly one of narrow elongate intermontane basins in the north-east trending Palaeo-Qilian Mountains, where a suite of coal-bearing clastics was deposited. The plant fossils of the early stage are quite different from those of South China, being monotonous and impoverished, similar to the contemporary plants of Central Asia, and Russia. Cycadophytes and pteridospermophytes are scarce, and *Coniopteris* has not been found. In the late stage, ginkgoales and conifers, such as *Ginkgoites huttonii, Baiera furcata, Phoenicopsis angustifolia*, and *Coniopteris*, increased rapidly. The freshwater bivalves include *Tutuella* and *Pseudocardinia*.

In Xinjiang, the depositional area was chiefly the Junggar Basin, Turpan Basin, and Tarim Basin, while the Palaeo-Tianshan Mountains were located in the central part. In these basins coal-bearing clastics were deposited. Plants flourished and are characterized by the *Phoenicopsis speciosa–Coniopteris hymenophylloides* assemblage, the principal elements of which include *Sphenobaiera longifolia, Pityophyllum nordenskioldi*, and *Ginkgoites sibirecus*. Freshwater bivalves were relatively developed, including *Sibireconcha, Ferganoconcha, Tutuella*, and *Cuneopsis*.

Gondwanan realm

The Gondwanan areas in China's territory belong to the Himalayan province, Gondwanan Tethyan region, the Gondwanan realm.

The Himalayan province

The Himalayan province formed the eastern rim of the Indian subcontinent and was located in the south temperate zone. It displayed mainly normal neritic environments with relatively abundant ammonites and bivalves. The ammonites are represented by the *Kondiloceras–Echioceras–Schlotheimia* assemblage and the bivalves by the *Meleagrinella ovalis–Entolium demissum* assemblage, together with the belemnite *Dicoelites*, gastropods, and forams. The general features of the biota indicate close affinities with Gondwana while being quite different from northern and central Tibet, which must reflect a major geographical separation.

Cluster analysis of Early Jurassic bivalves

Bivalves have been subjected to cluster analysis using the Otsuka coefficient. The results show both concordances and discordances with the provincialization indicated in the text (Fig. 10.7). On the one hand, areas attributed respectively to the Boreal and Laurasian-Tethyan realms can be separated readily, with the exception that the South China province shows a high similarity with the neighbouring Southwestern China province, which only implies close bivalve but not plant relations between the two. Another exception is the clustering of Himalaya with Guangdong–Taiwan, the only explanation of which is that both are marine biotas, while the North China terrestrial biota is different from the Guangdong–Taiwan marine biota although they belong to the same region.

10.2.3 MIDDLE JURASSIC PALAEOBIOGEOGRAPHICAL PROVINCIALIZATION OF CHINA

The distribution of mountains and oldlands was basically the same as in the Early Jurassic. However, the Middle Jurassic palaeogeography of China experienced a few changes compared with the Early Jurassic. The Guangdong–Jiangxi Bay disappeared, the Tibet–Qinghai–Yunnan Sea enlarged, a Boreal transgression formed the Wandashan Bay in north-eastern China, and the inland basins in south-eastern and north-western China shrank in size. The change in palaeoclimate was more remarkable, resulting in the northward expansion of the

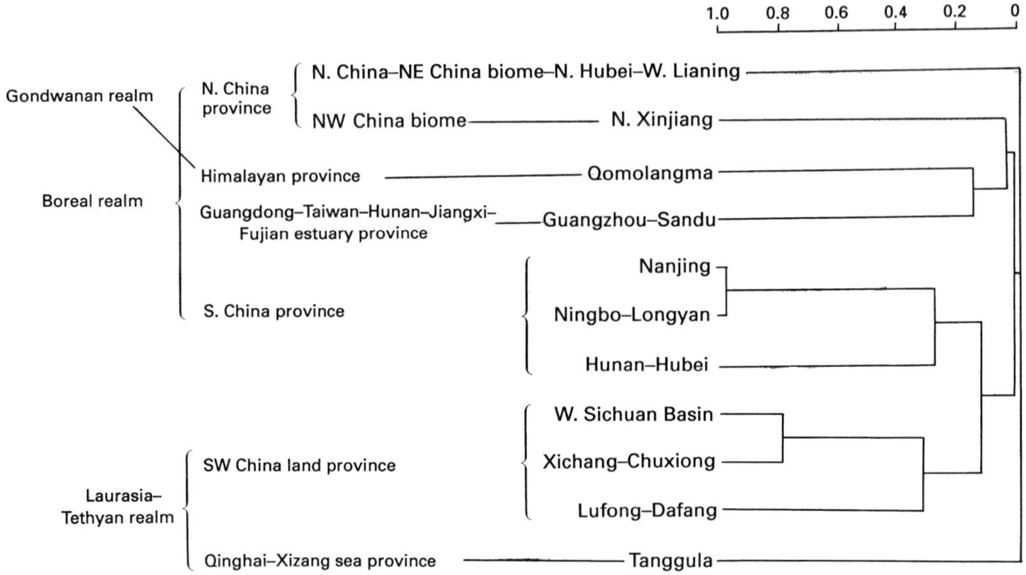

Fig. 10.7 Cluster analysis of the Early Jurassic bivalves of China.

subtropical arid belt. In the early Middle Jurassic (Bajocian) the boundary of the tropical–subtropical zone shifted northward to the Kunlun–Qinling–Dabie–Tianmoushan line, and in the late Middle Jurassic (Bathonian–Callovian) it further moved to the Erenhot (Inner Mongolia)–Qinhuangdao line. As terrestrial biogeographical provincialization was largely controlled by palaeoclimate, it showed quite different patterns between the early and late stages, necessitating separate accounts for these two stages.

Early Middle Jurassic palaeobiogeographical provincialization

The Laurasian Tethyan region of the Laurasian-Tethyan realm

The Tibet–Western Yunnan province

A notable transgression occurred in Middle Jurassic times, which expanded the pre-existing Tibet–Qinghai Sea and invaded western Yunnan. Marine biota of the province are characterized by ammonites and bivalves, remarkably similar to those of western Europe, and thermophilic *Burmirhynchia–Holcothyris* brachiopods typical of the eastern Asian part of Eurasian Tethys. The bivalves lack the important Ethiopian genus *Indogrammatodon* and cool-water *Buchia* typical of Gondwanan Tethys. These imply a wide separation of the two margins of Tethys. This province includes three biomes.

The Gangdise–Nianqingtanggula biome. This area consisted of a few islands and the Gangdise Lowland in the south and a deep-sea basin in the north, separating itself from Tanggula. The main groups are the *Burmirhynchia–Holcothyris* brachiopod assemblage, the *Trigonia–Goniomya* bivalve assemblage, the *Madracis–Dungulis* hexacoral assemblage, and plants such as *Ptilophyllum* and *Zamites* in terrestrial interbeds. The data available so far are not sufficient to clearly demarcate the early and late stages of the Middle Jurassic, so some of these assemblages may extend to the later stage. The faunas were evidently of tropical Tethyan type and quite different from those of the Himalayan province. To take bivalves, for example, the 18 species reported from the Qomolangma area of the Himalayan province have no elements in common with the contemporary 8 species from the Lhasa area

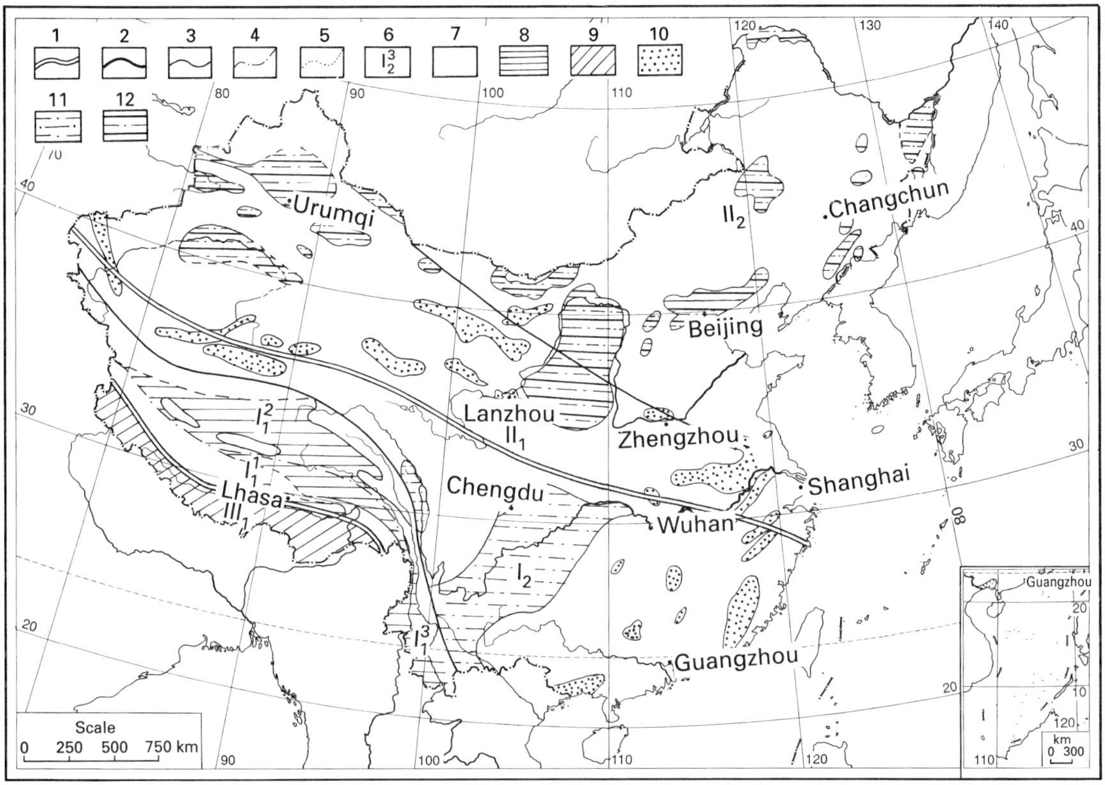

Fig. 10.8 Map of the early Middle Jurassic palaeobiogeographical provincialization of China. 1, Boundary of realms; 2, boundary of regions; 3, boundary of provinces; 4, boundary of biomes; 5, boundary of communities; 6, roman digit = realm mark, subscript = province mark, superscript = biome mark; 7, Oldland, 8–9, Marine and paralic regions: 8, South China; 9, North Gondwana. 10–12, Terrestrial regions (Late Permian and Mesozoic–Cenozoic): 10, intermontane basins; 11, fluvio–lacustrine basins; 12, coal-bearing basins. I, Laurasian–Tethyan realm; I_1, Tibet–Western Yunnan province; I_1^1, Gangdise–Nianqingtanggula biome; I_1^2, Tanggula biome; I_1^3, Western Yunnan biome; I_2, Kunlun–South China province; II, Boreal realm; II_1, Central–North-western China province; II_2, North–North-eastern China province; III, Gondwanan realm; III_1, Himalayan province.

of this biome, although the two areas are near to each other today.

The Tanggula biome. Here the Middle Jurassic sea covered a significantly larger area than in the Early Jurassic, and there was a thick accumulation of marine and alternating marine and non-marine deposits. The Yanshiping Group used to be taken to represent the Middle and Late Jurassic of this area. Because the age of its subdivisions are still under discussion, this book has adopted the viewpoint of *The Jurassic of China* (Wang Si'en *et al.* 1985) which regards the Seywa Formation of the Zhamnaqu Group as the deposits of early Middle Jurassic time, characterized by a *Dorsetenia–Oppelia* ammonite assemblage and a *Lopha–Anisocardia* bivalve assemblage. The contemporary non-marine strata contain a thick-shelled *Psilunio–Cuneopsis* bivalve assemblage typical for tropical Yangtze, which undoubtedly denotes that Tanggula had been incorporated with Eurasia by the Middle Jurassic. The red-bed Dabuka Formation in Qamdo area yields a terrestrial stegosaurid *Changdusaurus* assemblage, as well as a marine *Pseudotrapezium–*

Anisocardia bivalve assemblage, which lived in a water temperature of *c*. 24 °C (as determined by stable isotopic analysis), thus being tropical (Si Xiaoyin, personal communication).

The Western Yunnan biome. The Jurassic trangression reached a maximum during this epoch, and formed widespread marine–terrestrial alternating deposits which cannot be clearly subdivided into early and late stages. The marine *Burmirhynchia–Holcothyris* brachiopod assemblage and the *Camptonectes–Lopha* assemblage suggest affinity with Gangdise and Tanggula, while the freshwater *Psilunio–Cuneopsis* assemblage, together with gastropods, ostracods, and charophytes, denotes contact with Tanggula, Yangtze, and hence Eurasia. The purple and variegated clastics, dolomites, and gypsum deposits with rare plant remains suggest an arid, hot climate more similar to Tanggula and less so to Gangdise.

The Kunlun–South China province

During the early Middle Jurassic this land area was under a semi-arid tropical climate. The sedimentary deposits are purple and variegated clastics without coal seams, such as the Qianfoyan and Shaximiao Formations of Sichuan, the Upper Ziliujing Formation of Hubei, the Qijiang and Shigu Formations of Hunan, and Zhangping Formation of Fujian and Guangdong. Bivalves were the most abundant invertebrates, dominated by a thermophilic freshwater *Cuneopsis–Psilunio–Eolamprotula* assemblage including also *Pseudocardinia, Tutuella,* and *Ferganoconcha*. Although much impoverished compared with the Early Jurassic, the *Ptilophyllum–Coniopteris* flora persisted with moderate diversity, including *Cladophlebis, Brachyphyllum, Otozamites, Podozamites*, and *Dictyophyllum* (Mong Fansong 1985). The Sichuan Basin was the main habitat for dinosaurs such as *Omeisaurus* and *Sinoplisaurus*. Other fossils include the estheriid *Euestheria ziliujingensis* assemblage and the ostracod *Darwinula* assemblage.

The relief of this province consisted mainly of mountains and hills. It was especially so in Kunlun and western Sichuan, where only a few intermontane basins may have received deposits. These probably include the sediments of the Upper Yerqiang Group of Kunlun and corresponding volcanics and clastics in western Sichuan; the dating of both have not yet been determined. The main depositional area was a large inland basin covering the Sichuan Basin, eastern Yunnan, western Hubei, western Hunan, and parts of Guizhou and Guangxi. In south-eastern China there were, in addition, a few north-east trending intermontane basins.

The Kunlun–western Sichuan area is assigned to this province, together with South China, because it was also located south of the Kunlun–Qinling– Dabie line, but to the north and east of the Tibet–western Yunnan marine regime, and was also in the tropical zone, thus bearing the same, though little recorded, biota. However, influenced by the marine weather from the south, it was less arid than South China.

The Boreal realm

The early Middle Jurassic boundary between the Boreal and Tethyan realms in China basically lay along the Kunlun–Qinling–Dabie–Tianmoushan line. Areas to the north were under a subtropical–north temperate, humid to semi-humid climate symbolized by flourishing vegetation, often causing coal accumulation. Unlike the Late Palaeozoic, the Mesozoic coals were mainly formed by temperate plants, hence the southern Boreal boundary of the early Middle Jurassic is also marked by coal-bearing formations, beginning in the west from the Yerqiang Group of Gonggur Shan and the Yangye Formation of Tiklike in Xinjiang, the Xiaomeigou Formation in Qaidam, and the Yaojie Formation in eastern Qinghai and central Gansu, to the Yushanjian Formation of western Zhejiang and the Maolong Formation of central Zhejiang in the east, with the middle parts covering north-western Hubei, Dabie, the Funiu Mountains, and central Anhui lacking coeval deposits. This coal-bearing belt, in accordance with the palaeoclimate zone, displayed a west-northwest–east-southeast delineation of the temperate Boreal realm boundary. The flora of this stage inherited the Early Jurassic pattern and was still characterized by the *Coniopteris–Phoenicopsis* assemblage.

The composition is mainly of pteridophytes (49 species of 13 genera), ginkgoales (40 species, 8 genera), cycadophytes (27 species, 7 genera), conifers (21 species, 10 genera), and equisetales (10 species, 4 genera) (Wang Si'en *et al.* 1985).

In China two provinces can be recognized, namely the Central China–North-western China province and the North China–North-eastern China province.

The Central China–North-western China province

With its southern boundary coincident with the realm boundary, its northern boundary ran along the Xinjiang-Mongolia border–Yinchuan (Ningxia)–Lingfeng (Shanxi)–Xuzhou (Jiangsu). Located within the subtropical–north temperate humid belt, it was a principal area of early Middle Jurassic coal accumulation. Coal-bearing strata include, in addition to those mentioned above, the Xishanyao Formation of northern Xinjiang, the Kezilernur Formation of southern Xinjiang, and the Fanzhi Formation of Shandong. Ferns were most abundant in the flora, accounting for 32 per cent of the total species diversity; other important elements belonged to ginkgoales, cycadophytes, and conifers. The relatively high percentage (18 per cent) of cycadophytes indicates a quite warm climate. The whole flora is named the *Coniopteris–Phoenicopsis* flora. In inland lacustrine basins there flourished a *Psilunio–Eolamprotula* bivalve assemblage similar to that of South China, with major elements such as *Ferganoconcha*, *Tutuella*, *Pseudocardinia*, basically correlative within all basins although with endemic species. The *Darwinula–Timiriasevia* ostracod assemblage was the most flourishing and widespread in this province, commonly appearing in Kunlun–South China and Tibet–western Yunnan, but seldom occurring in North China–north-eastern China. Among the estheriids, *Euestheria ziliujingensis* still prospered, joined by an important newcomer *Qaidamestheria* in this province.

The North China–North-eastern China province

The North China Plateau, Palaeo-Yinshan, and North-eastern China Plateau occupied most of this area, disrupted by scattered intermontane or fault-block basins where early Middle Jurassic deposits were mainly coal-bearing clastics. These include the Taigou Formation of Yinshan, the Haifanggou Formation of western Liaoning, the Wanbao Formation of Inner Mongolia and Xingan, the Baichenzi Formation of Songliao region, and the Datong Formation of Shanxi. The flourishing vegetation yielded by these strata consists dominantly of ferns, with subordinate ginkgoales, cycadophytes, and conifers. In the Wanbao Formation of Jilin, the pteridophytes account for 19 species of 6 genera; ginkgoales 10 species, 5 genera; cycadophytes 10 species, 4 genera; and conifers 4 species, 4 genera. The Yaopo Formation in the western vicinity of Beijing yields pteridophytes (12 species of 7 genera), ginkgoales (15 species of 5 genera), conifers (8 species of 4 genera), and cycadophytes (6 species of 4 genera) (Chen Fen *et al.* 1984). The flora of this stage inherited its features from the Early Jurassic and was still represented by the *Coniopteris–Phoenicopsis* assemblage. Freshwater bivalves are mainly small-sized and thin-shelled, unlike the thermophilic large and thick-shelled *Psilunio–Eolamprotula* assemblage in the southern provinces. The foremost genus was *Ferganoconcha*, with subordinate *Sibireconcha*, *Pseudocardinia*, etc. The estheriids, with the persisting dominance of *Euestheria ziliujingensis*, are represented by the *E. ziliujingensis–E. haifanggouensis* assemblage (Chen and Shen 1979). Ostracods are lacking.

The Gondwanan realm

The Himalayan province

In the early Middle Jurassic this area, a marginal part of the Indian subcontinent, was about 30°S in the Gondwanan Tethys, with a subtropical– temperate climate. The facies signifies two kinds of marine environments. In the west, the Gamba–Nielam–Zanda area was neritic and rich in organisms, as shown by the lower Niehniehhiungla Formation of Nielam. The biota was dominated by ammonites and bivalves, followed by brachiopods, gastropods, and corals. These include the *Dorsetenia–Sonninia* ammonite assemblage, the *Indogrammatodon–Camptonectes* bivalve assemblage (with *Goniomya*,

Trigonia, *Parvamussium*, and *Praeexogyra* as the most important members), and the *Nerinea– Pictavia* gastropod assemblage. These three assemblages are most similar to assemblages in northern and eastern Africa, southern and western Asia, the Atlantic and western Europe, giving a typical Ethiopian aspect quite different from that of the Eurasian Tethys. The brachiopods include a number of endemics; *Burmirhynchia* and *Homoeorhynchia*, widespread in the Eurasian Tethys, also occur. In the east, the Jiangzi–Longzi area included a rift tensional marine trough with intermediate–basic volcanism, probably belonging to the marginal part of the expanding Neotethys. Few fossils have been reported apart from faunules containing *Garantiana*.

Late Middle Jurassic palaeobiogeographical provincialization of China

The Laurasian Tethyan realm

The Tibet–Western Yunnan province

The transgression that began at an early stage was still at its maximum in this late stage. Three biomes are recognized.

The Gangdise–Nianqingtanggula biome. This was still mainly an island in Tethys with similar environments as in the early stage. Most sediments accumulated in marginal geosynclines, including the Quosang Spring Formation of Lhasa, the Lagongdang Formation of middle Nujiang, and the upper Sangribaka Formation of upper Nujiang. The principal marine organisms were ammonites, brachiopods, bivalves, and hexacorals, represented by the *Macrocephalites–Indocephalites* ammonite assemblage, the taxonomy and biostratigraphy of which are now under revision (Westermann 1988); the *Burmirhynchia–Holcothyris* brachiopod assemblage, which is the same as in the early stage; and the *Trigonia–Goniomya* bivalve assemblage, containing *Liostrea birmanica*. The flora consists mainly of *Ptilophyllum* and *Zamites*.

The Tanggula biome. This area experienced an alternating marine and terrestrial environment, represented by the Wenquan (meaning 'hot spring') Formation of the Tanggula Mountains. Bivalves rank first in the biota, with a *Lopha–Pleuromya* marine assemblage and Eolamprotula freshwater assemblage (Wen Shixuan 1981) typical of tropical Tethys and Eurasia, respectively; the latter being indicative of a land connection with Eurasia. The *Burmirhynchia–Holcothyris* brachiopods also occur, as well as ammonites, gastropods, and echinoids. The Callovian *Reineckeia–Choffatia* ammonite assemblage in the Bagongbalansha Group of Karakorum is found in purple-red clastics intercalated with limestones, dolomites, and locally gypsum, with scarce plant remains, which also denote an arid tropical climate.

The Western Yunnan biome. Biotically this area is distinguished from Tanggula by the lack of brachiopods and ammonites. While in Tanggula there accumulated thick sequences of clastics and carbonates up to more than 2000 m thick, here the sediments are mainly clastics several hundreds of metres thick. Its biota mainly consist of bivalves, gastropods, ostracods, and charophytes, e.g. the *Protocardia–Isognomon* marine bivalve assemblage, the *Cuneopsis–Undulatula* freshwater bivalve assemblage, the *Viviparus–Amplovata* gastropod assemblage, and the *Euaclistochara lufengensis* charophyte assemblage.

Kunlun–South China province

This area was still within the arid tropical belt and was mostly covered by upland and mountains with a few intermontane and inland lacustrine basins, the former mainly in the eastern and western parts, the latter in the central part.

The Kunlun–Western Sichuan biome. The few records of sedimentation here denote an intermontane environment; there has been little palaeontological research. The rhyolites, purple to red tuffaceous conglomerates, and grey sandstones and tuffaceous sandstones in the Garze–Yushu area probably contain strata of this age but without definite confirmation from fossils. The upper part of the Yerqiang Group of Kunlun, consisting of light grey arkosic quartz-arenites, conglomerates, and siltstones, as well as the purple to red mudstones and siltstones in the Altag

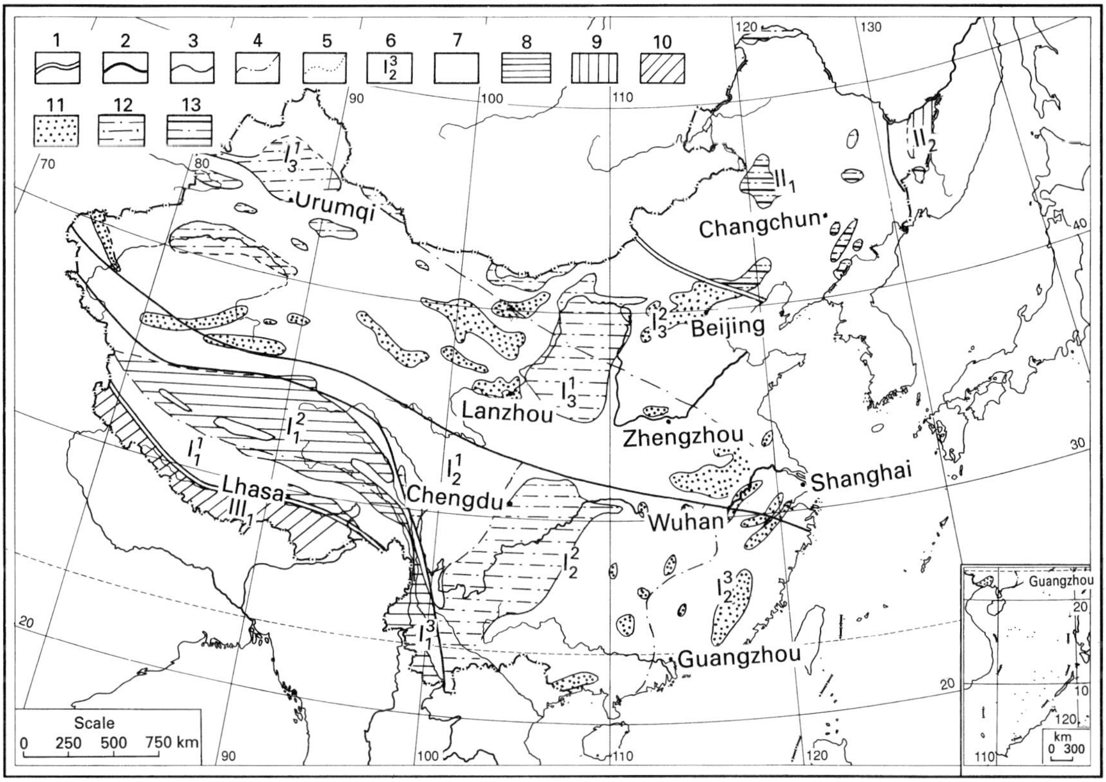

Fig. 10.9 Map of the late Middle Jurassic palaeobiogeographical provincialization of China. 1, Boundary of realms; 2, boundary of regions; 3, boundary of provinces; 4, boundary of biomes; 5, boundary of communities; 6, roman digit = realm mark, subscript = province mark, superscript = biome mark; 7, Oldland. 8–10, Marine and paralic regions: 8, South China; 9, North China; 10, Gondwana. 11–13, Terrestrial regions (Late Permian and Mesozoic–Cenozoic): 11, intermontane basins; 12, fluvio–lacustrine basins; 13, coal-bearing basins. I, Laurasian Tethyan realm; I_1, Tibet–Western Yunnan province; I_1^1, Gangdise–Nianqingtanggula biome; I_1^2, Tanggula biome; I_1^3, Western Yunnan biome; I_2, Kunlun–South China province; I_2^1, Kunlun–Western Sichuan biome; I_2^2, Central South China biome; I_2^3, South-eastern China biome; I_3, North–North-western China province; I_3^1, Central China–North-western China biome; I_3^2, Yinshan–Shandong biome; II, Boreal realm; II_1, North-eastern China province; II_2, Wandashan province; III, Gondwanan realm; III_1, Himalayan province.

area of Kunlun, may also belong to this age. As a whole they denote less arid circumstances than in South China and the area is thus differentiated as a separate biome.

The Yangtze biome. This includes the Sichuan Basin, eastern Yunnan, Guizhou, Guangxi, most of Hubei and Hunan, and was characterized by large lacustrine basins, among which the Sichuan–Yunnan one was the largest. Sediments are mainly purple-red clastics, such as the upper Shaximiao Formation of the Sichuan Basin, the upper Lufeng and upper Zhanghe Formations of Yunnan, the Yanglukou Formation of western Hunan, and the upper Ziliujing Formation of Hubei. Land vegetation was poor due to aridity, but charophytes, dominated by *Aclistochara*, were relatively abundant in the lakes. Among animals, freshwater bivalves and reptiles were the most important, represented by the *Eolamprotula–Cuneopsis* and *Mamenchisaurus–Omeisaurus* assemblages, respectively. Next in importance were gastropods and ostracods, especially *Darwinula*. The Sichuan Basin persisted as the main area for dinosaurs; 21 species of 19 genera of reptiles have been reported in this biome, plus two species undetermined. However, some specialists, such as Zhao Xijing, regard these vertebrates as Late Jurassic in age (Wang Si'en *et al.* 1985).

The South-eastern China biome. This covers southern Jiangsu, Zhejiang, Fujian, Guangdong, and eastern Hunan. Strata of this age include the upper Zhangping Formation of Guangdong and Fujian, and the upper Yushanjian Formation in the Jiande–Ganzhou area, mainly tuffs and grey to purplish red clastics with high contents of volcanic material. The biota was impoverished, with a few plants such as *Podozamites* and *Otozamites* and freshwater bivalves such as *Tutuella*, *Ferganoconcha*, and *Eolamprotula*. There is as yet no record of reptiles.

The North China–North-eastern China province

Due to northward expansion of hot arid climates, this area was already within the semi-arid tropical–subtropical climatic zone during the late Middle Jurassic, with poor vegetation incapable of giving rise to coal accumulation. Two biomes are recognized.

The Central China–North-western China biome. This large belt, extending from Xinjiang to the East China Sea was mainly occupied by mountains and highlands interspersed by basins of different sizes. Strata of this age are mainly grey to purplish red clastics, including the Toutunhe Formation of northern Xinjiang, the Qiketai Formation of southern Xinjiang, the Xinghe Formation of Qilian, Zhiluo Formation of Ordos, the Yungang Formation of Shanxi, the Ma'ao Formation of Henan, the upper Yuantongshan Formation of Anhui, and the upper Xiangshan Group of the Nanjing area. The *Coniopteris–Phoenicopsis* flora persisted into the late Middle Jurassic but became less diversified. In relatively less arid areas, such as the Xinghe Formation of Qilian, many bituminous mudstone and oil shale beds occur and yield a considerably diversified flora (19 species of 14 genera) in the lower part, which decreases rapidly in the succession, denoting subsequent climatic deterioration. As shown in the Ma'ao Formation of northern Henan, the bivalves were still represented by the *Eolamprotula–Psilunio* assemblage containing *Cuneopsis*, *Unio*, and *Pseudocardinia* as important members. Other groups were estheriids such as *Euestheria*, ostracods such as *Darwinula*, and charophytes such as *Euaclistochara*.

The Yinshan–Shandong biome. Volcanism was much stronger than in the previous biome, and the thick-shelled *Eolamprotula*, *Psilunio*, and *Cuneopsis* have not been reported here (RGLNIGP 1976). The topography was mainly highlands with scattered intermontane and fault-block basins. The Changhangou Formation of Yinshan consists of grey to greyish-green clastics and lacustrine limestones (Baotou area), or purple to brown mudstones interbedded by greyish green mudstones (Chaoge Qi to Baiyun'ebo). The Jiulongshan and Tiaojishan Formations of northern Hebei are dominated by greyish purple to brown conglomerates, ignimbrites, and other volcanics. The Lanqi Formation of western Liaoning contains purple to grey andesites, andesitic conglomerates, tuffaceous sandstones, and basalts. Fossils are poor. The *Coniopteris–Phoenicopsis* flora is less diversified, consisting also of *Neocalamites*, *Cladophlebis*, *Hausmannia*, *Eboracia*, and *Czekanowskia*. There are a few bivalves such as *Pseudocardinia*.

The Boreal realm

The North-eastern China province

This area remained temperate, humid to semi-humid, in the late Middle Jurassic, with a biota similar to the early stage but less abundant. Grey conglomerates, sandstones, mudstones, and locally coal beds, together with thick and widespread acidic–intermediate volcanics occupied scattered small basins. They comprise the Taonan Formation of the Songliao region, the Jubao Formation of Inner

Mongolia and Xingan, the upper Taiyangling Formation of central Jilin, and the Changantung Formation of the Zhangguangcai Ling region. The plants, as typified by fossils found in the Jubao Formation, constitute the *Coniopteris* flora, including *Coniopteris*, *Todites*, *Eboracia*, *Baiera*, *Czekanowskia*, *Raphaelia*, *Cladophlebis*, and *Hausmannia*. Among invertebrates were the *Ferganoconcha* bivalve assemblage and, in the Changantung Formation, insects such as *Somarura*, *Mesoneta*, and *Mesoleactra*.

The Wandashan province

This was formed by the transgression of the Boreal Sea into the north-eastern corner of Heilongjiang. Strata of this age are called the Dongshengcen (lower) and Peide (upper) Formations, featuring alternating terrestrial and marine coal-bearing clastics, accompanied by abundant volcanics and yielding a Boreal fauna. The *Arctocephalites* ammonite assemblage contains mainly *Stenocodoceras*, *Kosmoceras*, *Morphoceras*, and *Calliphylloceras*. The *Mesosascella morrisi–Entolium demissum* bivalve assemblage contains 39 species, including *Nuculoma maga*, *Palaeoneilo amygdala*, *Astarte extensa*, *Camptonectes lens*, and *Mytilus binificidi*, of which only a few are Boreal while the majority are also seen in Tethys, which reflects the obscurity of Jurassic provincialism. Abundant plants total 38 species of 33 genera. Pteridophytes dominate, including *Todites williamsoni* and *Cladophlebis nebbensis*; next in importance are cycadophytes, *Nilssonia compta*, *Pterophyllum*, etc.; ginkgoales, such as *Ginkgo digitata*, *Sphenobaiera*, and *Stenorachis lepida*; and conifers, *Pityophyllum lindstroemi*, *Schizolepis prynadae*, etc. This flora is not typically Boreal because *Phoenicopsis* is rare and many taxa are similar to the southern provinces. The Early and Middle Jurassic floras of these northern provinces of China bear more similarity with the Central Asian province of the Indo-European region than with the Siberian region. Thus the plants, as well as the bivalves, denote that this province was in the part of the Boreal realm that was at relatively low latitude with higher temperatures.

The Gondwanan realm

The Himalayan province

This was within the Gondwanan Tethyan region, with similar environments as in the early stage. The Larongla Formation in the Qomolangma area represents stable neritic facies with clastics and carbonates, whereas in the Jiangzi area geosynclinal clastics with a huge thickness and intermediate–basic volcanics predominate. The principal organisms are the *Macrocephalites–Indocephalites* ammonite assemblage, including mainly *Dolikephalites* and *Choffatia*, the *Belemnopsis lalonglensis–Hibolithes catlinensis* belemnite assemblage (Wu Shunbao 1982), bivalves such as *Entolium*, *Indogrammatodon*, and *Ctenostreon*, and brachiopods such as *Septaliphoria*. The ammonites have affinities with elements of corresponding age from Kutchi of India and Madagascar. As in the early stage, the bivalves show strong Ethiopian flavour, and the tropical *Burmirhynchia–Holcothyris* brachiopod assemblage, widespread in the eastern Eurasian Tethys (such as Pamir, Tibet, western Yunnan, and Burma), are not present. These denote a wide separation between Himalaya and Tibet–western Yunnan.

10.2.4 LATE JURASSIC PALAEOBIOGEOGRAPHICAL PROVINCIALIZATION OF CHINA

Compared with the Middle Jurassic, three remarkable changes took place in the Late Jurassic environments of China. First, along the coasts from south-east to north-east China there was intensive volcanic activity. Secondly, the large lacustrine basins in Sichuan–Yunnan, Ordos, and Tarim shrank in size. Thirdly, the Gangdise Lowland of Tibet was converted into a geosyncline. Kunlun, Bayan Har, and Qinling were still mountains crossing central China; the North and South China Highlands occupied a large part of continental China. The coal belt was still north of 40°N and east of 110°E, covering north-eastern China and eastern Inner Mongolia, reminiscent of the late Middle Jurassic. However, with the retreat of a hot, arid climate, most of North China and part of south-

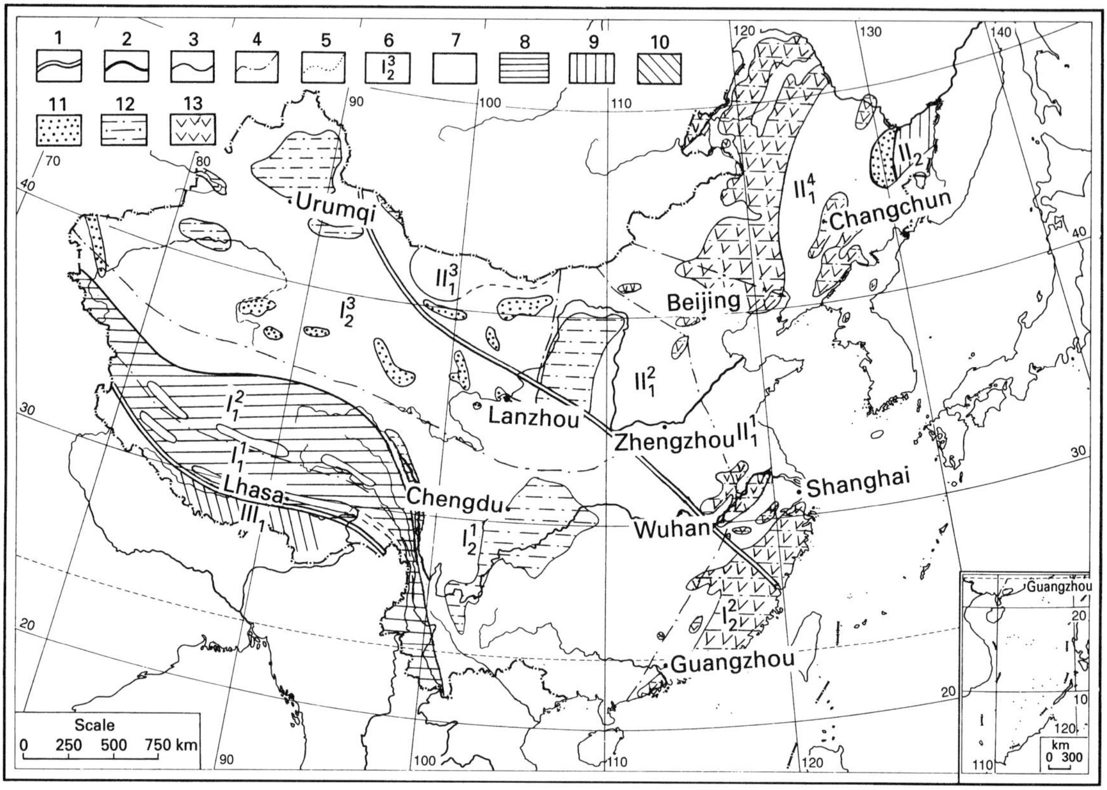

Fig. 10.10 Map of the Late Jurassic palaeobiogeographical provincialization of China. 1, Boundary of realms; 2, boundary of regions; 3, boundary of provinces; 4, boundary of biomes; 5, boundary of communities; 6, roman digit = realm mark, subscript = province mark, superscript = biome mark; 7, Oldland. 8–10, Marine and paralic regions: 8, South China; 9, North China; 10, Junggar–Xingan. 11–13, Terrestrial regions (Late Permian and Mesozoic–Cenozoic): 11, intermontane basins; 12, fluvio–lacustrine basins; 13, volcanics. I, Laurasian Tethyan realm; I_1, Tibet–Western Yunnan province; I_1^1, Gangdise–Nianqingtanggula biome; I_1^2, Tanggula–Western Yunnan biome; I_2, South–North-western China province; I_2^1, Kunlun–South China biome; I_2^2, South-eastern China biome; I_2^3, North-western China biome; II, Boreal realm; II_1, North–North-eastern China province; II_1^1, Fujian–Hebei biome; II_1^2, Anhui–Shaanxi biome; II_1^3, Northern Qilian–Gansu corridor biome; II_1^4, North-eastern China biome; II_2, Wandashan province; III, Gondwanan realm; III_1, Himalayan province.

eastern China turned from arid–semi-arid to semi-humid, with a flourishing of the Jehol fauna. Tethys was still under a tropical climate.

There is controversy concerning the Jurassic–Cretaceous boundary in China. This book has adopted the boundary suggested by Wang Si'en *et al.* (1985). Thus the J/K boundary is drawn between the Jiufutang Formation (lower) and Fuxing Formation (upper) in western Liaoning and between the Hepingxiang Formation (lower) and Jingxing Formation (upper) in western Yunnan. Five provinces of three realms are established within China (Fig. 10.10).

The Laurasian Tethyan realm

This realm covers areas north of Yarlung Zangbo and south of the Fuzhou–Wuhan–Gansu Corridor line. It consists of the Tibet–Western Yunnan province of the Laurasian Tethyan region and the North-western–South China province of the Central Laurasian region.

The Tibet–Western Yunnan province

This covers the marine and alternating marine and terrestrial regimes of northern and central Tibet, southern Qinghai, and western Yunnan. Bivalves were the dominant organisms, followed in importance by hexacorals, echinoids, gastropods, ostracods, and locally ammonites and vertebrates. Subdivision is made into two biomes.

The Gangdise–Nianqingtanggula biome

Gangdise was already a submarine rift-opened basin. The Duodigou and lower Lingbu Formations in the Qilin–Lhasa area of its southern part consist of limestones and clastics, with volcanics in the lower part. The biota includes the bivalve *Diozopteris*, gastropods *Ptygmatis* and *Nerinea*, hexacorals *Lochmaesmilia*, forams *Pseudocyclammina* and *Ammobaculites*, and calcareous algae *Salpingoporella*, *Sarfatiella*, and *Cladocoropsis*. In its northern part, the Naqu–Luolong area, outcropping clastic successions occur totalling more than 1000 m in thickness, with local coal beds, but without volcanics, and yielding abundant fossils, such the ammonites *Virgatosphinctes*, *Indosphinctes*, and *Prososphinctes*; the bivalves *Astarte*, *Goniomya*, and *Entolium*; hexacorals including *Thecosmilia*; the echinoid *Balanodicidaris*, and the gastropods *Nerinea* and *Ptygmatis*. All these common Tethyan fossils indicate a tropical seaway. This area did not incorporate with Eurasia and turn into land until the end of the Jurassic, when elements of the non-marine *Peregrinoconcha* fauna migrated from Yangtze to here. In the Early Cretaceous, the Eurasian non-marine bivalves *Trigonioides*, *Plicatunio*, and *Nippononaia* migrated into this region from the north.

The Tanggula–Western Yunnan biome

This includes the areas north and east of the Bangong Lake–Nujiang line. It was mainly littoral, and marine/terrestrial clastics dominate, intercalated with limestones. The *Virgatosphinctes–Perisphinctes* ammonite assemblage is typically Tethyan. The marine bivalves included *Myopholas*, *Pleuromya*, *Falcimytilus*, and *Quenstedtia*, while non-marine ones include *Eonippononaia* and *Peregrinoconcha* (Guo Fuxian 1985). There are also ostracods such as *Rhinocypris*, *Damonella*, and *Darwinula*, and reptiles such as *Lanpingosaurus*.

The North-western–South China province

This subtropical semi-arid area occupies nearly half of Chinese territory. Sediments are mainly purple, red, and variegated clastics that were deposited in open lacustrine, intermontane, or fault-block basins. Strong volcanism existed in the eastern part. Vegetation was poor and the major elements of the temperate Jehol fauna of the North-eastern–North China province, e.g. *Ephemeropsis*, *Eosestheria*, and *Lycoptera*, did not reach here. Three biomes are recognized.

The Kunlun–South China biome

The main sedimentary area was the Sichuan–Yunnan Basin. The dominant purple and red clastics yield few plant remains but rather abundant lacustrine organisms, such as the *Cuneopsis–Undulatula* bivalve assemblage, *Chuanestheria–Eosestheriopsis* estheriid assemblage, the ostracods *Damonella*, *Darwinula*, and *Cypridea*, and the vertebrates *Longguanichthys* and *Chienkosaurus*. Intermontane basins were scattered in western Kunlun and along the Hubei–Hunan–Guangxi border. The Kuzgongsu Formation of the former is made of red clastics yielding *Podozamites*; the Dongli Formation of the latter has a similar lithology.

The South-eastern China biome

This is an area with evidence of strong acidic–intermediate volcanism, accompanied by the deposition of conglomerates, sandstones, and tuffs, 1000–2000 m thick, reflecting the interaction between the Laurasian and Pacific plates. The biota includes the bivalves *Margaritifera*, *Ferganoconcha*, and *Mesocorbicula*, the estheriids *Eosestheria*, *Orthestheria*, and *Yanjiestheria*, the ostracods *Damonella* and *Cypridea*, and the plants *Cupressinocladus*, *Brachy-*

phyllum, and *Pagiophyllum*. Reptile remains are lacking.

North-western China biome

This consists of Xinjiang and most of Qinghai and was largely mountainous. The principal deposition took place in the Tarim, Jinggar and Qaidam basins, where purple, red, and variegated clastics predominate, with gypsum intercalations. The main fossil groups are the *Metacypris–Limnocythere* ostracod assemblage, including *Darwinula*, *Cypridea dzungarica*, and *Lycopterocypris*; charophytes, dominated by *Euaclistochara*; bivalves, such as *Margaritifera*, *Solenaia*, and *Sphaerium*; fish, including *Turfanichthys* and *Hunuanshania*; and reptiles such as *Tienshansaurus* and *Kelamayisaurus*.

The Boreal realm

The southern boundary with the Tethyan realm runs along the Fuzhou–Wuhan–southern Ordos–Gansu Corridor line. It covers northern and north-eastern China as well as part of south-eastern and north-western China. Biogeographically it was the distribution area of the Jehol fauna (Cao Meizhen *et al.* 1986) which was named after an old administrative region including northern Hebei and southern Inner Mongolia. The representatives of this fauna are the estheriid *Eosestheria*, the insect *Ephemeropsis*, and the fish *Lycoptera*, which, according to their distribution, reflect a warm temperate freshwater environment belonging to the Pacific drainage system. Palaeocimatically this was a warm temperate area, changing from semi-humid in the south-west to humid in the north-east, where coal accumulation proceeded. As in the North-western–South China province, its eastern part experienced extensive volcanism due to interaction with the Pacific Plate.

The North-eastern–North China province

This vast province displayed a variety of environments, including volcanic, fault-block, intermontane, and open lacustrine basins. Among animals, the Jehol fauna dominated. There were fewer plants than in the Middle Jurassic; they comprise pteridophytes, 28 species of 14 genera; conifers, 24 species, 15 genera; cycadophytes, 13 species, 10 genera; ginkgoales, 13 species, 6 genera; and a few equisetales, bryophytes, and lycopods, mostly persisting from the Early and Middle Jurassic and partly heralding the Early Cretaceous flora. Pteridophytes were principally distributed in the more humid coastal areas in the north-east.

The Fujian–Hebei biome

This biome occupies a belt that experienced strong volcanism, extending from northern Fujian, through Zhejiang, Jiangsu, and Shandong to Hebei. The scattered, small fault-block basins bear grey to purplish red clastics and acidic–intermediate volcanics. The biota includes estheriids, dominated by *Eosestheria* with subordinate *Nestoria* and *Orthestheria*; insects, dominated by *Ephemeropsis trisetalis* with subordinate *Archaeogomphus*, *Pseudosamarura*, and *Karataviella*; fish, such as *Lycoptera*; a *Ferganoconcha–Mengyinaia* bivalve assemblage; a *Mongolianella–Cypridea* ostracod assemblage; the plants *Otozamites* and *Pagiophyllum*; and some reptiles.

The Anhui–Ordos biome

The only major sedimentary basin is the Ordos Basin, with purplish red and grey clastics yielding abundant fossils, including the *Sphaerium–Ferganoconcha* bivalve assemblage, the *Cypridea–Lycopterocypris* assemblage (18 species, 13 genera), the estheriids *Yanjiestheria* and *Sinoestheria*, the fish *Lycoptera* (4 species), *Sinamia*, and *Ikechaomia*, the turtle, *Sinemys*, and other reptiles (*Eotomistoma* and *Huanghepterus*), and the plants *Sphenopteris*, *Pagiophyllum*, and *Zamiophyllum*.

The Northern Qilian–Gansu Corridor biome

This area includes northern Qilian, Beishan, and the Gansu Corridor (an elongate, lowland filled with oases between the former two mountains. The famous Silk Road led from Xi'an through this corridor to Xinjiang and further west). The prevailing purple-red clastics yield relatively abundant fossils, including bivalves, gastropods, ostracods, estheriids, insects, fish, reptiles and plants. The Late Jurassic Xinmingbao Group of the Gansu Corridor contains bivalves (37 species of 6 genera) (Ma Qihong 1986), including *Ferganoconcha* (10 species), *Danlengiconcha* (1 species), *Nakamuranaia* (3 species), *Gansuella* (4 species), *Neomiodonoides* (9 species), and *Limnocyrena* (10 species). Coeval with them are typical elements of the Jehol fauna: *Eosestheria*, *Ephemeropsis*, and *Lycoptera*. Other animals include the estheriid *Diestheria*, the ostracod *Cypridea*, the fish *Sinamia*, and the reptile *Chiayuesaurus*. Among plant remains there are pteridophytes (6 species of 4 genera), cycadophytes (3 species, 3 genera), ginkgoales (1 species, 1 genus), conifers (6 species, 6 genera), and the charophyte *Euaclistochara–Aclistochara* assemblage (Ma Qihong 1986).

The North-eastern China biome

The peculiar aspect of this area is the coal accumulation in the Late Jurassic, which denotes a warm and humid temperate climate. The strata consist of coal-bearing, variegated clastics and extensive volcanic rocks, containing abundant plants. The Xinganling Group of Daxinganling yields equisetales (2 species, 1 genus), pteridophytes (3 species, 2 genera), cycadophytes (5 species, 5 genera), ginkgoales (6 species, 4 genera), and conifers (11 species, 9 genera), including *Acanthopteris*, *Nilssonia*, *Coniopteris*, *Ginkgo*, and *Phoenicopsis*. *Ferganoconcha* is the most abundant bivalves, followed by *Sphaerium*; among estheriids, *Eosestheria*, *Keratestheria*, and *Nestoria* are common, as is the fish, *Lycoptera*.

The Wandashan province

The marine regime persisted from the Middle Jurassic into the Late Jurassic in this area, and coal-bearing marine–terrestrial alternating sediments containing Boreal biota were deposited. The bivalves were dominant, totalling 169 species of 37 genera, most of which occur in the Upper Jurassic of other regions. Important members are *Isognomon isognomonoides*, *Camptonectes shuguangensis*, *Quenstedtia qinglongshanensis*, *Arcomya*, *Buchia mosquensis*, and *B. rugosa*. A majority of this assemblage is of Boreal type but Tethyan and endemic elements are not uncommon, again showing the obscurity of provincialism as in the Middle Jurassic. In addition to marine bivalves, there are freshwater bivalves dominated by species of *Ferganoconcha*, such as *F. sibirica* and *F. tomiensis*, and the genus *Sibericoncha*, many of which are common in the Middle and Upper Jurassic strata in Siberia and adjacent areas. There are 15 species of 9 genera of brachiopods, including *Septaliphoria*, *Thurmannella*, *Daghanirhynchia*, *Kallirhynchia*, and *Yunshannella*, the assemblage of which shows considerable endemism. This province also had a flourishing Late Jurassic flora, with 81 species of 23 genera, including *Gleichenites nordenskioldi*, *Coniopteris*, *burejensis*, *Onychiopsis elongata*, *Acanthopteris onychioides*, and *Sphenolepis kurrianus*. The floral pattern shows a transition between the Siberian and Indo-European phytoregions, but with more affinities to the latter.

The Gondwanan realm

The Himalayan province

During the Late Jurassic this was a neritic sea along the Gondwanan Tethyan margin far away from Eurasia. The shelf deposits of clastics and carbonates reach 3262 m in thickness and contain a rich fauna, the main elements of which are the ammonites *Haplophylloceras*, *Phylloceras*, *Virgatosphinctes*, *Pterolytoceras*, and *Blanfordiceras*; the belemnites *Belemnopsis gerardi*, *B. uhligi*, and *Hibolithes* (Yang and Wu 1964); and the bivalves *Buchia blanfordiana*, *B. extensis*, *Astarte*, *Indogrammatodon*, and *Pseudolimea*. The ammonites are of Tethyan type. Among belemnites, 13 species and subspecies of *Belemnopsis* and 15 species of *Hibolithes* have been reported from the Upper Jurassic and lower Lower Cretaceous of this area; most are of Gondwanan Tethyan type and a few are endemic. The bivalves are also of Gondwanan Tethyan type, but there is an interesting occurrence of a monotonous *Buchia* assemblage in black shales which often show an alternation in sequence with *Indogrammatodon* assemblages in coarser clastics. This may imply a biotic submergence to deeper and cooler levels in the sea. The appearance of *Buchia* and other Boreal elements give another good example of bipolarity. The general features of the biota indicate little affinity with Gangdise, implying a wide Tethyan separation.

10.3 IMPLICATIONS OF JURASSIC PALAEOBIOGEOGRAPHY FOR PLATE TECTONICS, PALAEOCLIMATE, AND PALAEOECOLOGY

In the marine regime of China, the Boreal biota can be readily distinguished from the Tethyan biota. Within the latter a tropical or Eurasian Tethyan biota and a Gondwanan Tethyan biota can be discriminated, because the latter is less diversified and shows bipolarity with the Boreal biota. This division serves as a good criterion for marine provincialization.

Because most of China was continental in the Jurassic, the provincialization was mainly by phytogeography, and hence climate-controlled. Generally, the Kunlun–Qinling–Dabie line subdivided China into a southern arid tropical–subtropical biotic region and a northern semi-humid–humid, subtropical–temperate biotic region. The terrestrial provincialization was obscured by a majority of bivalves, estheriids, ostracods, and plants being common to both regions and showing transitional or mixed aspects. For terrestrial environments without ocean barriers, where land and drainage communications are feasible and climate changes are gradual, such minimal provinciality would be normal.

Two phenomena may be of significance to plate tectonics and palaeoclimate. The arid tropical zone (the red-bed zone), accompanied by the temperate coal-bearing zone immediately to its north, shifted northward consistently until the late Middle Jurassic, when it reached climax (Fig. 10.2). Then, after the Late Jurassic, it retreated, and meanwhile the Jehol fauna invaded southward. This is interpreted by some as evidence of clockwise and anticlockwise rotation of the Chinese continent with a range of 20 degrees in latitude (Fei Qi 1987). This can also be interpreted as the waxing and waning of the Pacific influence, as explained in Chapter 15.

Another remarkable phenomenon is the biotic development in the Tibet–Qinghai Plateau. The faunal disparity between the Himalayan province and areas north and east of it throughout the Jurassic is a strong argument for the existence of a vast Tethyan ocean. Brachiopods, bivalves, and other marine assemblages show that Gangdise and western Yunnan were already within the influence of tropical or Eurasian Tethys. The approaching and final contact of these blocks with the Eurasia continent can be traced by the occurrence of freshwater and terrestrial biota, epoch by epoch within the Jurassic, from Yangtze to Tanggula and then to Gangdise.

Biomes are subdivided by ecological and, indirectly, by lithofacies differentiation. The marine biomes, occupying smaller areas in the Jurassic, can be subdivided into normal neritic, such as the Himalayan province and part of Tibet–western Yunnan; rift-opening seaway, such as eastern Gangdise–Nianqingtanggula; alternating marine and terrestrial biomes, such as parts of Tanggula and western Yunnan, Guangdong–Hunan Bay (Early Jurassic), and Wandashan (Middle and Late Jurassic); and brackish estuaries, such as part of Jiangxi, Fujian, and Hunan (Early Jurassic). The terrestrial biomes, occupying much larger areas in Jurassic China, are mostly erosional, with records being preserved only in interspersed basins which include open lacustrine, intermontane, block-fault basins, and volcanic areas. These environments, together with the organisms that inhabited them, serve as the basis for biome subdivision, which in turn gives valuable information for the interpretation of Jurassic history.

REFERENCES

Cao Meizhen, Liu Qibin, and Chen Jinhua (1986). 'Jehol Fauna' discovered from Henan. *Acta Palaeontologica Sinica* **25**, (2), 211–13 (in Chinese).

Chen Fen, Dou Yawei, and Huang Qisheng (1984). *The Jurassic flora of West Hills, Beijing (Peking)*. Geological Publishing House, Beijing (in Chinese).

Chen Jinhua and Cao Meizhen (1986). Triassic and Jurassic coal-bearing measures and palaeoclimate zonation in China. In *Selected Papers from the 13th and 14th Annual Conventions of Palaeontological Society of*

China, pp. 205–21. Anhui Science and Technology Publishing House (in Chinese).

Chen Peiji and Shen Yanbin (1979). Mesozoic and Cenozoic estherian faunas and their distribution in the Red Beds of South China. In *The Mesozoic and Cenozoic Red Beds of South China*, pp. 9–97. Beijing Science Press (in Chinese).

Chen Zhonghui, Wu Shunbao, Lu Rulun, Ma Jinxian, and Liu Jinhua (1982). On conditions of deposition for Mesozoic strata and ages of Tanglong and Shigu Formations in Sandu, Hunan. *Earth Science–Journal of the Wuhan College of Geology* **3**, 143–53 (in Chinese).

Fei Qi (1987). A preliminary research into the plate collision, rotation and divergence pattern of China and its periphery since Mesozoic. *Earth Science—Journal of China University of Geosciences* **12 (5)**, 463–75 (in Chinese).

Guo Fuxian (1985). *Fossil bivalves of Yunnan*. Yunnan Science and Technology Press (in Chinese).

Hallam, A. (1971). Provinciality in Jurassic faunas in relation to facies and palaeogeography. In *Faunal province in space and time*, (ed. F. A. Middlemiss, P. E. Rawson, and C. Newell), pp. 129–52. Seel House Press, Liverpool.

Hallam, A. (1975). *Jurassic environments*. Cambridge University Press.

Hallam, A. (1977). Jurassic bivalve biogeography. *Palaeontology* **3**, 58–73.

Imlay, R. W. (1984). *Jurassic ammonite successions in North America and biogeographic implications*. Geological Society of America, Special Paper No. 27, pp. 1–12.

Kosygin, Y. A. and Parfenov, L. M. (1981). Tectonics of the Soviet Far East. In *The ocean basins and margins*, Vol. 5, *The Arctic Ocean*, (ed. A. E. M. Nairn, M. Churkin, and F. Stehli), pp. 377–412. Plenum Press, New York.

Ma Qihong (1986). Fossil lamellibranchs from the Late Jurassic Xinminpu Group in western Gansu, NW China. *Memoirs of Nanjing Institute of Geology and Palaeontology, Academia Sinica* **22**, 181–203 (in Chinese).

Mong Fansong (1985). Problems concerning the Jurassic of Hubei, Hunan, Guangdong and Guangxi. *Journal of Stratigraphy* **9**, (4), 317–21 (in Chinese).

Rage, J. C. (1988). *Gondwana, Tethys, and terrestrial vertebrates during the Mesozoic and Cainozoic*. Geological Society Special Publication No. 37, pp. 255–74.

RGLNIGP (Research Group of Lamellibranchiata, Nanjing Institute of Geology and Palaeontology, Academia Sinica) (1976). *Fossil Lamellibranchiata of China — Chinese Fossils of All Groups*. Science Press, Beijing (in Chinese).

Stevens, G. R. (1965). The Jurassic and Cretaceous belemnites of New Zealand and a review of the Jurassic and Cretaceous belemnites of the Indo-Pacific region. *Palaeontological Bulletin of Wellington* **36**.

Wang Hongzhen (1978). On the stratigraphic division of China. *Journal of Stratigraphy* **2**, (2), 81–104 (in Chinese).

Wang Neiwen (1984). A study of the paleobiogeography and plate tectonics of Xizang-Qinghai Plateau. *Bulletin of the Geological Institute, Chinese Academy of Geological Sciences* **9**, 1–28 (in Chinese).

Wang Si'en *et al.* (1985). *The Jurassic stratigraphy of China*. Geological Publishing House (in Chinese).

Wen Shixuan (1981). Palaeobiogeography of Qinghia-Xizang Plateau—evidence for continental drift. *Proceedings of the Symposium on the Geology of Qinghai-Xizang Plateau (Beijing, China)* Vol. 1, pp. 149–57.

Westermann, G. E. G. (1981). Ammonite biochronology and biogeography of the Circum-Pacific Middle Jurassic. In *The Ammonoidea* (ed. M. R. House and I. R. Senier). The Systematics Association Special Volume, No. 18.

Westermann, G. E. G. (1988). *Middle Jurassic ammonite biogeography supports ambi-Tethys origin of Tibet*. Geological Society Special Publication No. 37, 235–40.

Willeg, L. E. (1973). Belemnites from south-eastern Alexander Island. II, The occurrence of the family Belemnopsidae in the Upper Jurassic and Lower Cretaceous. *British Antarctic Survey, Bulletin* **28**, 33–59.

Wu Shunbao (1981). Jurassic and its bivalves of Fengjiachong, Lingling, Hunan. *Geological review* **27**, (5), 375–83 (in Chinese).

Wu Shunbao (1982). Characteristics of Early Jurassic–Early Cretaceous belemnoid assemblages from southern Xizang (Tibet). *Contribution to the Geology of the Qinghai-Xizang (Tibet) Plateau* **10**, 113–21 (in Chinese).

Yang Zunyi and Wu Shunbao (1964). Some Late Jurassic and Early Cretaceous belemnites of South Xizang. *Acta Palaeontologica Sinica* **12**, (2), 187–207 (in Chinese).

11 CRETACEOUS

Liu Benpei, Xu Yuling, and Chen Feng

11.1 CRETACEOUS PALAEOBIOGEOGRAPHICAL PROVINCIALIZATION OF THE WORLD

11.1.1 PLATE DISTRIBUTION IN THE CRETACEOUS WORLD

Pangaea experienced further disintegration during the Cretaceous. The main body of Laurasia approached its modern location but the Eurasian part deviated remarkably south-westward, forming an angle of several tens of degrees with modern latitudes. In the southern hemisphere, Gondwana underwent several rifting and dismembering phases, the most remarkable of which were the extensive rift-openning of the South Atlantic and rapid northward movement of the Indian plate. The Tethyan or Neo-Tethyan (*sensu* Sengör, 1984) Ocean existed between Eurasia and eastern Gondwana, but controversy persists as to its possible dimensions. Chinese palaeomagnetic data show that Dingri County of Himalaya on the northern margin of the Indian plate was 11.9°S, while the Gangdise–Lhasa block migrated from 1°N in the Late Jurassic to 20.3°N by the end of the Cretaceous, due to mid-ocean ridge expansion within Tethys. Considering that the Asian continent as a whole rotated clockwise during the Cretaceous, and that the Indian plate and Gangdise block may have drifted northward coevally (Sengör *et al.* 1988), a primary deduction on the width of Tethys along the Yarlung Zangbo suture estimates a span of no larger than 35° of palaeolatitude, which would be approximate the size of the modern Arctic Sea or medium ocean (Table 11.1).

11.1.2 CRETACEOUS PALAEOCLIMATE

The Cretaceous is known as a warm period in geological history, based on data from sedimentology, palaeontology, and especially from quantitative determinations using oxygen isotope analysis of sea-water palaeotemperature. The average Cretaceous sea-water temperatures in low latitudes were 15–20 °C, based on isotopic analysis of benthic foram shells, or 25–30 °C using planktonic forams. Even during the conspicuous end-Cretaceous cooling event, the average sea-water temperature was higher than 10 °C (by benthic forams) or 20 °C (by plankton). Evidently

Table 11.1 Palaegeomagnetic data of the Cretaceous of China (from the unpublished material of Ju Hong, 1986)

Localities	No. of magnetic poles (samples)	Present latitude	Average Cretaceous latitude	Difference between the Cretaceous and the present latitudes
Zhuzheng and Linqu	5	35.9°N	36.2°N	–0.3°
Zhejiang	4(19)	29.5°N	24°N	5.5°
Jiangxi	2	25.8°N	14°N	11.8°
Guangdong	1(348)	24.2°N	20.1°N	4.1°
Dukou, Sichuan	1(21)	26.6°N	29.7°N	–3.1°
Biru and Lhunzhub	11(159)	30.5°N	20.3°N	10.2°
Tingri, Xizang	2(24)	28.8°N	11.9°S	40.7°

there were no polar ice sheets, and only tropical, subtropical and warm temperate zones existed in the Cretaceous, although arid and humid belts are recognizable.

Three global cooling phases occurred in this period. The first was from the late Volgian in the Arctic (probably the Berriasian in Tethys) to the Barremian (Fig. 11.1, including B, V, H, B), followed by a global temperature increase after the Aptian, which reached a maximum for the whole Cretaceous period in the Albian. The second phase had begun by the end of the Albian (Fig. 11.1, Alb). This was relatively short-term and quickly turned back to a warm phase again, the effect of which was most evident in the surface horizon of sea water, where plankton suffered severely. The third phase, occurring during the end-Cretaceous (Maastrichtian), was the strongest in this period, and remarkably affected surface water, sea bottom and land; it has been widely postulated to be connected with an asteroid impact event.

There were also several phases of waxing and waning of arid and humid belts during the Cretaceous. The regional factors of palaeogeography and tectonics would have been influencing factors, but the effects of global warming and cooling were probably more important. Accumulating date show that in China during cooling phases the humid temperate zone expanded southward, while warming resulted in northward expansion of the hot arid belt. Such palaeoclimatic changes in connection with global temperature oscillations had a significant influence on organic evaluation and distribution, especially among terrestrial biotas (Liu Benpei and Chen Feng 1992).

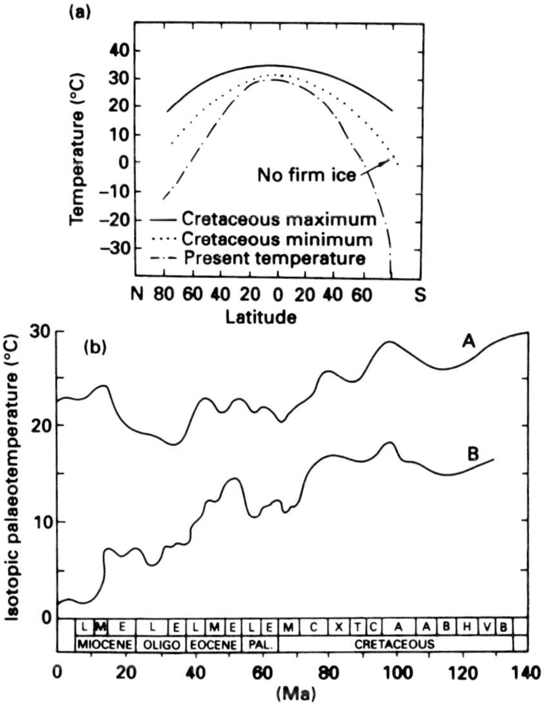

Fig. 11.1 Temperature curves of Cretaceous sea water. (a) Surface water temperatures of the Middle Cretaceous and Recent periods; (b) low-latitude water temperature from the Early Cretaceous to present, based on isotope analysis; A, plankton; B, benthos.

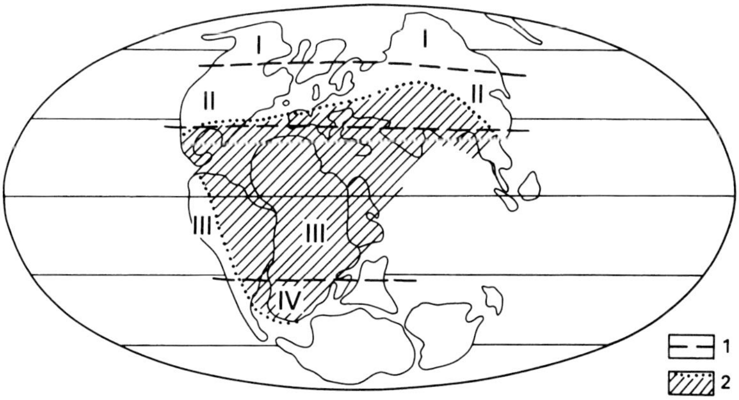

Fig. 11.2 Map of the early Early Cretaceous (Berriasian–Valangian) phytogeography and climate zonation of the world. I, Siberia–Canadian region (temperate); II, Europe–Cathaysian region (north subtropical); III, Equatorial region (tropical); IV, Australian region (south subtropical); 1, boundary of palaeobotanic regions; 2, distribution of semi-arid and arid climate.

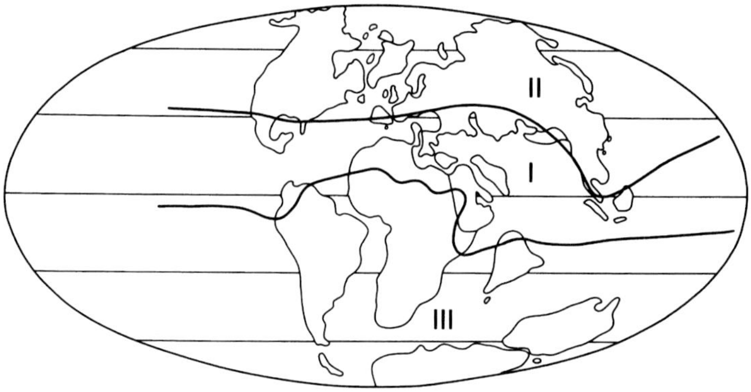

Fig. 11.3 Map of the Late Cretaceous bivalve zoogeography and climate zonation of the world oceans (after Kauffman 1981). I, Tethys; II, Boreal, III, Gondwana.

11.1.3 CRETACEOUS PALAEOBIOGEOGRAPHICAL PROVINCIALIZATION OF THE WORLD

Three biogeographic realms were established in the Cretaceous world, i.e. the tropical–subtropical Tethyan realm and approximately bipolarly distributed Boreal and Gondwanan realms (Figs 11.2, 11.3). However, these figures show that phytogeographical and zoogeographical boundaries of realms are not coincident, and non-coincidence may even exist among boundaries based on different animal groups, which is only natural because different taxa have their unique habitat, ecology and ability of migration.

11.2 EARLY EARLY CRETACEOUS PALAEOBIOGEOGRAPHICAL PROVINCIALIZATION OF CHINA

11.2.1 GENERAL INTRODUCTION

Concerning the determination of the Jurassic–Cretaceous boundary, extensive discrepancies remain

Table 11.2 Correlation chart of the Cretaceous of China

			Non-marine			
			E. Heilongjiang	Songliao	W. Liaoning	
Upper Cretaceous		K_2	Yilin Fm	Mingshui Fm Sifangtai Fm Nenjiang Fm Yaojia Fm Qingshankou Fm		
Lower Cretaceous	Albian and Aptian	K_2	Huashan Fm	Quantou Fm	Sunjiawan Fm	
	Neocomian	K_1^1	Muling Fm Chengzihe Fm (upper part)	Denglouku Fm Yingcheng Fm	Fuxing Fm Shahai Fm	

on its marine biostratigraphic criteria, on the isotopic dating, and on the correlation of terrestrial and marine strata. The early Early Cretaceous used here means the period from the Berriasian to the Barremian, i.e. the span of the Neocomian; Its isotopic dating is *c.* 130 ± 3–112 ± 2 ma. As for correlation with the terrestrial strata of eastern Asia, the base of the Cretaceous is placed within the Jehol group (*sensu lato*), between the Shahai Formation and the underlying Jiufutang Formation (Table 11.2), which is in accordance with the end-Jurassic global cooling and humidifying phase from the eventostratigraphic viewpoint (Liu Benpei *et al.* 1986).

It is known that a conspicuous end-Jurassic cooling phase occurred in the Palaeo-Arctic. Data also show that there was a coeval southward expansion of the humid belt (Chen and Cao 1986), a large-scale regression, and plate collision along the Mongolia–Okhotzk belt. These events together significantly influenced the biogeography of eastern Asia, causing some notable changes from the Late Jurassic pattern, although some relationships can still be discerned.

11.2.2 PROVINCIALIZATION

Eight provinces and four regions of three realms have been established for the early Early Cretaceous of China (Table 11.3).

The Laurasian Tethyan realm

Areas of China south of Tianshan–Qilian–Qinling and north of Yarlung Zangbo are attributed to the South China region of this realm, which can be further subdivided into four provinces, namely the South-eastern China, Sichuan–Yunnan, Tarim–Qaidam, and Tibet–western Yunnan provinces.

Within the South China region, the area between the Bangong Lake–Nujiang line and the Yarlung Zangbo line was the only area intermittently covered by the shelf sea of Tethys. It is thus established as the Tibet–Western Yunnan province, but is still related to this region because the intercalated terrestrial beds yield South China-type biota.

The rest of the whole region experienced continental conditions, with biota quite different from that

				Marine
Zhejiang	S. Hunan	M. Yunnan	Junggar	S. Tibet
Jujiang Gr.	Dongtang Fm Daijiaping Fm	Jiangdihe Fm		Jidula Fm Zongshan Fm Jiubao Fm Xiawuchubo Fm Lengqinre Fm
Fangyan Fm Chaochuan Fm Guantou Fm Shouchang Fm (upper member)	Shenwangshan Fm Dongjing Fm	Matoushan Fm Puchanghe Fm Gaofengsi Fm	Donggou Fm Tugulu Gr.	Chaqiela Fm Dongshan Fm

Table 11.3 Cretaceous palaeobiogeographical provincialization of China

Realm	Region	Province K_1^1	Province K_1^2–K_2	K_1^1	K_1^1–K_2
Boreal realm	N.China region	NE China–Inn. Mongol. province		*Acanthopteris–Nillsonia* flora *Classopollis* < 10% *Nippononaia*, *Protelliptio*	Mainly deciduous anglosperms and gymnosperms *Classopollis* < 10% *Aquillopreaidites*–form pollen. Endemic *TPN* Fauna, *Plesiolycoptera*, *Protelliptio*, *Euestherites*
				Kuntulunia	
Boreal realm	N.China region	N. China–Junggar province		Mixed flora *Classopollis* >10-<40%	Separately included in SE China province (L_1) and in SW China–NW China province (L_2) *Tsintaosaurus* (Pacific-type *TPN* fauna)
				Psittacosaurus	
	W. Pacific region	Wandashan province	Taiwan province	*Buchia* fauna	*Dufrenoyia*, *Epichelniceras*, *Holcoptylloceras*
Laurasian–Tethyan realm	S. China region	SE China province		*Manica*, *Weichselia* flora *Classopollis* > 40% *Migransia*, *Yanjiestheria* *Mesoclupea*	Pacific-type *TPN* fauna *Paralycoptera*. There were a few *Aquillopreaidites* along the sea.
Laurasian–Tethyan realm	S. China region	Sichuan–Yunnan province	SW China–NW China province	*Classopollis* > 50% *Orthestheria*, *Orthestheriopsis* *Peregrinoconcha*, *Cyotrigonioides*	Mainly evergreen trees and shrubs *Classopollis* + *Schizaeoisporites* >50%
Laurasian–Tethyan realm	S. China region	Tarim–Qaidam province	SW China–NW China province		Tethys-type *TPN* fauna *Aglestheria*
		Tibet–W. Yunnan province	Xizang–Tarim province	*Weichselia* There were few trisaccites, but no *Classopollis*	*Orbitolina* Hexacorals, rudists (Tethys-type *TPN* fauna)
Gondwanan–Tethyan realm	Indian region	Himalayan province		*Hibolithes*, *jiabulensis*–*Belemnopsis sinensis*	*Orbitoides* *Xenothyris tuilaensis* *Bourononia tibetica*

of the Boreal realm. Land plants were less diversified than in the latter. Deciduous ginkgoales were rare and czekanowskiales were lacking.

Most pteridophytes were equipped with small leaflets with split fringes and relatively thick membranes. The well-known *Weichselia*, adapted to arid climates, appeared. Conifers and cycadophytes, especially bennettitales, were very abundant; the former usually with thick cuticles and scale-shaped leaflets coherent with the branch, such as *Brachyphyllum, Frenelopsis,* and *Manica*. These evidently represent an aridity-tolerant tropical–subtropical flora.

Palynomorphs show the following features:

1. Gymnosperm pollen is dominant in the assemblages, usually accounting for 70–80 per cent of the total content, with diversified taxa of which the genus *Classopollis* has the highest content, approximately 40–60 per cent of the total. Other pollen came from ginkgoales, cycadales, bennettitales, *Ephedra,* and disaccate pines.

2. Among pteridophyte pollen, the Lygiduaceae remained dominant, *Schizaeoisporites* being the major component. This is an important characteristic of the sporo-pollen assemblage of the southern Tethyan realm.

The above-mentioned sporo-pollen assemblage obviously differed from that of the Boreal realm, indicating an arid climate of the tropical–subtropical zone. The lacustrine biota were also quite different from those of the Boreal realm, and exhibited conspicuous spatial differentiation, controlled by different drainage. This provides the basis for the subdivision of the region into three provinces, i.e. the South-eastern China, Sichuan–Yunnan, and Tarim provinces.

The south-eastern China province

This area covers the south-eastern coastal area of China to the east of Xuefengshan. In the early part of the Early Cretaceous there was a series of small inland basins, and continental volcanic eruptions have also been identified in the coastal area. The lacustrine biota displays certain differences from that of the Late Jurassic. It is now clear that the lower Shouchang Formation, which yields abundant *Ephemeropsis trisetalis* in Lin'an County, Huzhou County, and Ningbo County (Zhejiang) should be placed in the Upper Jurassic. In the Upper Member of the Shouchang Formation (Lower Cretaceous) there occurs a new insect *Solusipanorpa* assemblage, with *Ephemeropsis* no longer found. This change in insect fauna, which are sensitive to climatic change, indicates climatic change at the turn of the Jurassic and Cretaceous, and the increasing regionalism of insect distribution.

The endemic freshwater fish *Mesoclupea* continued to exist in the Upper Member of the Shouchang Formation, according to data from the Shouchang section in Jiande County, Zhejiang, which indicates that fish endemism existed in the south-eastern area at that time, with taxa quite different from those in the Boreal realm.

Though conchostracans were still characterized by the joint dominance of *Yanjiestheria* and *Orthestheria,* the widespread appearance of an African-type assemblage, marked by *Migransia* and *Congestheriella,* from the Upper Member of the Shouchang Formation in Zhejiang, Anhui, Fujian, and Jiangxi, reflected an invasion event of the Gondwanan fauna. This has great significance for research on the palaeogeographical evolution and the exchange of land fauna of both southern and northern sides of the Tethyan Ocean.

The freshwater bivalves were characterized by the presence of a great amount of *Nakamuranaia* spp. and the absence of *Ferganoconcha*. Whether or not the early *TPN* (*Trigonioides, Plicatounio, Nakamuranaia*) fauna occurred in this province remains to be clarified. The '*Trigonioides*' reported to have been found in the Shixi Formation in the Xinjiang basin of Jiangxi, and its juvenile form with distinctive radial sculpture on the umbo area occurring in the Upper Member of the Shouchang Formation, may be proved to be certain new types of that fauna, which again show some regionalism in the early Early Cretaceous.

Non-marine ostracods characterized by the *Cypridea morinia–Damonella–Mongolianella–Darwinula* assemblage are quite different from those of the Boreal realm, which were marked by *Cypridea* with many species and very complicated shell sculpture.

The Sichuan–Yunnan subprovince

This covers a vast area to the west of Xuefengshan and to the east of Bangong Lake–Nujiang River, but the depositional basins are mainly restricted to the Sichuan–Yunnan and Three Rivers (Jinshajiang, Lancangjiang, and Nujiang Rivers) areas. To the west of the Three Rivers area there were then high and steep mountain ranges on the northern side of Tethys, thus only a few intermontane basins with rare fossils exist. In western Yunnan the relief gradually became less uneven, and coastal basins appeared. There was a vast and flat denudation area of the Sichuan–Yunnan inland basin to the east of the Xuefeng water-divide that led the inland water system into Tethys, whereas the inland water system of the South-eastern China province ran into the Palaeo-Pacific Ocean. Such a palaeogeographic setting caused the distinct differentiation of lacustrine fauna between the Sichuan–Yunnan and South-eastern China provinces. Disagreements remain on the age of the Jingxing Formation in western Yunnan. The author suggests placing the Jurassic–Cretaceous boundary at the base of the Upper Member of the Jingxing Formation, based on the fact that the *Classopollis* content accounts for up to 78.8 per cent of the sporo-pollen assemblage in the Lower Member of the Jingxing Formation in Nandien of Lancang County (Hao *et al.* 1986, p. 215), whereas it dropped to 52.5 per cent in the Upper Member of the Hemiao section of Jiangchang County, where the Lygodiaceae were dominant in the sporo-pollen assemblage (Hao *et al.* 1986, p. 215). This fact, as in the case of insects, indicates an event of global temperature fall and humid-zone expansion. It should be noted that the abundant fossils of the Jingxing fauna occur in the Upper Member of the Jingxing Formation.

Conchostracans are represented mainly by the *Orthestheria–Orthestheriopsis* assemblage; a few African-type *Migransia* occur locally, but neither *Yanjiestheria* nor *Neodiestheria* is present. The fossils are very similar to those of the same age in the Wealden of southern England, and in the Hanover area of northern Germany, typically embodying the characteristics of conchostracans along the northern shore of Tethys.

Bivalves are characterized by a flourishing primitive or early *TPN* fauna. There have for a long time been diverse opinions on the origin of the *TPN* fauna, whether they were derived from freshwater unioniids or from marine trigoniids (Kobayashi 1968). The lowest horizon of the *TPN* genus, *Cyotrigonioides*, in western Yunnan is the Middle Jurassic Zhanghe Formation. Based on the similarity of dentition structure with *Nakamuranaia*, it can be inferred that *Cyotrigonioides* originated from ancient unioniids (Chen Jinhua 1982; Guo Fuxiang 1981, 1986; Chen Jinhua 1982; Yao and Yu 1985). However, western Yunnan was near Tethys and it was repeatedly subjected to transgression in the Late Mesozoic, illustrated by the presence of brackish bivalves such as *Mytilus* and *Modiolus* in the Jingxing Formation. There are numerous individuals of the famous *Peregrinoconcha* in the Jingxing Formation; the V-shaped sculpture on the shell surface remains unchanged, but the dentition structure varies so greatly within a population and among different species that they can be easily misidentified as different genera. Such a great variation of dentition is probably due to the rapid changes of geochemical conditions induced by the varied environments of the coastal plain. Possibly the earliest *TPN* fauna originated during the Middle Jurassic in western Yunnan. As for other freshwater bivalves, *Nakamuranaia* was also characterized by many diverse species and the endemic genus *Yunnanoconcha* appeared.

Among the ostracods, *Jingguella* occurs in great numbers and *Pinnocypridea* is common; *Cypridea* is monotonous in taxa and sculpture. They reflect an endemic character different not only from the Boreal realm, but also from the South-eastern province of the Circum-Pacific.

The Tarim–Qaidam province

This area covers the Tarim Basin of southern Xinjiang, the Qaidam Basin of Qinghai, and adjacent inland basins. The early Early Cretaceous is generally characterized by red beds with a few bluish-grey, green, and brownish-yellow intercalated beds. The fossils are usually so few that it is difficult to distinguish their biogeographical provincialization. According to data available for ostracods, in the Kapushaliang Group in the Kuqa district on the northern margin of the Tarim Basin *Jingguella*, a component of the Sichuan–Yunnan province, occurs; but in the Kezileisu Group of the south-western Tarim Basin some ostracods are very similar to those of the Boreal realm. Considering that this area in the Early Cretaceous was adjacent to the Pre-Aralsk– Fergana region (Kolesnikov 1980) to its west, thus showing some regional aspects, the author provisionally takes this area to be a province.

The Tibet–Western Yunnan province

The early Early Cretaceous transgression covered only a narrow zone between the Bangong Lake–Nujiang line and the Yarlung Zangbo line. To the west of Lhasa were shallow shelf deposits (Gangdise Sea) and, to its east, paralic coal-bearing deposits. In fact they comprised two biomes on the northern coast of Tethys. In the Gangdise Sea area hexacorals have been found in the region of Xainza, Baingoin, Gerze, and Sengge Zangbo (Shiquanhe River), among which *Actinastraea, Pseudominima major, Favia turbinata and Dermosmilia* cf. *laxata* were common taxa in Europe. They are associated with rudists, nerineids, echinoids, and a few large forams, indicating the tropical shallow-water environment of northern Tethys. In eastern Tibet, the paralic coal-bearing depositional area is represented by the Duoni Formation and Linbuzong Formation, yielding abundant flora besides marine bivalves and gastropods. The flora is dominated by Filicopsida with the thriving Weichseliacae, Schizacaceae, and Matoniaceae. The genera *Cladophlebis* and *Scleropteris* are marked by pinnacles and split fringes. Cycadales were next in importance, with members *Zamiophyllum*, *Ptilophyllum*, and *Zamites* predominant. Conifers usually possessed scale-like leaves, their main representatives being *Frenelopsis* and *Cupressiocladus*. Ginkgoales and undoubted pteridosperms have not been reported. The general appearance of the flora indicates affinities to the tropical and subtropical zone of northern Tethys. However, facing the Tethyan Ocean, this area was more humid than the South China province. To the west of the Nujiang River, the sporo pollen assemblage of the Duoni Formation was dominated by the pollen of disaccate Pinaceae and Podocarpaceae, and also the pollen of cycadales and pteridophytes. The absence of *Classopollis* also indicates a more or less wet climate. It is interesting that in the Duoni Formation some trisaccate pollen has been discovered, including *Podosporites* and *Microcachryidites*, showing certain similarities to the coeval assemblage of India.

The Boreal realm

It was mentioned in Section 11.1 that in the Cretaceous this realm lay within the warm temperate zone. The cooling event at the end of the Late Jurassic and accompanying humidity expansion in the northern hemisphere caused a remarkable southward shift of the southern boundary of this realm, approaching the line of Palaeo-Tianshan–Palaeo-Qilianshan–Palaeo-Yanshan. It should be noted that a transitional mixed zone of the Boreal and Tethyan realms existed between the southern boundary and that line. Because of the differences between various taxa in ecological habitat and migration ability, scientific specialists working on their own groups often come to different conclusions about the boundary between the two realms (Chen 1982; Li 1983; Hao *et al.* 1986). The boundary adopted in this book is a comprehensive one based on the majority of taxa. This line extends westward roughly along the Ili River, through the southern edge of Balchas Lake into central Asia, approximately fitting the southern boundary of the Europe Siberian region (Kolesnikov 1980); eastward it joins the south-eastern boundary of Japan's Inner Zone, and the western boundary of the Dongning–Suifen Basin at the border between the Russia and China (Kimura 1984) (Fig. 11.4).

The northern half of China belongs to the northern Laurasian region in the southern part of the Boreal realm. The sea that invaded eastern Heilongjiang for a short time in the earliest Early Cretaceous came

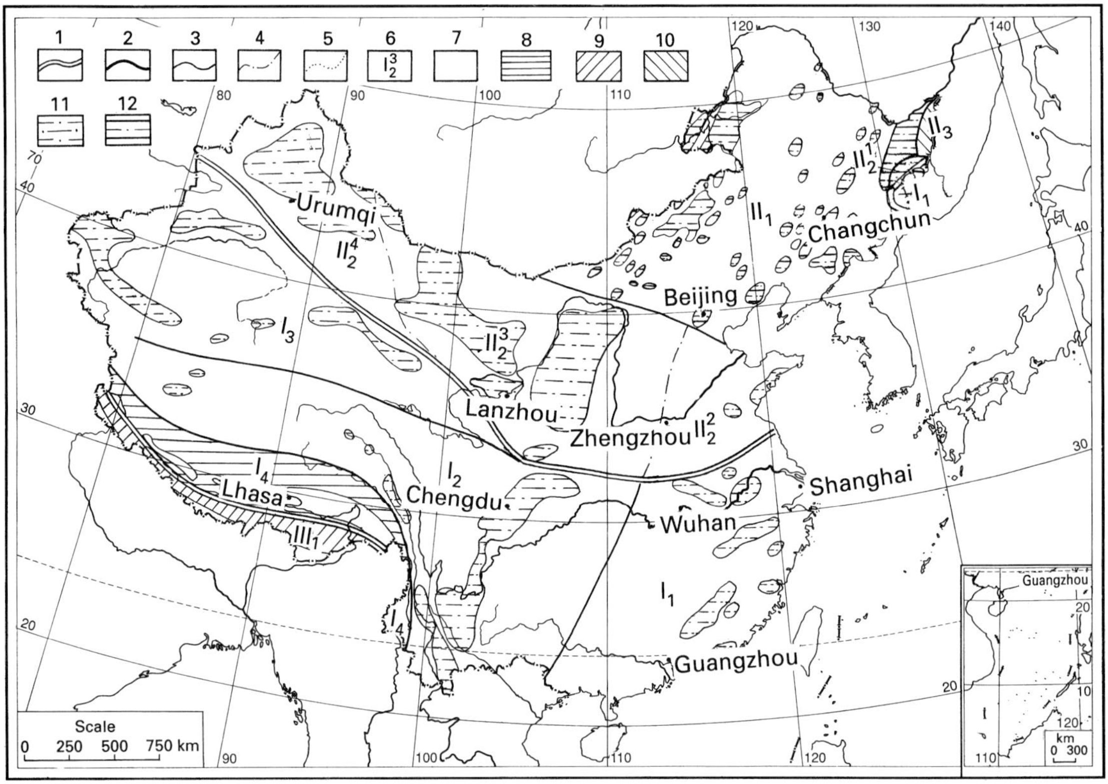

Fig. 11.4 Map of the early Early Cretaceous palaeobiogeographical provincialization of China. 1, Boundary of realms; 2, boundary of regions; 3, boundary of provinces; 4, boundary of biomes; 5, boundary of communities; 6, roman digit = realm mark, subscript = province mark, superscript = biome mark; 7, Oldland. 8–10, Marine and paralic regions: 8, South China; 9, Gondwana; 10, Junggar–Xingan. 11–12, Terrestrial regions (Late Permian and Mesozoic–Cenozoic): 11, fluvio–lacustrine basins; 12, coal-bearing basins. I, Laurasian–Tethyan realm, South China region; I_1, South-eastern China province; I_2, Sichuan–Yunnan province; I_3, Tarim–Qaidam province; I_4, Tibet–Western Yunnan province; II, Boreal realm, North-eastern China–Inner Mongolian region; II_1, North-eastern China–Inner Mongolia province; II_2, North China–Junggar province; II_3, North Pacific region, Wandashan province; III, Gondwanan Tethyan realm, India region; III_1, Himalayan province.

from the north Pacific. This transgressed area can be regarded as the edge of the north Pacific. Hence two regions, Northern China and North Pacific, are designated.

The Northern China region

This region can be subdivided into two provinces, i.e. North-eastern China–Inner Mongolia province and North China–Junggar province.

The Northeastern China–Inner Mongolia province

This is the area to the north of the Palaeo-Yinshan–Palaeo-Yanshan line. It possesses the following palaeobiogeographical features:

The land flora was dominated by the Filicopsida with Dicksoniaceae thriving; next in importance were ginkgoales, with broadleaf conifers and cycadales also common; this is called the *Acanthopteris–Nillsonia* flora. The general aspects coincide with those of the Siberia–Canadian 'Kingdom' established by Vakhrameev (1976), and are basically similar to those of the Japanese Inner Zone (Kimura 1984). The sporo-pollen assemblage belongs to the Disacciatrileti–*Cicatricosisporites* flora (Li Wenbeng 1983), with the following features:

1. It is a greatly diversified and abundant palynomorph flora. The main groups include Cyatheacea, Schizaeaceae, Osmunospermae, Lycopociaceae, Selaginellaceae, belonging to the pteridophytes, coniferales, ginkgoales, and cycadales. Angiosperm pollen is not reported.

2. Pteridophyte spores are characterized by the great diversity of Schizaeceae, the main genera of which were *Cicatricosisporites, Appendicisporites, Plicatella, Lygodiumsporites, Pilosisporites, Trilobosporites, Impardecispora,* and *Concavissimisporites,* among which *Cicatricosisporites* is highly diversified, with 15 species from the Chengzihe Formation in Jixi County and more than 15 species from the Shahai Formation in western Liaoning.

3. Gymnosperm pollen is characterized by the predominant disaccites of conifers, mainly pine-type pollen and *Piceaepollenites*. Cycadales are represented by tiny individuals of Cycadacean pollen. A small percentage (<10%) of *Classopollis* (of the Cheirolepicaceae) has been found in the assemblage. The characteristics of the sporo-pollen assemblage are compatible with the warm temperate climatic belt to which this province belongs. A difference in the sporo-pollen assemblages has been noted between the southern and northern parts of the province; this is basically reflected in the content of *Classopollis*. In the Jixi Basin in the northern part the content is less than 1 per cent, and in Guyang and Fuxin in the south it is up to 5 per cent, indicating a slightly higher aridity in the south than in the north.

The lacustrine fauna was very different from that of the Late Jurassic. The freshwater *Kuntulunia–Kuyangichthys* fish group appeared. As regards the insects, the *Ephemeropsis* assemblage was extinct. Freshwater bivalves were characterized by virtually new species on *Nipponoaia, Protelliptio (Plesielliptio),* and the thriving *Neomiodon*? The Jurassic survivor *Ferganoconcha* was still present and abundant in local horizons (Yu Xihan 1982) and can be correlated with the eastern Siberian region of Russia (Kolesnikov 1980). As for the conchostracans, an association of *Yanjiestheria* with the remaining *Eosestheria* has been found in horizons corresponding to the Shahai Formation in many localities of Liaoning and Jilin, but no conchostracans have been reported from the overlying Fuxing Formation, so far, which must reflect their rapid decline.

Among terrestrial vertebrates the thermophilic reptile *Psittacosaurus* has been found from the lower horizon of the Shahai Formation at the southern edge of this province, whereas from the Fuxing Formation only the primitive mammals *Endotherium* and *Manchurodon* have been recorded (Zhou Mingzhen 1953), indicating a tendency toward a more humid and less warm climate.

The North China–Junggar province

This is the zone of mixed fauna between the line of Palaeo-Yinshan–Palaeo-Yanshan and the line of Palaeo-Tianshan–Palaeo-Qilianshan–Palaeo-Qinling. The fauna has basically been attributed to the Boreal realm, but the flora has a mixed and transitional nature. In sporo-pollen assemblage the content of *Classopollis* ranges from 10 to 40 per cent, indicating a transitional state.

The transitional nature of land flora from north to south is notable from the fossil assemblages and their spatial distribution. In the south, the lower part of the Donghe Group in the Huicheng Basin of Gansu is dominated by red beds; the upper part bears coal seams (locally minable), the flora of which is dominated by filicopsidales and ginkgoales, consisting of both *Acanthopteris* and *Ginkgoites* of the North China type and *Brachyphyllum* and *Manica* of the South China type (Qi Hua 1984). Northward, thin coal seams in the Lamawan Formation at the top of the Zhidan Group in the north-eastern Ordos Basin again show the coexistence of *Acanthopteris* and *Brachyphllum*. The floral assemblage of the Laiyang Formation in the Laiyang Basin of Shandong is considered to belong to the South China type, but it occurs essentially in the horizons of members 1–5 of the Laiyang Formation, which are attributed to the

Late Jurassic in this book. For the early Early Cretaceous the character of the Laiyang flora is not yet known clearly. Westward to the Gansu Corridor and Junggar area of northern Xinjiang plant fossils occur only rarely.

The sporo-pollen assemblages nevertheless depict the transitional nature between the southern and northern regions. Li Wenben (1983) designated a transitional zone, which he set in general within the southern region. According to the principles adopted in this book for dividing the continental Jurassic and Cretaceous strata, the uppermost member of the Laiyang Formation of Shandong, the Tuoli Group of the Western Hills of Beijing, the lower Hekou Group at the Gansu–Qinghai border, and the Xinminpu Group of the Gansu Corridor are regarded as early Early Cretaceous, hence the transitional zone should be attributed to the North China region on the basis that the *Classopollis* content is less than 40 per cent in these strata. It must be explained that at the southern margin of this region, the content of *Classopollis* is up to 57 per cent in the Huaya Formation of the Tanba district, Chengxian County within the Huicheng Basin of Gansu (Zhang Zhifu 1982), which might be related to periodic climatic change in the drier zone adjacent to the southern region.

The lacustrine biota also display spatial differentiation within the North-eastern China–Inner Mongolia province. The freshwater fish *Kuntulunia* has been discovered in the Liwaxia Formation and Madongshan Formation in the upper Liupanshan Group of the Ordos Basin (Liu Xianting *et al.* 1985); the specimens collected are more abundant than those from the Guyang Basin of Inner Mongolia. The fish from the Xinminpu Formation of the Gansu Corridor include both *Qilianichthys* (Kuyangichthyidae) and *Coccolipis* (=*Sunolipis* Coccolipidae). Further westward to the Junggar and Turpan Basins in northern Xinjiang the freshwater fish were dominated by a more or less specialized siyuichthyid fauna (Su Dezhao 1985). This spatial discrimination of freshwater fish groups may have been caused by the restricted distribution of fluvial–lacustrine systems, so that they may be taken for independent biomes of this province. Such isolated inland basins formed by the inland system and separated from the outside world might be the reason why the siyuichthyid fauna exhibits comparatively primitive features from an evolutionary point of view.

The insects were greatly reduced in number at the change from Jurassic to Cretaceous, which corresponds with the climatic change clearly indicated in this province. The reappearance of a great number of endemic taxa in the Cretaceous demonstrates the difference between this province and the North-eastern and the China–Inner Mongolia province.

As regards the freshwater bivalves, the early *TPN* fauna has been found in this area. It has moderately diversified genera and species; besides *Nipponanaia* it includes *Plicatounio* spp., and *Nakamuranaia* has more species than in the North-eastern China–Inner Mongolia province. In the upper horizons of the Jingchuan Formation, Huatin County, Gansu and Longxian county, Shaanxi at the southern edge of the Ordos Basin, *Nakamuranaia* has more than 10 species, of which *Nakamuranaia yongkangensis, N. angulata, N. subcentralis, N. subequilateralis,* and *N. elliptica* are common South China species, and the *Yunnanoconcha? chuxiongensis* endemic to south-western China also appears. The ostracods in the Hekou Group of the Minhe Basin and in the upper Liupanshan Group at the southern edge of Ordos Basin (Qi Hua 1984) also show a degree of mixing. Obviously these basin lay where the southern and northern regions joined up.

Unlike the north-eastern China–inner Mongolia province, where the vertebrate *Psittacosaurus* and its allies are restricted to the lower horizons of the Shahai Formation, in this province they occur in the whole Lower Cretaceous (Tugulu Group, Qinshan Formation). The diachroneity from north to south of these thermophilic vertebrates is evidently related to the fact that this province was more influenced by the hot, arid climate from the south. The famous *Dsungaripterus* fauna from the Tugulu Group of the Junggar Basin in northern Xinjiang indicates a unique inland lake basin environment.

A similar transitional fauna in this province has been noted in the Jixi district of eastern Heilongjiang. The flora of the Chengzihe Formation *(sensu stricto)* and Muling Formation of the upper Jixi Group are characterized by a great number of *Gleichenites* spp. and *Frenelopsis*, commonly seen in the southern region, in addition to the general features of the *Acanthopteris–Nillsonia* flora. *Gleichenites* accounts for 10 per cent of the composition of the sporo-pollen assemblage in the Muling Formation. Based on this, the Jixi area is treated as an extension of the North China–Junggar province (Fig. 11.4). Provisionally this area is taken to be an independent biome—the Jixi biome. North Xinjiang is also considered to be an independent biome—the Junggar biome—due to the endemic siyuichthyid fishes and the absence of early *TPN* bivalves. Two other biomes designated provisionally for the remaining area embracing the Taihangshan Mountains, i.e. the Shangdong and Ordos biomes. However, the difference of terrestrial faunas between the two remains to be further studied.

The Wandashan province of the North Pacific region

The data collected in recent years are concerned only with the Donganzhen district, Raohe county, eastern Heilongjiang (Sha and Yuan 1985). Although the stratigraphy must be further studied, the discovered buchiids, such as *Buchia fischeriana*, *B. unschensis*, *B. subokensis*, etc. obviously indicate a transgression in the North Pacific region in the early Early Cretaceous.

The Gondwanan Tethyan realm

This realm covers the area to the south of the Yarlung Zangbo River. It belongs to the northern edge of the Indian plate. Only one palaeobiogeographical province is designated for the limited area in China, i.e. the Himalayan province. An ammonite assemblage containing *Sarasinella*, *Necopnites*, and *Spiticeras*, very similar to the Valanginian fauna in the Salt Range of Pakistan, has been found in the Gamba and Gyangze districts. In the Gabula Formation of Gyangze county there has also been discovered a *Hibolithes jiabulensis–Belemnopsis sinensis* belemnite assemblage, most closely related to the Salt Range and next to the Neocomian belemnites of Indonesia and Kutch (Catch) of India. However, in Gabula *Cylindroteuthes porrecta* has also been found recently. This is characteristic of North America and Europe, and may have infiltrated from outside Gondwanan Tethys (Yang *et al.* 1985). It should be noted that the *Weichselia* flora has been found in the continental strata of northern India, suggesting the location of northern India at low latitudes with a warm climate at that time.

11.3 LATE EARLY CRETACEOUS TO LATE CRETACEOUS PALAEOBIOGEOGRAPHICAL PROVINCIALIZATION OF CHINA

11.3.1 OUTLINE

In the late Early Cretaceous (Aptian–Albian) the global climate became warmer. This accompanied the change of palaeogeographical and palaeotectonic patterns in eastern Asia, resulting in a reorganization of land communities and a shift in the boundaries of biogeographical provincialization. The global temperature rise began after the Aptian. The tropical–subtropical arid zone expanded northward, with the southern boundary of the Boreal realm moving northward to northern Heilongjiang province, and covering the vast area of north-eastern, north, and northwestern China. At the end of the Albian the global temperature fell (Hallam 1985). This short cooling event had a significant influence on marine plankton and a lesser one on benthic organisms. On land it considerally influenced the sporo-pollen compositions (Yu Jiangxian *et al.* 1983), but similar information has yet to be reported for other terrestrial organisms. The strongest Cretaceous cooling event on a global scale occurred during the Maastrichtian. It is distinctively demonstrated by isotopic data on the shells of both oceanic plankton and benthic or ganisms (Douglas *et al.* 1981), by the change in angiosperm composition (Krassilov 1975), and by the size of the spiracles on dinosaur eggshells (Zhao Zikui 1978). This cooling event and the following probable extraterrestrial impact event at the end of Cretaceous had great significance in the famous catastrophic event causing the mass extinction of marine and land faunas. It is necessary to state that the northern boundary of the tropical–subtropical zone extended to the northern border of China after the late Early Cretaceous. From then on the vast territory of China as a whole experienced a tropical–subtropical climate. Even north-eastern China was under a subtropical (indicated by the sporo-pollen assemblage) or warm temperate climate (indicated by the megaplant fossils). Even though China underwent two events of temperature fall at end of the Albian and Maastrichtian, a temperate climate did not return. However, some fluctuations of warmth or aridity took place in time and space from the late Early Cretaceous to the Late Cretaceous, which had a significant influence on the patterns of biogeographical provincialization.

11.3.2 PROVINCIALIZATION

According to the scheme of Zaklinskaia (1963), who designated the Eastern Siberia–Circum-Pacific phytogeographical realm and Atlantic phytogeographical realm, north-eastern China must obviously belong to the Eastern Siberia–Circum-Pacific, due to the abundant *Aquilapollenites* angiosperm pollen, although from the late Early Cretaceous to Late Cretaceous the sporo-pollen composition in north-eastern China indicates a tropical–subtropical climate (Yu Jingxian *et al.* 1983). According to data from a few localities of megaplants, the north-eastern China flora was composed of deciduous angiosperms and gymnosperms. Thus, north-eastern China or the area to the north of Palaeo-Yinshan–Palaeo-Yanshan is assigned to the Boreal realm. For the whole period from late Early to Late Cretaceous three realms, four regions, and six provinces are established (Table 11.3, Fig. 11.5).

The Laurasian-Tethyan realm

The extent of transgression coverage in this realm increased from the late Early Cretaceous to the Late Cretaceous so that the western boundary of Tethys within the South China region moved northward to include the western edge of the Tarim Basin.

The South China region

The global temperature rise after the Aptian pushed the northern boundary of this province to the north of the Palaeo-Yinshan–Palaeo-Yanshan line. The vast area of northern China and north-western China lay within the arid tropical–subtropical zone and belonged to the same palaeobiogeographical region, together with the South China province. Because the fossils are distributed unevenly, provinces can only be designated roughly, i.e. the South-eastern province to the east of the Xuefengshan–Taihangshan line and the South-western–North-western province to the west of the same line. Each of the two provinces is again subdivided into a southern and a northern area along the Ili River–Xining City–Xi'an City–Huaihe River line. The southern areas of each province were arid zones, with *Classopollis* and *Schizaeoisporites* accounting for more than 50 per cent of the total flora; in the northern areas *Classopollis* and *Schizaeoisporites* have less than 50 per cent, but *Classopollis* accounts for more than 10 per cent of the flora signifying a transitional area from arid to humid. These four areas correspond to subprovince or biome but they have not been named so because the data are insufficient.

The South-eastern province

The land plant assemblages did not change in their general features during the Early Cretaceous in the south-eastern area. Only scattered fossils have been discovered in Yining, Guangxi in Late Cretaceous deposits. Besides the gymnosperm *Brachyphyllum* in the Bali Formation, there were the tropical and subtropical lauraceae evergreen trees and bushes *Cinnamomum* and *Nectandra*, which still exist today. Usually they had small leathery leaves, with entire margins, indicating a warm and arid environment. Sporo-pollen assemblages, according to available data, were widely distributed. The lower part of the late Early Cretaceous Lushangfen Formation in the western hills of Beijing has a *Classopollis* content of less than 10 per cent, which rapidly increases to more than 40 per cent in the upper part. Moreover, gypsum deposits 0.3–3 cm thick occur at the top of the overlying Xiazhuang Formation. In the Zhejiang-Fujian

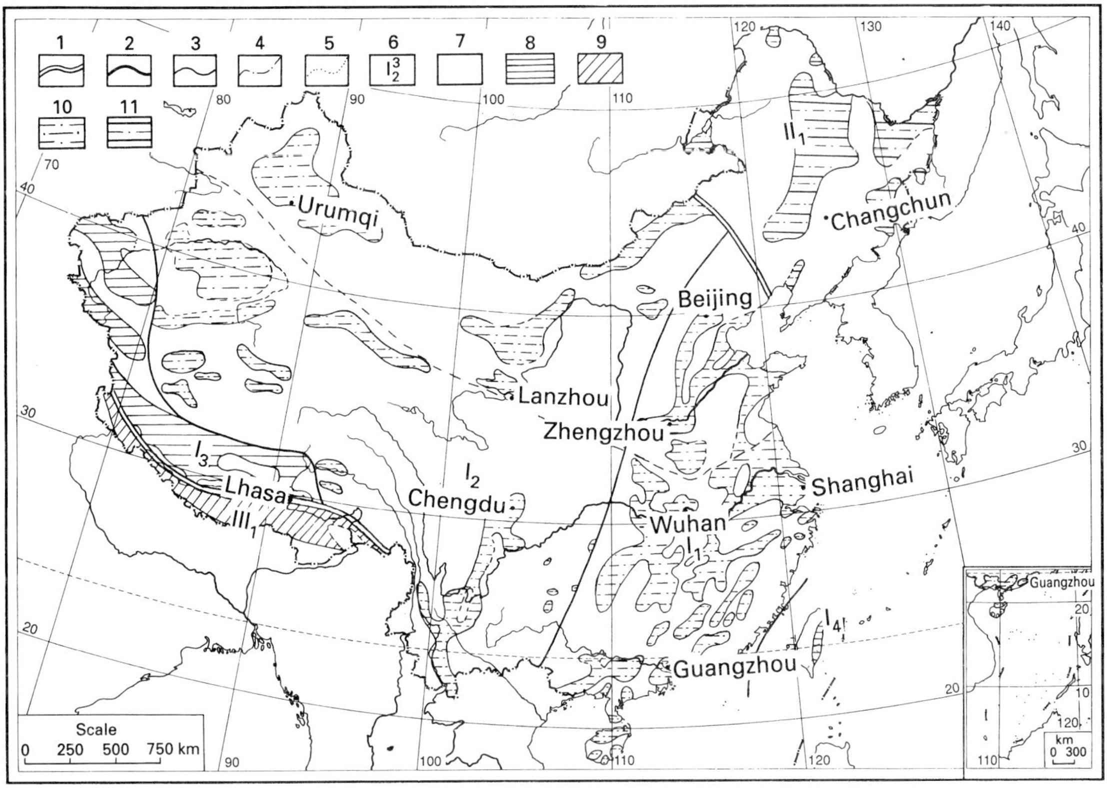

Fig. 11.5 Map of the Late Cretaceous palaeobiogeographical provincialization of China. 1, Boundary of realms; 2, boundary of regions; 3, boundary of provinces; 4, boundary of biomes; 5, boundary of communities; 6, roman digit = realm mark, subscript = province mark, superscript = biome mark; 7, Oldland. 8–9, Marine and paralic regions: 8, South China; 9, Gondwana. 10 11, Terrestrial regions (Late Permian and Mesozoic–Cenozoic): 10, fluvio–lacustrine basins; 11, coal-bearing basins. I, Laurasian–Tethyan realm, Southern China region (areas north to the dashed line are transitional; *Classopollis* > 10 per cent, *Classopollis* + *Schizaeoisporites* < 50 per cent; areas south to the dashed line are arid: *Classopollis* + *Schizaeoisporites* > 50 per cent; I_1, South-eastern China province; I_2, South-western–North-western China province; I_3, Tibet–Tarim province; I_4, West Pacific–Taiwan province; II, Boreal realm, North China region; II_1, North-eastern China–Inner Mongolia province; III, Gondwanan Tethyan realm, India region; III_1, Himalayan province.

area along the south-eastern coast, the *Classopollis* content of the Aptian strata, such as the Guantou Formation of eastern Zhejiang, the Gecun Formation of Jiangsu, and the Second Member of the Zhuxiang Formation in central Anhui, is generally ±30 per cent, whereas the total content of *Classopollis* and *Schizaeoisporites* is up to 55 per cent or more than 80 per cent, with the latter greatly exceeding the former in the west, i.e. in the Albian Wulong Formation of western Hubei and the Shenwanshan Formation of southern Hunan (Li Wenbeng 1983). These facts denote either further enhancement of the Albian arid zone, a decrease in humidity following the increase in distance from the Palaeo-Pacific Ocean, or a difference between south and north. In the Late Cretaceous, rocksalt was deposited in the Jianghan Basin, the Huicheng Basin of southern Jiangxi, and the northern Jiangsu depression, and the total content of *Classopollis* and *Schizaeoisporites* ranged from more than 50 per cent to ±70 per cent, indicating a continuously arid tropical–subtropical climate. The Maastrichtian cooling event acounts for the increase of temperate deciduous broadleaf plants in the Yuyang Formation of the Jianghan Basin, and the decrease in the total content of *Classopollis and Schizaeoisporites* to less than 27 per cent (Wang and Zhao 1980). In the inland area, there were only individuals of *Aquilapollenites* in the Jianghan Basin in the Late Cretaceous. However, eastward along the same latitude, the *Aquilapollenites* from the Taizhou Formation in northern Jiangsu increased notably, both in quantity and in variety (Zhou and Wang 1982), which indicates a tendency towards increasing humidity as the Pacific was approached. To sum up, the change of sporopollen assemblages in south-eastern China reflects not only periodic pulsations of global temperature, but also the distance and influence of the Pacific, and is thus significant for palaeoclimatic and palaeogeographical interpretation.

Among lacustrine organisms, there is still a distinct diversity of conchostracans with provincial distributions. A late Early Cretaceous conchostracan *Cratostracus* assemblage appears in the Guantou Formation of eastern Zhejiang. This assemblage still includes *Orthestheria* and rare *Migransia*, and even, problematically, *Yanjiestheria*. This can be regarded as the continuous development of the Early Cretaceous *Yanjiestheria–Migransia* assemblage. The South-eastern province in the Late Cretaceous was characterized by the universal appearance of a *Tenuestheria–Sinoestheria* assemblage. A great number of 'African' conchostracans, such as *Zhestheria, Feiyunella,* and *Linhaiella* occurred, in addition to the *Orthestheria* that had survived from the Early Cretaceous. The above-mentioned fossils appeared in the Congo Basin of Africa in the latest Early Cretaceous. This implies that they may have migrated from south to north into Eurasia (Chen and Shen 1977). More African conchostracans invaded during the Late Cretaceous, demonstrating the even more frequent migration of land fauna between southern and northern continents. However, the present data show only migration from Gondwana northward. It is still to be clarified whether some other genera and species of conchostracans migrated in the opposite direction (Wang Si'en 1983).

The freshwater bivalves were characterized by the great abundance of the *TPN* fauna, among which *Trigonioides* and *Plicatounio* were predominant, with *Nippononaia* and *Nakamuranaia* relatively subordinate. Such a composition is different from the *TPN* composition in the Boreal realm and is most similar to the *TPN* fauna of coeval examples in Japan and South Korea. *Pseudohyria* was already present in the late Late Cretaceous, but such special members of the Boreal realm as *Protelliptio (Plesielliptio)* were absent.

As for ostracods, in the late Early Cretaceous there occurred the *Cypridea–Darwinula* (small-sized)–*Eucypria (=Lycopterocypris)* assemblage, found in the Guantou Formation of eastern Zhejiang and the Huizhou Formation of southern Anhui. Late Cretaceous ostracod assemblages occur in the Qujiang Group of western Zhejiang and the Wangshi Group of eastern Shandong, and are characterized by the decline of *Cypridea* and the appearance of new genera such as *Candona, Candoniella,* and *Limnocythere*, but they exhibit no obvious differences from the faunas of the Boreal realm.

In the late Early Cretaceous there occurred such special fish genera as *Paralycoptera, Yungkangichthys* and *Huashia,* quite different from the *Manchurichthys* assemblage of the Boreal realm. *Paralycoptera* of this region and *Plesiolycoptera* of the Boreal realm are both referred to as Osteoglossomopha, but they evolved from *Asiatolepis (='Lycoptera sinensis')* and *Lycoptera* respectively, and thus record different lineages. This indicates pronounced biogeographical provincialization between the two realms.

The South-western–North-western province

There is no formal report on any discovery of land plant fossils in this area, and so far deduction can be made only on the basis of some sporo-pollen material. The late Early Cretaceous may be represented by the Wulong Formation of western Hubei province where the *Schizaeoisporites* content is up to 24–45 per cent and *Classopollis* is 3–10 per cent. The total sum of both is up to more than 50 per cent indicating an arid climate. In the contemporary strata of the Ailaoshan Mountains of Yunnan the contents of *Schizaeoisporites*, *Classopollis*, *Exesipollenites*, and *Ephedripites* are all high, indicating an even more arid tropical–subtropical climate than the South-eastern province.

The lacustrine fauna also displays obvious biogeographical differentiation. Among conchostracans, *Orthestheria* survived in the Mengong Formation of the late Early Cretaceous; some of its species are very similar to those found in the coeval Wealden Group in England. The Late Cretaceous is characterized by the *Aglestheria* assemblage, including *Aglestheria, Halysestheria, Nemestheria*, plus a few *Orthestheria* and *Jilinestheria*. Compared with the *Euestherites* assemblage of the North China Region, both diversity and abundance are low and most of the shells are thin, indicating a more or less arid tropical–subtropical climate. The bivalves indicate a flourishing *TPN* fauna. In the Shiwandashan Mountains of Guangxi in the southern part of the province, the Napai and Xinlong Formations of the late Early Cretaceous age yield more or less abundant *TPN* fauna; their individuals are big, and many special members have been found in the synchronous strata of Muong Phalane in Laos and in the Khorat Plateau of Thailand. This fauna displays a certain differentiation from the *TPN* fauna of Zhejiang–Fujian, South Korea, and Japan of the Circum-Pacific zone. The Bali Formation of the early Late Cretaceous contains a *Pseudohyria* assemblage with a species composition quite different from those found in the northern region. Data of late Late Cretaceous bivalves are few. In the Sanshui Formation of Guangdong the *TPN* fauna had already disappeared; only *Sphaerium*, *Eupera*, and *Pisidium* were left (Huang and Zhang 1982), obviously different from the coeval *Protelliptio (Plesielliptio), Pseudohyria, Sphaerium* assemblage of the North China region.

With regard to ostracods, from the late Early Cretaceous to the Late Cretaceous the assemblages are more or less uniform; no more notable differentiation between south and north or between east and west existed. The Late Cretaceous ostracod assemblage of the Junggar and Turpan Basins in northern Xinjiang might be correlated to the fauna, dominated by *Cristocypridea*, of the Songliao Basin, eastern Shandong, the Jianghan Basin, and the Nanxiong Basin of eastern Guangdong. In Xinjiang and Qinghai, the other lacustrine and plant fossils of this age are rare; the sporo-pollen assemblages also display similarities with those in South China, so that they are incorporated into the South-western province.

The Tibet–Tarim province

With the expansion of transgression in this stage the northern boundary in the western sector extended northward to include the north-western edge of the Tarim Basin. However, it remains to be further studied whether the northern boundary of the transgression in the eastern sector reached the Golmud area in Qinghai. The southern boundary was still restricted north of the line of the Yarlung Zangbo River. In North Tibet, in the area of Baingoin, Xianza, Gerze, Rutog, and the Shiquanhe River, among the Aptian–Cenomanian hexacoral assemblage there occurred relatively more European elements, such as *Aulastraeopora deangelisi*, or their allies, like *Montlivaltia mayaoensis*, so the assemblage as a whole shows affinities to Eurasian Tethys, although the South Indian species *Hydnophora insignis* occurred as well. The large benthic foram *Orbitolina* was the typical representative of tropical subtropical shallow-water regimes. It is distributed mainly in Xigaze to the north of the Yarlung Zangbo River suture, in Lhasa and Lhunzhub of central Tibet, in eastern Karakorum, and on the southern edge of the Kunlun Mountains, and the containing deposits have been referred to the Aptian–Cenomanian. In the Kashi–Yengisar area of the western Tarim Basin in Xinjiang, the Late Cretaceous deposits are rich in small benthic foraminifers, associated with the oyster *Turkostrea*, the rudist *Osculigera*, and the trigoniids *Trigonia ferganensis* and *T. turkestanensis*, which in general corresponds with the Eurasian Tethyan fauna as in northern Afghanistan, north-eastern Iran and central Asia. There is a high proportion of oysters, which might be related to a specific gulf environment. In the area along Baingoin and the Shiquan River of Gangdise there occur bivalves including *Horiopleura haydeni*, *Requienia*, and

Radiolites, very similar to the Albian–Cenomanian faunas in Afghanistan and Kashmir.

Six gastropod species of the Late Cretaceous Xigaze Group were discovered on the northern bank of the Yarlung Zangbo River, three of which are nerineids commonly seen in the Late Cretaceous of Europe.

The ammonite *Deshayesites* was found in Niumagou of Lhunzhub County to the north of Lhasa. This genus has been discovered in the Aptian of England, North America, and South America. The early Turonian *Thomasites koulabicus* and late Santonian–early Campanian *Placenticeras khareamensis* occur in the Kukbai area, Wuqia County, Xinjiang, also indicating an affinity with Eurasian Tethys.

Brachiopods did not have a notable provincialization; especially in the Late Cretaceous most were cosmopolitan. In Xainza, Baingoin, and Shiquan River of Gangdise the Aptian–Cenomanian brachiopod fauna contains *Orbirhynchia* of the Pugnaxidae, commonly seen in north-western Europe; some species of the terebratulid *Praelongithyris* are very similar to those in the late Early Cretaceous strata in France and Italy. The brachiopods in general signify an occurrence on the northern side of Tethys.

The West Pacific province

Cretaceous ammonites and bivalves have been discovered from boreholes in the Beigang area, Yunlin County and the Penghu archipelago in western Taiwan. However, the Palaeocene forams *Globigerina spiralis* and *Globorotalia pseudobulloides* from the matrix of the mollusc fossils proved that these samples were the products of reworking. None the less, the transportation distance should not have been very great, so they can still indicate the presence of the Cretaceous transgression in the Taiwan area. On this basis a Taiwan province in the West Pacific region is tentatively suggested. The ammonites consist of the late Aptian genera *Dufrenoyia* and *Holcophylloceras*. The bivalves include *Misosacella? taiwanensis*, *Costocyrena paikongensis*, *Tetoria (Paracorbicula)* sp., *Nemocardium* aff. *yetshiraensis*, and *Amygdallum* aff. *ishidoensis*, the dating of which is Cretaceous but which might extend to the Palaeogene. The fauna contains Pacific, Tethyan and endemic members. Available data show that at least during the late Early–Late Cretaceous the Palaeo-Pacific transgression influenced the eastern margin of the Asian continent.

The Boreal realm

After the late Early Cretaceous the palaeogeographical and palaeotectonic patterns in north-eastern Asia were greatly changed. Though normal marine transgression invaded Chinese territory no more, the Late Cretaceous global transgression still brought about intermittent ingression effects into the NNE-trending depression basin in eastern China, creating brackish environments and special biotic communities. Thus, within Chinese territory for this realm there have been designated one region (the North China region), one province (the North-eastern China–Inner Mongolia province), and two biomes (the Inland biome and the Paralic Basin biome).

The North-east Asia region

The North-east Asia region covers the Lena-Koluma and Baikal-Amur provinces of Russia (Kolesnikov, 1980), and the North-eastern China–Inner Mongolia province. The latter two most probably belong to the same province. They were all transitional areas from the warm temperate zone to the subtropical zone in the northern hemisphere. The land flora was dominated by warm temperate plants, with some elements of warm temperate–subtropical type. The flora of the late Early Cretaceous Dalazi Formation in the Yanbian region of Jilin is poorer than the early Early Cretaceous Fuxing flora, and is dominated by conifers with scale-like or needle-shaped leaves, followed next in importance by angiosperms and bennettitales, lacking ginkgoales, and having only a few filicopsid specimens. Such a flora was the product of the post-Aptian temperature rise and aridity expansion. The North-eastern China–Inner Mongolia province had since been deprived of humid coal-forming environments, except in a few localities. Some drought-enduring plants, such as *Brachyphyllum*, *Manica*, *Frenelopsis*, and *Cupressionocladus*, thrived in South China in the early Early Cretaceous, and entered this area at a late stage, demonstrating the diachroneity of the drought-enduring plants during their northward migration. The flora from the Albian to Cenomanian or Turonian is generally difficult to subdivide (Vakhrameev 1952). The Quantou and Qingshankou Formations of the Songliao Basin contain a flora containing *Trapa* and *Platanus* which probably extends from Early to Late Cretaceous. *Trapa* is a genus living at present in the temperate zone; *Platanus* is now thriving in the warm temperate–subtropical zone. It seems that the short cooling event at the end of the Albian did not interrupt the development of flora in this area, but brought some changes, however, in sporo-pollen assemblages. The middle and late Early Cretaceous flora in Hunchun County, Jilin and in Jiayin County, Heilongjiang still indicate a warm temperate climate, most plants being deciduous, but a few, such as *Myrica*, being evergreen. The Late Cretaceous angiosperm plants usually have leaves with dentate leaf margins, occasionally thin and leathery, comparable with those of the modern plants of the warm humid zone between the Yellow River and Yangtze.

In the sporo-pollen assemblage the *Classopollis* content began to rise again in the late Early Cretaceous. In the Fuxin Basin it began to increase from the upper Fuxing Formation (Banlashan Formation), and in the Sunjiawan Formation it was up to *c.* 13 per cent (Pu and Wu 1985). In the Songliao Basin this content also increased notably from the Denglouku Formation to the Quantou Formation, and it was up to 38 per cent in the upper part of the Quantou Formation. Other drought-enduring thermophilic plants, such as *Schizaeoisporites* and *Ephedripites*, also display synchronous increases. Therefore, in the Albian, this area reached the maximum dry, hot palaeoclimate, being in fact a subtropical arid climate, coincident with the palaeoclimate reflected by the above-mentioned flora of the Dalazi Formation and also with the conclusion that the Albian was the warmest stage in the Cretaceous, according to isotopic thermometry of marine foram shells (Hallam, 1985).

The content of *Classopollis*, *Schizaeoisporites*, and *Ephedripites* dropped rapidly at the base of the Qingshankou Formation, indicating the sudden change of climate from dry–hot to wet–cool (Fig. 11.6). This is most probably related to the global cooling event at the end of the Albian and may serve as an effective index in correlating synchronism between different lithofacies and biofacies. In sporo-pollen assemblages from the Qingshankou Formation to the Mingshui Formation, the *Classopollis* content fluctuated below 10 per cent and some *Aquilapollenites*, *Pyroteacidites*, and *Beaupreaidites* existed. As a whole this indicates a more-or-less humid tropical–subtropical climate in the Late Cretaceous, which is slightly different from the conclusion drawn by analysing the megaplant fossils.

In the sporo-pollen assemblage of the Mingshui Formation, temperate deciduous broadleaf plants increased significantly, whereas the tropical and subtropical Palmaceae, Myrtaceae, Loranthaceae, and Proteaceae decreased; the content of *Aquilapollenites* was up to 12.5 per cent but *Classopollis* only 0.5–4 per cent; *Schizaeoisporites* was still dominant, but its abundance and diversity declined, which undoubtedly indicates the notable cooling event at the end of the Cretaceous period.

As for the lacustrine fauna, the conchostracans also reflect a close relationship to climatic change. *Yanjiestheria* and *Orthestheria* of the Dalazi Formation had replaced the *Eoestheria* assemblage. In recent years the 'African' genus *Migransia* has been found in the Denglouku Formation. This again indicates the diachroneity during which the southern fauna invaded northward following the expansion of the hot arid belt. The Late Cretaceous *Euestherites* assemblage is characterized by its high diversity (4 orders, more than 21 genera, more than 100 species), its close lineage relationships, and strong endemism; these phenomena may be related to the peculiar environment of the paralic Songliao Basin.

After the late Early Cretaceous the eastern Asian freshwater bivalves evolved into the acme stage of the *TPN* assemblage. In this area *Plicatounio* was most common; *Nippononaia* and *Martinsonella* had their representatives too, while *Trigonioides* occurred only in the Yanbian region in the eastern part of this province but not in the Songliao Basin.

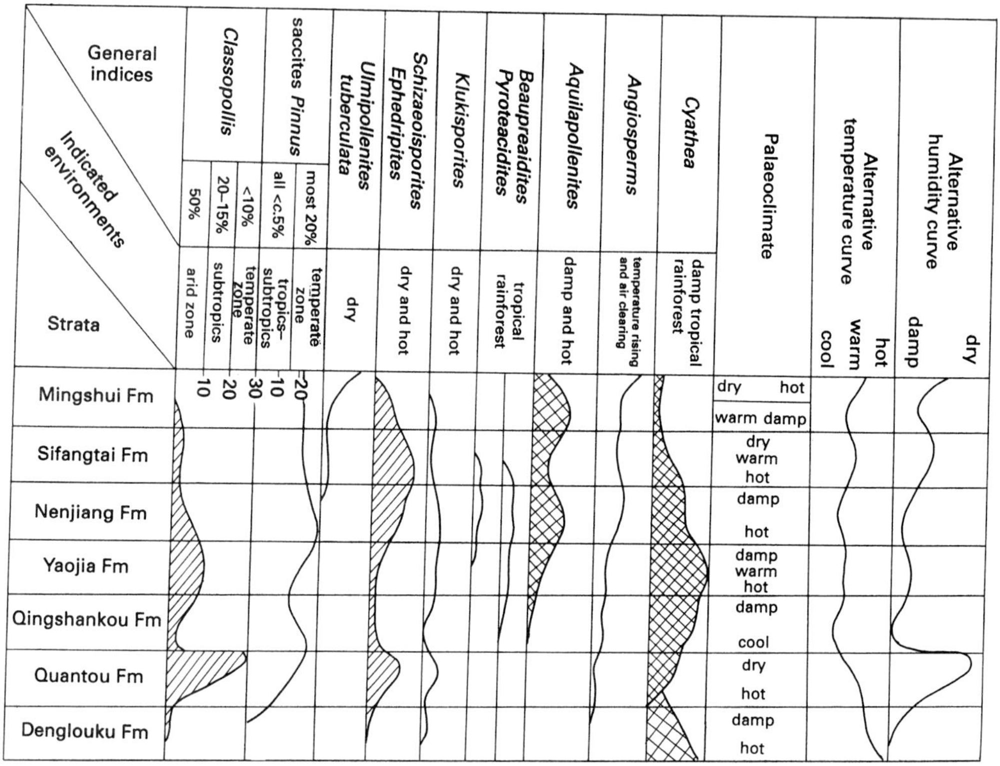

Fig. 11.6 The relationship between Cretaceous sporo-pollen assemblages and palaeoclimate changes (after unpublished material of Zhang Ying 1985).

In the late Late Cretaceous a relatively monotonous *Pseudohyria–Protelliptio (Plesielliptio)* assemblage appeared in this province; *Trigonioides, Plicatounio, and Nippononaia* all disappeared, and the formerly diminished *Sphaerium* and *Margaritifera* reappeared. In brief, the Middle and Late Cretaceous bivalves of this province are similar to those of Siberia and Kazakhstan, but more or less different from those of South China. A number of Late Cretaceous brackish bivalves, including *Striarca, Mytilus*, and *Musculus*, have been found in the Nenjiang Formation of the Songliao Basin. These represent the special biofacies under the circumstances when the paralic basin was invaded by the sea. The late Early Cretaceous ostracod assemblage exhibits reasonably close similarities with that in South China (DRIDPF 1976; Hao *et al.* 1974), reflecting a tendency towards northward migration due to climatic warming. The early Late Cretaceous and late Early Cretaceous ostracod assemblages in this province are similar to contemporary ones in Mongolia.

The *Manchurichthys* fish assemblage from the late Early Cretaceous Dalazi and Baojiatun Formations and the *Sungarichthys* fish assemblage from the Late Cretaceous Nenjiang Formation are obviously different from the freshwater fish groups in South China. It should be noted that the Qingshankou and Nenjiang Formations of the Songliao Basin in Late Cretaceous underwent marine ingression, and in consequence many brackish *Hama, Jilinichthys*, and shark's teeth occur in these deposits (Chang and Chou 1976).

The Gondwanan Tethyan realm

This realm in China covers only the Himalayan province on the northern margin of the Indian plate.

The upper Zongshan Formation in the Gamba region on the northern side of Qomolangma is rich in fossils. There are 6 genera and 8 species of hexacorals, but most of them are new or indeterminate species with a very strong endemism. Fifteen species of bivalves have been reported, including the well-known rudist assemblage in carbonates. Among them the rudists *Bournonia tibetica, B. haydeni* and *Plagioptychus tibeticus*, and the non-rudists *Inoceramus (Eudocostea) haydeni* and *Aetostreon zongshanensis* are endemic species. Only five of them can be correlated to other parts of Tethys. However, the fauna as a whole approximates that of the late Late Cretaceous Baluchistan fauna in Pakistan (Wen *et al.* 1976). As regards the brachiopods, the unique *Xenothyris tuilaensis*, which belonged to a lineage that underwent parallel evolution with the world-wide Magathyrid lineage, occurs here. The large benthic foraminifers are characterized by the presence of the distinctive *Orbitoides* assemblage, an index of the Campanian to Maastrichtian (Hao Yichun *et al.* 1986). It is possible that in the Maastrichtian an important cooling and regression event occurred (Kauffmann 1981), accompanied by other events in Tethys, which enhanced the endemism of many organic groups.

REFERENCES

Chang, M. M. and Chou, C. C. (1976). Discovery of *Plesiolycoptera* in Songhuajiang-Liaoning basin and origin of Osteoglossomorpha. *Vertebrata PalAsiatica* **14**, (3), 146–53 (in Chinese).

Chen Jinghua (1982). Mesozoic transgressions, regressions and bivalve provinces in China. *Acta Geologica Sinica* **56**, (4), 334–46 (in Chinese with English abstract).

Chen Peiji and Shen Yanbin (1977). On the discovery of Afrograptidae (Conchostraca) in Zhejiang with its significance. *Acta Palaeontologica Sinica* **16**, (1), 81–94 (in Chinese with English abstract).

Douglas, R. G. and Woodruff, F. (1981). Deep sea benthic foraminifera. In *The sea*, (ed. C. Emiliani), Vol. 7. Wiley-Interscience, New York.

DRIDPF (Development Research Institute of Daqing Petroleum Field) (1976). *Cretaceous ostracoda from the Songliao basin*. Science Press (in Chinese).

Guo Fuxiang (1981). Bivalves from the Jingxing Formation (Cretaceous) in western Yunnan with notes on the origin of the trigonioids in Asia. In *Selective Papers of the 12th Annual Conference of the Palaeontological Society of China*, pp. 61–79. Science Press, Beijing (in Chinese with English abstract).

Guo Fuxiang (1986). *On Trigonioidaceans (non-marine Cretaceous bivalves) and Asian non-marine Cretaceous System*. Yunnan Science and Technology Publishing House, Kunming (in both Chinese and English).

Hallam, A. (1985). A review of Mesozoic climates. *Journal of the Geological Society* **142**, (3), 433–45.

Hao Yichun, Shu Deyin, Li Yougui, Yuan Peihua, and Yuan Fengtian (1974). *Cretaceous–Tertiary ostracoda fossil from Songliao plain*. Geological Publishing House, Beijing (in Chinese).

Hao Yichun *et al.* (1986). *The Cretaceous system of China—stratigraphy of China*, Vol. 12. Geological Publishing House, Beijing (in Chinese).

Huang Baoyu and Zhang Xianqiu (1982). Late Cretaceous and Early Tertiary nonmarine lamellibranchs from the Pearl River Delta. *Acta Palaeontologica Sinica* **21**, (3), 267–79 (in Chinese with English abstract).

Kauffman, E. G. (1981). Plate tectonics and global sea level changes: A comparison of the global Cretaceous record with that of Tibet. *Geological and Ecological Studies of Qinghai–Xizang Plateau* **1**, 413–26.

Kennedy, W. J. and Odin, G. S. (1982). The Jurassic and Cretaceous time scale in 1981. In *Numerical dating in stratigraphy* **I**, 557–92.

Kimura, T. (1984). Mesozoic floras of east and southeast Asia, with a short note on the Cenozoic floras of southeast Asia and China. *Geology and Palaeontology of Southeast Asia* **25**, 325–50.

Kobayashi, T. (1968). The Cretaceous non-marine pelecypods from the Nam Phung Dam Site in the Northeastern part of the Khorat Plateau, Thailand with a note on the Trigonioidae. *Geology and Palaeontology of Southeast Asia* **4**, 109–38.

Kolesnikov, C. M. (1980). Classification, stratigraphic distribution and biogeography of Mesozoic limnic bivalvia from USSR. In *Limnobios of ancient lake-basins of Eurasia*, pp. 9–65. Nauka, Leningrad (in Russian).

Krassilov, V. A. (1975). Climate changes in eastern Asia as indicated by fossil floras. 2. Late Cretaceous and

Danian. *Palaeogeography Palaeoclimatology Palaeoecology* **17**, 157–72.

Li Wenbeng (1983). Microfloral areas of Early Cretaceous in China. In *Palaeobiogeographic provinces of China*, pp. 142–51. Science Press, Beijing (in Chinese).

Liu Benpei, Chen Feng, and Wang Wuli (1986). On the non-marine Jurassic–Cretaceous boundary in eastern Asia in the light of event stratigraphy. *Earth Science Journal of Wuhan College of Geology* **11** (5), 465–72 (in Chinese with English abstract).

Liu Benpei and Chen Feng (1992). Temperature drop-events in the Cretaceous and their stratigraphic significance. In *Aspects of nonmarine Cretaceous geology* (ed. N. J. Mateer and P. J. Chen), pp. 41–8. China Ocean Press.

Liu Xianting, Ma Fengzhen, and Liu Zhichen (1985). Discovery of *Kuntulunia* in Shaan-Gan-Ning basin and its stratigraphic significance. *Vertebrata PalAsiatica* **23**, (4), 255–63 (in Chinese with English abstract).

Pu Ronggan and Wu Hongzhang (1985). Mesozoic sporopollen assemblages in western Liaoning and their stratigraphic significance. In *Mesozoic stratigraphy and palaeontology of Western Liaoning*, Vol. 2, pp. 121–212. Geological Publishing House, Beijing (in Chinese with English abstract).

Qi Hua (1984). Non-marine Jurassic–Cretaceous stratigraphy of Northwest China. *Bulletin of Xi'an Institute of Geology and Mineral Resources, Chinese Academy of Geological Sciences* **7**, 1–28 (in Chinese with English abstract).

Sengör, A. M. C. (1984). *The Cimmeride orogenic system and the tectonic of Eurasia* Geological Society of America, Special Paper No. 195, pp. 1–82.

Sengör, A. M. C., Altiner, S., Cin, A., Ustaomer, T., and Hsü, K. J. (1988). Origin and assembly of the Tethyside orogenic collage at the expense of Gondwana. In *Gondwana and Tethys*, (ed. M. G. Audley-Charles and A. Hallam). Geological Society of London, Special Publication No. 37, pp. 119–82.

Sha Jingeng and Yuan Fushen (1985). *Buchia* fauna from Dong'anzhen Formation near Dong'an of Raohe county, Heilongjiang. Acta *Palaeontologica Sinica* **24**, (6), 651–62 (in Chinese with English abstract).

Su Dezhao (1985). Late Mesozoic fish faunas from Xinjiang. Memoir of Institute of Vertebrate Palaeontology and Palaeoanthropology. *Academia Sinica* **17**, 61–136 (in Chinese with English abstract).

Vakhrameev, V. A. (1952). Cretaceous stratigraphy and fossil flora of western Kazakhstan. *Regional Stratigraphy USSR*, Vol. 1 (in Russian).

Vakhrameev, V. A. (1981). Pollen *Classopollis*: indicator of Jurassic and Cretaceous climates. *Palaeobotanist* 1981, 28–9.

Vakhrameev, V. A. (1984). Early cretaceous flora and climate of the earth. *Soviet Geology* 1, 41–49 (in Russian).

Wang Daning and Zhao Yinniang (1980). Late Cretaceous–Early Paleogene sporopollen assemblages of the Jianghan basin and their stratigraphical significance. *Professional Papers of Stratigraphy and Palaeontology* **9,** 88–110 (in Chinese).

Wang Si'en (1983). Some Jurassic–Cretaceous conchostracans from Qinghai. *Acta Palaeontologica Sinica* **4**, 465–71 (in Chinese with English abstract).

Wen Shixuan, Lan Xiu, Chen Jinghua, Chen Chuzhen, and Gu Zhiwei (1976). *Lamellibranch fossils from the Mt. Qomolangma area,* Science Research Report of the Mt. Qomolangma area, Palaeontology, No. 3, pp. 1–152. Science Press, Beijing (in Chinese).

Yang Zunyi, Nie Zetong, Guo Tieying, Liang Dingyi, and Chang Yizhi (1985). Mesozoic biotic characteristics and their relationship to crustal evolution in the Ngari Region. Tibet (Xizang), China. *Earth Science—Journal of Wuhan College of Geology, Special Papers on Stratigraphy and Paleontology* **3**, 1–8 (in Chinese with English abstract).

Yao Peiyi and Yu Jingshan (1985). *Plicatounio (Guangxiunio)*, a new Lower Cretaceous non-marine bivalve subgenus from Fusui County, Guangxi. *Professional Papers of Stratigraphy and Palaeontology* **14**, 227–36 (in Chinese with English abstract).

Yu Jingxian, Guo Zhenying, and Mao Shaozi (1983). Cretaceous palynological assemblages from the district south of the Songhua River. *Professional Papers of Stratigraphy and Palaeontology* **10**, 1–86 (in Chinese with English abstract).

Yu Xihan (1982). Freshwater bivalvia (mollusca) from Upper Jaurassic Fuxin Formation, western Lianoning. *Bulletin of Shenyang Institute of Geology and Mineral Resources, Chinese Academy of Geological Sciences* **4**, 185–94 (in Chinese with English abstract).

Zaklinskaya, E. D. (1963). *Pailtza pokraitocemennaikh i eë zhachenie dlyr obosnovania stratigraphii verkhnevo mila*. Trudai Geologicheskii Instituta, AN SSSR, 74.

Zhang Zhifu (1982). Sporo-pollen assemblage of the Donghe Group in the Huixian-Chenxian basin and their stratigraphic significance. *Bulletin of Xi'an Institute of Geology and Mineral Resources, Chinese Academy of Geological Sciences* **5**, 101–112 (in Chinese with English abstract).

Zhao Zikui (1978). A preliminary investigation on the thinning of the dinosaurian egg-shells of the Late Cretaceous and some related problems. *Vertebrata Pal Asiatica* **4**, 213–21 (in Chinese with English abstract).

Zhou Mingzhen (1953). Discovery of Mesozoic mammal fossils from Northeast China and their significance. *Acta Palaeontologica Sinica* **1,** (3), 150–6 (in Chinese with English abstract).

Zhou Shanfu and Wang Lianyuan (1982). Two new genera from the Taizhou Formation of the Rutung Region, North Jiangsu. *Selective Papers of 1st Symposium of Palynological Society of China,* pp. 89–92. Science Press, Beijing (in Chinese).

12 TERTIARY

Tong Linfeng

12.1 TERTIARY PALAEOBIOGEOGRAPHICAL PROVINCIALIZATION OF THE WORLD

12.1.1 TERTIARY PALAEOGEOGRAPHICAL OUTLINE OF THE WORLD

The Tertiary palaeogeography is quite different from that of the Cretaceous. Following Ziegler *et al.* (1983), the Atlantic and Indian oceans increased in width, whereas the Pacific Ocean shrank during the Tertiary period. The African and Indian plates came into collision with the Eurasian, and the Tethys or Neotethys became a series of mountain ranges. Meanwhile huge quantities of sedimentary and volcanic rocks were deposited along the margin between the Eurasian continent and Pacific Ocean.

In the Palaeogene, Africa, India, South America, Australia, and Antarctica were separated from each other and approached their modern positions; the ancient continent of Gondwana had disintegrated completely. The Indian plate was moving northward and united with the Eurasian plate by the end of the Eocene. In the Late Palaeocene, the Australian plate was separated from the Antarctic plate and moved northward. The Antarctic plate moved southward to a higher latitude. Meanwhile, South America separated from Africa. The African plate was so closed up to the north that the western Tethys seemed like a trough. The North American and European continents were separated from each other and the North Atlantic Ocean attained its basic form as it is now. By the end of the Oligocene, there was a general uplift again, but the southern Kazakhstan and southern Tianshan Mountains still experienced a Tethyan transgression, and there was also a Palaeogene transgression in the Tarim Basin of China (Zhou 1982).

Neogene palaeogeography is similar to that of modern times. All the continents turned into land. Marginal seas were formed on the East Asian continent facing the Pacific Ocean, and then the gulf of California opened. Meanwhile Africa was so closed up to Europe that the southern and northern Tethys seas disappeared except for the remaining Mediterranean Sea, which was sealed up during the Late Miocene (Yuan and Du 1984).

12.1.2 TERTIARY CLIMATE AND BIOGEOGRAPHICAL PROVINCIALIZATION IN THE WORLD

In the Tertiary the global seasonal climate changes and the latitudinal climatic zonation were more clearly differentiated than in the Cretaceous (Brenchley 1984). On the whole, the climate tended to become cool. However, it was still warmer in the Palaeogene than at present. The northern boundary of the subtropics lay at a latitude of about 42°N, which is over 10–13°N of the present boundary. In polar regions the climate was temperate. The Palaeogene transgression enlarged the Atlantic, Indian, and Pacific Oceans and linked them together.

The nummulitids which represented warm climate distribution spread widely in geosynclinal areas from the Alps to the Himalayas, and further to Australasia and the eastern shore of Africa. The nummulitid belt coincides with the present-day Agulhas warm current. In brief, although in the Palaeogene there existed a phytogeographical and climatic zonation between the equator and the poles, the cool temperate to cold zone was restricted and the poles lacked ice caps; the tropical and subtropical zones shifted to high latitudes (CPGC 1979).

Biogeographical provincialization was controlled not only by the climatic zonation but also by natural barriers affecting the distribution of organisms. Taking the above-mentioned factors into consideration, the Tertiary can be divided into five biogeographical realms, i.e. the Boreal realm, South Asia

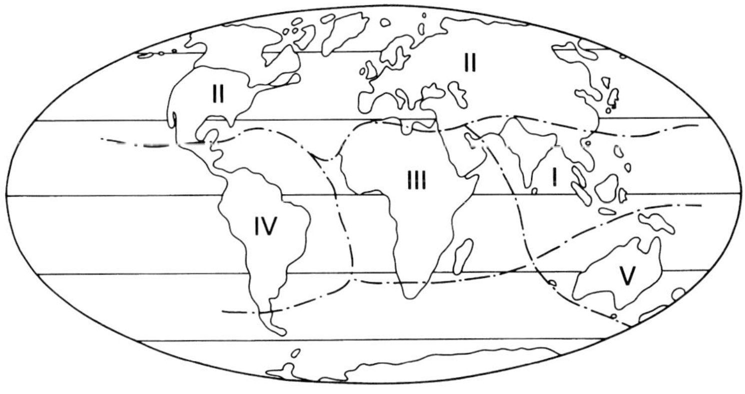

Fig. 12.1 Map of the Tertiary palaeobiogeographical provincialization of the world. I, South Asia realm; II, Boreal realm; III, Palaeoequatorial Africa realm; IV, Palaeoequatorial America realm; V, Australia realm.

realm, Palaeoequatorial Africa realm, Palaeoequatorial America realm, and Australia realm (Fig. 12.1, Table 12.1).

The Boreal realm

This includes Europe, North America, and Asia to the north of the Arabian peninsula and the Himalayas. The general features of the realm are as follows. At the early stage the southern part possessed obvious characteristics of subtropical phytogeographical provincialism, while the northern part was in the temperate zone with a warm and humid climate. At the late stage, temperate climate was predominant in the whole realm, with rather uniform plant communities. Temperature and humidity fell and it became cooler at an even later stage, and the hygrophile plants migrated southward. The previously rather uniform plant communities began to differentiate following the uplift of huge mountain ranges in Tethys, the Andes, and other regions. Xerophile plant communities appeared in North America. Because Europe and Asia had been connected by corridors with each other several times since the late Cretaceous, the mammals of these two continents mutually interchanged to give rise to many common genera and species, such as the primitive carnivore creodontids, the majority of clasper carnivores, proboscideans, perissodactyls, and artiodactyls. However, each continent possessed its own endemic taxa. In short, the advanced representatives of the mammals were dominant in this realm and the primitive marsupials only occurred in the Early and middle Tertiary in Europe, with only one or two remaining in North America.

In this realm, the Neogene marine biota in marginal areas of the Mediterranean contained planktonic forams like *Praeorbulina*, endemic molluscs, and nannoplankton.

The Palaeogene in this realm possesses the following characteristics:

1. North America and southern Eurasia belong to the northern subtropical phytogeographical realm. The flora contains hygro-thermophilic Lauraceae and Fagaceae, and evergreen Moraceae, Meliaceae, Proteaceae, Myrtaceae, and Palmae. They are mainly distributed in South China and the south-eastern coast of China. From there the distribution area ran obliquely to the north-west and extended its northernmost boundary to England (500°N) and Alaska (620°N), forming a Tertiary Palaeo-Mediterranean phytogeographical realm (not completely coinciding with the boundary of the faunal realm in Fig. 12.1). In the London Clay a great deal of Palmae, Burseraceae, Sapotaceae have been found, the tropical content of which accounts for 32 per cent; among them *Nipa truticans* spread widely along the sea shore of the Palaeo-Mediterranean, but today it remains only in the tropical areas of south-eastern Asia. The northern part of Eurasia is attributed to the temperate zone where deciduous needle-leaf trees developed.

Table 12.1 Tertiary palaeobiogeographical provincialization of the world

<table>
<tr><th colspan="3"></th><th>Palaeocene</th><th>Eocene</th><th>Oligocene</th><th colspan="2">Miocene–Pliocene</th></tr>
<tr><td rowspan="3">N. temperate–N. subtropical</td><td rowspan="3">Boreal realm</td><td>Asia</td><td>Anagalida
Perissodactyla
Rhinocerotidae</td><td rowspan="2">Propachynolophus
Creodonta
Lauraceae
Meliaceae
Moraceae
Myrtaceae
Proteaceae
Fagaceae</td><td>Didymoconidae</td><td colspan="2" rowspan="3">Temperate deciduous broadleaf flora is represented by *Picea*, *Abies*, *Larix*, *Alnus*, and *Betula* Foraminifer *Praeorbulina*

Marginal Mediterranean endemic mollusc faunas and nannoplankton</td></tr>
<tr><td>Europe</td><td rowspan="2">Perissodactyla
Condylanthra
Rhinocerotidae</td><td>Cainotheriidae
Ardynomys</td></tr>
<tr><td>North America</td><td>Anthracotheriidae
Ursidae
Canidae</td><td>Merycoidontidae
Desmatolagus</td></tr>
<tr><td rowspan="3">Torrid zone</td><td>Asia–meridional realm</td><td>South Asia</td><td colspan="2">Bontang fauna is characteristic of anthrocotheres</td><td></td><td>Siwalik forest marsh fauna</td><td rowspan="2">Typical tropical forms *Carnivora* Redentia and a few *Gazella* migrated from the Boreal realm</td></tr>
<tr><td>Palaeoequatorial Africa realm</td><td>Central Africa</td><td colspan="2">*Nummulites* and thick-shell mollusc fauna
Cinnamomus, *Sabalites* tropical rainforest flora
Discocyclina, *Lepidocyclina*, Alveolinidae formed nummulitic limestone</td><td>Fauna represented by Hyracoidea and Embrithopodae</td><td>*Mastodon*–*Hipparion*–Primates and older fauna</td></tr>
<tr><td>Occidental Palaeoequatorial realm</td><td>S. America</td><td colspan="2">Ungulates, carnivorous marsupials and endentates and other old faunas</td><td>Floras characteristic of general plant families
Cinnamomus
Sabalites</td><td colspan="2">Faunas represented by South American apes and monkeys as well as Rodentia

No ungulates and carnivorous marsupials</td></tr>
<tr><td>S. subtropical–S. temperate zone</td><td>Australia realm</td><td></td><td colspan="3">There were south temperate floras and marsupials during the early Eocene</td><td colspan="2">Flora represented by *Zucalyptus* *Dingo*, *Thylacoleo*, *Diprotodon*</td></tr>
</table>

2. In the Early Palaeocene animal migration was obstructed between Asia and North America. The Palaeo-Mediterranean Sea connected with the Arctic Ocean through the north–south trending Turgai Strait, interrupting the land animal interchange between Europe and Asia. In Late Palaeocene the interchange increased; with the help of the Bering land-bridge the mammal genera common to Eurasia and North America increased to no less than 40 per cent, whereas in the Late Palaeocene taxa common to Europe, Asia and America were very few, such as the Arotostylopidae, *Coryphodon* and *Microparamylus* which originated from Asia and probably passed through the Bering land-bridge into North America (Yuan and Du 1984).

3. Some endemic taxa were developed for each continent because of the geographical separation and the differentiated adaptation of animals to various environments and climates. For example, *Heptodon* (Perissodactyla) and hypodontids (Condylanthra) were restricted within Asia and North America, whereas *Propachynolophus* (Perissodactyla) was confined to Europe and Asia. The Anagalida were unique in the Palaeocene of Asia (Table 12.1).

4. From the Late Eocene to the Oligocene, the Bering land-bridge and Turgai Strait, which extended from the Palaeo-Mediterranean Sea to the Arctic Ocean, disappeared, so that the European and Asian fauna could communicate with each other. Many mammals that appeared in South China, such as the Chalicotheriidae (Perissodactyla), Anthracotheriidae (Artiodactyla), and Unsudae, Felidae, and Canidae (Carnivora), are also found in southern and western Europe.

During the Middle to Late Eocene, due to the collision of the Indian and Asian plates, the animals of these two lands began to interchange. Hence the fauna on both southern and northern sides of the Himalayas were quite similar to each other in the early Neogene, which also reflected similar palaeogeographic environments in both areas at that time. For example, *Hipparion xizangense* found in the Bulong Basin in northern Tibet possesses the same features as *H. primigenius* from the European Vallesian, while *Chiotherium tanggulaense* of the Bulong Basin is close to *C. intermedia* from Nagri, in the Siwalik Group of northern India. At that time the uplifting of the Himalayas was not high enough to obstruct the faunal communication between the two places. However, in the latest Miocene, the Himalayas were uplifted about 3000 m in altitute, high enough to form a barrier, and thus the Jielong mammal fauna became different from the southern Asian Siwalik fauna, while remaining similar to the *Hipparion* fauna from the Pontian in western Asia (NIGPAC 1982).

In the Neogene the climate was arid in this realm; the temperature dropped rapidly after the beginning of the Oligocene. The average annual temperature along the north-western coast of North American dropped from 27 °C to 12 °C within 1–2 ma, so that the temperate phytogeographical region expanded southward, and the previously evergreen plants were gradually replaced by the psychro-hygrophilic *Picea*, *Abies*, and *Larix*, with a certain amount of *Betula*, *Corylus*, and *Alnus*.

The South Asia realm

This realm borders the Boreal realm along the Hindukush, Himalayas and Nanling Mountains. It includes India, Pakistan, southern Iran, Malaysia, Indochina, and the Chinese territory south of Nanjing. After the Eocene, India integrated with Asia, so that the animals of this realm intermixed with those of the Boreal realm. However, they still have specific features of this realm, e.g. the Siwalik fauna was outstanding in many varieties of elephant, and the Eocene Pangdang fauna in Burma was characterized by a flourishing of the Anthracitheriidae. In the Pliocene there was a typical tropical and subtropical mammal fauna quite different from that of the Boreal realm.

The Palaeoequatorial Africa realm

This includes the major part of Africa and the Arabian peninsula and has the following characteristics.

1. During the Early Palaeogene there was a tropical fauna consisting of *Nummulites* and an equatorial rainforest flora. Many nummulitids, thick-shelled molluscs, and hermatypic corals were widely distributed in the warm currents of the equatorial ocean. The nummulitids were distributed from the Alps to the Himalayas. They went so far as to reach Australasia, central America, and eastern Africa, forming nummulitic limestone with large forams such as *Discocyclina*, *Lepidocyclina*, and *Alveolinidae*.

2. The tropical rainforest flora was mainly thermophilic with *Cinnamomus*, *Magnolia*, *Sabalites*, *Quercus*, and *Fagus*, as well as large gymnosperms and ferns.

3. During the early stages there were four to six orders of mammals in Africa, including elephants, leporids, and orycteropids. Later, the number of mammal orders gradually increased to about 16; all originated in Eurasia and North America, including Carnivora, Vespertilionida, Primates, Ungulata, Rodentia, and Lagomopha. In the Oligocene, Africa migrated northward and integrated with Europe. The interchange of animals between Africa and Europe resulted in the dispersal of many mammals such as elephants and mastodons.

4. In the Neogene, a *Mastodon–Hipparion–*Primates fauna developed in Africa; in the Miocene, there appeared the earliest *Ramapithecus*.

In brief, this realm was connected with the Boreal realm in the Early Tertiary, but in the Late Tertiary animal communication with areas elsewhere decreased because of the existence of barriers such as vast deserts.

The Palaeoequatorial America realm

This includes South America, central America, and southern Mexico, and has the following features.

1. Together with the Palaeoequatorial Africa realm and South Asia realm it belonged to the tropical zone. They had similar floral aspects. Proteaceae, Dilleniacene, and Restionales which developed in South America were also seen in Africa, indicating an interconnection between them. South America also shared with southern Africa many pan-tropical plant families and genera such as *Cinnamomus* and *Sabalites*.

2. Because South America had been separated from the other continents for a long time, in the Tertiary it gave rise to a unique fauna. About six orders of endemic mammals evolved and occupied different habitats, including placentals and marsupials, such as ungulates, carnivorous marsupials, and edentates in the early stage. Rodentia of a middle stage immigrated from Europe through North America and central America.

3. In the Late Pliocene, with the help of the central American land-bridge, the South American *Choloepus dedaclylus*, *Hystrix*, and *Armadillo* migrated to North America, whereas North American Equidae, Cervidae, Canidae, Felidae and Ursidae migrated southward. As a result of competition after the intermingling of these mammals, among the mammals that originated in South America there remained only a few edentates.

The Australia realm

This includes Australia and its surrounding islands. At the beginning of the Eocene Australia was separated from Antarctica and drifted northward, and in the Late Tertiary it gradually approached South Asia. Australian marsupials and South Asian Vespertilionidae, Rodentia, etc. mutually migrated, so giving rise to a mixture by limited interchange. During the northward drifting, the realm contained the southern temperate flora in the Palaeogene. In the Miocene, the major part of Australia was already situated in the high-pressure zone near the tropic of Capricorn, forming a vast area of desert. In its northern part there existed some South-East Asian tropical flora, while in its south-eastern part there still remained some Antarctic plants. On the whole,

Australia has been long isolated until today. The Tertiary biota is not known in detail, but the modern biota is quite different from that of Indonesia which is only separated from Australia by a narrow seaway. This reflects the fact that its Tertiary biota did not migrate far and the land has been isolated since then.

12.2 TERTIARY PALAEOBIOGEOGRAPHICAL PROVINCIALIZATION OF CHINA

12.2.1 INTRODUCTION

After the Cretaceous phase of Yanshan movement, the Palaeogene was a long-term stable stage (Zhou 1982; Li *et al.* 1984). The Chinese continent was surrounded by warm seas, its relief was gradually evened up and its temperature and humidity were higher at that time than at present. Except for the seas in Tibet, Himalaya, 'Keshi Bay' to the west of Tarim and Taiwan, the vast area of China was terrestrial. However, a few basins, including the North China depression, north Jiyang depression, Jianghan Basin, north Jiangsu depression, and the Baise Basin, experienced marine transgression and were intermittently connected with the open sea, and consequently contain brackish biotas. By the end of the Eocene, that part of Tethys between China and India disappeared and these two landmasses were integrated. In the Late Neogene, the Qinghai–Tibet plateau and the vast area of north-western China was uplifted. The high mountains within this area obstructed the development of high humidity, so that grassland-desert belts appeared. The increased difference in temperature between sea and land produced a monsoonal circulation system and caused a universal temperature drop. At that time the sea in the west of the Himalayas and Tarim basin disappeared, and all China became land with stable basins, except for a marine zone in Taiwan and at both sides of the Qiongzhou Strait.

The Tertiary biota was diversified and abundant. It is characterized by thriving angiosperms and mammals. The terrestrial fauna includes mammals, fish, reptiles, gastropods, bivalves, ostracods, estheriids, and insects. The marine organisms were forams, ostracods, gastropods, and bivalves.

In the Palaeogene the fauna within China, including areas of the Boreal realm and South Asia realm which lay to the south of the Yarlung Zangbo River, was not well differentiated. Climatically it was a middle latitude area which was very warm. The climatic zonation is classified from north to south into a humid subtropical zone, an arid subtropical zone, and a wet tropical zone (CVC 1980). However, in the late Oligocene most areas turned cooler. In the Neogene, China was divided into two realms, namely the Boreal realm and the South Asia realm lying to the south of the Hengduan and Nanling Mountains. The climate at that time turned cool universally. Regression took place in the west. Because Europe was integrated with Asia, the continental climate was enhanced inland. The monsoon circulation system led to a humid climate in the east and an arid one in the west. Thus the Palaeogene climatic zonation changed. The climatic zonation and differences in palaeogeographical conditions and lithofacies controlled the Tertiary palaeobiogeographical provincialization of China, which is listed in Tables 12.3 and 12.4. The features of each realm are as follows (CCBC 1978; Guo 1983).

The Boreal realm in China covers the area to the north of the Himalayas and Huaihe River. It is different from the South Asia realm in the following respects:

1. The Boreal realm is characterized by a distinctive subtropical flora different from the tropical flora of the South Asia realm.

2. The Early Tertiary, especially Palaeocene, fauna possessed unique Asian features (CPCC 1979), e.g. the flourishing of *Prodinoceras* mammals and the existence of endemic Asian Anagolida. All these became extinct In the Palaeogene and had no direct relationship with modern Asian mammals. Most palynomorphs were adapted to a northern subtropical humid climate, e.g. *Betulaepollenites plicoides* and *Aquilapollenites*, and the hygrophilic gymnosperm *Taxodiaceaepollenites*. The tropical–subtropical evergreen plants *Cinnamomus* and *Sabalites* are also notable (NIGPAC 1978).

Table 12.2 Correlation chart of the Tertiary of North China

	NW China Province					NE China Province		N. China Province
	Northern Inner Mongolia	Southern Xinjiang	North-eastern Xinjiang	Western Xinjiang	Xining, Qinghai	Eastern Liaoning	Southeast Heilongjiang	Northern China
Pliocene	Pliocene	Kuche	Putaugou	Atushi	Linxia	Quijiatun	Gangtou	Minghuazhen
Miocene	Tonggur Damiao	Jidike Suweiyi	Taoshuyuanzi	Wuqia Gr	Xiejia		Ma'anshancun	Guantao
Oligocene	Naogangdai Eerdengaobao			Wulagen and others	Mahalagou	Yanghe	Meihe	Dongying
Eocene	Salamulun Yrdingmanha Ashanton Bayanwulan	Xiaokuzibai	Liankan ? Dabu		Honggou	Gengjiajie Xilutian Jijuntun Guchengzi		Shahejie Kongdian
Plaeocene	Naomugen	Talake	Taizichun	Qimge Aertashi	Qijiachuan	Lizigou Lachutai		

Table 12.3 Tertiary palaeobiogeographical provincialization of China

	Palaeocene	Eocene	Oligocene	Miocene	Pliocene
S. Asia realm					
W. Pacific region					
Taiwan province	*Globigerina triloculinoides* *Globorotalia pseudobulloides*	*Discocyclina*		*Miogypsina-Pulleniatina* fauna Lauraceae flora	*Pulleniatina obliqueloculata*
S. Lingnan province	*Orbiculapollenites* *Aquilapollenites* *Gothanipollis*	*Cinnamomus* *Paramynodon* *Malatelmatherium*	*Anthracokeryx* *Sabalites*	*Ammonia-Uvigerina-Globigerina* assemblage	*Pseudorotalia-Asterorotalia-Pulleniatina* assemblage
Himalayan province	*Nummulites canalifera*	*Nummulites larvigatus* *Discocyclina* *Orbitolites* *Araucaria*		Included in Qinghai–Tibet province	
Hengduan province		Undivided		*Primates, viverra* *Dorcatherium minus* *Dryopithecus kaiyuanensis*	
Boreal realm					
China region					
S. China province					
S. China biome	Anagalidae Bemalambdidae *Subtriporopollenites*	Mesonychidae Anthracotheriidae Deperatellidae Magnoliaceae Proteaceae	S.China province:	*anchitherium* *Brachypotherium* *Hyoherium* *Dicrocerus*	Forest-type faunas
W. China biome	*Atlantipollis* *Nudopollis* *Triprojectacites*	*Palibina, Trudopollis* *Embolotherium graugeri* *Amynodon alxaensis*	Qinghai–Tibet province:	Betulaceae–Ulmaceae–*Quercus* *Quercus semicarpifolla*	
Kashi biome	plankton: *Globigerina triloculinoides*; *Globorotalia pseudobulloides* benthas: *Nonion, Cibicides*; *Anomalinoides* Marine bivalves and gastropods			Included in Junggar-Qillian biome	

Table 12.3 (*cont.*)

	Palaeocene	Eocene	Oligocene	Miocene	Pliocene
Boreal realm China region NW China province Junggar–Qilian biome	*Piceaepollenites* *Phenacolophus* Proteaceae *Ephedripites*	*Piceaepollenites* *Pinuspollenites* *Dzungariotherium turfanensis*		*Eucricetodon youngi* *Oioceros xijiaensis* *Rhinoceros pollenites*	Chenopodiaceae, Gramineae, Compositeae *Mastodon*, *Rhinoceros* and others formed wilderness prairie
Erenhot biome	Included in N. China province			*Platybelodon grangeri* *Listriodon mongoliensis* *Pinus* Chenopodiaceae	*Ochotona* *Anatolomys*
Altay biome	*Taxudium* *Alnus* *Myrica* *Juglaus* *Pterocarya*	Undivided			
N. China province Shanxi–Shaanxi biome and N. China biome	*Betulaepollenites* *Taxodiaceaepollenites* *Pantodonta* No *Phenacolophusnor Prodinoceras* Undivided	*Artiodactyla* *Derissodactyla* *Rodentia* No *Anthracokeryx*		*Hipparion, Palaeomeryx* fauna *Piceaepollenites* *Chenopodiaceae* *Stephanocemas thomsoni* *Aceratherium* (abundant)	*Hipparion* spp., *Antilospira licenti*, *Gazella*, *Ephedripites* formed dry prairie *Chilotherium* *Dipoidesmajera* *Nyctereutes sinensis*
NE China province				*Quercoidites* Betulaceae *Ilexpollenites* Pinaceae Mixed coniferous with broadleaf forest	Betulaceae Pinaceae Mixed coniferous with broadleaf forest

3. The mammals were all of advanced types, such as Carnivora, Perissodactyla, and Artiodactyla, quite different from those of rather primitive type in the South Asia realm (IVP 1979).

In the Neogene, the boundary between the Boreal realm and South Asia realm shifted south to the Himalayas. This is reflected by the *Hipparion* fauna in the Jilong Basin which is similar to that in North

Table 12.4 Occurence of the main Tertiary fossils of China

		NE China Pr.				N. China Pr.				NW China Pr.			W. China biome			S. China biome			SW China Pr.	S. Nanling Pr.		Taiwan Pr.	Himalayan Pr.
	E. Heilongjiang	E. Liaoning	Songliao Basin	Linqu	Cixian	Bohai Sea and coast	N. Hebei	C.-S. Shanxi	Erenhot	Junggar–Palaeo–Qilian	Xining	Gyrong Basin	Bilong Basin	N. Qomolangma	Jianghan Basin	Nanxiong Basin	Jiangsu–Zhejiang–Anhui	Sichuan Basin	Lufeng	Nanning Basin	Bibu Bay	Taiwan	Tingri–Gamba
Quercus	M			M	M									E									
Betulaceae	M			M	M		M		P			E											
Salix	M			M																			
Juglans	M			M										E									
Podocarpidites	M																						
Abiespollenites	M																						
Piceaepollenites							M																
Glyptostrobus europaeus		E																					
Metasequoia disticha		E																					
Taxodiaceae	A	E	E			A																	
Stephanocemas				M	M																		
Carpinus subcordata				M																			
Liquidambar				M		M																	
Diatomys shantungensis				M																			
Suidae				M																			
Cervidae					M		M			P													
Chinocythere					E																		
Quercoidites microhenrici		E			O	O											O						
Dorgyirgia					O	O																	
Ulmipollenites undolosus					O	O											O						
Aquilapollenites	A					E										A							
Chenopodiaceae							M	M	M	M	M												
Palaeomeryx							M	M															
Hipparion								M		M		M	M	M					M				
Gazella gaudryi												P											
Elasmotherium								M															
Platybelodon grangeri									M														

Table 12.4 (*cont.*)

	E. Heilongjiang	NE China Pr. E. Liaoning	Songliao Basin	Linqu	Cixian	N. China Pr. Bohai Sea and coast	N. Hebei	C.-S. Shanxi	Erenhot	NW China Pr. Junggar–Palaeo–Qilian	Xining	Gyrong Basin	W. China biome Bilong Basin	N. Qomolangma	Jianghan Basin	S. China biome Nanxiong Basin	Jiangsu–Zhejiang–Anhui	Sichuan Basin	SW China Pr. Lufeng	S. Nanling Pr. Nanning Basin	Bibu Bay	Taiwan Pr. Taiwan	Himalayan Pr. Tingri–Gamba
Anchitherium									M						M		M						
Ochotona									P														
Anatslomys									P														
Eucricetodon youngi										M													
Mastondon										M	M												
Rhinoceros										M	M												
Spirocerus										P													
Quercus semicarpifolia															M								
Hipparion tanggulaense													M										
H. guizhongensis												P											
Chilotherium xizangensis												P											
Brachypotherium																	M						
Dicrocerus																	M						
Hyoherium																	M						
Polypodiaceae						O											M						
Pterisporites																	M						
Myrtaceidites															E								
Magnolipollis															A		A						
Primates													M						M				
Viverra																	*M*						
Dorcatherium minus																	M						
Dryopithecus kaiyuanensis																		M					
Brachyrhizomys naguensis													M										
Miogypsina																						M	
Pulleniatina																					P	P	
P. obliqueloculata																						P	
Ammonia																					M		
Uvigerina																					M		

Table 12.4 (*cont.*)

		NE China Pr.				N. China Pr.				NW China Pr.			W. China biome			S. China biome			SW China Pr.	S. Nanling Pr.		Taiwan Pr.	Himalayan Pr.
	E. Heilongjiang	E. Liaoning	Songliao Basin	Linqu	Cixian	Bohai Sea and coast	N. Hebei	C.-S. Shanxi	Erenhot	Junggar–Palaeo–Qilian	Xining	Gyrong Basin	Bilong Basin	N. Qomolangma	Jianghan Basin	Nanxiong Basin	Jiangsu–Zhejiang–Anhui	Sichuan Basin	Lufeng	Nanning Basin	Bibu Bay	Taiwan	Tingri–Gamba
Globigerina																					M	A	
Pseudorotalia																					P	P	
Asterorotalia																					P	P	
Bemalambda															A				A				
Linnanialofoensis																			A				
Anthracokeryx																	E						
Archaeolambda																A							
Sinocypris																A							
Parailyocypris																A							
Eucypris																A							EO
Cypridea																						EO	
Nummulites larvigatus																							EO
Discocyclina														A									EO
Orbitolites														A									EO
Discorbis															O								
Miliodidae							O																
Triloculina							O																

A, Palaeocene; E, Eocene; O, Oligocene; M, Miocene; P, Pliocene.

China, thus being Eruasian, while different from the Siwalik fauna. To the east, the boundary between the two realms was situated along the Nanling Mountains; the latitude-controlled phytogeographical differentiation was clear: subtropical and temperate zones were to the north of the boundary, and the tropical zone was to the south.

12.2.2 PALAEOGENE PALAEOBIOGEOGRAPHICAL PROVINCIALIZATION OF CHINA

There are two realms and two regions of the Palaeogene in China (Fig. 12.2). Five provinces controlled by the different climatic zones are identified. Five biomes are designated due to the different biofacies.

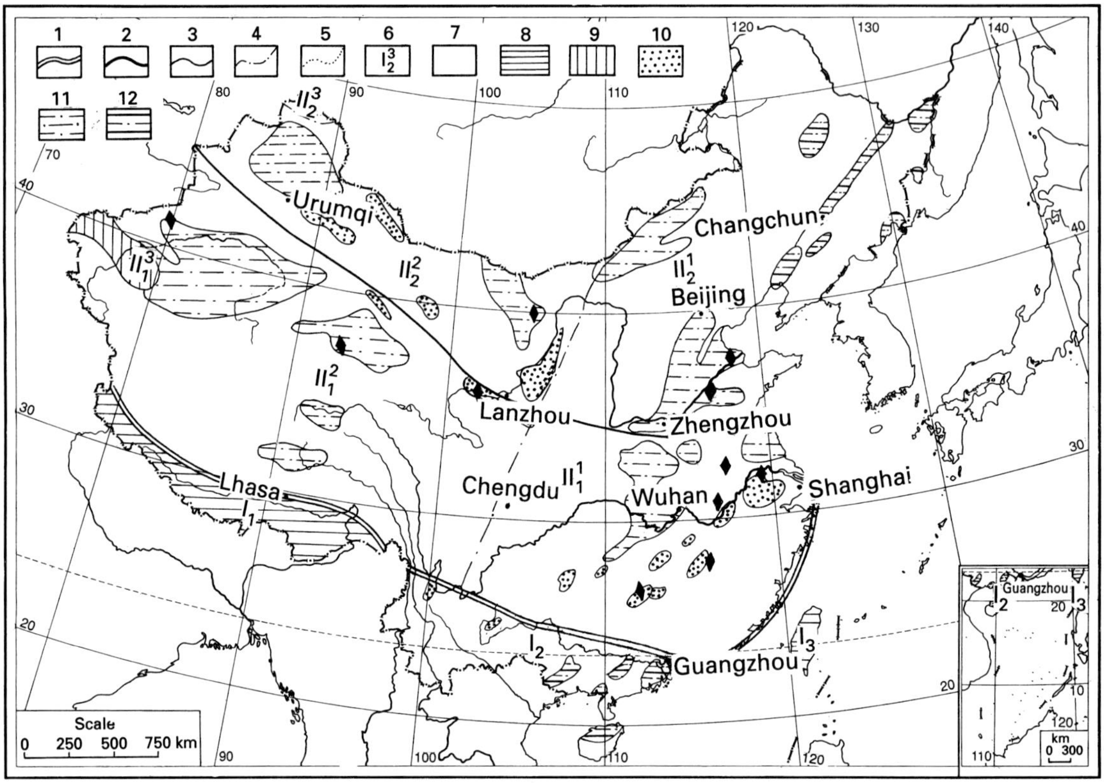

Fig. 12.2 Map of the Palaeogene palaeobiogeographical provincialization of China. 1, Boundary of realms; 2, boundary of regions; 3, boundary of provinces; 4, boundary of biomes; 5, boundary of communities; 6, roman digit = realm mark, subscript = province mark, superscript = biome mark; 7, Oldland. 8–9, Marine and paralic regions: 8, South China; 9, North China. 10–12, Terrestrial regions (Late Permian and Mesozoic–Cenozoic): 10, intermontane basins; 11, fluvio–lacustrine basins; 12, coal-bearing basins. I, South Asia realm; I_1, Himalayan province; I_2, Lingnan province; I_3, West Pacific region, Taiwan province; II, Boreal realm; II_1, Southern China province; II_1^1, South China biome; II_1^2, Western China biome; II_1^3, Kashi biome; II_2, North–North-western China province; II_2^1, North China–Shanxi–Shaanxi biome; II_2^2, Junggar–Qilian biome; II_2^3, Altay biome. Black rhomboids, gypsum deposits.

The Boreal realm

For the Palaeogene three provinces are recognized. From north to south they are: the North China–North-western China province, and the South China province.

The North China Province

This covers north-eastern China, North China. The biota was dominated by spores and pollens, ostracods, gastropods, charophytes, and mammals, among which spores and pollens and ostracods were especially highly diversified.

The North China–Shanxi–Shaanxi biome

Palaeocene strata are found in the Lachutai Formation, Fushun region, eastern Liaoning. Angiosperm pollen predominated and may exceed 80 per cent. Especially the primitive triporate pollen was prominent, e.g. *Betulaepollenites*, representing a subtropical hygrophilous flora (NIGPAC 1978). The representative locality for Palaeocene mammals in this biome

Figure 12.3 Reconstruction of the Early Tertiary community of China. 1, dinoceratids; 2, creodonts; 3, *Eohippus*; 4, *Osmunda*; 5, *Nelumbo*; 6, *Glyptostrobus*; 7, *Dryophyllum*; 8, *Cinnamomus*; 9, *Sabalites*; 10, *Metaseguoia*.

is the Naomugen Formation of Siziwangqi in Inner Mongolia. Fifteen species of 11 genera, nine families, and six orders have been found altogether. Of this fauna more than 90 per cent were Multituberculata, being very few in Mongolia and absent in Xinjiang. Members of this fauna, such as *Pantodont*a, lived in the shallow, marginal areas of lakes and low, wet lake-beach grasslands. This biome lacks the *Prodinoceras* that occurs in Xinjiang, and the *Mongolotherium* of Mongolia (Yuan and Du 1984).

From the Eocene to Oligocene subtropical deciduous trees and shrubs still predominated, mixed with tropical evergreen plants. Most of these are also commonly seen in Europe and North America, including *Glyptostrobus europaeus*, *Metasequoia disticha*, *Sequoia chinensis*, *Dryophyllum dewalquei*, *Alnus*, and *Betula*. Palynomorph floras are marked by the high content of Taxodiaceae, *Quercoidites*, and *Ulmipollenites*, and a few *Aquilapollenites* remained. Along the coast of the Pohai Sea (Chihli Bay) a brackish fauna developed, e.g. in the Jiyang depression, the Shahejie Formation of which yields many *Trilculina*, together with *Discorbis*, *Nonion*, and *Florilus*. Moreover, biolithites from Oligocene strata yield the fossil worm tubes *Iguiliosia* and *Hydroides*, and codiacean fossils of the siphonales of chlorophytes, typical of warm subtidal zones (Li Yuntong *et al.* 1984). These marine fossils in intercalated horizons indicate that the North China plain underwent marine ingressions in the Early Tertiary. The mammal fauna was different from that of the Palaeocene. They include still-living taxa of Perissodactyla, Artiodactyla, and Rodentia. There were also *Eomoropus*, *Caenolophus*, and *Anthracothema* as in the South Asia realm, but *Anthracokeryx*, which is abundant in the South Asia realm, is lacking.

North-western China province

This covers the rest of Qinling and the area to the north of Tianshan.

The Junggar–Qilian biome

The spores, pollen, and ostracods in this biome are less diversified than those in the North China–Shanxi–Shaanxi biome. The main characteristics of this biome are as follows:

1. Palaeocene palaeobotanical data are few. Taking eastern Qinghai and the intermontane basins of Gansu as examples, in its early stage the sporo-pollen assemblage comprised mainly angiosperm pollen with triporate pollen predominant. *Nomopolles* may account for up to 5 per cent of the sporo-pollen assemblage. In the late stage there predominated gymnosperm pollen, of which that of the xerophite *Ephedripites* reaches as high as 60 per cent; disaccites increased also (CVC 1980). As regards Palaeocene mammals, *Prodinoceras* and *Phenacolophus* of the ancient Artiodactyla occurred (being absent from the North China–Shanxi–Shaanxi biome). Differentiation was caused by the differences between sedimentary environments and the lithological and ecological characters of the two biomes. In this biome the mammals are preserved in sediments of fluvial facies.

2. In the Eocene shrub prairies developed, characterized by the drought-enduring *Palibinia*. The sporo-pollen assemblage characterized by abundant tricolpate and tricolporate angiosperms, such as *Celtispollenites*, *Melilaceoidites*, *Quercoidites*, and *Sapindaceidites*, while *Nomopolles* disappeared. Herbaceous plants increased and there was a high content of *Ephedripites*. All these reflect an overall aridity. According to the climatic zonation, the North China–Shanxi–Shaanxi biome together with this biome should be attributed to the subtropical zone. The difference lies in that the former was more humid, while this realm was rather arid.

The Altay biome

This biome is situated in the north-western corner of Xinjiang. The plant fossils in the Wulunguhe Formation of Aletai consist of the gymnosperms *Taxodium*, *Glyptostrobus*, and *Sequoia* and the angiosperms *Alnus*, *Pterocarya*, *Juglans*, and *Myrica*, all being thermophilous and hygrophilous genera. This represents a subtropical–warm-temperate mixed forest of deciduous and broadleaf trees (Li Bai 1986). This biome is evidently different from the Junggar–Qilian biome.

South China province

Its northern boundary neighbours the North China–North-western China province and the southern boundary is along the line of the Nanling and Gangdise Mountains. The mega-plant fossils are poor in the province. Spores and pollens, ostracods, and mammals are very abundant except in the western part. In the eastern part the climate was more humid than that in the west, due to the influence of the Pacific Ocean. Except for areas along the eastern coast red sediments developed over the whole province, containing gypsum and rock salt, indicating an overall dry and warm climate, and the province belonged to the arid subtropical zone. Bordering on the Helan and Hengduan Mountains, the province is divided into three biomes due to the difference in vegetation and in mammals.

The South China biome

This is located to the east of Helan and Hengduan Mountains. The vegetation belonged to the mixed forest of broadleaf and coniferous trees, mainly of subtropical type. The Palaeocene sporo-pollen assemblage included *Ulmipollenites minor* and *Ulmoideipites krempii* (Ulmaceae) as predominant members, followed in importance by *Subtriporopollenites* and *Plicapollis*. From the Eocene until the Oligocene there existed subtropical vegetation mingled with tropical evergreen elements, such as Magnoliaceae, Proteaceae, and Aquifoliaceae. At a later stage hygrophilous trees increased in importance. This biome lacks typical tropical plants (Sun and He 1980).

The Palaeocene mammals show clear endemism. Anagalidae, Esthongebidae, and Bemalambdidae from the Shanhu and Nonshan Formations were all East Asian Palaeocene elements (Zhou *et al.* 1977; Lei 1984). Most of them lived in the water margins or swamps. There were no representatives of the Primates, Carnivora, or Insectivora that typically inhabited the plains and forests. The sediments associated with the fauna were of static lake facies.

During the Eocene and Oligocene, there appeared, in addition to the swamp animals, plain-and-forest-inhabiting primates, mesonychids (Condylarthra), deperatellids (Perissodactyla), anthracotheres (Artiodactyla), and *Amynodon*, quite different from those mammals of the North China–North-western China province.

The southernmost part of the biome was a subtropical swamp-forest area where tropical taxa were more important. The animals living here included *Propalaeotherium* in the forests and *Elasmotherium*, alligators, and tortoises in the lakes.

The Western China biome

This is located to the west of the previous biome and bordered along the Gangdise Mountains in the south. It was difficult for moisture from the Pacific Ocean to enter here because it was very much inland and barred by high mountains. North Tibet was within the subtropical high-pressure belt so that the vegetation was of an arid savanna–shrub type. Mega-plant fossils are rare in this biome. In the Eocene, xero-thermophile *Palibinia* was widespread. There were also xero-thermophilic representatives among the spores and pollen, such as *Atlantipollis*, *Trudopollis*, and *Nudopollis*, occurring in the upper Tarak Formation. Extratriporopollenitids and *Ephedripites* are also present. In the Oligocene, the climate cooled gradually and the mammals of the Chagangulage Formation were dominated by forest–prairie types, such as *Embolotherium graugeri*, *Paracaducodon suhaituensis*, and *Amynodon alxaensis* (Perissodactyla). In the early Oligocene mammal herbivores increased, e.g. in the arid prairie of south-western Alashan the predominant elements were small rodents, insectivores, and perissodactyls, which were different from the anthracotheriids and amynodontids occurring in the eastern biome in the Oligocene.

The Kashi biome

Lying in the northern foothills of the Kunlun Mountains, in a narrow belt, this is located in the arid subtropical area of this province. It is distinguished as a biome because during the Palaeocene and Eocene there was shallow sea or lagoonal sedimentation, different from that of other biomes. There were abundant molluscs, such as *Brachidontes jeremejewi* and *Potamides*. Forams were diversified; in addition to thriving small benthic forams such as *Nonion*, *Cibicides*, and *Anomalinoides*, there were tropical–subtropical plankton of Early Palaeocene age, such as *Globigerina triloculinoides* and *Globorotalia pseudobolloides*. This marine area was a branch of the northern Tethys (Hao *et al.* 1980).

The South Asia realm

This consists of the Himalayan, Lingnan, and Taiwan provinces.

The Himalayan province

Lying to the south of the Yarlung Zangbo River this province is characterized by shallow shelf marine facies belonging to Tethys (NIGPAC 1982). Distributed in the Dingri and Gangba areas, the deposits are shallow marine carbonates. The *Nummulites* fauna was dominated by large benthic forams without planktonic forams, accompanied by bivalves and gastropods. The main nummulitids were the Palaeocene *N. canalifera* and the Eocene *N. lavigatus*, *Discocyclina*, and *Orbitolites*, all common taxa in southern Asia, the Middle East, and the Palaeo-Mediterranean. In addition, there were a few small forams (mainly miliolids) similar to those of the Kashi biome in Xinjiang.

In the Early Palaeogene the province was the westward extension of the humid tropical zone of Lingnan province. The mega-plant and sporo-pollen fossils were only distributed in southern Tibet. After the Eocene the deposition changed into lacustrine facies. Dominant plants included the tropical–subtropical *Eucalyptus* and *Ficus*. Elements of India–Malaysia evergreen rainforests were also common, such as *Araucaria* and *Altingia* excelse. There were no typical temperate elements.

The Lingnan province

Lingnan literally means areas south of Nanling. Actually this province covers areas to the south of the Shiwandashan Mountains including small faulted basins in Nanning, Baise, and Maoming, as well as the Zhanjiang, Leizhou, and Qiongzhou areas. Situated in low latitudes, with high temperatures and moderate precipitation, this province belongs to the humid tropical climatic zone, yielding abundant organisms. The major characteristics of the province are as follows:

1. In the Palaeocene there appeared tropical *Gothanipollis* and the primitive angiosperm pollen *Aquilapollenites*, and locally tropical *Feneserites*.

2. There were abundant mammals adapted to nearshore swamp and forest environments in the humid tropical–subtropical belt, such as *Anthracokeryx*, *Paramynodon*, and *Malatelmatherium*, in contrast to the fauna of the North China–North-western China province, which lacked *Anthracokeryx*. Around the basins there lived a forest–shrub fauna, such as carnivores and *Meryx* (Zhou *et al.* 1977).

3. The Nadu Formation of the Baise Basin contains an abundant *Anthracokeryx* fauna very similar to the Pangdang fauna of Burma, and thus belongs to the same biogeographical region. Both faunas have the same animals, such as *Anthracokeryx birmanica*, *A. gohice*, *Paramynodon birmanicus*, and *Malatelmatherium birmanicus* (M. cf. birmanicus in Baise) (Yuan and Du 1984).

The Taiwan province

In Early Tertiary times this province was on the margin of the western Pacific. It was intensely deformed and uplifted during the early Neogene, and thick, marine–terrestrial alternating, coal-bearing deposits accumulated in the depressions on the western side of the central mountain range. The Palaeogene was characterized by nummulitids. The Palaeocene is represented by the planktonic forams *Globigerina triloculinoides* and *Globorotalia pseudobulloides*, the former being the Palaeocene index member for south Asia, the Middle East, South America, and Central America. The Eocene–Oligocene nummulitid fauna were dominated by *Discocyclina*, comparable with the Tibetan assemblage. In the Palaeogene this province was in a marine regime, as indicated by marine facies without continental deposits, located within the mobile zone of the Pacific, and containing a Pacific warm-water fauna. In this respect it is quite different from the Nanling and other provinces.

Cluster analysis of Palaeogene mammals and plants

According to the published data there were 268 species of Palaeogene mammals; 75 genera of plants; 82 genera of spores and pollen; 58 genera of ostracods; 63 genera of gastropods; and 29 genera of charophytes. The mammals and plants were chosen for cluster analysis, deleting some generalized taxa and taking the stratigraphical units demarcated by the National Chart of Stratigraphic Regionalization as samples. Using the Otsuka coefficient, the results of cluster analysis we obtained were coincident with the traditional provincialization.

Figure 12.4a is the cluster analysis diagram for Palaeogene mammals. The calculation covers 110 genera. It is seen from the figure that if Otc = 2 then three groups can be made. The first group consists of the Hexi Corridor, Qaidam, Xining–Linxia, Alxa, and Erenhot Qog, forming the North China–North-western China province. The second group consists of Markang and east Yunnan, forming the West China biome of the South China province. The third group consists of Hefei, Jiangxi–Hunan–Guangdong and Zhujiang River, forming the South China biome of the South China province. Zhujiang River in palaeogeographic provincialization belongs to the Lingnan province, but the cluster analyses show that it has more similarity to the South China biome. It can be interpreted that in both the South China biome and Lingnan province the mammals of wet swamp type were so abundant that the cluster analysis failed to discriminate between them. However, as a whole, the cluster analysis shows compatibility with palaeobiogeographical provincialization.

Figure 12.4b is the cluster analysis diagram of Palaeogene plants, calculated from 75 genera. The

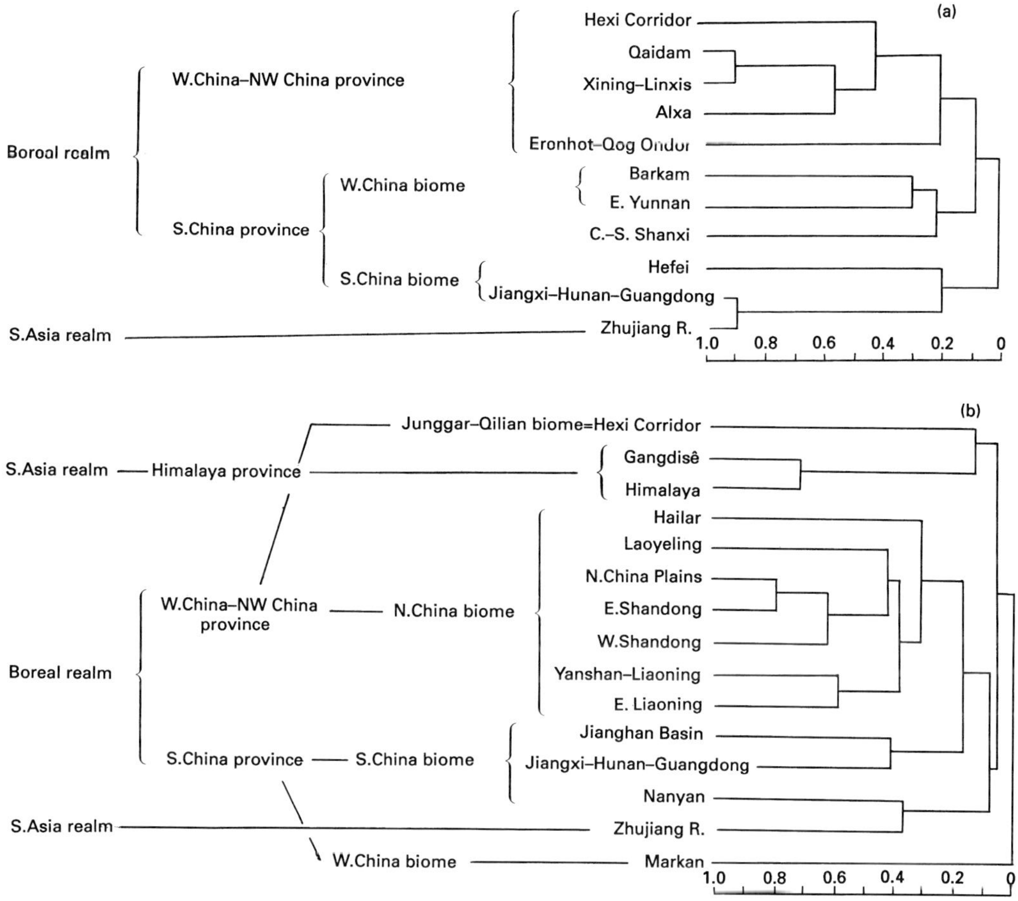

Figure 12.4 Cluster analysis of the Early Tertiary fossils of China. (a) Early Tertiary mammals; (b) Early Tertiary plants.

figure shows that the Hexi Corridor and Markan each constituted an independent group. Gangdise and Himalaya constitute the Himalaya Province. Jianghan and Jiangxi–Hunan–Guangdong constitute the South China biome. Zhujiang River represents the South Asia realm. The other units of stratigraphical provincialization within the North China–Shanxi–Shaanxi biome of North China–Northwestern China province aggregate into one group. The cluster analysis indicates coincidence with palaeobiogeographical provincialization.

12.2.3 NEOGENE PALAEOBIOGEOGRAPHICAL PROVINCIALIZATION OF CHINA

The Neogene realms in China are identical with those of the Palaeogene, but the Neogene regions and provinces are more varied than those of the Palaeogene. This is because the Himalayan movements resulted in tectonic differentiation; the western plateau and high mountains uplifted rapidly, while the eastern relief became gradually subdued, so that the monsoon circulation system strengthened in the east, while the continental climate was enhanced in the west. Two realms, eight provinces, and four biomes are designated within the territory of China (Figs 12.5, 12.6).

The Boreal or Palaearctic realm

The area of this realm has been indicated earlier. In the late Neogene the interchange of animals between the European and Asian continents was blocked. The distribution pattern and provincialization of Neogene mammals gradually approximates to those of today. In the whole Palaearctic realm there developed the *Hipparion* fauna. It was a temperate zone with vast areas of prairie and forest. The prairie elements, such as *Equus* and *Rhinoceros*, were rather abundant, so this realm is different from the South Asia realm which was dominated by forest fauna and lacked a prairie fauna.

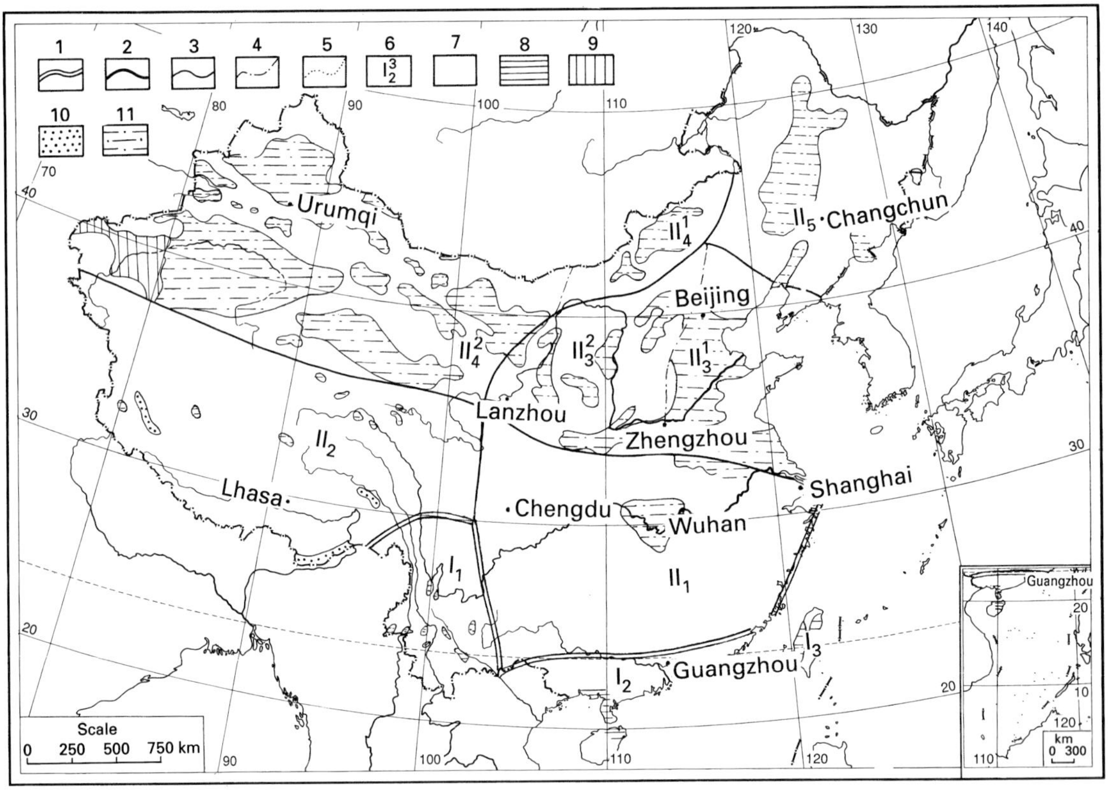

Figure 12.5 Map of the Neogene palaeobiogeographical provincialization of China. 1, Boundary of realms; 2, boundary of regions; 3, boundary of provinces; 4, boundary of biomes; 5, boundary of communities; 6, roman digit = realm mark, subscript = province mark, superscript = biome mark; 7, Oldland. 8–9, Marine and paralic region: 8, South China; 9, North China. 10–11, Terrestrial regions (Late Permian and Mesozoic–Cenozoic): 10, intermontane basins; I_1, fluvio–lacustrine basins. I, South Asia realm; I_1, South-western China region, Hengduan Mountains province; I_2, Lingnan province; I_3, West Pacific region, Taiwan province; II, Boreal realm, China region; II_1, South China province; II_2, Tibet–Qinghai province; II_3, North China province; II_3^1, North China biome, II_3^2, Shanxi–Shaanxi biome; II_4, North-western China province; II_4^1, Erenhot biome; II_4^2, Junggar–Qilian biome; II_5, North-eastern China province.

Fig. 12.6 Reconstruction of the Neogene community of China. 1, *Rhinoceros*; 2, *Gazella*; 3, *Palaeotragus*; 4, *Drypithecus*; 5, *Mastodon*; 6, *Palaeomeryx*; 7, *Hipparion*; 8, *Machairodus*; 9, *Alnus*; 10, *Lindera*; 11, *Hydrangea*; 12, *Ceratophyllum*; 13, *Pteris*; 14, *Cyperacites*; 15, *Carya*; 16, *Pterocarya*.

The China region

Is subdivided into North-eastern, Northern, and North-western China, Qinghai–Tibet, and South China provinces.

The North-eastern province

This borders on the Daxingan Mountains in the west, at a latitude of 42°N it demarcates itself from the North China province in the south, and its eastern and northern boundaries may go beyond the boundary line of China. Under a temperate climate the province is characterized by a mixed forest of coniferous and broadleaf type, with a few subtropical members. No mammals have been reported.

Along the Daxingan Mountains fossils are rare. The mixed coniferous–broadleaf forest was dominated by deciduous trees with some ferns. The Songhuajiang–Liaohe plain and the eastern mountainous areas of Jilin and Heilongjiang have more or less abundant plants and sporo-pollen fossils. The Miocene Ma'anshancun Formation yields a similar flora to the Shanwang Formation of Shandong, North China Province. Both were dominated by warm temperate components. The deciduous broadleaf elements of this formation are *Quercus*, *Zelkova*, *Acer*, *Betula*, *Salix*, and *Juglans*, mingled with a few subtropical genera such as Ilex and *Lequidambar* (CCBC 1978). Locally relatively abundant coniferous pollen, such as *Podocarpidites*, Abiespollenites, *Piceaepollenites*, and *Ketelleria*, has also been found.

The North China province

This is located to the south of the North-eastern China province, adjacent to the Helan Mountains in the west and bordering on the Huihe River in the south; it lay in the warm temperate–subtropical climatic zone. Because of the influence of the east Pacific monsoon, the climate of the eastern part was rather humid, and the dominant facies is lacustrine, whereas the western part was a loess plateau belt, less influenced by monsoons and drier than the eastern part. Consequently, the vegetation of the two parts was different. On the other hand, the mammals in both parts were mainly the same *Hipparion* fauna, although there is a small difference in content due to the different environments. Taking into consideration the mammals, plants, climate, and lithofacies, the province may be subdivided into the North China and Shanxi–Shaanxi biomes.

The North China biome borders on the Taihan Mountains in the west and differs from the Shanxi–Shaanxi biome in the following ways.

1. The North China biome had a rather humid climate due to the influence of the west Pacific monsoon. In Miocene times diatomite, oil shale, and lignite were deposited. The plants and sporo-pollen assemblages are dominated by deciduous broadleaf forest and herbaceous plants. The mammals were mainly the common genera and species of the North China *Hipparion* fauna, represented by the Middle Miocene Shanwang Formation, containing much *Aceratherium*, plus *Stephanocemas thomsoni* and *Diatomys shantungensis*, but without herbivorous grassland hypsodont bovoids, thus indicating a forest–swamp environment. Similar faunas are also found in Cixian County, southern Hebei, and Zhangbei County, northern Hebei. The plant and sporo-pollen assemblages indicate a corresponding deciduous forest vegetation. The temperate Betulaceae, Ulmaceae, and *Quercus* were mixed with the subtropical Fagaceae (evergreen type), *Rhus*, *Liquidambar*, and Meliaceae. In the Late Miocene, Zhangqiu County of Shandong was the easternmost locality of the north China *Hipparion* fauna.

2. The mammal fauna of the Jiulongkou Formation in Cixian County, which is located at the western boundary of this biome, includes typical grassland cervids besides the forest–swamp mammals similar to those of the Shanwang Formation. This indicates that in its bordering area in the west with the Shanxi–Shaanxi biome the faunas were well mixed with each other.

The Shanxi–Shaanxi biome. This is adjacent to the North China biome in the east and borders on the Helanshan Mountains in the west; its southern part belongs to the eastern sector of Qinling, and it is adjacent to Inner Mongolia in the north. It is already too far from the Pacific Ocean for the monsoon to reach, so the climate was more arid than that of the North China biome; both biomes belong to the warm temperate–subtropical zone. The deposits of this biome are dominated by red earthy sediments quite different from those of the North China biome. In the Early Miocene basalt (Hannuoba Formation) was erupted in northern Hebei, covering Wanquan, Shangyi, Youyu, and Zuoyun Counties. *Piceaepollenites* predominated in sporo-pollen assemblages, followed in importance by angiosperm pollen of the Betulaceae and Fagaceae, and pollen of the herbaceous Chenopodiaceae, *Potomageton* and *Convolvulus*. This vegetation was similar to that in the North China biome, being warm temperate mixed coniferous–broadleaf forest. Middle Miocene mammals consisted of the swamp inhabitants Elasmotheriinae and grassland–forest Cervidae, Proboscidea, and *Palaeomeryx*. The Lower Pliocene Baode Laterite of Shanxi contains many species of *Hipparion*, the large hypsodont *Gazella*, the drycodont *Palaeomeryx*, and Suidae, indicative of a grassland habitat. Pollen of the drought-enduring herbaceous *Ephedripites*, Chenopodiaceae, and the grassland *Hipparion weihonese*, *H. chiai*, and the relatively hypsodont *Antilope* occur southward in Yushe, Huoxian, Yingchi, Sanmenxia, Lushi, and Lantian Counties of southern Shanxi and northern Henan.

In summary, during the Early Miocene the climate was warm and humid, indicated by the presence of swamp-living *Rhinoceros*. In the Late Miocene, however, the climate became semi-arid and warmer and grassland animals and drought-enduring plants appeared.

The North-western China province

This includes the major part of Inner Mongolia and Xinjiang and the northern parts of Qinghai and Gansu. It is adjacent to the Northern province and the North-eastern province in the east, and in the south borders on the Kunlun Mountains. With the uplift of the Xingan, Taihang, and Lueliang Mountains to the east, and the Tibet–Qinghai Plateau in the south, neither Pacific nor Indian Ocean monsoons could reach this province, so it experienced an inland arid climate with inland grassland and forest–grassland vegetation (CVC 1980). On the basis of biological characteristics and sediment characteristics, two biomes are designated.

The Inner Mongolia–Erenhot biome. This is the area to the north of Yinshan and Daqingshan, with an arid climate, piedmont deposits being predominant.

Plant fossils of the Miocene Erenhot Tonggur Formation represent a forest–grassland vegetation, dominated by *Pinus*, *Alnus*, Fagaceae, and a large number of herbaceous plants, such as Compositeae, Liliaceae, Chenopodiaceae, and *Liquidamber*. The fauna is represented by Proboscidea such as *Platybelodon* and also *Anchiterium*, but without the Ellasmotheriinae of the Shanxi–Shaanxi biome, thus reflecting a grassland habitat. Other differences from the Shanxi–Shaanxi biome are that Pliocene grassland rodents, such as *Meryx*, *Lepus sinensis*, and *Circelulas teilhardi*, occurred here, and that the flora was dominated by herbs and savanna-type trees, such as Alnus and Betula; no forest-type plants of the Shanxi–Shaanxi biome have been found.

The Junggar–Qilian biome. This is a vast area to the west of the Helan Mountains and to the north of the Kunlun Mountains, including the large basins of Tarim, Junggar, and Qaidam, as well as mountain basins with inland deposits of fluvial–lacustrine facies. The biome is characterized by arid grassland mammals of the family Muridae, such as *Eucricetodon youngi* of the early Neogene Xiejia Formation. The vegetation was of grassland type with forests. Unlike the Inner Mongolia–Erenhot biome, a desert–grassland landscape developed around Qaidam, Jiuquan, and Hami in the late Neogene. The plant communities were mainly composed of Gramineae, Chenopodiaceae, Compositeae, Liliaceae, and Leguminosae. The fauna included *Hipparion*, *Mastodon*, and *Rhinoceros*. In the eastern Qaidam Basin, the Pliocene deposits contain *Cervus*, *Machairodus*, and *Spirocerus*. This desert–grassland landscape extended to the eastern Junggar Basin. The drought-ensuring *Ulmus* was found in the brownish-red argillite of the Tianshan piedmont along the southern margin of the basin. In north Xinjiang *Mastodon*, *Hipparion*, and *Gazella* also occurred.

The Qinghai–Tibet province

This province lies to the south of the Kunlun Mountains, to the north of the Gangdise Mountains and the Yarlung Zangbo River, and to the west of the Hengduan Mountains. The fauna being poor, its establishment was based on plant and sporo-pollen data. Because the Qinghai–Tibet Plateau was being uplifted continuously, the palaeobotanical zonation in this province was obvious. The early Neogene was dominated by Betulaceae. In the middle and late stages *Quercus semicarpifolia* appeared. The early floral assemblages can be regarded as temperate forest, consisting of Betulaceae–Ulmaceae–*Quercus* and Juglandaceae. The climate in the late stage turned cooler and more arid; the main floral representatives were *Quercus semicarpifolia* and Salicaceae. This province is unique not only in plant assemblages but also in palaeogeography and palaeoclimate, differing from other provinces. The Bulong Formation of the Bulong Basin in northern Tibet contains the *Hipparion* fauna and flora representing a deciduous broadleaf forest vegetation type (SEQX 1980).

The South China province

This occupies a vast area to the north of Nanling and to the south of Qinling and the Huihe River. In the Miocene the province was in the southern subtropical zone (southern South Tropical–Subtropical phytoprovince, CVC 1980) with well-developed deciduous broadleaf and evergreen forest. In the early stage the climate was arid, and in the late stage, warm and humid. Much red, weathered crust was formed. Around Fangshanpu town east of Nanjing city a Miocene forest fauna, including *Anchitherium*, *Brachypotherium*, *Hyoherium*, and *Dicrocerus*, lived, being quite different from the grassland or forest–grassland fauna in the western part of North China. The sporo-pollen assemblage of the Yancheng Formation in this province is characterized by the abundance of hygrophilous deciduous broadleaf varieties with a few thermophilous members. The pteridophyte spores of hygrophilous Polypodiaceae and *Pterisporites* thrived and are associated with swamp thrived and are associated with swamp Nymphaeaceae. On the whole, the climate in the east was more or less humid, with well-developed subtropical evergreen forests and swamps. The western part of this province lacks fossils. In Neogene times the Sichuan Basin was an area surrounded by mountain ranges. A great thickness of clastic deposits accumulated, and the red and greyish-green argillites contained gypsum beds in the west. The climate was evidently arid and hot.

The South Asia realm

This realm covered the area to the south of Hengduan and Nanling Mountains. It was in the tropical wet climatic zone, quite different from the Boreal or Holarctic realm. According to the different biotic characteristics, this realm may be subdivided into three provinces:

The Hengduan Mountains province

This province is short of fossil data. The mammals from the Xiaolongtan Formation of Lufeng and Kaiyuan in Yunnan were the representative fauna, which show a basically South Asian but relatively endemic *Hipparion lufengensis* assemblage with primates, *Viverra*, *Dorcatherium minus*, and *Dryopithecus kaiyuanensis*. *Hipparion lufengensis* is similar to *H. nagriensis* of the Siwalik fauna. Though there were elements in common with the North China-type *Hipparion* fauna and Middle Miocene Eurasian members such as *Macrotherium*, *Aceratherium*, and *Potamochoerus*, on the whole the *Hipparion* fauna of Lufeng is basically similar to the Siwalik fauna and much different from that of the Boreal realm, so this province should be attributed to the South Asia realm. Lithologically the Lufeng Group is a suite of fluvial–lacustrine facies of sandstones, conglomerates, siltstones, and lignite deposits. The flora and sporo-pollen assemblages all indicate a humid environment with an abundance of Lauraceae.

The Lingnan province

This belongs to the tropical zone lying to the south of the South Asian tropical zone, with the dominant plant family being the Annonaceae, but lacking a temperate flora. Sporo-pollen data collected from boreholes in Weizhou Island, Leizhou Peninsula, Beichangpo Formation of Hainan Island, and the Pearl River mouth, show that there were abundant tropical *Trochodendron*, Theaceae, Hamamelidaceae, and Flacourtiaceae, with subordinate *Polypodium*, *Gothanipollis*, and *Sonneratia*, but few Pinaceae. Hence the climate was undoubtedly tropical (SSBPC *et al.* 1981).

Forams have been collected from clastic rocks of littoral facies in this province. They comprise the Miocene *Ammonia–Urigerina–Globigerina* assemblage and the Pliocene *Pseudorotalia–Asterorotalia–Pulleniatina* assemblage, both belonging to a warm-sea benthic type; there are very few plankton (SSBPC *et al.* 1981).

The Taiwan province of the West Pacific region

In Taiwan province, the Miocene consists of huge thicknesses of sand and mud deposits, bearing coal seams, and with an abundance of forams, gastropods, and bivalves. The forams are characterized by the *Myogypsina–Pulleniatina* assemblage, with many tiny plankton. This assemblage consists of more than

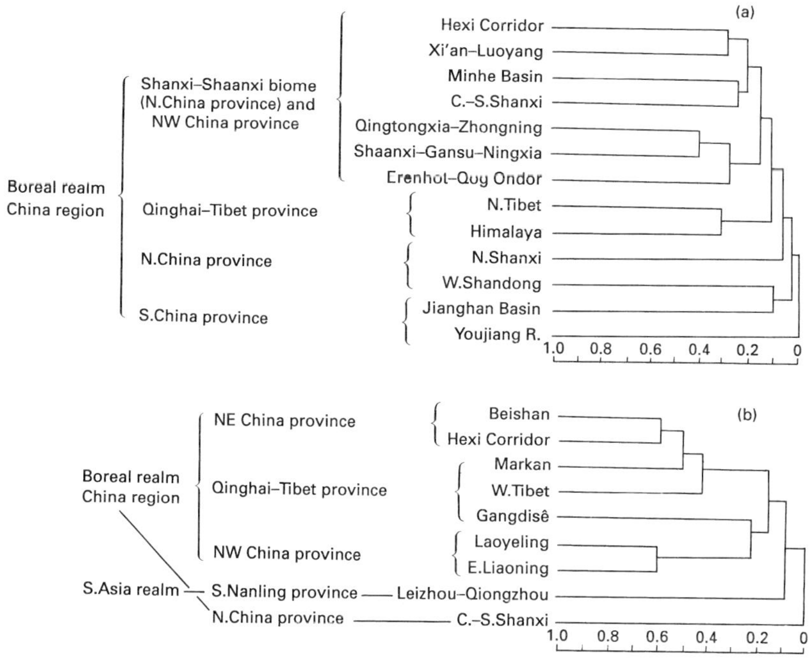

Fig. 12.7 Cluster analysis of the Late Tertiary fossils of China. (a) Late Tertiary mammals; (b) Late Tertiary plants.

300 species, the principal members being open-sea warm-water genera such as *Miogypsina*, *Asterorotalia*, *Pseudorotalia* and *Bulimina*, belonging to the tropical zone. The Sidi Formation of Middle Miocene age yields abundant lauraceans and *Pasania* and *Castanopsis* of the evergreen type of Fagaceae, but without typical temperate-type plants. The assemblage also yields *Coniogramme*, *Bambusa*, and *Phyllostachys*, which live at present in areas of low latitude. All these facts indicate that this province belonged in the Neogene to the tropical zone, climatically identically to the Lingnan province. However, considering the palaeogeographical features, the sediment environments and foraminiferal assemblages, it should be designated as an independent province.

Cluster analysis of Neogene mammals and plants

There have been reports of 168 genera and species of Neogene mammals, 70 genera of plants, 63 genera of ostracods, 80 genera of sporo-pollen, 40 genera of gastropods, and 40 genera of forams. The mammals and plants were chosen for cluster analysis. As in the Palaeogene, generalized genera and species were omitted, and the same kind of stratigraphic units were used.

Figure 12.7a is the diagram for the cluster analysis of Neogene mammals based on calculations using the data of 168 genera and species of mammals. It shows that, if Otc = 2, northern Shanxi, western Shandong (both belonging to the Northern province), Jianghan and Youjiang (both belonging to the South China province) are each an independent group. The Gansu Corridor (Hexi), Xi'an-Luoyang, the Minhe Basin of eastern Qinghai and south-central Shanxi, all belonging to the Shanxi–Shaanxi Biome of the North China province, aggregate into one subgroup; but another subgroup of the same biome, consisting of Qingtongxia–Zhongning and Ordos, are mixed with Erenhot Qog of the North-western China province into one group. Such an unexpected categorization results from the fact that the Miocene mammals, mainly the northern *Hipparion* fauna, covered both areas, so the boundary between the two provinces cannot be delineated on this basis.

Figure 12.7b is the diagram of Neogene plant cluster analysis, using the data of 70 genera for the calculations. If Otc = 2.5, Leizhou–Qiongzhou and south-central Shanxi each form a group. The rest of the areas make the China region, within which Laoyeling and eastern Liaoning form the North-eastern China province, the other five together constituting the North-western China and Tibet–Qinghai provinces. Here the demarcation of North and North-western China is clear, reflecting climate discrimination. Clusterings of mammals and plants, considered together, can be of value as a source of reference in studies of provincialization.

REFERENCES

Brenchley, P. J. (1984). *Fossils and climate*, pp. 174–9. Wiley, New York.

CCBC (Compiling Committee of the Cenozoic Botany of China, by the Beijing Institute of Botany and Nanjing Institute of Geology and Palaeontology, Academia Sinica) (1978). *Botanic fossils of China*, Vol. 3, *Cenozoic Botany of China*. Science Press, Beijing (in Chinese).

CPGC (Compiling Committee of the Physical Geography of China, Academia Sinica) (1979). Zoogeography of China. In *Physical Geography*, pp. 71–106. Science Press, Beijing (in Chinese).

CVC (Compiling Committee of the Vegetation of China) (1980). *The vegetation of China*. Science Press, Beijing (in Chinese).

Guo Shuangxing (1983). Note on phytogeographic provinces and ecological environment of Late Cretaceous and Tertiary floras in China. In *Palaeobiogeographic provinces of China*, (ed. Compiling Committee of Basic Theories in Palaeontology), pp. 164–75. Science Press, Beijing (in Chinese).

Hao Yichun, Qiu Songyu, Lin Jiaxing, and Zheng Xuelu (1980). *Forminifera*. Science Press, Beijing (in Chinese).

IVP (Institute of Vertebrate Palaeontology and Paleoanthropology, Academia Sinica) (1979). *Handbook of vertebrate fossils of China*. Science Press, Beijing (in Chinese).

Lei Yizhen (1984). On the subdivision of the Early Tertiary mammalian fauna in Central–South China. *Bulletin of Yichang Institute of Geology and Mineral Resources, Chinese Academy of Geological Sciences* **8**, 41–51 (in Chinese).

Li Bai (1986). Climatic implication of Paleocene flora in Altay, Xinjiang Autonomous Region. In *Selected Papers from the 13th and 14th Annual Conventions of the Palaeontological Society of China*, pp. 249–57. Anhui Science and Technology Publishing House (in Chinese).

Li Yuntong *et al.* (1984). *The Tertiary system of China*. Geological Publishing House, Beijing (in Chinese).

NIGPAC (Nanjing Institute of Geology and Palaeontology, Academia Sinica and Institute of Petroleum Exploration and Development, Ministry of Petroleum Industry) (1978). *Early Tertiary spore and pollen grains from the coastal region of Bohai*. Science Press, Beijing (in Chinese).

NIGPAC (Nanjing Institute of Geology and Palaeontology, Academia Sinica, Institute of Botany, AC) (1982). *Palaeontology of Xizang*, Vol. 5. Science Press (in Chinese).

Qian Nai (1967).

SEQX (Scientific Expedition to the Qiughai-Xizang Plateau) *et al.* (1980). *Palaeontology of Xizang*, Vol. 1, pp. 1–105. Science Press, Beijing (in Chinese)

SSBPC (South Sea Branch of Petroleum Corporation of the Peoples Republic of China) *et al.* (1981). *Tertiary palaeontology of the north continental shelf of the South China Sea*. Guangdong Science and Technology Press, Guangzhou (in Chinese).

Sun Xiangjun and He Mingyue (1980). *Palaeocene spores and pollen assemblage in Jiangxi*. Science Press, Beijing (in Chinese).

Yuan Fuli and Du Henjian (1984). *Cenozoic biostratigraphy in China*. Geological Publishing House, Beijing (in Chinese).

Zhou Mingzhen, Zhang Yuping, Wang Banyue, and Ding Suyin (1977). *Mammalian fauna from the Palaeocene of Nanxiong Basin, Guangdong*. Science Press, Beijing (in Chinese).

Zhou Tingru (1982). *Palaeogeography*. Publishing House of the Beijing Normal University (in Chinese).

Ziegler, A. M., Scotese, C. R., and Barret, S. F. (1983). Mesozoic and Cenozoic paleogeographyic maps. In *Tidal friction and the earth's rotation*, II (ed. Brosche and Sundermann).

13 QUATERNARY

Cao Bexun

13.1 QUATERNARY PALAEO-BIOGEOGRAPHICAL PROVINCIALIZATION OF THE WORLD

In the Cenozoic, the continents of the northern hemisphere shifted towards the North Pole, and Antarctica occupied the south polar section. This situation, plus climatic changes, were favourable for the formation of the permanent polar ice sheets of Arctica and Antarctica in the Pliocene and Miocene, respectively. World temperature further declined after the Pliocene and the Earth has been experienced the late Cenozoic ice age. Quaternary biogeographical distributions were controlled by the above-mentioned plate-tectonic and palaeoclimatic changes. Thermophilous faunas migrated to low latitudes, temperate floras shifted southward and psychrophilous herbs and coniferous trees significantly increased in numbers. The modern provincialization therefore formed through addition of these alterations to the Tertiary biogeographical provincialization.

The Quaternary terrestrial provincialization mainly consisted of a Holarctic realm and an Austral realm (Fig. 13.1). The former corresponds to the Palaeo-Arctic and the Neo-arctic of modern zoogeography or the Panarctic of modern phytogeography, consisting of a vast continuous terrestrial area surrounding the Arctic Ocean. The geographical and ecological differentiation was relatively monotonous. From north to south there were regions of tundra, coniferous forest, temperate deciduous forest and grasslands, and, correspondingly, tundra and temperate forest and grassland fauna, showing conspicuous latitudinal climatic–biotic differentiation. The North China region belongs to this realm. Fossil records demonstrate this realm to be the origination area of many Cenozoic faunas, including the primates. The Austral realm includes the Ethiopian, Neotropical, Oriental, and Australian realms of modern zoogeography, or tropical or subtropical realms of modern phytogeography. The South China region belongs to this realm. Due to

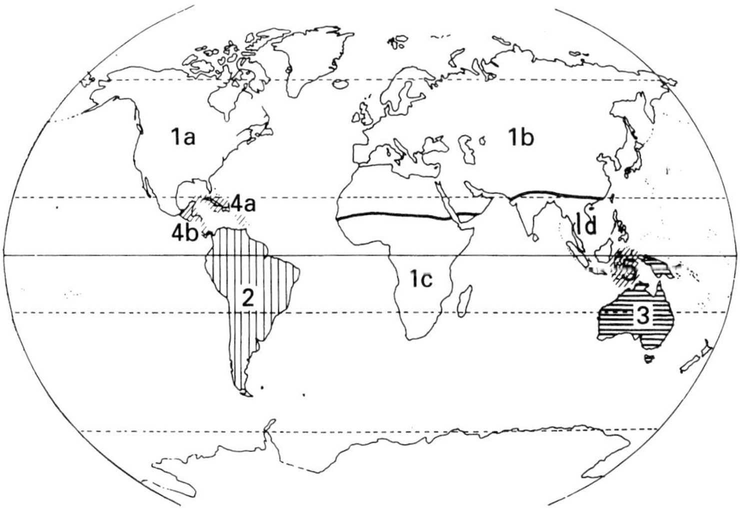

Fig. 13.1 Map of Recent terrestrial zoogeographical provincialization of the world (after Ross 1974*b*; for reference see Chapter 1 of this book). 1, Eurasian–African–North American realm (Holarctic realm); 1a, Nearctic region; 1b, Palaearctic region; 1c, Ethiopian region; 1d, Oriental region; 2, Neotropical realm, region and subregion; 3, Australian realm and region; 4, Caribbean transitions; 4a, West Indian transition; 4b Middle American transition; 5, Celebesian transition.

separation by oceans in this realm, provincialization caused by geographical barriers is more striking than that caused by latitudinal differentiation, and the biotas were more diversified than in the Boreal realm (Yuan and Du 1984). Affected by the alternation of glacial and interglacial climates, both marine and terrestrial biotas migrated to and fro between lower and higher latitudes, a phenomenon more notable in the northern hemisphere.

13.2 QUATERNARY PALAEO-BIOGEOGRAPHICAL PROVINCIALIZATION OF CHINA

The factors controlling the formation and development of the Quaternary biogeographical provincialization of the world certainly influenced China, as was reflected in the latitudinal climatic–biotic zonation and back and forth longitudinal biotic migration during the Quaternary in the low-mountain, hill, and plain areas of eastern China. However, the unique factor by which China is distinguished from the rest of the world is the uplift of the Tibet–Qinghai Plateau. In the Pliocene when the Tibet–Qinghai Plateau was about 1000 m in altitude (Li *et al.* 1979), the east–west difference in the relief of the Chinese landmass was far less than it is today. The palaeoclimate of of China was then under the influence of a planetary wind system which was obviously latitudinally zoned and dominated by warm temperate and tropical–subtropical weather (Zhang Linyuan 1984), and the whole of China belonged basically to the same zone of *Hipparion* fauna (Chow 1964). In the Quaternary the situation changed. In addition to the world-wide temperature decrease, the compression of the Indian plate against Eurasia continued. Consequently, the Tibet–Qinghai Plateau was rapidly uplifted and enlarged in size. Thus the Himalayas are estimated to have been elevated by as much as 500 m since the climatic optimum of the Holocene (Hsu Ren *et al.* 1973). This uplift directly interrupted the latitudinal climatic–biotic zonation in western China and replaced it with vertical zonation. During the middle and late part of the Middle Pleistocene, the altitude of the plateau reached *c.* 3000 m and that of the Himalayas was above 5000 m, forming a barrier to the south-west Indian monsoon. As a result, not only the Tibet–Qinghai Plateau suffered cold and arid weather, but also north-western China, now inaccessible to humid winds from all directions, became progressively more arid, which inevitably had a major effect on the biotic environments. In addition, the upheaval of the Qilian–Qinling Mountains was another factor in the development of Quaternary biogeographical provincialization (Cao 1988). All these tectonic, geomorphological, and climatic changes altered the Tertiary provincialization, which was characterized by a close relationship between north and south and east–west intercommunication, into the Quaternary provincialization, characterized by north–south differentiation and restricted intercommunication, and were, to a large extent, responsible for the eastward shift of human history in China and East Asia.

Based on modern zoo- and phyto-geographical provincialization (ECPGAS 1979, 1983), and also taking other data into account (Pei 1957; Chow 1964), North China and South China biogeographical regions have been established for the Quaternary of China. Each is, in turn, subdivided into a few provinces. Boundaries of these regions and provinces are similar to, but not always identical to, both the modern ones and the corresponding boundaries of physical environments (Fig. 13.2).

13.2.1 THE NORTH CHINA REGION

This region includes the land areas of North China, Inner Mongolia, north-eastern and north-western China to the north of the Qinling–Dabie line, together with most of the Tibet–Qinghai Plateau and the shelf areas north of Taiwan, forming a southern part of the Palaeo-Arctic zoo-realm and the Panarctic phyto-realm. According to differences in biotic distribution and ecological environments, it is subdivided into five provinces: North China, North-eastern China, North-western China, Tibet–Qinghai Plateau, and Northern Shelf province.

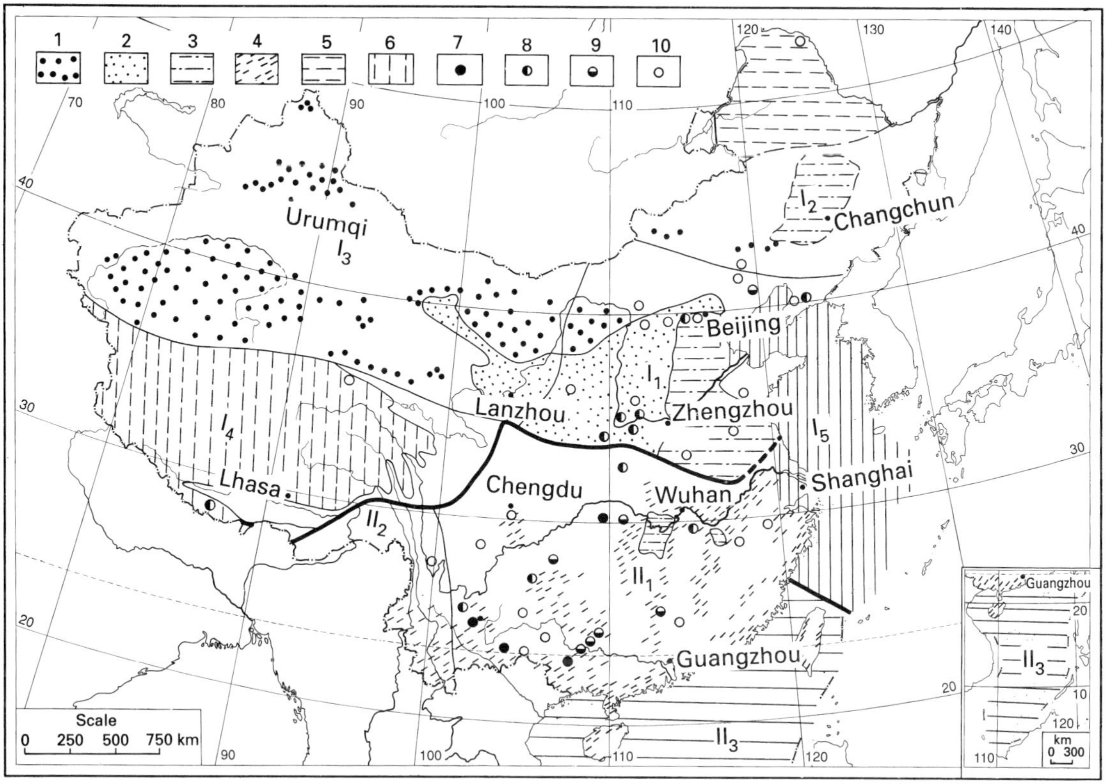

Fig. 13.2 Map of the Quaternary palaeobiogeographical provincialization of China. I, Boreal region; I_1 North China province; I_2 North-eastern China province; I_3, North-western China province; I_4, Tibet–Qinghai Plateau province; I_5, Northern Shelf province; II, South China Region; II_1, South China province; II_2, Hengduan Mountains province; II_3, Southern Shelf province; 1, Desert, 2, loess; 3, alluvium; 4, red earth; 5, latitude frozen soil; 6, highland frozen soil, 7, anthropoid apes; 8, *Homo erectus (Pithecanthropus)*; 9, *Homo sapiens* (Neanderthal man); 10, *Homo sapiens* (modern man).

The North China province

The western half of this province is the East Gansu–Shanxi–Shaanxi loess plateau; its eastern part is the North China alluvial plain. It is thus the main loess are (including secondary loess) of China (Liu 1985). The Quaternary mammal fauna of this province evolved from the warm temperate forest–grassland *Hipparion* fauna of the Pliocene (Yuan and Du 1984; Xue 1985). Representatives of its Quaternary faunas are: the Early Pleistocene *Proboscidipparion–Equus* fauna, or the Nihewan fauna; the Middle Pleistocene *Sinanthropus–Megaceros pachyosteus* fauna, or the Zhoukoudian fauna; the Late Pleistocene *Cervus elaphus–Crocuta ultima* fauna, or the Sara'ushu fauna; and the Holocene *Elaphurus davidianus* fauna. They were all dominated by Perissodactyla, Artiodactyla, Carnivora, Proboscidea, and Rodentia. However, the Early and Middle Pleistocene faunas were of forest–grassland type with a number of forest animals, such as *Machairodus*, *Cynailurus*, and *Cervus*, while the Late Pleistocene fauna belonged to temperate grassland type with more grassland inhabitants, especially rodents. The terrestrial molluscs comprised both xerophilous *Cathaica* and hygro-thermophilous *Metodontia* (Chen *et al.* 1982). By the end of the Late Pleistocene, not only the few South China elements such as proboscideans,

Table 13.1 Correlation chart of the Quaternary of China

	Age 10^4 a BP	Polarity epoch	Mammal faunas (f.) N. China	Mammal faunas (f.) S. China	Palaeoanthropology	Cave deposits (cd) & formations (fm) of Zhoukoudian
Holocene		Brunhes n	Banpo f.	Shanxiandong f.	Banpo man	
Late Pleistocene	1.1		Upper cave f.	Zhiyang f.	Zhiyang man	Upper cd; Donglinzhi cd
	2		Sara'-ushu f.; *Coelod–Mammuthus f.*	Liujiang f.	Liujiang man	New cd (or No. 14 site)
	5	n	Dingchun f.	Maba f.	Maba man	
Middle Pleistocene	10 20			*Ailuropoda–Stegodon*	Dali man	Peking man cd (No. 1 site)
				Yanjinggou f.	Peking man	Zhoukoudian Fm
	50	Brunhes	Zhoukoudian f.			
				Guanyingdong f.		
Early Pleistocene	73	r Jaramillo	Gongwangling f.	Biejiashan f.	Lantien man	Taipenshan Fm & No. 12 site
		ma Olduvai	Nihewan f.	Gaopen f. Liucheng f.		
					Yuanmou man	
	240	Matuya		Yuanmou f.		
			Youhuo f.			
Pliocene		Gauss n			*Sivapithecus*	No. 14 site

Rock-stratigraphic unit		Glacial (g.) & interglacial (ig.) ages		Quaternary transgressions (t.)
N. China (l. loess)	S. China	Himalaya	Other regions	
Holocene l.	Recent deposits	Postglacial period		Holocene t. (1 cycle)
Malan l.	Xiasu or Chendu Clay	Qomolangma g.	Dali g.	Late Pleistocene t. (2 cycles)
			L–D ig. Lushan g.	
Upper Lishi l.		Jiabla ig.	D–L ig.	Middle Pleistocene t. (2 cycles)
	red earth with netted texture			
		Nieniexiongla g.	Dagu g.	
Lower Lishi l.				
		Pali ig.	B–D ig.	
	Bosajin red gravels		Boyang g.	Early Pleistocene t. (3 cycles)
Wuchen l.		Xixiabangma g.		
	Yuhuatai gravels			
Baoda red clay				

rhinocerids, and *Taipirus*, but also the *Coelodonta antiquitatis*, which migrated into the region, disappeared from North China. The Holocene and modern mammals of North China are mere relicts of the flourishing Pleistocene fauna.

Warm temperate deciduous broadleaf forest was already established at the beginning of the Quaternary in the North China province. The main elements were *Quercus*, *Ulmus*, and *Carpinus*, mixed with subtropical conifers such as *Podocarpus*, *Cedrus*, *Kelelecia*, and broadleaf trees such as *Liquidambar*, *Carya*, and Fagaceae. However, the subtropical elements diminished spatially from south to north, and temporally from the Early to the Late Quaternary, and are now on the verge of total disappearance. Meanwhile, psychrophilous conifers, such as *Abies* and *Picea*, and herbs, such as Chenopodiaceae and *Artemisia*, increased in quantity and area following the climatic drying and cooling. The dimensions of grassland in North China enlarged considerably in the Late Pleistocene Malan Loess time (*c.* 100 ka BP). Biotic distributions changed accompanying the alternation of glacial and interglacial ages. Records indicate that the North China biota migrated southward and from mountain to plain when cold air invaded from the north. During the time of high-frequency alternations of cold and warm climates in late in the Middle Pleistocene, *Megaceros pachyosteus* migrated southward to Hexian County on the northern bank of the Yangtze, Anhui (Huang Wanpo and Huang Cixuan 1985). At the peak of the Dali ice age (*c.* 15 ka BP) the Boreal element *Equus przewalsky* migrated to Penhu Island (at about 24°N) (Huang 1986). Also, at about 15–30 ka BP the coniferous forest of the North China mountains descended to river valleys and plains, e.g. the Abies–Picea forest of about 23 ka BP in Beizhuang Village, Weinan County (Shaanxi) (Cao *et al.* 1966; IBAS 1966).

The North-eastern China province

This includes areas of north-eastern China and eastern Inner Mongolia north of 42°N. There is now a permafrost in its northern part (the north Daxingan Mountains), whereas in the southern part there are low mountains, hills, and the Songliao Plain.

Records of its Early and Middle Pleistocene mammal fossils are fewer than in North China. However, abundant discoveries of *Dicerorhinus merki*, *Pseudaxis grayi*, *Macacus robustus*, and *Hyaena sinensis* from Liaoyang County and Yinkou district, Liaoning (Zhang Zhenhong *et al.* 1980) denote a similarity with North China. During the Dali (or Wurm) ice age of the Late Pleistocene (*c.* 11–40 ka BP), this province experienced the Siberian ice-age climate (Sun Jianzhong *et al.* 1985); areas north of 43°N were covered by permafrost; there were isolated masses of a permafrost in areas between 40°N and 43°N; and glacial moraines and other glacigenic phenomena were common at this time (Cui and Xie 1985). At this time the biota of the North-eastern China province was characterized by the *Mammuthus–Coelodonta antiquitatis* fauna, including *M. primigenius*, *M. (Parelephas) sungari*, *C. antiquitatis*, and accompanying *Bison exiguus*. This fauna lacked elements typical of those living within the Arctic Circle, and is considered as a division of the Siberian tundra fauna which migrated southward into northern China and Japan (Chow 1978). Areas north of 43°N were mainly inhabited by *Mammuthus* adapted to a permafrost and tundra; south of 43°N the faunal elements shifted from a coexistence of *Mammuthus* and *Coelodonta* to the dominance of *Coelodonta*, which migrated to Shouxian County, Anhui (*c.* 34°N) and the vicinity of Shanghai. The *Mammuthus–Coelodonta* fauna is roughly synchronous with the Late Pleistocene *Cervus elaphus– Crocuta ultima* fauna of North China, but they belonged to different provinces.

In the North-eastern China province, the temperate mixed forest of deciduous conifers and broadleaf trees had become established since the beginning of the Quaternary. In the Late Pleistocene, during the Dali ice age, the vegetation changed to that of a dry and cold grassland, composed of *Chenopodia* and *Artemisia*, with tundra in the north and *Abies–Picea* forests in parts of the valleys and plains; while during interglacial ages there grew mixed coniferous and broadleaf forests, dominated by *Betula*. In the Holocene, the vegetation consisted of psychrophilous forests and grasslands, similar to the modern situation (Qiu *et al.* 1985).

The North-western China province

This province consisted of Xingjiang, western Gansu, and part of Qinghai, i.e. the main area of deserts in China today.

Being located near to the centre of the Eurasian continent and far away from the sea, this area had been arid since the Pliocene. The situation deteriorated further in the Quaternary because the accelerated uplift of the Tibet–Qinghai Plateau formed a barrier to the Indian Ocean monsoon and the neotectonic block-fault movements of this area created an intercalated series of snow-covered high mountains and enclosed arid intermontane basins. Such deteriorated environments greatly impeded the development of Quaternary biota in this province.

There are few records of Quaternary faunas in this province. A few fossils of Middle to Late Pleistocene *Equus sanmeniensis*, *E. hemionus*, *Hyaena sinensis*, and *Rhinoceros* have been discovered in the Gonghe Basin, Qinghai. *Equus sanmeniensis* is also reported from the Xiyu Conglomerate of Anjiechang, south Junggar, Xinjiang (Hong 1981) and the Wuquanshan Conglomerate of Lanzhou City, Gansu, together with rodents (Guo 1987); the Late Pleistocene *Mammuthus–Coelodonta* fauna is reported to have migrated to the vicinity of Urumqi City, Xinjiang (Han and Ye 1986). These limited records reveal to a certain extent the intercommunication and differentiation between this province and North China. Geographical and climatic difficulties apparently restricted the faunal abundance and diversity.

Within the enclosed basins and lakes there lived ostracods adapted to varying salinity. A common feature is the presence of euryhaline Mediterranean elements such as *Cyprideis littoralis* and *C. torosa*, as well as endemic *Qinghaicypris crassa* (Huang Baoren 1985). Among terrestrial molluscs the Palaeo-Arctic xerophilous *Cathaica* had the widest distribution (Chen *et al.* 1982).

Early–Middle Pleistocene forest vegetation still existed in the basins of the North-western China province, e.g. a palynomorph assemblage of broad-leaf forest has been reported from lacustrine deposits of that age around the Qinghai Lake. In the Late Pleistocene, the land turned into desert, in accordance with the consistent widening of the Taklamagan (Tarim) desert since 63 ka BP (Li Baojing 1988). Meanwhile, in the mountainous areas there still developed a few alpine conifer forests due to an increase of precipitation; however, they were quite different from the mountainous vegetation on the southern slopes of the Himalayas which were under the influence of an oceanic climate.

The Tibet–Qinghai Plateau province

This includes most of the plateau north of the Himalayas and south of Golmud (Qinghai). With an average altitude above 4500 m and a plateau permafrost area of 1.5 million km^2, it is the highest cold plateau of the world (Tong 1982).

During the Pliocene this area was a high plain with an altitude of about 1000 m, covered by subtropical vegetation and inhabited by the *Hipparion* fauna, and karst phenomena were commonplace. In the Early Pleistocene it reached 2000 m in altitude, and was not yet a serious barrier to the humid air from the Indian Ocean. During the global ice age of the Middle Pleistocene, there developed in this area the largest Quaternary mountainous glaciation of China—the Nieniexiongla glaciation. After this ice age there was a relatively lengthy Jiabla interglacial age, when a crust of residual red earth was formed on the plateau (Zeng 1976). There is not yet any record on the Early and Middle Pleistocene mammal fauna of this area, probably due to scarcity of such fauna caused by the large-scale glaciation.

There have been scattered discoveries of Late Pleistocene *Cervus*, *Bos*, *Ovis*, *Equus*, etc. (Huang Wanpo, 1980) which, together with a number of modern Palaeo-Arctic animals on the plateau, such as *Lynx*, *Meles*, *Cervus elaphus*, *Mustela sibirica*, and *Equus hemionus*, indicate that this province still belonged to the Palaeo-Arctic region in the Quaternary. However, animal diversity and abundance were greatly reduced and there remain only a few species specialized to the plateau environment, such as *Bos grunniens*, *Paoutholops hodgsoni*, *Yulpea dweeilata*, and *Altirara pareri* (SETQP 1982).

Vegetation development on the Tibet–Qinghai Plateau was under constant degeneration during the

Quaternary. In the Early and Middle Pleistocene there were still mixed forests of coniferous and broadleaf trees. From the Late Pleistocene on, the altitude of the plateau gradually exceeded the precipitation limit, and consequently only bushes and moorland remained except for a few coniferous trees in the valleys. Such a situation was in remarkable contrast to the southern slopes of the Himalayas where the oceanic influence dominated and, accompanying the uplift of the plateau, a flourishing vertical zonation of vegetation developed during the Quaternary.

The Northern Shelf province

This embraces the ancient shelves of the Bohai Sea (Chihli Gulf), Yellow Sea, and East China Sea north of Taiwan, including the inundated areas of North China during Quaternary marine ingressions.

In its northern part, the Bohai Sea, Yellow Sea, and surrounding area experienced frequent transgressions during the Quaternary. The areas surrounding Bohai underwent eight such episodes in the Quaternary, during which there became established a warm temperate brackish foraminiferan fauna—the Palaeo-Bohai foram assemblage—with nearly 100 species (Lin 1982; Wang *et al.* 1985). The most abundant species are *Ammonia beccarii*, *A. tepida*, and *Spiroloculina communis*. There were, in addition, occasionally thermophilous elements such as *Asterotalia* and *Pseudorotalia*, and endemic elements such as *Stomoloculina multiangula*. This assemblage occurs repeatedly in the sequence and appears as far westward as 100 km into the North China Plain accompanying the transgression. Marine bivalves of the northern part of this province were dominated by the *Potamocorbula* assemblage, which was a widespread group in the temperate zone of the northern hemisphere, adapted to clastic substrates in shallow seas (*c.* 20–60 m in depth) (Huang Baoren 1985).

In the southern part, the foram diversity of the ancient East China Sea reached nearly 200 species (Lin 1982), twice that in the northern part. In addition to the taxa of the northern part, the forams here included diversified littoral–neritic elements and relatively more deep-water members, such as *Bulimina* and *Uvigerina*, and a few planktonic forams, such as *Globigerina* and *Globorotalia*. Thermophylous *Pseudorotalia* and *Asterotalia* occur more frequently here than in the northern part, and *Globorotalia tumida* made its appearance here, being typical of tropical–subtropical water associated with the Kuroshima Current. Marine bivalves were dominated by the *Paphia* (*Paratapes*) *undulata–Timoclea minuta* assemblage, together with abundant oysters, which could build up coquina bars in river outlets. This bivalve assemblage belongs to the mixed South and North China type (Huang Baoyu 1982).

13.2.2 THE SOUTHERN CHINA REGION

This region covers areas south of the Qinling–Dabie line and east of the eastern margin of the Tibet–Qinghai Plateau, including most parts of Yangtze, south-eastern and south-western China and the Hengduan Mountains, and shelf areas south of the Taiwan strait. It belongs to the tropical–subtropical zone and is the main distribution area of Quaternary laterites in China. Its Pliocene Hipparion fauna was similar to the synchronous one of the Indian subcontinent while somewhat different from the temperate Palaeo-Arctic *Hipparion* fauna. This region is subdivided into the South China province, the Hengduan Mountains province, and the Southern Shelf province.

The South China province

This province consists of Yangtze, south-eastern China and Lingnan (Guangdong and Guangxi).

The earliest Quaternary mammal fauna of this province is called Yuanmou fauna. Dated back to 3.1 ma, this fauna contains many Pliocene elements, some elements of the Oriental realm, such as *Hystrix*, *Viverricula malaccensis fossilis*, *Cynailurus*, *Hyaena*, *Rhinoceros*, *Rusa*, and *Bibos*, and a few taxa common to both South and North China, such as *Microtus*, *Panthera tigris*, *Vulpes* and *Bos*. As a whole, the Early Pleistocene mammals of South China show a certain differentiation, though not distinctive, from the synchronous ones of North China in its Oriental aspect.

Slightly later, there emerged in the South China region the *Ailuropoda–Stegodon* fauna, with distinctive Oriental aspects (Chow 1964; Li 1981; Ji 1985). The principal elements of this fauna include *Ailuropoda*, *Stegodon*, *Tapirus*, *Rhinoceros sinensis*, *Sus*, *Bibos gaurus grangeri*, *Gigantopithecus blacki*, *Rhinopithecus*, and *Hylobates*, the last three being primates. Their distribution extends from South China eastwards to Taiwan, Japan, and the Philippines. Several temporal assemblages can be subdivided within this fauna. The early assemblage, as a rule, had *Gigantopithecus blacki* as its most important element, for example the *G. blacki* assemblage of Liuchen, Guangxi. In the middle Middle Pleistocene assemblage or the *Ailuropoda–Stegodon* fauna (*sensu stricto*), there was no *G. blacki*; an example is the Yanjinggouo assemblage of Wanxian County, Sichuan. The Late Pleistocene assemblage was usually accompanied by *Homo* species. The general trends seen in this fauna from the Early to the Late Pleistocene were the increase in size of the main elements (such as *Ailuropoda*, *G. blacki*, and *Tapirus*) and the reduction in their area of distribution, which reflected temperature fluctuations of Quaternary climate in the South China province. *Ailuropoda* is now restricted to mountainous areas bordering Sichuan, Gansu, and Shaanxi; elephants only survived within Chinese territory in the Xishuanbanna area of southern Yunnan, and *Tapirus* has migrated to equatorial regions. Quaternary terrestrial molluscs of this province are represented by those adapted to humid, hot habitats, such as *Cyclophorus*, *Macrochlamys*, *Hemiphaedusa*, and *Vitrea* (Chen *et al.* 1982).

The Quaternary flora of the South China province was inherited from the Pliocene tropical–subtropical flora without essential change. The Yangtze region was dominated by mixed forests of subtropical evergreen broadleaf and deciduous broadleaf trees. In the Nanning–Guangzhou area of Guangdong and Guangxi there were tropical–subtropical evergreen broadleaf trees, whereas Hainan Island and archipelagos in the South China Sea were occupied by tropical rainforests. However, on the Yunnan–Guizhou Plateau and in mountainous areas of Yangtze and south-eastern China, the percentage of temperate coniferous forest increased during the Quaternary ice ages. For example, during the Dali ice age, there was a temporary occurrence of *Abies–Picea* forest in Lichuan (western Hubei), Tianmushan (Zhejiang), and a few localities in Guizhou (Kong 1977). Even further southward, in Yuanmou (Yunnan), together with the dominance of tropical–subtropical Early Pleistocene palynomorphs, such as *Magnolia*, *Stauntonia*, and Celastraceae, Moraceae, Myrtaceae, and Palmae, there was a period showing a percentage increase of the mountainous coniferous elements *Abies* and *Picea* (Sun 1984). These phenomena, like those in the faunas, demonstrate the temperature fluctuations of Quaternary climate in the South China province, which, overall, were on a much smaller scale than those of the North China Province.

The Hengduan Mountains province

This province covers the Hengduan Mountains bordering eastern Tibet, western Sichuan, and western Yunnan, as well as the Tibetan areas south of Lhunzhub, including Yadong.

This region was marked by great altitude differences and a strong oceanic influence brought about by the south-western monsoon from the Indian Ocean, causing its humid climate. The vegetation types had not experienced significant changes since the Tertiary, the most conspicuous aspect being their vertical zonation. Along the southern slopes of the Himalayas, the zonation in ascending order is: tropical forest (elevation below 1600 m) → mountainous evergreen broadleaf forest (elevation 1600–2500 m) mountainous warm temperate mixed forest of coniferous and broadleaf trees (elevation 2500–3100 m) → mountainous cool temperate coniferous forest (elevation 3100–3900) → subalpine cold bushes and moorlands (elevation 3900–4700) → alpine cold moorlands (elevation above 4700 m) (SETQP 1982). The situation is similar with the Hengduan Mountains region, within which, however, the percentage of Palaeo-Arctic animals and Panarctic plants increases at higher elevations due to the more northerly location. Such climatic and geomorphological diversity renders this province the reservoir of organic life in China. Quite a few relict plants, such as *Ginkgo*, *Pseudolarix*, Eucommiaceae, and Sapindaceae took refuge here when the cold climate invaded, and then flourished and expanded outward when the climate turned warm again. This province also became the habitat of some precious and protected Oriental animals, such as *Rhinopithecus roxellanae*, *Ailurus fulgens*, *Hylobates*, and *Nycticebus coucang* (ECPGAS 1979, 1983).

Intermigration between the southern and northern mammals of eastern China had already begun in its western parts during the late Early Pleistocene. About 0.8 ma ago, the major elements of the southern or *Ailuropoda–Stegodon* fauna, *Stegodon* and *Tapirus*, reached to the vicinity of Lantian County, Shaanxi, about 35°N, as discovered in the Gongwangling fauna (Xue 1985). Between 0.46 and 0.23 ma ago, the main arena of intermigration shifted eastward, with *Ailuropoda* reaching Zhoukoudian to the south-west of Beijing, 39°48'N and becoming a member of the Zhoukoudian fauna (Wu *et al.* 1985). During the Holocene warming episode 4000–5000 years ago, elephants and tapirs crossed Qinling once more and appeared in Anyang, Henan at the middle reaches of the Yellow River (Zhou 1973; Yang 1985). The eastward shift, evidently effected by the upheaval of the Tibet–Qinghai Plateau and Qinling, ultimately led to freer intermigration and intermixture between southern and northern mammals along the Huihe River drainage, which was more evident in the Middle and Late Pleistocene than in the Early Pleistocene (Pei 1957).

The South China province also ranks first among the Chinese regions for primate fossils. *Ramapithecus lufengensis* and *Sivaopithecus lufuengensis* have been discovered in Lufeng County, Yunnan (Qi 1979). In Yuanmou County, Yunnan, *Homo erectus yuanmouensis* was found initially (Zhou and Zhang 1984), followed by the discovery of another *Homo erectus* even older than *yuanmouensis*, (Nengren 1987). A survey of palaeo-anthropological localities and horizons of China shows that anthropoids and early hominids appeared more to the west, while middle and late hominids appeared more to the east, so the anthropological development also reveals an eastward shift.

The Southern Shelf province

This province, corresponding to the ancient South China Sea, covers the shelf area south of the Taiwan Strait. During the Quaternary in this area there thrived tropical–subtropical littoral–neritic biotas, consisting of forams, corals, bivalves, and mangrove forests. The highly diversified forams amounted to more than 300 species and are called the Palaeo-South China Sea foram assemblage (Lin 1982). Among them are large-sized forms, such as *Amphistegina* and *Peneroplis*, small benthics, such as *Asterotalia* and Pseudorotalia, whose quantities far exceeded those in the Northern Shelf province, as well as relatively abundant plankton, such as *Globigerina*, *Globigerinoides*, *Orbulina*, and *Globorotalia*. *Globorotalia truncatulinoides* occurs in massive quantities in the Pleistocene of Taiwan, and its first appearance is taken as the beginning of that epoch (Huang 1962). Bivalves of this province are dominated by the *Tridacna–Hipopus* assemblage, signifying a warm-water fauna coexisting with tropical coral reefs (Huang 1982). Of course, the composition of the Palaeo-South China Sea fauna, altered with palaeoclimatic changes.

13.3 BRIEF CONCLUSIONS

In the marine regime of the Quaternary of China, a biogeographical provincialization between the southern and northern shelves has not been dramatically changed since the Pleistocene. The terrestrial biotic provincialization between north and south was only initially displayed in the Early Pleistocene. In the Middle Pleistocene the South and North China regions became firmly established and a diversity distinction between eastern and western China emerged. This situation replaced the Tertiary pattern of biogeography, which featured similarity between north and south and intermigration between east and west. The pronounced upheaval of the Tibet-Qinghai Plateau during the Middle and Late Pleistocene not only interrupted the latitudinal climatic and biotic zonation in western China, but also caused the aridity and coldness there. As a consequence, the biotas of western China degenerated. In north-western China, survivors tended to be drought resistant; in the Tibet–Qinghai Plateau, they tended to be psychrophilous; while most organisms shifted eastward, and eastern China became the intermigrating and intermixing arena of southern and northern biotas. The development of mankind in China also experienced an eastward shift and associated northward radiation under the influence of the Tibet–Qinghai uplift, which had an important bearing on the cultural, economics, and social development of Chinese society.

REFERENCES

Cao Boxun (1988). Quaternary climate of China. In *Palaeoclimatology*, (ed. Zhao Xiwen), Chapter 17, pp. 1–24 China University of Geoscience Press, Wuhan (in Chinese).

Cao Boxun, Zhao Boyi, Liu Shiyun, and Guan Kangnian (1966). A preliminary research on the Cenozoic of Youhe River, Weinan, Shaanxi. In *Symposium of Cenozoic Field Meeting, Lantian, Shaanxi*, pp. 37–47. Science Press, Beijing (in Chinese).

Chen Deniu, Lu Yanchou, and An Zhisheng (1982). The *Cathaica* fossil assemblage in the loess strata and their significance. In *Symposium of the 3rd National Conference on Quaternary*, pp. 7–15. Science Press, Beijing (in Chinese with English abstract).

Chow Benshun (1978). The distribution of the woolly *Rhinoceros* and woolly mammoth. *Vertebrata PalAsiatica* **16**, (1), 45–57 (in Chinese with English abstract).

Chow Mingchen (1964). The Quaternary fauna province change in China. *Dongwuxue Zazhi* (*Journal of Zoology*) **6**, (6), 274–8 (in Chinese).

Cui Zhijiu and Xie Youyu (1985). On Late Pleistocene periglacial environment in the north part of China. *Quaternaria Sinica* **6**, (2), 115–19 (in Chinese with English abstract).

ECPGAS (Editorial Committee of Physical Geography, Academia Sinica) (1979). *Physical geography of China* (*Zoogeography*). Science Press, Beijing (in Chinese).

ECPGAS (Editorial Committee of Physical Geography, Academia Sinica) (1983). *Physical geography of China* (*Phytogeography*). Science Press, Beijing (in Chinese).

Guo Zhugang (1987). Mammal fossils found at Wuquanshan Conglomerate. *Vertebrata PalAsiatica* **25**, (2), 70–9 (in Chinese).

Han Shuti and Ye Wei (1986). Studies of the Late Pleistocene strata and the sedimentary environment in Cangfanggou near Urumuqi. *Quaternaria Sinica* **7**, (2), 87–95 (in Chinese with English abstract).

Hong Li (1981). The outline of Quaternary strata of Xinjiang. In *Symposium of Quaternary geology and Glacial Geology of Xinjiang*, pp. 36–54. Xinjiang People's Press, Urumqi (in Chinese).

Hsu Ren, Kong Zhouchen, Sun Xiangjun, Tao Junrong, and Du Naiqiu (1973). Quaternary palaeobotanical studies on the Mount Jomolangma Region. In *Report on the Scientific Expedition in Mount Jomolangma Area (1966–1968), Quaternary Geology*, pp. 30–63. Science Press, Beijing (in Chinese).

Huang Baoren (1985). Quaternary ostracod biogeographical province in China. *Quaternaria Sinica* **6**, (2), 87–8 (in Chinese with English abstract).

Huang Baoyu (1982). Quaternary lamellibranch fauna provinces in East China and their characteristics. In *Quaternary geology and environment of China*. China Ocean Press, Beijing (in Chinese).

Huang Dunyou (1963). Study of planktonic foraminifera fossils from the No. 3 drilling well, Bokong, Yunlin County, Taiwan. *Petroleum Geology of Taiwan* **2**, 153–82 (in Chinese)

Huang Wanpo (1980). The sites of Quaternary mammal faunas in Xizang. In *Paleontology of Xizang*, Vol. I, pp. 35–41. Geological Publishing House, Beijing (in Chinese with English abstract).

Huang Wanpo (1986). Quaternary mammalian faunas and climate variations in East China. *Quaternaria Sinica* **7**, (2), 54–60 (in Chinese).

Huang Wanpo and Huang Cixuan (1985). Mammal fossils and sporopollen at Hexian Man locality and their composition. In *Symposium of National Conference on Quaternary Glaciers and Periglacial Deposits*, pp. 180–2. Science Press, Beijing (in Chinese).

IBAS (Institute of Botany, Academia Sinica) *et al.* (1966). Study on Cenozoic palaeontology from Lantian Area, Shaanxi. In *Symposium of Cenozoic Field Meeting, Lantian, Shaanxi*, pp. 157–82. Science Press, Beijing (in Chinese).

Ji Hongxiang (1985). Zoogeographical distribution and living environment of Quaternary mammalian faunas in China. In *Symposium of National Conference on Quaternary Glaciers and Periglacial Deposits*, pp. 138–42. Science Press, Beijing (in Chinese with English abstract).

Kong Zhaochen (1977). Quaternary plants discovered in east Yunnan and west Guizhou and preliminary approach on palaeofloras and palaeoclimate. In *Symposium on Quaternary glacial geology of China*. Geological Publishing House (in Chinese).

Li Baojing (1988). A preliminary research on the Pulu section in the southern margin of Takalamakan Desert. *Kexue Tongbao* **33**, (2), 140–4 (in Chinese).

Li Jiejun, Wen Shichang, Zhang Qinsong, Wang Fubao, Zhen Benxing, and Li Bingyuan (1979). A discussion on the period, amplitude and type of uplift of the Qinghai–Xizang Plateau. *Scientia Sinica* **22**, (1), 1314–28 (in Chinese with English abstract).

Li Yanxian (1981). On the subdivision and evolution of the Quaternary mammalian faunas of South China. *Vertebrata PalAsiatica* **19**, (1), 67–75 (in Chinese with English abstract).

Lin Jingxin (1982). Quaternary foraminifera faunas of China. In *Symposium of the 3rd National Conference on the Quaternary*, pp. 31–3. Science Press (in Chinese).

Liu Tungsheng (1985). *Loess and environment*. Science Press, Beijing (in Chinese).

Nengren (1987). *Yunnan—one of the human race source provinces*. Zhong Guo Dizi Bao.

Pei, W. C. (1957). The zoogeographical divisions of Quaternary mammalian fauna in China. *Vertebrata PalAsiatica* **1** (1), 9–24 (in Chinese).

Qi Guoqin (1979). Pliocene mammalian faunas of Lu Feng, Yunnan, *Vertebrata PalAsiatica* **17**, (1), 14–22 (in Chinese with English abstract).

Qiu Shanwen, *et al.* (1985). Basic characters of natural environment on the Northeast Plain in Pleistocene. In *Symposium of National Conference on Quaternary Glaciers and Periglacial deposits*, pp. 208–11. Science press, Beijing (in Chinese).

SETQP (Scientific Expedition of the Qinghai–Xizang Plateau) (1982). *The physical geography of Xizang*. Science Press, Beijing (in Chinese with English abstract).

Sun Jianzhong *et al.* (1985). Paleoenvironment of the last glacial stage in Northeast China. *Quaternaria Sinica* **6**(1), 82–8 (in Chinese)

Sun Mongrung (1984). Preliminary palynological analysis of the data from Yuanmou Basin. In *Yuamou Man*, (ed. Zhou Guoxing). Yunnan People's Press, Kunming (in Chinese).

Tong Boling (1982). Some features and influencial elements of permafrost in Qinghai–Xizang Plateau. In *Symposium of Permafrost Studies in Qinghai–Xizang Plateau*, pp. 70–9. Science Press, Beijing (in Chinese).

Wang Pinxian, Min Qiubao, and Bian Yunhua (1985). Quaternary stratigraphy of marine-continental transitional facies in East China. *Quaternaria Sinica* **6**, (1), 35–48 (in Chinese with English abstract).

Wu Rukang, *et al.* (1985). *Multi-disciplinary study of the Peking Man site at Zhoukoudian*. Science Press, Beijing (in Chinese with English abstract).

Xue Xiangxu (1985). Mammalian faunas in the loess distributive province of China. In *The new developments of loess studies in China*, pp. 115–33. Shaanxi People's Press (in Chinese).

Yang Huairen (1985). *Quaternary geology*. Higher education Press, Beijing (in Chinese).

Yuan Fuli and Du Hengjian (1984). *Cenozoic biostratigraphy of China*. Geological Publishing House (in Chinese).

Zeng Benxing (1976). Quaternary glaciation on the Mount Qomolangma Area. In *Report on the Scientific Expedition (1966–1968) in the Mount Qomolangma Area, Quaternary Geology*. Science Press, Beijing (in Chinese).

Zhang Linyuan (1984). Quaternary ice ages and evolution of monsoon climate in China. In *Evolution of mountain glaciers and the Quaternary glaciation*, pp. 30–4. Lanzhou University Press (in Chinese).

Zhang Zhenhong, Zou Baoku, and Zhang Likai (1980). The discovery of fossil mammals at Anping, Liaoyang. *Vertebrata PalAsiatica* 28, 154–61 (in Chinese with English abstract).

Zhou Guanxing and Zhang Xingyong (1984). *Yuanmou Man*. Yunnan People's Press, Kunming (in Chinese).

Zhu Kechen (1973). A preliminary study on the climatic change fluctuation during the last 5,000 years in China. *Scientia Sinica* **2**, 168–88 (in Chinese).

14 IMPLICATIONS OF THE PALAEOBIOGEOGRAPHICAL PROVINCIALIZATION OF CHINA FOR PLATE TECTONICS

Yin Hongfu

14.1 FOUR MAJOR GEOHISTORICAL–TECTONIC UNITS CONTROLLING THE PALAEOBIOGEOGRAPHICAL PROVINCIALIZATION OF CHINA

The Central Asia–Mongolia palaeo-ocean separated Siberia and China proper during the Late Proterozoic and Palaeozoic, and had mainly closed up by the end of Permian, although a welding of the Mongolo– Okhotsk suture was not completed until Late Jurassic to Early Cretaceous times (Kosigen and Parfenov 1981). It included the vast northern Chinese borderland, Xingan–Junggar province. The Xar Moron River–Gaxunnur Lake–Ebinur Lake subduction zone is one of the major remains of this vast ocean.

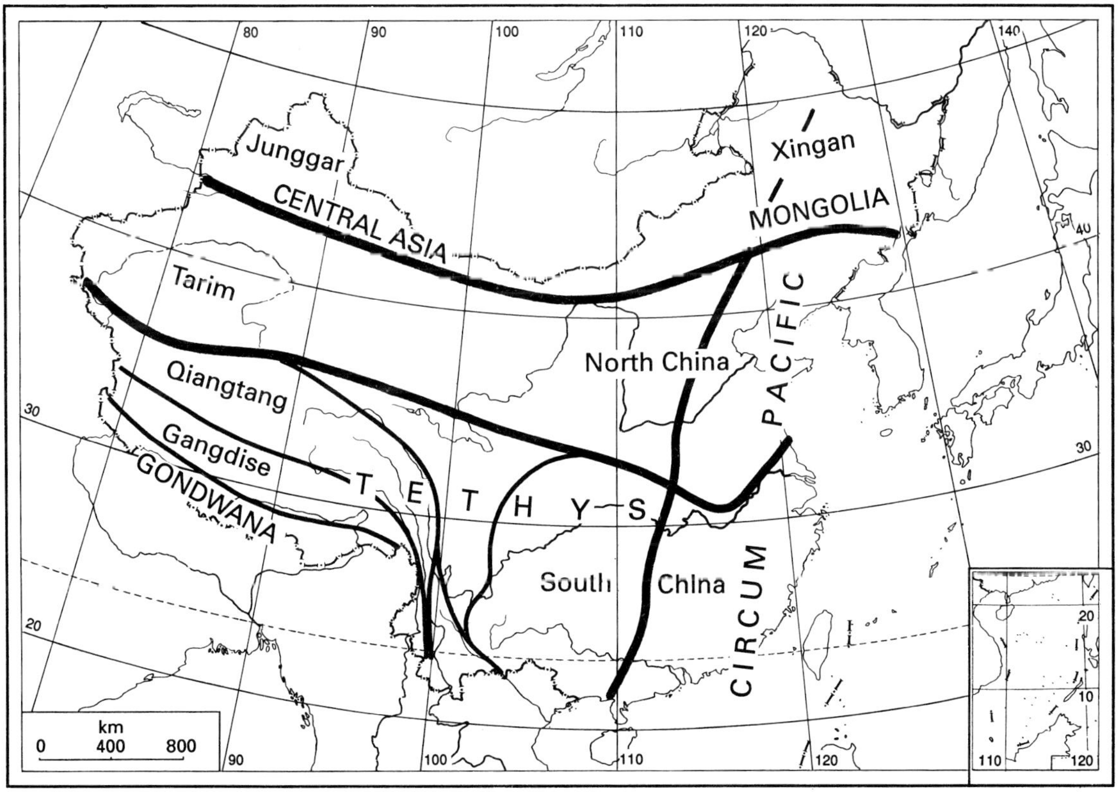

Fig. 14.1 Major geohistorical–tectonic units controlling the palaeobiogeographical provincialization of China.

Most of China proper belongs to the tropical–subtropical Tethyan (*sensu lato*) Ocean region for a large part of the Phanerozoic. Tarim, North and South China were microplates drifting in Tethys (*sensu lato*) since at least the Silurian. Tethys (*sensu lato*) is by far the most significant essential unit in the palaeobiogeography of China.

Gondwana is now restricted to the south of the Yarlung Zangbo subduction zone. However, it is likely to have played a more important role in the geological past. Qiangtang (including Tanggula), Gangdise (including Nianqingtanggula), South China, Tarim, and maybe even North China were probably once within its realm. These pieces have progressively been sutured to Eurasia throughout the Phanerozoic.

The Pacific Ocean (*sensu lato*) had a strong influence on Chinese palaeobiogeography after the Late Triassic, resulting in the formation of a Mesozoic–Cenozoic biogeographical belt superimposed on pre-existing units.

14.1.1 TETHYS

The concept of Tethys

Sengör (1984) has made a general review of the Tethys concept. Names cited in literatures include Proto-Tethys, Palaeo-Tethys, Meso-Tethys or Mesogee, and Neo-Tethys. It seems reasonably safe to say that a Proto-Tethys existed in the Devonian and Early Carboniferous (Boucot and Gray 1986; Talent *et al.* 1986). The Tethys (*sensu lato*) is now applicable to a wide tropical sea or ocean that existed between a southern (Gondwana) and a northern (Siberia or Eurasia) landmass. In this sense we have extended the usage of Proto-Tethys back to the Silurian, although there are few precedents. This is because, from our viewpoint, the Silurian world was already climatically differentiated. In the southern hemisphere the Malvinokaffric realm provides good evidence for a moderately high global climatic gradient (Boucot 1990). Its northern counterpart is not so clear because Siberia and its shallow benthic communities had not reached such high latitudes as Gondwana. Still, we have recognized a *Tuvaella* biota as non-tropical (for discussion see Chapter 1, p. 7 and Chapter 5, p. 104). In between these two there existed a low-latitude ocean (Scotese and McKerrow 1990), inhabited by typical tropical biota, for example in Yangtze. It is the latter that we have assigned to the Proto-Tethys.

Various usages have been applied to Palaeo-, Meso-, and Neo-Tethys. This book follows the common usage, treating Palaeo-Tethys as existing between Laurasia and Gondwana during the Carboniferous and Permian. Beginning from the later Permian and during the Mesozoic, Turkey, Iran, Afghanistan, North Tibet, and other microplates disintegrated from Gondwana to form the intermediate Cimmerian Continent (Sengör 1984) which delineated Palaeo-Tethys to its north and Meso-Tethys to its south. Palaeo-Tethys shrank and finally disappeared in the Triassic due to the successive accretion of Cimmerian blocks onto Eurasia. Neo-Tethys is used sometimes as a synonym of Meso-Tethys, and sometimes to represent the remnant 'Mesogee' in the Cenozoic. Moreover, there were also Para-Tethys and southern Para-Tethys, the former existed to the north of Tethys during the Late Mesozoic and Cenozoic, roughly along the line from central Europe–Don River–North Caspian–Central Asia to western Tarim; the latter corresponds to the Siwalik depression of India. In this book we have used Palaeo-Tethys for the tropical realm in the Carboniferous and Permian and Tethys for its equivalent in the Mesozoic and Cenozoic, including both Meso-Tethys and Neo-Tethys.

The contents of Tethys kept changing throughout geological history, due to plate movements. There are two Chinese models of plate movement in Tethys. One is the one-way conveyor model in which the Gondwanan margins kept breaking away and spreading, compelling Tethys—the conveyor itself—to subduct underneath Eurasia and join its loading plates onto this supercontinent. The other is the accordion model, according to which the two supercontinents, as well as microplates within Tethys, migrated to and fro, not directionally. Most Chinese geologists favour a basically conveyor-type, but partly accordion-type model. Examples of the accordion-type model, following which plates rifted from Eurasia and moved southward, are riftogenesis and southward movement of the South China Sea (Miocene) and Andaman Sea (Recent) (Tapponier *et al.* 1982), possible southward migration of Qiangtang during the Permo-Triassic, and possible movement of Junggar away from Siberia during the Ordovician, as noted later in this chapter.

In China, the Proto-Tethys was more or less defined during the Silurian and Devonian, between Gondwana and southern Kazakhstan–North China (Figs 5.1, 6.1). In the Carboniferous and Permian, the dispersion of Iran, Afghanistan, Qiangtang, Indochina, and South China from Gondwanaland delineated the Palaeo-Tethys to their north (Figs 7.1, 8.1), with the Qinling Geosyncline (between South and North China) and Jinshajiang River–Red River Geosyncline (between South China and Indochina) as two eastward seaways connecting with Panthalassa. Palaeo-Tethys was eliminated during the Triassic Indosinian Orogeny. Meso-Tethys bifurcated in China, the main branch being represented by the Yarlung Zangbo suture formed in the Triassic, and a second one by the Bangong Lake–Nujiang River suture between Qiangtang and Gangdise, formed at the end of the Jurassic.

Palaeobiogeographical significance of Tethys

Albeit wedge-shaped towards the west, Tethys was an ocean extending roughly parallel to latitude, with its northern margin located near the equator. This situation greatly enhanced the latitudinal distribution of palaeobiogeographical realms after the Carboniferous. Generally four realms can be subdivided:

Northern or Eurasian Tethys

Considered together with its leading microplates, such as Armorica, the Central Massif of France, Bohemia, Panon, South, and North China. Usually it exhibited tropical biotas which may be divided into a western Hercynian–Mediterranean region and an eastern Cathaysian region, which was especially conspicuous during the Permian and Triassic.

Southern or Gondwanan Tethys

This included Madgascar, the Persian Gulf, the Salt Range, Kashmir, the Himalayas, and parts of Indonesia and Australia. Usually it belonged to the temperate, or at most subtropical, zone.

The intermediate Cimmeride microplates

These included Qiangtang, Gangdise, Helmand, Central Iran, etc. During their northward migration toward Eurasia, they experienced changing climate and transitional biotic characters, from temperate to tropical.

The abyssal part of Tethys

This was mostly subducted except for a few localities near sutures. Due to the frequent disintegration and migration of microplates from Gondwana toward Eurasia, and partly vice versa, Tethys exhibited a complex interlacing pattern of microplates and micro-oceanic basins throughout its whole existence. This situation greatly facilitated biotic migration and fossil transportation, by island hopping, Noah's Arks, or 'Viking Funeral Ships', terrain accretion, and the formation of olistoliths and melanges, so geographically or temporally mixed biotas became commonplace. This, in turn, has given rise to various interpretations with regard to Tethyan plate movement and climate development, which are now difficult but fascinating 'hot spots' in the palaeobiogeographical research of China.

The shallow Tethys concept (Sonnenfeld 1981) has been carefully examined, but we have found the concept adopted herein more appropriate in interpreting our data.

14.1.2 GONDWANA

The Gondwana biotic realm was formed after the late Early Palaeozoic and was best defined during the Late Palaeozoic and Triassic. It has always been a first-order palaeobiogeographical unit throughout its existence, representing non-tropical biota. It has been called the Malvinokaffric realm for Silurian and Devonian times, representing southern temperate marine biota. Later, it included the Permo-Carboniferous *Glossopteris* flora and glacio-marine fauna, signifying cool temperate to cold subantarctic and Antarctic realms, and also the Triassic *Dicroidium* flora and Gondwanan Tethyan fauna, denoting austral biotas grading from temperate to subtropical. Gondwanaland underwent considerable disintegration after the Jurassic and hence gradually lost its biotic distinctiveness. The modern Australian terrestrial and circum-Antarctic marine biotas may be regarded as remnants of the Gondwanan biotic realm. It is to be noted, however, that the Gondwanan realm does not completely coincide with Gondwanaland. Microplates such as South China may once have belonged to the Gondwanan realm but were not necessarily merged with Gondwanaland.

Chinese areas supposed to have been once part of the Gondwanan realm include Qiangtang (including Tanggula), Gangdise (including Niangqingtanggula), and South China. Tarim is also suggested by some authors because it closely resembles South China in Sinian tillites and Cambro-Ordovician biotas. North China may also possibly have departed from Gondwana in earlier geohistory.

Qiangtang and Gangdise

The suggested northern margin of the Gondwanan realm on Qinghai–Tibet plateau (Fig. 14.2) has been progressively shifted northward during the past two decades. Glacio-marine fauna and *Glossopteris* flora were first discovered south of the Yarlung Zangbo suture. Later, the Bangong County–Nujiang suture was widely recognized and in Xainza, an area between the two sutures, Palaeozoic strata and fauna corresponding to those from the Qomolangma area were discovered. Hence the northern margin was moved to the Bangong County–Nujiang suture. Still later, Gondwanan facies were recognized to the north and east of that suture, for example Duoma, Ritu district south of Longmuco (*Eurydesma*, *Oriocrassatella*, and *Subansiria*); Gangmacuo, north of Gerze (*Cyathaxonia* and *Amplexus*); and Baoshan, western Yunnan (*Eurydesma*). To the west of Ritu, Norin (1946) established the Horpa-tso Group which has yielded typical diamictites with a cold-water fauna. In all these localities, plus a few more areas, diamictites have been found that are regarded as tilloids or Blaini-type deposits, although some recent researches considered these pebble-bearing slates to be deposits of massive subaqueous gravity flows (Yin and Wen 1988). A considerably deformed ophiolites belt has been recognized along the Longmuco–Shuanghu–Lancanjiang line, equally dividing Qiangtang, so the northern margin was transferred northward to this line (Fan 1985; Huang Zhixun 1989).

This, however, is not a final answer. Huang *et al.* (1984) have insisted that a Hohxil–Jinshajiang suture is by far the most significant line in geohistory demarcating Gondwanan and tropical Tethys. This view is supported by palaeomagnetic data. Table 14.2 clearly shows that from the Early Palaeozoic to the Triassic, areas east of the Jinshajiang tectonic zone were always near the equator, while the Tibet–Qinghai regions south-west of this zone were around 20–30°S. The Triassic latitudinal difference of 30–40° should imply an oceanic barrier between the two. Rao (1989) reported a cool-water solitary coral fauna including *Tachylasma*, *Bradyphyllum*, and *Timorphyllum* from the Permo-Carboniferous of the Qamdo area, and *Atomodesma* from the Permian of Jinggu and Lancang. These localities overstep the Lungmu County line, so the Hohxil–Jinshajiang suture is again called to attention. One paper even reported the discovery of a cold-water fauna in Tsaidam. It should be noted that the northern margin of Gondwana is basically delineated according to the Late Carboniferous–Early Permian situation. The Middle Permian *Verbeekina–Iranophyllum* fauna (tropical Tethyan) and Late Permian Cathaysian flora overlapped the cold-water fauna from Qiangtang all along to the southern side of Yarlung Zangbo. This has been explained as being due to both the spreading of the tropical climate and the northward movement of the relevant microplates. In the Mesozoic, Qiangtang and

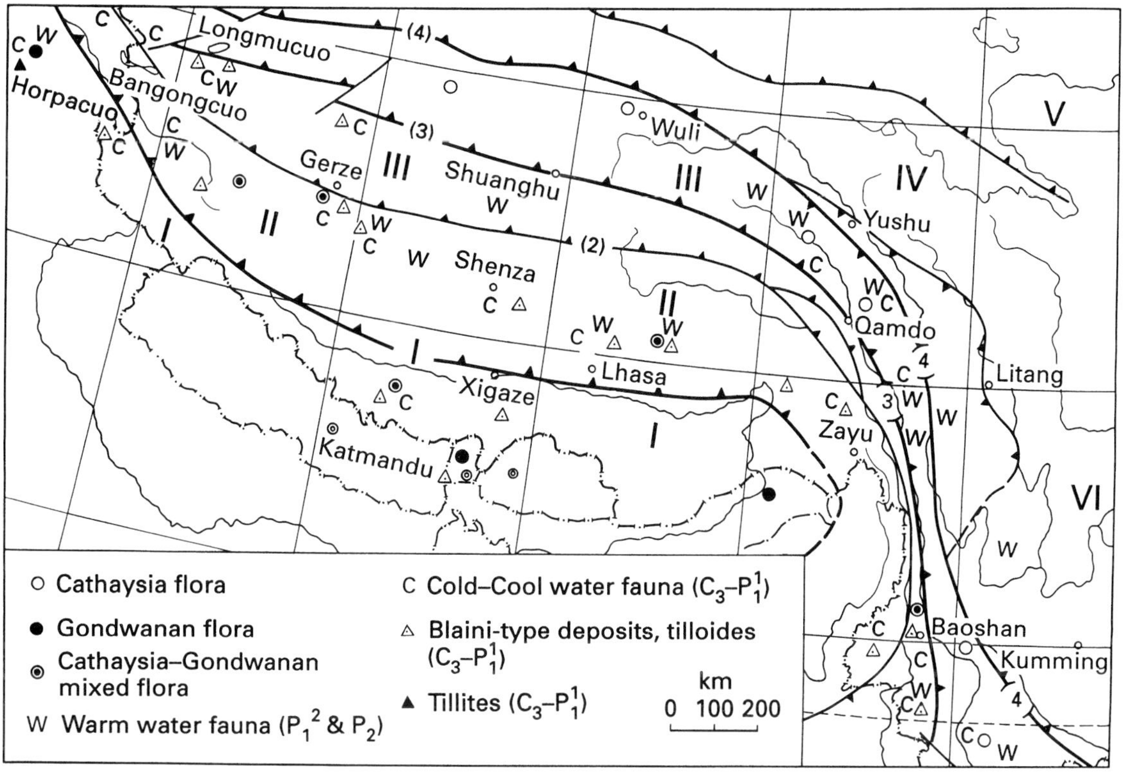

Fig. 14.2 Suggested northern border of Gondwana in view of palaeobiogeography (compiled from Huang and Chen 1987; Yin and Wen 1988; Rao 1989; and other references). Tectonic units: I, Himalaya; II, Gangdise–Nianqingtanggula; III, Qiangtang–Tanggula; IV, Bayan Har; V, Qinling; VI, South China. Sutures: 1, Yarlung Zangbo River; 2, Bangong Lake–Nujiang River; 3, Longmuco Lake–Shuanghu–Lancanjiang River; 4, Hohxil Mountains–Jinshajiang River–Red River.

Gangdise moved successively into tropical Tethys, so the northern margin has to be drawn first along Bangong County–Nujiang, then Yarlung Zangbo (Table 14.1).

Thus delineated, a large part of the Tibet–Qinghai plateau is attributed to the Gondwanan realm or, more strictly, the Gondwanan Tethyan region, during the Late Palaeozoic and Mesozoic. The biogeographical differences between Cathaysian (eastern Eurasian) Tethys and Gondwanan Tethys, or their substitute names of the Early Palaeozoic, largely based on the Yangtze and Qomolangma biotic provinces, are shown in Table 14.2. This table reveals an initial differentiation in the Ordovician, presumably reflecting the departure of South China from Gondwana. Biotic differences strengthened after the Silurian, continued to the end of the Cretaceous, and disappeared in the Tertiary (Eocene), which was evidently due to collision of the Indian Plate.

Table 14.1 Palaeolatitudes of tectonic units on both sides of the Jinshajiang suture (after Rao 1989). The double line demarcates southern and northern palaeolatitudes.

Unit	West of Jinshajiang suture				East of Jinshajiang suture					
	Himalaya	Gangdise	Qiangtang	Yangtze margin	Yangtze					
Period	Nyalam Dingri Lhaze Gyangze Pagri	Doilung-deqen Lhasa Sumdo Baxoi Bomi Nagqu Shenza	Qamdo Tuoba Amdo Mangkang	Yanyuan Lijiang	Xiaguan	Panxi	Meigu	Emei	Qujing	Changsha
Tertiary	24.4°N									
Cretaceous	38°S	22.8°N	31.5°N			28.7°N		31.8°N		36.1°N
Jurassic	22.8°S	10°S	8.5°N		23.6°N	20.0°N	26.6°N	26.1°N		26.5°N
Triassic	24.1°S	28.0°S	17.4°S	30.0°N	14.0°N	16.7°N	19.3°N	15.9°N		23.0°N
Permian						5.4°S	4.5°S	6.5°S		1.6°S
Carboniferas	24.0°S	21.0°S								
Devonian	33.0°S								0.6°S	
Silurian		27.0°S							0.1°S	
Ordovician										
Cambrian									2.0°S	

Table 14.2 Comparison of biotic provincialization between Cathaysian Tethys (South China type) and Gondwanan Tethys

	Cathaysian Tethys (South China type)	Gondwanan Tethys
Tr	Provincialization disappeared	
K	Forams: *Orbitolina*	*Globotruncana*
	Marine bivalves: Hippuritidae	*Inoceramus* spp. (*I. xizangensis*)
	Non-marine bivalves: *Trigonioides*	No *Trigonioides*
	Plants: Eurasian Tethyan flora (*Weichselia, Onychiopsis*)	No Eurasian Tethyan flora
	Ammonites	Southern ammonite fauna (Wang Naiwen 1984)
J_{2-3}	Eurasian Tethyan fauna (*Birmirhynchia–Holcothyris*)	Ethiopian fauna (*Indogrammatodon*)
	Marine bivalves: *Liostrea birmanica*	*Buchia*
	Freshwater bivalves: *Eolamprotula–Psilunio* assemblage	No *E.–P.* assemblage
	Belemnites: rare	*Belemnopsis, Hibolites*
T_3–J_1	Provincialization not remarkable	
T_{1-2}	Coral–lithistid sponge–red algae (reef-building assemblage) (T_2)	No such assemblage
	Bivalves: *Claraia aurita–Eumorphotis inaequicostata* assemblage (T_1)	*Claraia decidens* assemblage
	T_2 ammonites: *Paraceratites–Progonoceratites* assemblage	*Keyserlingites* assemblage
	T_1 *Tirolites*	*Otoceras, Svalbardiceras*
	Plants: *Voltzia–Pleuromeia* flora	*Dicroidium* flora
P	Plants: *Gigantopteris* flora (*Gigantonoclea, Lobatannularia, Rhizomopsis*)	*Glossopteris* flora (*Raniganjia, Dizeugotheca, Dichotomopteris*)
P_2	Fusulinids: *Palaeofusulina*	Few
	Brachiopods: *Lyttonia–Cathaysia–Peltichia*	This assemblage not abundant
	Bivalves: *Tambanella–Hunanopecten*	*Atomodesma*
	Ammonites: *Pseudotirolites–Pleuronodoceras* assemblage	No such assemblage
P_1	Fusulinids: *Verbeekina–Neoschwagerina* assemblage	*Monodiexodina* (seen also along Tethyan margin)
	Corals: *Iranophyllum–Ipciphyllum* assemblage	Small solitary corals (*Lytvolasma, Wannerophyllum*)
	Brachiopods: *Tyloplecta yangtzeensis, Cathaysia*	*Costiferina, Calliomarginatia*
	Bivalves: *Gujocardita, Pernopecten*	*Eurydesma*
	Ammonites: *Shouchangoceras, Pseudohalorites*	*Uraloceras*

Table 14.2 *(cont.)*

	Cathaysian Tethys (South China type)		Gondwanan Tethys
C_{2-3}	Forams: *Fusulinella, Triticites*		Rare
	Corals: *Kepingophyllum, Caninia*		Small solitary corals (*Empodesma, Cyathaxonia*)
	Brachiopods: *Axolithophyllum, Choristites, Natannella*		*Stepanoviella, Attenuatella, Tregonotreta*
C_1	Fusulinids: *Eostaffella*		Rare
	Corals: *Kueichouphyllum Cystiphrentis, Pseudouralinia*		*Ekvasophylloides, Parahumboldtia Siphonophyllia*
	Brachiopods: *Gigantoproductus, Eochoristites, Martiniella*		*Syringothyris, Pseudosyrinx, Fusella*
D	Corals: Disphyllidae, etc (*Hexagonaria, Keriophyllum, Zaphrentis*)		Small solitary corals (*Metriophyllum*)
	Tentaculites		*Nowakia acnaria, Guerichia xizangensis*
	Ammonites: clemeniids		*Imitoceras* (South Tibet)
	D_1 pisces: *Polybranchiaspis–Galeaspis–Yunnanolepis* assemblage		
S	Corals: *Idiophyllum, Shensiphyllum*, Stauriidae		*Triplophyllum* (Qomolangma)
	Brachiopods: *Pentamerus, Nalivkinia*		
	Nautiloids: *Sichuanoceras, Yunnanoceras, Yichangoceras*		*Michelinoceras*, M. (*Kopaninoceras*)
	Trilobites: *Coronocephalus, Ptillillaenus*, etc.		*Leonaspis, Primaspis* (western Yunnan)
O		Initial differentiation	
	Trilobites: Yangtze type		Yangtze type
	Blastoids: unknown		Yes (western Yunnan)
	O_3 Corals: *Boreolasma*		*Boreolasma* unknown
	O_2 Nautiloids: Yangtze type		North China type (Qomolangma)
	O_1 *Brachiopods*: *Yangtzeella*		*Yangtzeella* unknown

Note: The Cathaysian Tethys and Gondwanan Tethys were established only since the Late Palaeozoic. The Early and Middle Palaeozoic data listed in Tables 14.2 and 14.3 should in fact belong to their predecessors—please refer to the relevant chapters for the specific names. Moreover, the boundary between the Cathaysian Tethys and the Gondwanan Tethys shifted in time along the Jingshajiang line, Bangong Lake–Nujiang line, and Yarlung Zangbo Line, for details of which please also refer to the relevant chapters.

South China

South China consists of two geotectonic units; the Yangtze Platform and south-eastern geosynclines. The Yangtze Platform is surrounded by geosynclinal zones but most of the northern and western geosynclines have been subducted. The south-eastern geosynclines may be regarded as better preserved parts of the surrounding zones. Accumulating evidence has shown that South China was once part of eastern Gondwana (Fig. 14.3).

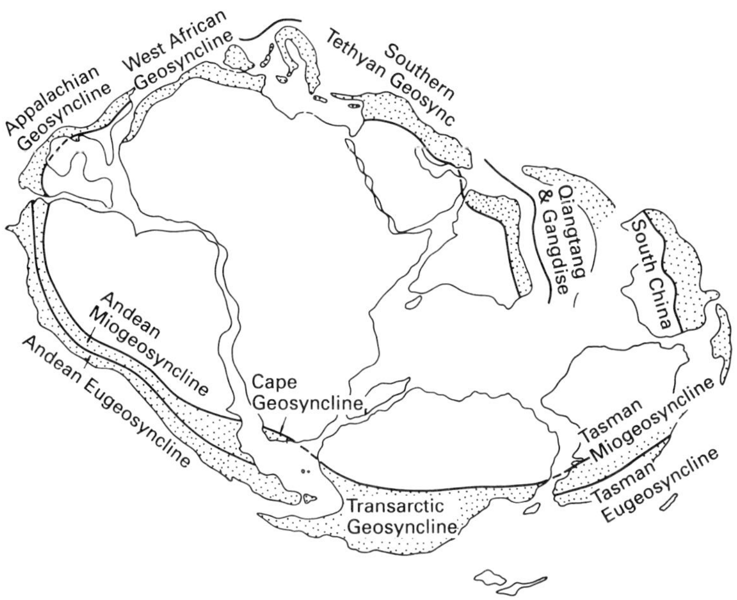

Fig. 14.3 Gondwanaland and its marginal geosynclines (after Seyfert and Sirkin 1973 adding Tibet–Qinghai and South China).

Tectonic and mineralogical analogy with Gondwana

The basement complex of the Yangtze Platform was formed during the Jingning Orogeny (850 ma), but the Central Sichuan Nucleus may be older. Similarly, the basement complex of the Central Australia Carpentarian Group formed at about the same time (867 ma). Both the Yangtze and Australia Platforms are accompanied by Caledonian orogenic belts—the south-eastern China and Tasmanian Geosynclines or Lablan Fold Belt. The stratigraphic sequence and type of phosphorus mines in Damao, Hainan Island are almost identical with those in Australia. The ages of and horizons of Early Palaeozoic copper and phosphorus minerogenesis closely correspond in Yangtze, Australia, Africa, and India.

Sedimentary covers

The earliest sedimentary covers of both platforms, Sinian and Adelaidean, have similar sequences. They begin with the clastic Liantuo and Callana Formations at about 850 ma, respectively; followed by tillites, the Nantuo Tillites (728 ma) and Lower Umberatana Tillites (738 ma); and ending up with the Ediacara fauna (570–640 ma) of the Middle Dengyin Formation and Pound quartzite.

Analogous palaeobiogeographical aspects during the Cambrian and Ordovician

Yangtze and Central Australia yield similar Cambrian faunas. The Early Cambrian trilobite *Redlichina* group combined Australia, Antarctica, South and North China and South Asia (Salt Range) into one realm. Many Middle and Late Cambrian agnostids and other trilobites (for example *Histridura* and *Balangia*) are common to both. Jell (1974) clustered Middle Cambrian trilobite faunas of China, Siberia, northern India, Australia, New Zealand, and Antarctica into a single Tolchuticook realm. On the other hand, Early Cambrian archaeocyathids of Siberia, South China, and Australia are quite distinct from each other, and the earliest skeletal fossils (small shelly fossils) show a cosmopolitan pattern. During the Ordovician, the Australian shallow-water faunas were highly endemic, especially at the species level, but the graptolite faunas of western Europe, South China, and Australia are similar enough to permit zone-by-zone correlation between these areas. There are also characteristic trilobites (*Asaphopsis*, *Taihungshania*, and *Ninkianolithus*) and cephalopods (e.g. *Manchuroceras*) common to South China and Australia. The latest Ordovician *Hirnantia–Dalmanitina* fauna, as well as the trace fossil *Cruziana*, spread from western Europe and North Africa to northern Tibet and South China, presumably via a peri-Gondwanan belt. Whittington and Hughes (1973) suggested a similar Late Ordovician trilobite region for South China and Australia, and Cocks and Fortey (1990) put the two into one Arenigian Dikelokephalinid region. South China also shows analogy with other Gondwanan areas, e.g. Qomolangma (Early Ordovician *Wutinoceras* and Middle Ordovician *Sinoceras*), Qiangtang (Early Ordovician graptolites), the Qamdo area of Tibet and Baoshan area of western Yunnan (Early Ordovician *Ordosoceras*).

It is thus concluded that South China belonged to the Gondwanan realm in the Cambro-Ordovician and may even have been incorporated with East Gondwanaland earlier in Earth history. The biotic differentiation from Tibet–western Yunnan and Australia was initiated after the Ordovician, and became evident in the Silurian when the endemism of Yangtze reached 42.7 per cent (Table 1.4). Meanwhile, marine eastern Australia formed an independent region of the Old World realm (the Tasman region), whereas terrestrial central and western Australia formed a part of the Old World floral realm (the *Baragwanathia* flora (latest Silurian–Pragian)), and an independent fish province or region (the wuttagoonaspid–phyllolepid fauna (Early Devonian)) (Veevers 1984). It has been suggested (Wang and Liu 1979) that riftogenesis may have occurred between northern Australia and the Red River–Eastern Yunnan–Yuanjiang region, western Yunnan. Associated with the latter are Palaeozoic ophiolites.

14.1.3 CENTRAL ASIA–MONGOLIA PALAEO-OCEAN

This was a Late Proteozoic–Palaeozoic ocean between Siberia and North China–Tarim. It extended from the Lake Balkhash area, through northern Xinjiang and Mongolia to Daxingan and the Xiaoxingan Mountains. Further westward it stretched through Central Asia to connect with the Urals, and further eastward it may have extended through Primoria to areas between Honshu and Hokkaido, but this segment was deformed by the superimposition of the Mesozoic and Cenozoic circum-

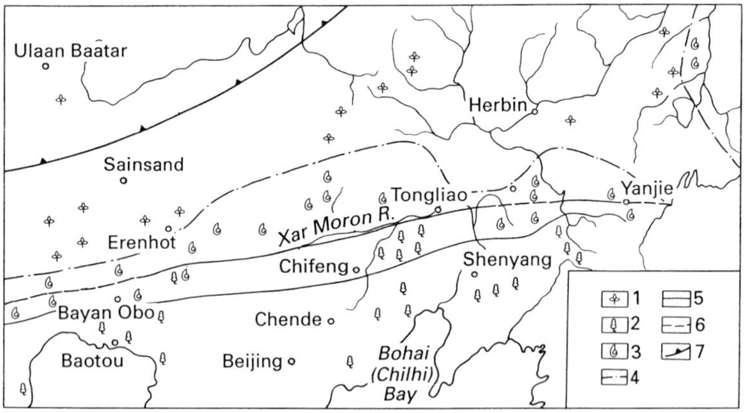

Fig. 14.4 Distribution of Late Carboniferous fossils in the eastern segment of Central Asia–Mongolia. 1, Angaran floral region and localities; 2, Cathaysian floral region and localities; 3, marine faunal localities; 4, northern limit of the sea; 5, southern limit of the sea, also northern border of the North China microplate; 6, Xar Moron–Solon Mountains suture, limiting the distribution of Cathaysian flora; 7, demarcation line suggested by Guo 1986.

Pacific Fold Belt. Palaeomagnetic data show that by the end of the Palaeozoic the southern margin of Siberia was about 55°N (Khramov *et al.* 1981), whereas the north margin of North China was about 20°N, leaving a distance of 3000–4000 km in between. This distance is now 1000–1500 km; the difference being due to subduction of the Central Asia–Mongolia paleo-ocean and subsequent crustal shortening by folds and nappes.

As regards the Mongolia–Xingan segment, many authors believe that its spreading mid-ocean ridge was located along the Xar Moron River–Solon Mountains line (Fig. 14.4). A trough–arc–basin system reminiscent of the modern western Pacific margin developed in the vast area between this line and the Siberian plate. There the Xingkai'an (Early to Middle Cambrian), Caledonian, and Hercynian Orogenies formed a succession of fold belts accreting onto the Siberian plate, namely, from north to south: Late Proterozoic Yenesey-Baikal, Early Palaeozoic Sayan–Mongolia–Argun (Eergun'nahe), and Late Palaeozoic Tianshan–Xingan Fold Belts. Within the three belts there are nearly ten underthrust– subduction zones signified by ophiolites and melanges, most of them extended in an arcuate east–west direction, convex southward, and underthrusting northward. The area between the Xar Moron–Solon line and North China–Tarim is much narrower, presumably being an old continental margin of Andean type. Along a narrow belt 50 to 100 km wide south of the line, there are Early and Late Palaeozoic subduction zones underthrusting southward (roughly along the line represented by 5 in Fig. 14.4).

Guo (1986) proposed a different view of the eastern segment. He maintained that collision of Siberia and North China began in the Middle–Late Devonian and ended in the Early Permian rather than in the Late Permian. The suture is the Solon Mountains–Hegenshan Mountains line, roughly along the eastern section of the China–Mongolia boundary.

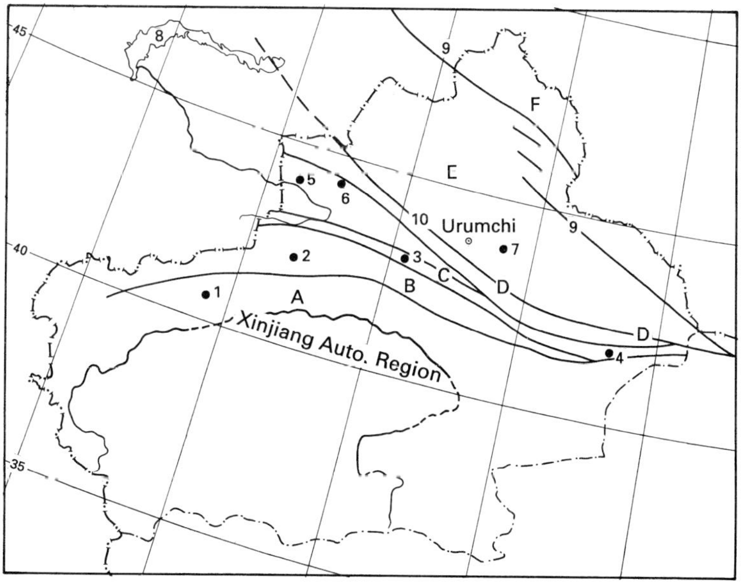

Fig. 14.5 Distribution of biogeographically important fossils in the western segment of Central Asia-Mongolia. A, Tarim; B, southern Tianshan; C, central Tianshan; D, northern Tianshan; E, Junggar; F, Altay. 1, Kalpin; 2, Haliktau; 3, Balentai; 4, Kawabulak; 5, Hochen; 6, Borohoro Mountains (Poluokenushan); 7, Bogda Mountains; 8, Balkhash Lake; 9, Zajsan–Erbix–Karameili Line; 10, Ebinur–Gaxunnur Line, in eastern Xinjiang, this line may be along the Aidinnur–Tudeng underthrusting zone.

The western extension of the Central Asia–Mongolia Palaeo-ocean in northern Xinjiang is still under discussion. Wang (1981) chose Ebinur Lake–Gaxunnur Lake as the demarcation line between Siberia and North China–Tarim. Li *et al.* (1982), on the other hand, maintained that Junggar is a portion of the Kazakhstan plate wedged between Siberia and North China, its northern boundary with Siberia being the Zajsan Lake (Ozera)–Ertix River–Karameili Mountains suture, and its southern boundary with Tarim being southern Tianshan. The key point of dispute is the attribution of Junggar.

From the palaeobiogeographical viewpoint (Fig. 14.5), three lines are important. The Zajsan–Ertix–Karameili line was evident in the Ordovician because Late Ordovician strata in Junggar yield endemic heliolitids as well as *Agetolites* coral faunas, the latter being widespread in South and North China. Both were lacking in the Altay–Mongolia–Xingan region. The Ebinur–Gaxunnur line was important throughout the Early Palaeozoic: the Cambrian trilobites of southern and northern Kazakhstan resemble those of South China and Siberia, respectively. *Damesella* (Middle Cambrian) occurs in the Korgas and Borohoro (Poluokenu) Mountains of west Central Tianshan, and *Dideroceras* and *Sinoceras* (Early and Middle Ordovician) in Keping and Kawaburak, indicating that these fossils of South and North China type did not cross the Ebinur line. Early Ordovician trilobites display a benthic → transitional → planktonic tendency from Tarim northwards to central Tianshan, implying a continental slope. Late Ordovician trilobites from north and south of this line show a low similarity coefficient (0.17, Fig. 4.10) and, finally, the Silurian *Tuvaella* fauna, characteristic of the Mongolia–Xingan region, was discovered in Hongliuyuan on the northern slope of the Bogda Mountains. Junggar is thus postulated to have been a microplate waxing and waning alongside the realm boundary and finally colliding with Siberia in the Devonian (see NK in Figs 3.1, 4.1, 5.1, 6.1). This postulation is supported by tectonic analysis (Wu 1987). A third line between central and southern Tianshan became significant in the Late Palaeozoic. For example, the South-China-type *Yunnanella* and *Stringocephalus* (Devonian) were restricted to southern Tianshan; Tethyan-type Permian fusulinids reached Balentai but not Kawaburak; on the other hand, the Carboniferous and Early Permian Angaran flora only reached northern Tianshan. These lines represent sutures of different periods. Extending eastward, they connected with the Xar Moron–Solon line in the Permo-Carboniferous. It is generally accepted that the eastern segment of Central Asia–Mongolia had substantially closed by the end of the Permian. However, the Lower Triassic in nearby Changchun, Jilin province, is reported to be a coal-measure series, compatible with contemporaneous deposits in Siberia and different from the red beds immediately south of the line. This fact suggests the possibility that the timing of collision may have been delayed until the Early Triassic.

Palaeomagnetic data have revealed that the relative location, with Siberia in the north and North China–Tarim to its south, was established after the Silurian. The latitudinal distance between the centres of the two was usually about 30°. Therefore, palaeobiogeographical differentiation between the two reflected not only an oceanic barrier but also climatic divergence. As a rule, Siberia has yielded temperate–cool temperate biotas, whereas Tarim–North China has tropical–subtropical fossils. Such a differentiation continued until the Late Triassic. A comparison between the two plates is shown in Table 14.3.

14.1.4 PACIFIC OCEAN

Although the oldest known deposit of the Pacific dates back only to the Jurassic, many have suggested an earlier date for the ocean's origin (Hilde *et al.* 1977; Tozer 1984). From the palaeobiogeographical viewpoint we believe that the influence of the Pacific has been significant at least since the Late Triassic. The influences are more climatic and physicogeographic than tectonic. Through formation of a South-East Asian monsoon area and a Pacific drainage system (Yangtze River, Yellow River, etc.), the Pacific has greatly affected the distribution patterns of plants and non-marine

Table 14.3 Comparison of biotic provincialization between Cathaysian Tethys (North China–Tarim type) and Boreal (Mongolia–Xingan type)

	Cathaysian Tethys (North China–Tarim type)	Boreal (Mongolia–Xingan type)
T_{2-3}	Plants: *Danaeopsis fecunda–Bernoullia zeilleri* flora (warm temperate)	*Sphenopteris* (temperate Uralian flora)
T_1	Plants: *Voltzia–Pleuromeia* flora (subtropical–warm temperate)	*Pseudoaraucarites*, Angaran flora (temperate)
Permo-Triassic transition	Plate collision	
P_2^2	Plants: North margin mixed flora and Late Catahysian flora (N. China type)	Angaran flora (*Comia, Iniopteris, Callipteria*)
P_1–P_2^1	Plants: Middle–Late Cathaysian (N. China type) flora: *Emplectopteris–Gigantonoclea–Lobatannularia*	Angaran flora: *Zamiopteris–Walchia* (west); *Z.–Nephropsis Crassinervia* (east)
P_1	Fusulinids: *Monodiexodina* unknown	*Monodiexodina*
	Corals: *Lytvolasma* unknown	*Lytvolasma* and other small solitary corals
	Brachiopods: *Leptodus, Meekella*	*Yakovlevia, Horridonia, Stepanoviella*
	Ammonoids: *Uraloceras*, etc. unknown	*Uraloceras, Paragastrioceras*
C_3	Plants: Early Cathaysian (N. China type) flora: *Neuropteris ovata–Lepidodendron posthumii*	Angaran flora: *Angaropteridium, Angaridium, Noeggerathiopsis Crassinervia, Zamiopteris*
C_1^2–C_{2-3}	Biotas of the two regions somewhat mixed	
C_{2-3}	Rugose corals: mainly large, solitary, e.g. *Bothrophyllum*	Mainly small, solitary, e.g., *Amplexocarinia Cystodendrophora*
	Tabulate corals: *Donetzites*	
	Fusulinids: *Pseudoschwagerina moelleri* assemblage (C_2)	Unknown
	Brachiopods: South China type	Cosmopolitan
C_1	Rugose corals: *Yuanophyllum, Pseudouralinia, Kueichowphyllum*	*Enygmophyllum, Sugiyamaella Zaphriphyllum*
	Tabulate corals: *Fuchungopora, Fuchungoporella*	Few
	Brachiopods: *Gigantoproductus, Striatifera, Vitilliproductus*	*Syringothyris, Rotaia, Marginatia*
	Plants: Euramerican flora *Triphyllopteris, Cadiopteridium*	Angaran flora: *Chacassopteris, Caenodendron, Angaropteridium*
D_3	Corals: South China type (*Disphyllum, Hexagonaria*, etc.)	*Nalivkinella*
	Brachiopods: *Yunnanella–Yunnanellina* assemblage	*Y.–Y.* assemblage unknown *Mucrospirifer*
	Ammonoids: unknown	*Sporadoceras* (Onor)
	Plants: *Lepidophloeum*	*Archaeopteris*

Table 14.3 (*cont.*)

	Cathaysian Tethys (North China–Tarim type)	Boreal (Mongolia–Xingan type)
D_2	Brachiopods: *Athyris–Athyrisina* assemblage and *Stringocephalus* etc.	No *A.–A.* assemblage and *Stringocephalus*
	Corals: S. China type: Disphyllidae, etc.	*Turgantites*
D_1	Corals: aff. S. China type, with N. American elements	Mainly small, solitary (*Siphonophrentis*, etc.)
	Brachiopods: aff. S. China type: *Orientospirifer–Dicoelostrophia* assemblage, with Boreal elements	No *O.–D.* assemblage, *Brachyprion*, *Coelospirifer*
	Plants: probably tropical flora	*Sciadophyton* (Junggar)
S	Corals: Stauriidae, Fletcheriidae	*Kodonophyllum, Kyphophyllum*
	Brachiopods: Pentamerus	*Tuvaella–Tannuspirifer* assemblage
	Nautiloids: *Sichuanoceras, Yunnanoceras*	*Trocholites*
	S_{2-3} trilobites: *Coronocephalus, Acanthopyge*	*Encrinurus, Cheirurus*
O_3	Corals: Agetolitids	*Calapocia-Sibiriolites* ass.
O_{1-2}	Cephalopods:	
	Lituites–Sinoceras assemblage	No *Lituites* and *Sinoceras*
	Actinoceratids abundant	*Kionoceras* (Inn. Mongol., Jilin)
	Trilobites: *Eoisotelus*, Nereinidae	*Sphaerexochus* (Luohe, Xingan area)
Cm	Trilobites: Asia–Australia type Redlichina, *Damesella*, etc.	Siberia (transitional) type *Erbia* assemblage (Yichun in eastern section), north Kazakhstan assemblage (western section)

faunas, including Chinese bivalves, ostracods, and conchostracans and has thus established a circum-Pacific unit at the rank of realm or region since the Early Mesozoic, the boundary of which oscillated around the north-east trending Xuefeng Mountains–Lueliang Mountains line (Fig. 14.1). Details will be given in Chapter 15.

As a source region of terrains, the Pacific margin is also significant for Asian accretion. The Nadanhada Mountains in north-eastern China are now recognized as one of the terrains, including Sikhote-Alin, Mino, and Hida, derived from the Pacific and accreted to Asia during the Late Jurassic–Early Cretaceous time interval, and the Jiamusi block may be another terrain accreted earlier (Kojima 1989; Mizutani *et al.* 1989; Fig. 14.6). Many discussions on south-eastern China tectonics refer to terrains from the Pacific. Ji and Coney (1985) postulate that most parts of China were terrains accreted to Eurasia, implying a Pacific source for many of them. Radiolarians play an important role in the identification of these terrains.

14.2 IMPLICATIONS OF PALAEOBIOGEOGRAPHY FOR PLATE TECTONICS OF CHINA

Palaeobiogeography contributes to plate tectonic reconstructions in two ways. First, palaeobiogeographical provincialization reflects palaeoclimatic zones, palaeogeography and palaeolatitude, which help in locating plate margins. The second is concerned with biotic distance between plates.

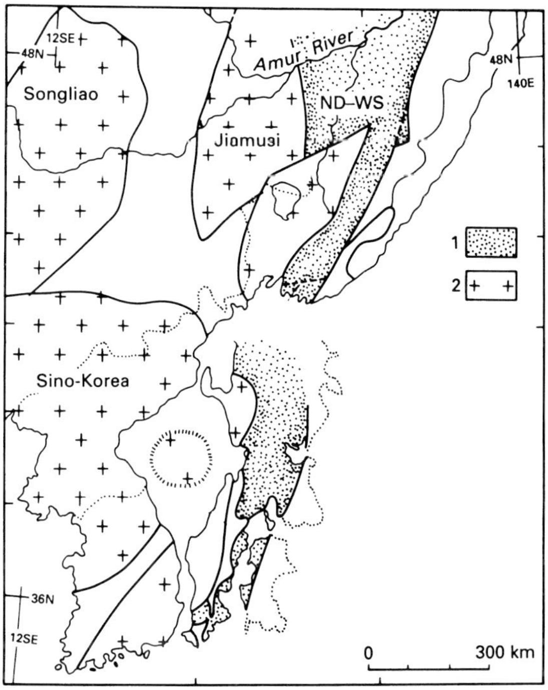

Fig. 14.6 Terrains of north-eastern China, Sikhote–Alin, and Japan prior to the opening of the Japan Sea (after Kojima 1989). 1, Mino-type disrupted terrains characterized by Upper Palaeozoic limestone, Triassic bedded chert, and Jurassic clastic rocks; 2, Continent-type stratigraphic terrains. ND–WS, Nadanhada–Western Sikhote–Alin terrain; other terrain names omitted.

Biotic distance is a relative and qualitative concept derived from similarity coefficient values and the presence or absence of index elements between the two plates under comparison. This concept reflects the degree to which the two plates are separated, and thus indirectly helps in determining the plate location.

Three biogeographical units persisted in China; namely, the Boreal (or Siberia, Junggar–Mongolia–Xingan), Tethys (or Proto-Tethys, Palaeo-Tethys, Eurasian and Cathaysian Tethys), and Gondwana (including Gondwanan Tethys). Tables 14.2 and 14.3 show the long-term development of their origination, establishment, and disappearance. A fourth unit, the circum-Pacific, superimposed itself on the previously formed units and has generated north-east trending demarcating lines since the Late Triassic.

14.2.1 PLATE MOVEMENT OF EASTERN CHINA IN VIEW OF PALAEOBIOGEOGRAPHY

Sinian, Cambrian, and Ordovician

Provincialization

The major boundary line was the Xar Moron–Solon line. The Xingan area yields archaeocyathids and trilobites of Siberian type and has therefore been attributed to the Siberian realm. South and North China belonged to the same region of the Asia–Australia (Cambrian) or Austral (Ordovician) realm. North China shows a resemblance to North America–Siberia whereas South China resembles Australia; they are thus subdivided into two provinces. Differences between the two are as follows:

1. Sinian. In North China, macroscopic algae mainly occurred in the Early Sinian and the metazoans were pre-Ediacaran, represented by the *Pararenicola* assemblage (Xuzhou–Huainan fauna). In South China, macroscopic algae occurred in the Late Sinian, and the metazoans belonged to the Ediacara fauna characterized by *Paracharnia*, annelids, and sabelliditids.

2. Cambrian. The North China–Qilian province has yielded only 70 Early Cambrian genera, with a few bradoricopids and earliest skeletal small shelly fossils, and lacking archaeocyathids. In the Upper Yangtze biome of South China alone 274 genera have been found, with abundant archaeocyathids, bradoricopids, and earliest skeletal fossils. In the Middle and Late Cambrian, the North China–Qilian province had eight dominant families, whereas South China had 14, and each province possessed a dozen or so endemic genera.

3. Ordovician. Regarding Ordovician nautiloids, North China was dominated by armenoceratids,

Australia	South China	North China	Siberia (Mongolia–Xingan)		
E. Sinian (Australia–South China)		U. Sinian		macroscopic algae	Sin
yes (Australia–South China)		pre-Ediacara		Ediacara fauna	Sin
Redlichina (Australia–North China)			Transitional	Cm_1 trilobites	Cm
endemic	endemic	unknown	endemic	archaeocyathids	Cm
abundant (Australia–South China)		few		bradoricopids & earliest skeletal fossils	Cm
similar (Australia–South China)		rare		Cm_{2-3} agnostids	Cm
Agetolites ass. (Australia–North China)			*Calapocia* etc.	O_3 corals	O
orthoceratids (Australia–South China)		armenoceratids (North China–Siberia)		O_{1-2} nautiloids	O
Taihungshania, Ninkianolithus, Yangtzella (Australia–South China)		*Favistella, Amsassia* (North China–Siberia)		other fossils	O
Hirnantia–*Dalmanitina* ass. (Australia–South China)				O_3^2 fauna	O
Tasman fauna →	*Pentamerus* (marginal sea)		*Tuvaella*	brachiopods	S
	Coronocephalus (marginal sea)		*Encrinurus*	trilobites	S
	Sichuanoceras (marginal sea)		*Trocholites*	nautiloids	S
	endemism 42.9%	(land)			S
	South China type (marginal sea) (see Tables 14.2,14.3)		Boreal type	brachiopods & corals	D
Wuttagoon-aspid, etc.	Polybranchiaspid, etc. endemism *c.* 20–30%	(land)		pisces	D
Glossopteris flora →	Euramerican (South China–North China)		Angaran	flora	C_{1-2}
	Tethyan (South China–North China)		Boreal	fauna	C_{1-2}
Gondwana fauna →	Cathaysian (South type)	(North type)	Angara	flora	C_3–P
	Cathaysian (South China–North China)		Boreal	fauna	C_3–P
Dicroidium →	*Dictyophyllum*–*Clathropteris* ass.	*Bernoullia*–*Danaeopsis* ass.	Uralian type	flora	T

Fig. 14.7 Palaeobiogeographical comparison of eastern China and adjacent areas.

South China by orthoceratids. The similarity coefficient of Early Ordovician nautiloids between the two areas barely reached 0.19. As to Ordovician trilobites, North China was characterized by the *Eoisotelus* assemblage (Early Ordovician) with many North American elements, whereas South China was characterized by *Taihungshania* (Early Ordovician) and *Nankinolithus* (Late Ordovician) assemblages. The similarity coefficients are only 0.12 (Middle Ordovician) and 0.06 (Late Ordovician). South and North China possessed the same Late Ordovician *Agetolites* coral assemblage, but North China–Qilian formed an *Agetolites–Favistella* subgroup. The latest Ordovician *Hirnantia–Dalmanitina* assemblage characterizes South China, Tibet, and other peri-Gondwanan areas, but is unknown in North China–Qilian (except for a few *Dalmanitina*) and the Boreal realm.

4. Affinities with other plates. The Cambrian trilobites, Ordovician nautiloids (except for Tremadocian), trilobites, and corals (*Favistella, Amsassia*) of North China–Qilian have affinities with the Siberian or North America–Siberian type, or include their important elements. On the other hand, Cambrian trilobites, Ordovician nautiloids and trilobites of South China, and Tarim show affinities with Australia and southern Europe (then part of Gondwana); the trace fossil *Cruziana* exhibits a similar pattern, and the graptolite zonation also correlates with that of Australia.

Based on biotic distance, palaeoclimatic and palaeomagnetic data (Figs 14.8–14.9), we have located North China–Qilian as a microplate between Siberia and Gondwana, with South China near Australia.

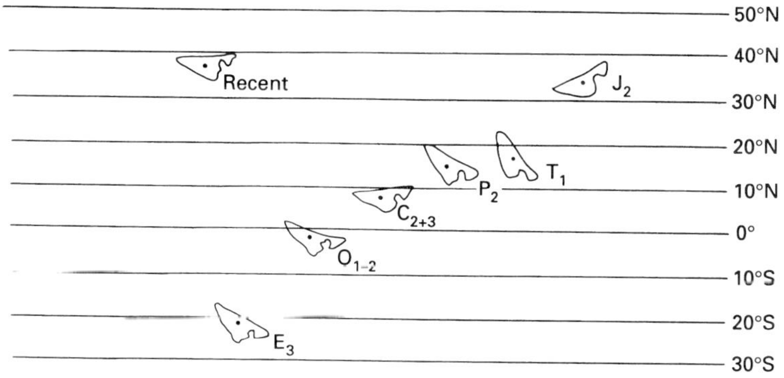

Fig. 14.8 Preliminary orientations of North China during the Phanerozoic (synthesized by Zhu Hong, unpublished). C_3, Late Cambrian; O_{1-2}, Early and Middle Ordovician; C_{2+3}, Late Carboniferous; P_2, Late Permian; T_1, Early Triassic; J_2, Middle Jurassic.

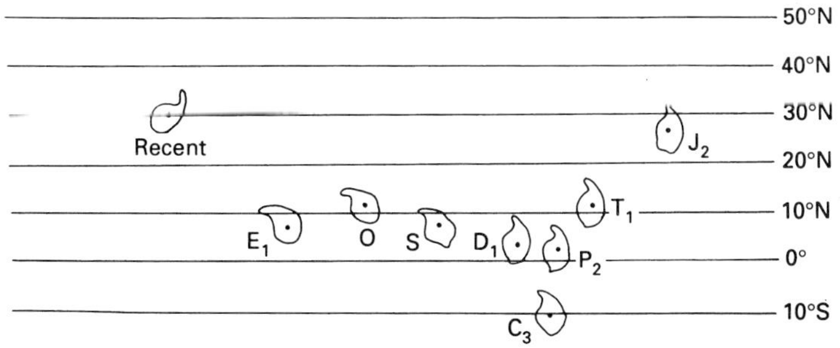

Fig. 14.9 Preliminary orientations of South China during the Phanerozoic (synthesized by Zhu Hong, unpublished). C_1, Early Cambrian; O, Ordovician; S, Silurian; D_1, Early Devonian; C_3, Late Carboniferous; P_2, Late Permian; T_1, Early Triassic; J_2, Middle Jurassic.

Plate rotation

As depicted by Scotese and McKerrow (1990) and most other reconstruction maps, the Siberia–Australian axis displayed clockwise rotation during this time interval, resulting in the northward movement of Siberia and the southward migration of Australia. Accordingly, North and South China underwent relative rotation. Based on faunal diversity, phosphorus distribution, and sedimentary facies, as indicated in Section 3.3 and Table 3.8, from the Early to the Late Cambrian, North China shifted from a zone of dry tropical climate to a zone of temperate climate, implying northward movement, while South China moved from temperate to dry tropical and thus southward. This is in accordance with the palaeomagnetic data, so towards the end of the Ordovician North China lay to the north-west, whereas South China lay to the south-east.

Some authors (Berry and Wilde 1990; Burrett *et al.* 1990) have suggested a location of South China in the southern middle latitudes, and an anticlockwise rotation of Gondwanaland from the Cambrian to the Ordovician to account for the alleged Cambrian similarity and Ordovician disparity between South and North China. This idea probably came from the interpretation that the Early Ordovician *Didymograptus bifidus* assemblage and latest Ordovician *Dalmanitina–Hirnantia* assemblage of cool-water origins, thus signifying a location of high latitudes in the southern hemisphere. However, the character of the *D. bifidus* assemblage is still in dispute (Chapter 4), and the *D.–H.* assemblage has a palaeolatitudinal distribution extending to low latitudes (Cocks and Fortey 1990). As explained in Chapter 4, South China was mainly under a tropical–subtropical climate and was at a low latitude (Fig. 14.9), and the biotic difference between South and North China was strikingly more reduced, rather than enlarged, in the Ordovician than in the Cambrian. Therefore is no need to accept the above suggestions. We have preferred the pattern adopted herein instead of the scenario suggested by these authors.

Silurian and Devonian

The distribution pattern of plate tectonics and the palaeobiogeography of China changed significantly from Silurian times. Further clockwise rotation caused Gondwana and Siberia to enter southern high and northern middle latitudes, respectively. Subsequently, realms were climatically differentiated, namely the Boreal, Malvinokaffric and Proto-Tethys. Simultaneously, bipolarity appeared in biotic provincialism.

Silurian

The Junggar–Xingan province of the Siberian region was clearly demarcated from Proto-Tethys by brachiopods (*Tuvaella*, *Tannuspirifer*, and *Stegerhynchella*), nautiloids (*Trocholites* and *Phragmoceras*), trilobites (*Encrinurus* and *Cheirurus*), and corals (*Kodonophyllum* and *Entelelasma*). The whole biota displayed low diversity, somewhat cosmopolitan but with more affinities to North America, western Europe, and Siberia, having a few typical Boreal elements but lacking typical tropical elements. These characteristics denote its temperate Boreal nature.

North and South China, as well as Tarim, remained in low latitudes and are attributed to the East Proto-Tethyans realm, showing similarities with Kazakhstan and Europe. Although there were a few Boreal elements (Late Silurian corals *Evenkiella* and *Altaia* from Inner Mongolia) and Australian elements (Early Silurian coral *Mucophyllum* from Inner Mongolia), the general fauna differed considerably from these two realms. Albeit that North China proper remaired terrestrial, its western margin (Qilian) shared many common elements with South–Central Tianshan and Kazakhstan. Yangtze was left alone in low latitudes when Australia moved towards the south pole. Its high endemism (42.7 per cent) and high diversity reflects its situation as an isolated migrating tropical microplate.

Devonian

The Devonian pattern inherited many Silurian features. The Junggar–Xingan province exhibited low diversity (e.g. Late Devonian ammonoids), a few Boreal elements within a generally cosmopolitan assemblage, lack of tropical Yangtze-type elements (especially in brachiopods), and typical small, solitary corals. North China remained an ancient massif; despite the fact that the Qilian seaway was closed, another marginal sea in western Qinling introduced tropical Yangtze elements into southern Tianshan. Yangtze was still highly diversified and endemic (27 per cent and 28 per cent respectively in Early Devonian corals and brachiopods, as well as in fish).

Carboniferous and Permian

Two major events controlled Chinese palaeobiogeography in this time interval.

Rotation of the Siberia–Australia axis

This axis had rotated through 90° since the Cambrian, and both were then in their maximal high-latitudinal positions. Consequently, the three realms—Boreal, Palaeo-Tethys, and Gondwana—were most clearly demarcated. Likewise in China, Junggar–Xingan, Cathaysian, and Gondwanan Tethys were distinctive. Palaeo-Tethys, owing to its tropical nature, showed high endemism, hence an eastern Cathaysian region was definitely established. The widely recognized Cathaysian flora started its existence in the Late Carboniferous and persisted to the end of the Permian. We suggest that a Cathaysian fauna formed at least in the Permian (see Section 8.2). Its distinctiveness has been indicated for all major Permian groups, some already quoted as Cathaysian (see Chapter 8, pp. 168–172, 179–181).

The gradual eastward closing of the Central Asia Mongolia Paleo-ocean

This controlled the distribution and exchange of Boreal and Palaeo-Tethyan (Cathaysian) biotas. The western segment, from Borohoro (Poluokenu) in northern Tianshan to the Beishan area of Gansu, shrank greatly in the Early Carboniferous. Its Xinjiang segment closed during the Carboniferous, whereas the Gansu and western Inner Mongolia segments closed during the Early Permian. Within this segment the Early Carboniferous marine fauna was a mixture of Boreal (*Syringothyris* and *Pseudosyrinx*) and Yangtze (*Kueichowphyllum*, *Yuanophyllum*, and *Gigantoproductus*) elements. In the Carboniferous the Angaran and Cathaysian floras were separated, while in the late Early Permian, the Angaran flora and Boreal marine elements crossed the Ebinur–Gaxunnur line into the Beishan area (Shuangbaotang Formation); in the Late Permian, Angaran elements invaded southern Tarim. Because no global or regional cooling occurred during that time, this southward invasion of temperate plants should be interpreted as due to northward movement of Tarim-North China.

Timing of the North China–Siberia collision

In the eastern segment, Gaxunnur–Xar Moron was still a demarcation line between the Angaran and Cathaysian floras up to the end of the Permian. Nevertheless, the marine fauna between these two floras showed mixed aspects in both the Carboniferous and Permian, which means that the sea was no longer broad and deep enough to bar the exchange of Boreal and Palaeo-Tethys faunas. Some recent Permian reconstruction maps place North China far away from Siberia, separated by a vast sea (Sengör *et al.* 1988; Scotese and McKerrow 1990), claiming support from palaeomagnetic (Lin *et al.* 1985) and phytogeographical (Chaloner and Creber 1988) differences. This is, however, incompatible with existing geological data, which evidently show that North China mainly collided with Siberia in the Late Permian or, at the latest, Early Triassic (Li *et al.* 1982; Nie *et al.* 1990), although complete closure may have been later. Three points should be made here.

1. By Late Permian times only a shallow sea existed between North China and eastern Siberia, i.e. in the Inner Mongolia–Xingan area. There is no sign of any marine incursion in this area during the Triassic, although the terrestrial basin deposits still show a strong distinction between the southern (south Liaoning) and northern (north of Changchun) sides of the Xar Moron Line.

2. A mixture of Angaran and Cathaysian floras occurred in the Late Permian in the western part of Tarim–Qilian province (see Chapter 8, p. 182). Although Chaloner and Creber (1988, p. 208) cast doubt about the existence of a mixed flora as suggested by Li and Yao (1983), it was backed by detailed research (Wang Dexu *et al.* 1984, see reference in Chapter 7). Immediately to its north, the Beishan area displayed an overlapping of a late Early Permian Tethyan fauna with an early Early Permian Boreal fauna. To its east the Early and Late Permian sea in Inner Mongolia–Xingan had a marine fauna similar to Tethyan type (see Chapter 8). Further east, in the Upper Permian Hongshan Formation of the south-eastern Xiaoxingan Mountains, Cathaysian floral elements are also mixed in with the Angaran flora (Huang 1983). These are best interpreted as indicating the geographical proximity of the North China and Siberia Plates.

3. Palaeomagnetic data show a rapid northward shift of North China after the late Early Permian, as required by the Siberia–North China collision (see following data from Taiyuan, North China, by McElhinny *et al.* 1981); meanwhile Siberia shifted southwards (Khramov *et al.* 1981):

Lower Shihezi Formation	
(upper L. Permian)	9.5°N
Upper Shihezi Formation	
(lower U. Permian)	
lower member	12.6°N
middle member	14.2°N
upper member	16.4°N
Shiqianfeng Formation	
(upper U. Permian)	18.8°N

On the basis of this evidence we have placed North China much nearer to Siberia than others have done.

Triassic

Timing of South China–North China collision

There have been three major viewpoints concerning the timing of the North China–South China collision: Jingningian (Late Proterozoic), Caledonian, and Indosinian (Triassic). The two microplates did undergo a Late Proterozoic opening–closing cycle, as demonstrated by orogeny, a blueschist–kyanite–talc-schist–ecologite belt (Sichuan, Hubei), and an ophiolite zone (northern Hubei). However, this did not end the geosynclinal history of Qinling.

Caledonian contact of the two continental blocks may well have happened. The latest ophiolite suite in Qinling is located along the Shangnan–Danfeng fault in central Qinling, dated to about 400 ma by many reports, and with a composition similar to that of an island-arc type. There was an overall uplift at the end of the Early Palaeozoic, resulting in the general absence of Upper Silurian and Lower Devonian rocks. In terms of palaeobiogeography, the discovery of ichthyofauna and plants from the Upper Devonian Zhongning Formation of Ningxia, along the margin of North China, is of considerable importance. The fish include *Sinolepis* and *Remigolepis*, which are only known elsewhere from South China and New South Wales (Australia). The flora includes such taxa as *Eolepidodendron wusihense*, *Hamatophyllum verticallatum*, and *Sphenopteridium taihuensis*, which are unique to the Wutong Group of Jiangsu and Zhejiang, South China (Zhao *et al.* 1986). Such communication of land biota should reflect the geographical proximity of the two blocks. In the Dabie area the basal conglomerate pebbles of the North China-type Carboniferous Yangshan Coal Series contain Silurian fossils of Yangtze type, implying that the two blocks had already been in contact before the Carboniferous.

However, the contact between the Silurian and Devonian, during which time the suggested collision should have taken place, is a disconformity instead of an unconformity, except in the Yunjiashan–Xieyuguan zone of northern Qinling, where metamorphosed Lower Palaeozoic rocks are unconformably overlain by Carboniferous rocks. There was not an overall Caledonian Orogeny in Qinling.

The real orogeny occurred in the Triassic, i.e. the Indosinian Orogeny. Before it, large-scale riftogenesis took place along the central Qinling 'suture' in the Permian and Triassic, which may have been due to northward shift of North China during the Permian. A vast Late Permian–Triassic deep sea existed in the west Qinling Mountains region, the width of which is estimated to have reached 1000 km (Yin *et al.* 1989). Olistostromes, thick turbidites, and a few contourites were deposited in this basin, together with radiolarians and nannofossils. It narrowed eastward, but a cool and deep-water Triassic fauna was distributed along Qingling–Middle and Lower Yangtze Valley, denoting a deep seaway connecting with the Boreal realm (Section 9.2). The Middle and Lower Yangtze Valley was the northern margin of the Yangtze Platform during the Permian and Triassic, receiving Early Triassic turbidites. However, deep-sea deposits in Qinling and Yangtze Valley are miogeosynclinal, without volcanism and ophiolites, so there was no oceanic development and suturing event in the Triassic.

The great Indosinian Orogeny after the Ladinian caused the final retreat of the sea in eastern China and the mountain-building of Qinling. The sea retreated once and for all during the Indosinian Orogeny, in the Ladinian in the Yangtze Valley and Norian in western Qinling. This appears to indicate the final welding (from east to west) of South and North China, for after the Indosinian Orogeny, South and North China moved as one plate. No more marine deposition and tectonics-induced biogeographical differentiation occurred between the two after the Late Triassic. However, the Triassic suturing is only supported by orogeny but without a corresponding ophiolites suite and volcanism, and is thus inferred by some to be intracratonic rather than intercratonic. Chinese geologists talk of this tectonic dilemma as 'collision (Caledonian) without orogenesis, followed by orogenesis (Indosinian) without collision'; this will be discussed further in Section 14.2.3.

The 'Xiangganzhe Triassic suture'

Hsü *et al.* (1988) and Sengör *et al.* (1988) advocated a Xiangganzhe Triassic suturing event when south-eastern China merged with Yangtze. There have been strong arguments against their attempt to regard the widespread Precambrian Banxi Group (cited by them as Banxi melange) as an overthrusted allochthon. For a brief review of the arguments for and against, reference shall be made to Rowley *et al.* (1989) and Hsü *et al.* (1989). Here we only discuss the timing from the palaeogeographical point of view. During the Late Palaeozoic and Early–Middle Triassic the whole of southern China, including Yangtze and south-eastern China, was covered by basically one epicontinental sea. Their litho- and biofacies changes in space are readily traceable. The biotas, as indicated in the text, have never been differentiated to the provincial level. Late Early Permian and Late Permian radiolarian-bearing siliceous rocks, as cited by Sengör *et al.* (1988) from Yang *et al.* (1981), were, in fact, deposited in extensional troughs within Bahama-type carbonate platforms (Yang *et al.* 1987), and are thus intracratonic. They are always of small size, reaching at most the upper bathyal zone in peripheral parts of southern China. So the palaeogeography and palaeobiogeography of this region do not support a Xiangganzhe Triassic suture.

Jurassic and later

After the Indosinian Orogeny the mainland of China, exclusive of Tibet, welded with Eurasia. From the Early Triassic to Recent, Siberia migrated southward by about 20° and meanwhile rotated clockwise (Khramov *et al.* 1981). Palaeomagnetic (Figs 14.8–14.9) and palaeoclimatic (Fig. 15.3) data show a similar scope of migration and direction of rotation. Moreover, judging from the palaeoclimate, the migration may have been episodic, with the main southward movement occurring in the Jurassic. This process was accompanied by intraplate crustal shortening, notably in the Kunlun–Qinling and Mongolia–Xingan mountain belts.

14.2.2 PLATE MOVEMENT OF WEST CHINA IN VIEW OF PALAEOBIOGEOGRAPHY

Sinian, Cambrian, and Ordovician

There is little information about Altay. However, it is generally accepted that areas north of the Zajsan–Ertix Line belonged to Siberia and thus to the Boreal realm. Junggar was a microplate wandering between Siberia and Asia–Australia (Cambrian) or Sino-European (Ordovician) realms. Tarim, together with south and central Tianshan, displays biotic similarities with South China. For example, about two-thirds of its Cambrian trilobites and about 50 per cent of its Middle Ordovician trilobites were the same as those of South China; and the similarity coefficient between Early Ordovician nautiloids of the two reached 0.31. Tarim's palaeoclimate changed in accordance with that of South China (Table 3.8). It thus seems that Tarim was associated with Gondwana just like South China. The differences between the two were:

1. The diversity of Tarim was much lower than that of South China (Tables 3.2, 3.5), but this may be due mainly to insufficient research.
2. Tarim was more related to Europe and Kazakhstan, e.g. in the Middle Ordovician the former possessed abundant lituitid nautiloids, as well as the trilobites *Ampyxinella*, *Cyclopyge*, and *Paracybeloides*. Taking account of this we have placed Tarim together with South Kazakhstan and near Europe.

Few Cambrian data are available for Tibet and western Yunnan. Their Ordovician faunas basically resemble the Yangtze province. However, lack of *Boreaolasma* (Late Ordovician) and *Yangtzella* (Early Ordovician), both typical of Yangtze, and the appearance of Cystoidea (*Sinocystis*, etc., western Yunnan) imply initial differentiation between the two. Chen (1983) reported differences between Silurian nautiloids north and south of the Yarlung Zangbo line.

Silurian and Devonian

With Junggar approaching and accreting to Siberia, the Ebinur–Gaxunnur line evidently defined the boundary between the Boreal and Proto-Tethyan realms, as shown by the distribution of the *Tuvaella* fauna. North Tianshan was then part of the Central Asia–Mongolia Palaeo-ocean, and the Devonian Tethyan faunas were basically restricted to its south, although there may be a few elements in common with the South China fauna.

The South Tibet (Himalaya)–Western Yunnan province was clearly distinguished from South China during the Silurinan. It is characterized by *Kopaninoceras* and peculiar fishes and corals, and lack of the *Coronocephalus–Pentamerus–Sichuanoceras* fauna which is typical of South China. *Sichuanoceras* also occurs in Xainza of the Gangdise–Nyianqingtanggula area. Nevertheless, in the Devonian, the Bangong Lake–Nujiang line seemed to be of more significance. It served as a demarcation line for fish, brachiopods, corals, and tentaculites. In addition, there is a belt of Lower Palaeozoic along the Jinshajiang and Lancanjiang Rivers. Its fauna is basically of Yangtze type, but possesses a few peculiar elements, such as the Late Silurian *Stortophyllum*. Jinshajiang may be another demarcation line during the Early Palaeozoic. Because most Silurian–Devonian information about Gangdise is based on data from Xainza, and little is known about Qiangtang, their attribution to the Gondwanan realm requires more evidence. The Silurian differentiation coincides with the departure of eastern Gondwana from South China.

Carboniferous and Permian

From the Early Carboniferous to the early Early Permian (Qixia'ian) Gangdise experienced several cold episodes, marked by diamictite deposits together with cold-water fauna, and thus belonged to the Gondwanan realm (Yin and Wen 1988). South Qiangtang (south of the Longmuco line) also has diamictites and cold-water fauna, but northern Qiang and western Yunnan produced a mixture of Cathaysian and psychrophilous biotas, thus indicating a transitional zone. During the late Early Permian (Maokouan) and Late Permian, Qiangtang and western Yunnan were dominated by Cathaysian flora and warm-water faunas, and the transitional zone shifted southward to Gangdise. The latter was then mixed with cold-water (*Lytvolasma*, *Wannerophyllum*, *Costiferina*, and *Monodiexodina*) and warm-water (*Verbeekina–Neoschwagerina* assemblage, *Iranophyllum–Ipciphyllum* assemblage) elements. The Himalayas area persisted as a Gondwanan type throughout the Permo-Carboniferous, except for local invasion of Maokouan Tethyan elements. Such a biogeographical situation is interpreted as due to the successive dispersion of Qiangtang and Gangdise from Gondwana. The former departed earlier (Carboniferous to early Early Permian), and the latter later (late Early Permian to early Late Permian). This timing is mainly based on biotic data, because volcanic data from the Jialashankou Formation shows that its riftogenesis began in the early Early Permian. They migrated northward and entered the tropical Tethys during the Middle and Late Permian and Early–Middle Triassic respectively. This explains the pattern shown in Figs 14.2 and 14.10.

Mesozoic and Cenozoic

In the southern hemisphere, the India plate was at first located between 40° and 60°S, with its northern side then heading east. It rotated anticlockwise and migrated northward during the Late Cretaceous and Early Palaeogene, as assumed by many authors. There is a consensus on the timing of collision between the eastern Cimmerides (Qiangtang and Sibumasu) and South China–Indochina in the Late Triassic (Mitchell 1981; Li *et al.* 1982; Sengör 1984; Huang and Chen 1987; Sengör *et al.* 1988), which is also supported by palaeobiogeographical data in this book. It is noted, however, that the

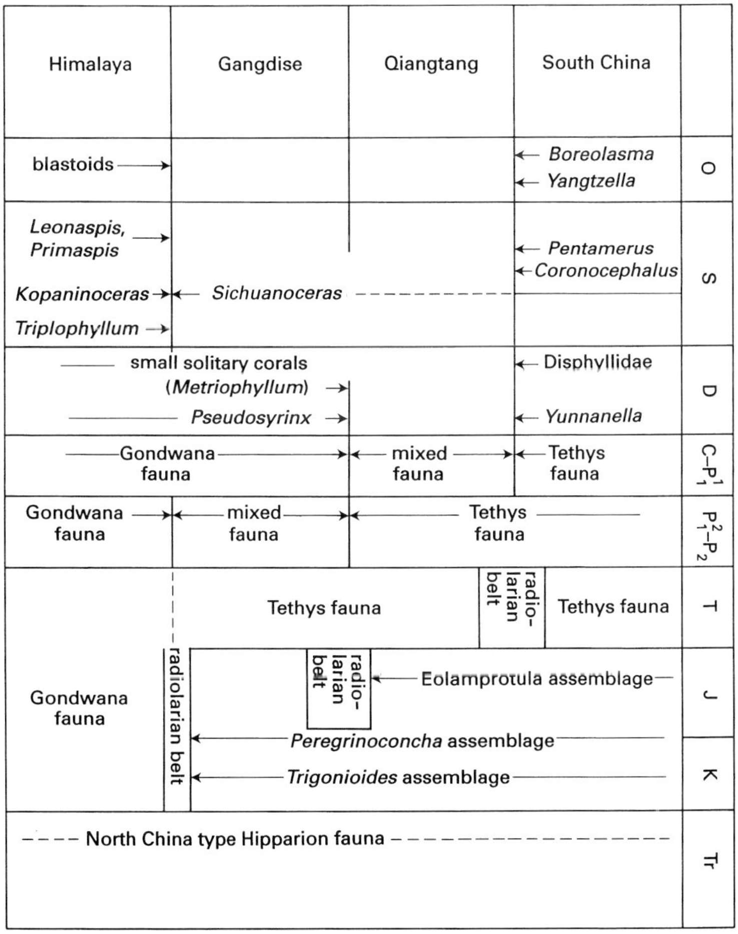

Fig. 14.10 Palaeobiogeographical comparison of plates of south-western China.

Gangdise block is exceptional. Its Jurassic history clearly shows that it was not sutured with Eurasia until the end of the Jurassic (Chapter 10, p. 241). The Triassic–Jurassic deep sea along the Bangong Lake–Nujiang line and associated ophiolites suggest that it was part of a branch of Tethys extending from Afghanistan to Burma (Sengör's Waser–Bangong–Nujiang–Mandalay suture).

During the Mesozoic, the Tibetan section of Neo-Tethys or Tethys can be subdivided into a Gondwanan Tethys, along the northern margin of the Indian subcontinent, and a Cathysian Tethys, along the newly accreted southern margin of Eurasia. The former was usually temperate but may have become subtropical–tropical, for instance in the Late Triassic–Middle Jurassic. The latter was always tropical–subtropical. The Mesozoic biotic disparity on both sides of Yarlung Zangbo is shown in Table 14.2. The two Late Palaeozoic transitional biogeographical units, Qiangtang and Gangdise, accreted into Eurasia at different times during the Mesozoic, as demonstrated by three deep-sea radiolarian belts (Wang Naiwen 1984; Wu 1988) and three cross-border events.

The three radiolarian belts are the Early–Middle Triassic belt of Litang, western Sichuan (*Xiphostylus*, *Cenellipsis*, *Sphaeroum*, and *Coenosphaera*); the Jurassic belt of Dongqiao, eastern Tibet (*Sethocyrtis*, *Criptocapsa*, and *Dicolocapsa*); and the Late Jurassic–Early Cretaceous belt of Xigaze, Xialu and Lhaze, southern Tibet (*Acanthocircus*, *Mirifusus* and *Cecrops*). The first belt represents the Ganze–Litang fault, supposed to be a back-arc, small oceanic branch of the Jinshajiang suture, which itself certainly contains Triassic radiolarians, as reported by many geologists but not yet adequately investigated. The second and third belts represent the Bangong Lake–Nujiang and Yarlung Zangbo sutures. The successively younger order of the three belts, supported by ophiolites and other indicators of ocean crust, indicate the successive separation between Yangtze, Qiangtang, Gangdise, and India.

As to the three cross-border events, the first event happened on the Tanggula area of the Qiangtang microplate. Middle Jurassic freshwater unioniids of the Yangtze province crossed the Jinshajiang suture and entered Tanggula; *Eolamprotula guangyuanensis* was discovered in the Yanshiping Group (Middle–Late Jurassic), reflecting the collision of Yangtze and Qiangtang before the Middle Jurassic, probably by the end of the Triassic. The second event occurred on the Gangdise microplate during the Jurassic–Cretaceous transition. Jurassic freshwater bivalves of Eurasian type were restricted to the north of the Bangong Lake–Nujiang line. The end-Jurassic–early Early Cretaceous *Peregrinoconcha–Sinonaia–Eonippononaia* assemblage, typical non-marine bivalves of Yangtze's western margin, crossed the border and appeared in the Lagondang Formation (Baxoi County) of Gangdise. Later, the Early Cretaceous *Trigonioides–Plicatunio–Nippononaia* freshwater bivalve assemblage also appeared in this area. None of them succeeded in crossing the Yarlung Zangbo. This second event reflects the collision of Qiangtang and Gangdise. The third event took place in the Himalayas during the Tertiary. The youngest marine deposit there is of the Eocene. Later, a *Hipparion* fauna of North China type (*H. guizohongensis*, *Palaeotragus microdon*, and *Gazelia gaudryi*) is found in the terrestrial Miocene (Gyirong County), which indicates a land connection between India and Eurasia.

These interesting phenomena, in accordance with other geological data, support the general idea that the Qiangtang microplate accreted to Eurasia during the Indosinian Orogeny, closing the Jinshajiang micro-ocean while extending the Bangong Lake–Nujiang rift into the ocean crust in the Jurassic (geosynclinal deposits of the latter were laid down after the Triassic). The Gangdise microplate accreted to Eurasia at the end-Jurassic, closing the Bangong Lake–Nujiang micro-ocean while extending Yarlung Zangbo into the ocean crust in the Late Jurassic and Cretaceous (geosynclinal and volcanic deposits of the latter were laid down after the Early Triassic). The collision of India and Eurasia led to land connection after the Eocene and before the Miocene. The general pattern clearly demonstrates that Neo-Tethys was contained by micro-oceans and microcontinents dispersing from Gondwana and accreting to Eurasia.

14.2.3 A MULTI-PHASIC INCORPORATION MODEL FOR THE MIGRATING MICROCONTINENTS

The Chinese blocks pose a tectonic dilemma un resolved by traditional plate tectonics. What we are concerned with is a series of northward migrating blocks, moving at different rates. Their collision was often due to a quicker block catching up with the slower one in front of it, and thus it was irregularly pulsatory. Because both were moving in the same direction and were small in mass, the first contact of two blocks was usually at one point rather than along a line, and therefore usually did not produce significant orogenys, and the continental crust thus created was not consolidated. Contact does not mean welding. Because of unequal velocities and direction in their constant northward migration, a long time interval may have existed between their first contact and final welding. For example, in the case of South and North China, although they may have made contact during the Caledonian Orogeny, a final welding may not have been achieved until the Triassic, from a biogeographical view. The first contact would give the northerly block an impetus to move faster, which would result in extensional riftogenesis along or near the weak, previously formed suture. Such riftogenesis may have led to the formation of deep-sea turbidites, basalts, spilites, or even ophiolites. When the northerly block accreted to Eurasia and thus became unmovable, then the southerly one would collide with it head-on. This last procedure was not suturing because the suture or continental crust had already been inititated, but its orogenic strength may well have exceeded that of the previous collision and led to the formation of a highly consolidated crust through strong crustal shortening and welding. Such a complex procedure of plate incorporation, including point contact (suturing), riftogenesis, and final welding, is called multiphasic incorporation. This event occurred between North and South China. The disappearance of Early Palaeozoic oceanic crust (ophiolites) was supposed to be end-Silurian–Devonian, which only produced epiorogenic rather than orogenic effects. Beginning from Carboniferous times, and especially in the Permian and Early and Middle Triassic, there was rift opening along central Qinling which was more of less parallel to the Early Palaeozoic suture. After the collision of North China with Siberia in the Late Permian, there came the head-on collision of South China which caused the significant mountain-building of the Indosinian Orogeny in the Ladinian. This was followed by the even stronger Yanshanian Orogeny accompanied by great crustal shortening.

A similar event happened between South China and Indochina. The first suturing, at a point, was probably during the Carboniferous, with Late Palaeozoic ophiolites along the Song Ma suture, as revealed by palaeomagnetic data (Gatinsky *et al.* 1984; Metcalfe 1988). The Late Carboniferous and Permian Cathaysian flora of South China type (Li and Yao 1983), the Early Permian *Neochwagerina–Lepidolina* foram province, and the *Monticulifera–Urushtenoidea* brachiopod subprovince (Nakamura *et al.* 1985) covered both areas. South China drifted northward after the mid-Permian, forming a wide-spread rift opening and Omeishan basalt eruptions on its then southern (now western) side, followed by thick Early–Middle Triassic turbidites in Guizhou and Guangxi along its southern border. In northern Indochina a huge sequence of pre-Norian turbidites with spilites and basalts accumulated along the Song Da suture, parallel to Song Ma. South China collided and became 'immovable' after the Ladinian–Carnian time interval, then the Norian Indosinian Orogeny took place between Indochina and South China, which was not suturing but nevertheless was much stronger than the Carboniferous suturing event. Such a unique situation of suturing history in unidirectional, small-block accretion has usually led to controversy about the timing of collision, which might have been multiple, as are the cases for the Qinling and Song Ma–Song Da lines. I suggest that the above model be considered in interpreting the unique situation in eastern Tethys and similar settings.

REFERENCES

Berry, W. B. N. and Wilde, P. (1990). Graptolite biogeography: implications for palaeogeography and palaeoceanography. *Memoir of the Geological Society of London* **12,** 129–38.

Boucot, A. J. (1990). Silurian biogeography. Memoir of *the Geological Society of London* **12**, 191–6.

Boucot, A. J. and Gray, J. (1986). The Tethyan concept during the Palaeozoic. In *Shallow Tethys,* Vol. 2, (ed. K. G. McKenzie), pp. 31–58. A. A. Balkema, Rotterdam.

Burrett, C., Long, J., and Stait, B. (1990). Early–Middle Palaeozoic biogeography of Asian terranes derived from Gondwana. *Memoir of the Geological Society of London* **12,** 163–74.

Chaloner, W. G. and Creber, G. T. (1988). Fossil plants as indicators of Late Palaeozoic plate positions. In *Gondwana and Tethys.* (ed. M. G. Audley-Charles and A. Hallam), Geological Society, Special Publication No. **37**, 201–10.

Chen Junyuan (1983). Silurian rugose coral assemblages and palaeobiogeography of China. In *Palaeobiogeographic provinces of China*, (ed. Editorial Committee of the Treatise of Basic Theories of Palaeontology), pp. 43–52. Science Press, Beijing.

Cocks, L. R. M. and Fortey, R. A. (1990). Biogeography of Ordovician and Silurian faunas. *Memoir of the Geological Society of London*, ***12,*** 97–104.

Fan Yinnian (1985). Palaeobiogeography of the Carboniferous–Permian rugose corals of Tibet, China. *Contribution to the Geology of the Qinghai–Xizang Plateau* **16**, 105–12. Geological Publishing House, Beijing (in Chinese).

Gatinsky, Y. G., Hutchison, C. S., Minh, N. N., and Tri, T. V. (1984). *Tectonic evolution of Southeast Asia.* 27th International Geological Congress, Moscow, Report No. **5**, pp. 225–39.

Guo Shengzhe (1986). On determination of convergence line between Siberian Plate and Sino-Korea Plate and its biostratigraphic evidence. *Bulletin of Shengyang Institute of Geology and Mineral Resources, Chinese Academy of Geological Sciences* **14**, 127–36 (in Chinese with English abstract).

Hilde, T. W. C., Uyeda, S., and Kreenke, L. (1977). *Evolution of the western Pacific and its margin.* United Nations ESCAP, CCOP Tech., Bulletin 10, Tectonophysics 38.

Hsü, K. J., Sun Shu, Li Jiliang, Chen Haihong, Pen Haipo, and Sengör, A. M. C. (1988). Mesozoic overthrust tectonics in south China. *Geology* **16**, 418–21.

Huang Benhong (1983). On late Late Paleozoic palaeophytogeographic regions of eastern Tianshan–Hingan foldbelt and its geological significance. *Contribution on the Project of Plate Tectonics of Northern China* **1**, 138–54 (in Chinese with English abstract).

Huang Jiqing and Chen Bingwei (1987). *The evolution of the Tethys in China and adjacent regions.* Geological Publishing House, Beijing (in Chinese and English).

Huang Jiqing, Chen Guoming, and Chen Bingwei (1984). Preliminary analysis of the Tethys–Himalayan tectonic domain. *Acta Geologica Sinica* **58**, (1), 1–17 (in Chinese with English abstract).

Huang Zhixun (1989). Paleobiogeography and paleostructure of Carboniferous–Permian in north Qinghai–Xizang (Tibet) Plateau. *11th Congrès International de Stratigraphie et de Geologie du Carbonifere Beijing 1987* **4,** 197–208.

Jell, P. A. (1974). Faunal provinces and possible planetary reconstruction of the Middle Cambrian. *Journal of Geology* **82,** 319–50.

Ji Xiong and Coney, P. J. (1985). Accreted terranes of China. In *Tectonostratigraphic terranes of the Circum-Pacific region* (ed. H. G. Howell). Earth Science Series, Circum-Pacific Council for Energy and Mineral Resources, Houston, No. **1**, pp. 349–61.

Khramov, A. N., Petrova, G. N., and Pechersky, D. M. (1981). Paleomagnetism of the Soviet Union. In *Paleoreconstruction of the continents,* (ed. M. W. McElhinny and D. A. Valencio), pp. 177–94. American Geophysical Union & Geological Society of America.

Kojima, S. (1989). Mesozoic terrane accretion in Northeast China, Sikjote-Alin and Japan regions. *Palaeogeograpohy, Palaeoclimatology, Palaeoecology* **69**, 213–32.

Kosygin, Y. A. and Parfenov, L. M. (1981). Tectonics of the Soviet Far East. In *The ocean basins and margins,* Vol. 5, *the Arctic Ocean*, (ed. A. E. M. Nairn, M. Churkin, Jr., and F. G. Stehli), pp. 377–412, Plenum Press, New York.

Li Chunyu, Wang Quan, Liu Xueya, and Tang Yaoqing (1982). *Explantory notes to the tectonic map of Asia.* Cartographic Publishing House, Beijing (in Chinese and English).

Li Xingxue and Yao Zhaoqi (1983). Carboniferous and Permian floral provinces in East Asia. In *Palaeobiogeographic provinces of China,* (ed. Editorial Committee of the Treatise of basic Theories of Palaeontology), pp. 74–82. Sciences Press, Beijing (in Chinese).

Lin Jinlu, Fuller, M., and Zhang, W. Y. (1985). Preliminary phanerozoic polar wander paths for the North and South China Blocks. *Nature* **313**, 444–9.

McElhinny, M. W., Embleton, B. J. J., Ma Xinhua, and Zhang Zhenkun (1981). Fragmentation of Asia in the Permian. *Nature* **293 (17)**, 212–16.

Metcalfe, I. (1988). Origin and assembly of south-east Asian continental terranes. In *Gondwana and Tethys*, (ed. H. G. Audley-Charles and A. Hallam). Geological Society of America, Special Publication No. 37, pp. 101–18.

Mitchell, A. H. G. (1981). Phanerozoic plate boundaries in mainland Southeast Asia, the Himalayas and Tibet. *Journal of the Geological Society of London* **138**, 109–22.

Mizutani, S., Shao Ji'an, and Zhang Qinglong (1989). The Nadanhada terrane in relation to Mesozoic tectonics on continental margins of East Asia. *Acta Geologica Sinica* **63** (1), 204–16 (in Chinese with English abstract).

Nie Shangyou, Rowley D. B., and Ziegler A. M. (1990). Contraints on the location of the Asian microcontinents in Palaeo-Tethys during the Late Palaeozoic. *Memoir of the Geological Society of London* **12**, 397–410.

Norin, E. (1946). *Geological explorations in western Tibet*. Sino-Swedish Expedition Publication No. **20**, pp. 1–214, Stockholm.

Rao Rongbiao (1989). On the north boundary of Gondwanaland in Tethys-Himalaya tectonic realm. *Acta Geologica Sichuan* **9**, (1), 18–26 (in Chinese).

Rowley, D. B., Zielgler, A. M., and Nie Shangyou (1989). Comment on Mesozoic overthrust tectonics in South China. *Geology* **17**, (4), 284–386.

Scotese, C. R. and McKerrow, W. S. (1990). Revised world maps and introduction. Memoir *of the Geological Society of London*, **12**, 1–24.

Sengör, A. M. C. (1984). *The Cimmeride orogenic system and the tectonic of Eurasia*. Geological Society of America, Special Paper No. **195**, pp. 1–82.

Sengör, A. M. C., Altiner, S., Cin, A., Ustaomer, T. and Hsü, K. J. (1988). Origin and assembly of the Tethyside orogenic collage at the expense of Gondwana. In *Gondwana and Tethys*, (ed. M. G. Audley-Charles and A. Hallam). Geological Society of London, Special Publication No. **37**, pp. 119–82.

Seyfert, C. K. and Sirkin, A. (1973). *Earth history and plate tectonics*. Harper & Row, New York

Sonnenfeld, P. (ed.) (1981). *Tethys, the ancestral Mediterranean*. Benchmark Paper in Geology No. **53**. Hutchinson & Ross.

Talent, J. A., Gratsianova, R. T., and Yolkin, E. A. (1986). Prototethys: Fact or phantom? Palaeobiogeography in relation to the crustal mosaic for the Asia–Australia hemisphere in Devonian–Early Carboniferous times. In *Shallow Tethys*, Vol. 2, (ed. K. G. McKenzie), pp. 87–112. A. A. Balkema, Rotterdam.

Tapponnier, P., Peltzer, G., Ledain, A. Y., Armijo, R., and Cobbold, P. (1982). Propagating extrusion tectonic in Asia, new insights from simple experiments with plasticine. *Geology* **10**, 611–18.

Tozer, E. T. (1984). Marine Triassic faunas of North America: their significance for assessing plate and terrane movements. *Geologische Rundschau*, **71**, (3), 1–22.

Veevers, J. J. (1984). *Phanerozoic earth history of Australia*. Oxford Geological Science, Series **2**. Clarendon Press, Oxford.

Wang Hongzhen (1981). Geotectonic units of China from the viewpoint of mobilism. *Earth Science – Journal of the Wuhan College of Geology* **1981**, (4), 42–66 (in Chinese with English abstract).

Wang Naiwen (1984). The rise of Gondwana research in China and some key problems on paleobiogeography related to plate tectonics. *Bulletin of the Chinese Academy of Geological Sciences* **10**, 103–15 (in Chinese with English abstract).

Wang Quan and Liu Xueya (1979). The drift of Yangtze continental segment and the early action of Gondwana Oldland—an important tendency of dynamic action of the Phanerozoic globe. *Journal of the Changchun Geological Institute* **1**, 1–7 (in Chinese).

Whittington, H. B. and Hughes, C. P. (1973). *Ordovician trilobite distribution and geography*. Special Papers in Palaeontology, Palaeontological Association of London, No. **12,** pp. 235–40.

Wu Haoruo (1988). Jurassic and Cretaceous Radiolaria from Xialu cherts in the Yarlung Zangbo ophiolitic zone, South Xizang and their geological significances. *Bulletin of the Institute of Geology, Academia Sinica* **3**, (2), 3–222 (in Chinese with English abstract).

Wu Qingfu (1987). Evidence for Junggar Massif and its position in the evolution of Kazakhastan plate. *Contributions to the Project of Plate Tectonics in Northern China* **2**, 29–38. Geological Publishing House, Beijing (in Chinese with English abstract).

Yang Zunyi, Wu Shunbao, and Yang Fengqing (1981). Permo-Triassic boundary of South China. *Earth Sciences – Journal of Wuhan College of Geology* **1981**, (1), 4–15 (in Chinese with English abstract).

Yang Zunyi, Yin Hongfu, Wu Shunbao, Yang Fengqing, Ding Meihua, and Xu Guirong (1987). *Permian–Triassic boundary stratigraphy and fauna of South China*. Geological Memoirs, Series B, No. 6 (in Chinese with English summary).

Yin Hongfu, Yang Fengqing, Lai Xuelong, and Yang Henshu (1989). Tectonic zonation of the Triassic of Qinling. *Modern Geology* **1**, (2), 108–18 (in Chinese with English abstract).

Yin Jixiang and Wen Chuanfen (1988). Diamictites of the Carboniferous and Lower Permian, their stratigraphical characteristic and origin. *Bulletin of the Institute of Geology, Academia Sinica* **3**, 26–55.

Zhao Xiuhu, Wu Xiuyuan, and Gu Qichang (1986). Late Devonian flora from Sou*thern Ningxia. Acta Palaeontologica Sinica* **25**, 544–60 (in Chinese with English abstract).

15 IMPLICATIONS OF PALAEOBIOGEOGRAPHICAL PROVINCIALIZATION FOR THE MESOZOIC AND CENOZOIC PALAEOCLIMATE AND PALAEODRAINAGE OF CHINA

Yin Hongfu

15.1 GENERAL INTRODUCTION

Figure 15.1 demonstrates an important phenomenon of palaeoclimate distribution in the geological history of China. There are two distribution patterns of tropical, arid tropical, humid subtropical, and humid temperate zones in the northern hemisphere. In the central and western portions of the Afro-Eurasian supercontinent and western North America, where the humid NE trade winds from the ocean cannot reach, the four belts are arranged in a regular south–north order (A, B, C of Fig. 15.1). In eastern Asia and eastern North America, however, the wet trade winds from the Pacific and Atlantic oceans have pushed the arid belt further westward, so that here, from south to north, are arranged wet tropical, humid subtropical, and humid temperate zones. Consequently, north of the tropical zone the humid subtropical–temperate and the arid subtropical (sometimes even temperate) zones may be arranged in an east–west direction instead of south–north. During Mesozoic and Cenozoic times in China, when arid weather prevailed, as T_{1-2}, J_2^2, and K_1^2–K_2 of Fig. 15.2, the first pattern was displayed; when humid weather pushed the aridity westward, there prevailed the second pattern, as in T_3, J_3–K_1^1, and N–R of Fig. 15.2.

The mainland of China became a part of the Asian continent after the Indosinian Orogeny, after which climatic differentiation caused by varied latitudes, relief and precipitation became the dominant controller of biotic provincialization. Four climatic belts can be recognized through palaeobiogeographical analysis, based mainly on floras but also with some consideration of faunas, (Figs 15.2 and 15.3, Table 15.1).

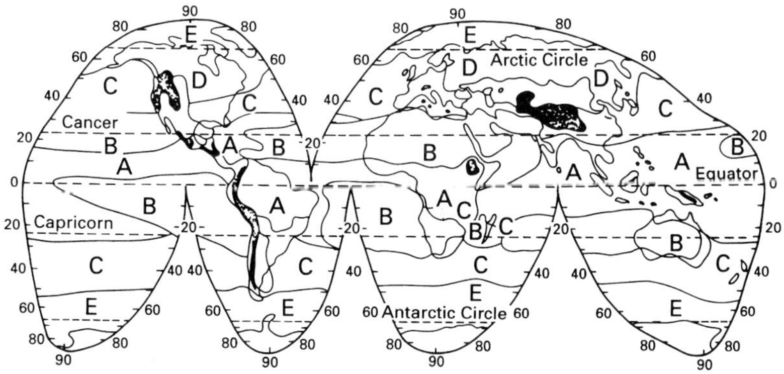

Fig. 15.1 Climatic regions of the world (after Strahler 1954). A, Tropical; B, dry; C, mesothermal, corresponding to humid subtropical and temperate in this book; D, microthermal; E, ice climates.

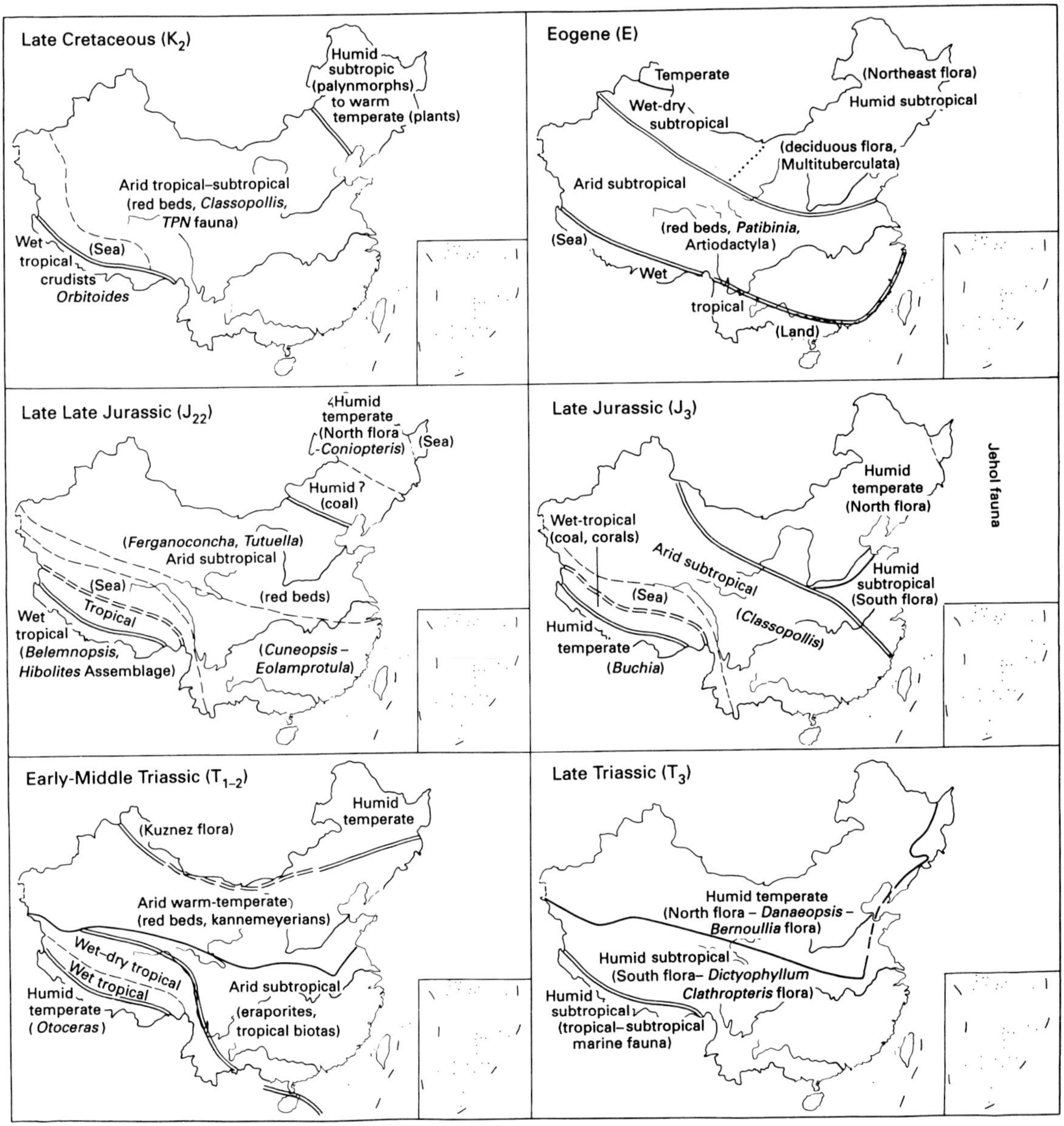

Fig. 15.2 Mesozoic and Cenozoic climatic zonation based on palaeobiogeographical data.

15.1.1 WET AND WET–DRY TROPICAL

This belt is located to the south of the arid and humid subtropical belt and is marked by marine tropical faunas including hermatypic corals, rudists, nummulitids, etc., and tropical floras as shown in Table 15.1. Its temperature may have corresponded to modern tropical rainforest temperatures, or may have been even higher. For example, the estimated air temperature of the Triassic of central and western USA

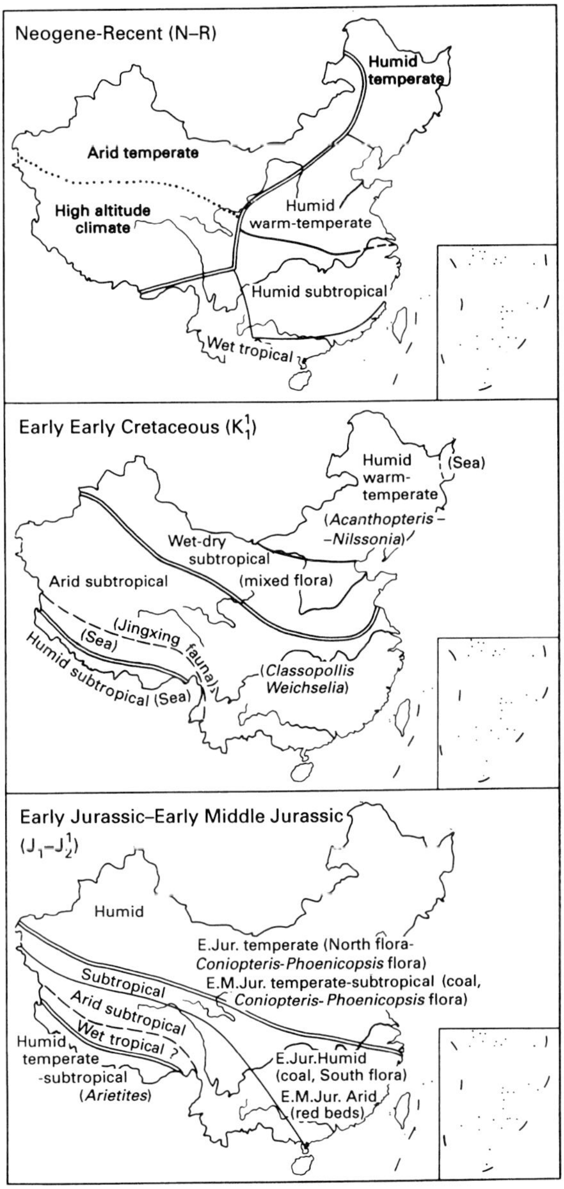

reached 35–40° C; the equatorial surface water of the Cretaceous was 7° C higher than now (Flakes 1979).

15.1.2 ARID SUBTROPICAL

Characteristic deposits of this belt are widespread terrestrial red beds intercalated with gypsum and other evaporites. Floras are rare or lacking; microfloras are dominated by *Classopollis*. Faunas are marked by red-bed reptiles typical of East Asia and non-marine, mostly freshwater, unioniid bivalves. The latter includes the Middle Jurassic *Eolamprotula–Cuneopsis–Psilunio* (ECP) assemblage, the end-Jurassic–earliest Cretaceous *Peregrinoconcha–Sinonaia–Eonippononaia* assemblage, the Early Cretaceous *Trigonioides–Plicatunio–Nippononaia* (TPN) assemblage and the Late Cretaceous *Pseudohyria*. When humid weather prevailed (T_3, J_3–K_1^1, and N–R), this belt was pushed westward, sometimes even beyond China (T_3). In the times of aridity expansion (T_{1-2}, J_2^2, K_1^2–K_2), it extended eastwards and northwards, reaching the eastern coast of China and invading the temperate belt (T_{1-2}), sometimes squeezing the humid subtropical belt to a very limited size or even eliminating it (T_{1-2}, J_2^2). Areas southwest of the Kunlun Mountains–Qamdo–central Yunnan belonged to the transgressive margin of tropical Mesotethys and were located on the western side of the Chinese continent, so they were generally under arid tropical–subtropical conditions before the India–Eurasia collision. Along the coast lived a shallow Tethyan fauna, such as the Eocene *Turkostrea* assemblage of western Xinjiang.

15.1.3 HUMID SUBTROPICAL

According to the definition of the Editorial Board of '*Physical Geography of China*', Academia Sinica (EBPGC 1981), the subtropical zone today denotes areas between 25°N and 35°N (or S) which, in China, consists of areas north of the Tropic of Cancer and south of the Qinling Mountains–Huaihe River; accumulated temperature degrees (>10° C), 4500–8000° C; days with temperature >10° C, 218–365; annual precipitation, 1000–1800 mm. On the other hand, the warm temperate zone in China consists of North China and south Xinjiang accumulated temperature degrees (>10° C), 3200–4800° C; days with temperature >10° C, 171–218; annual precipitation, varied.

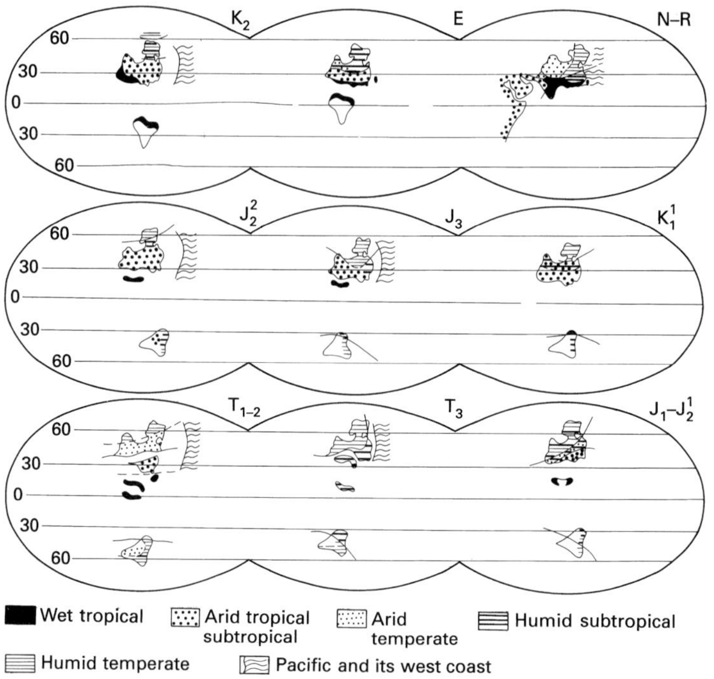

Fig. 15.3 Mesozoic and Cenozoic climatic zonation based on palaeobiogeographic data (reconstructed). T_{1-2},Early and Middle Triassic; T3, Late Triassic; J_1–J_{12}, Early to early Middle Jurassic; J_2^2, late Middle Jurassic; J_3, Late Jurassic; K_1^1 early Early Cretaceous; K_2, Late Cretaceous; E, Early Tertiary (Palaeogene); N–R, Neogene to Recent.

The above criteria are generally unavailable in geological history, so we have to recognize them by comparative methods. A basically tropical biota with temperate elements may be tentatively assigned to subtropical, and a basically temperate biota with tropical elements to warm temperate, especially when these biotas were arranged south–north and graded into each other. We are aware of the inaccuracy of this method, but we think that tentative labelling may express our inclination better than ambiguous formulation, and may form an object of study facilitating future examination.

The southern-type floras of Mesozoic periods are taken as tropical–subtropical. The Late Triassic *Dictyophyllum–Clathropteris (D.–C.)* flora and Early–Middle Jurassic *Ptilophyllum-Coniopteris (P.–C.)* flora have been regarded by many as tropical–subtropical Indo-European floras (see Chapters 9 and 10; and Wu Shunqing 1983). The Late Cretaceous *Glyptostrobus–Metasequoia* assemblage in the Songliao Basin, Jiayin (north-eastern China) and the Palaeogene plant assemblage north of the Yellow River are attributed to a humid subtropical regime (Chapter 12; Guo Shangxing, 1983). Among the floras contemporary with the Jehol fauna[1] (J^3–K_1^1), the *Onychiopsis–Ruffordia* flora from south of the Yellow River and Qinling Mountains, typified by the Jiande assemblage in Zhejiang, has a southern-type character but has been regarded as an arid flora because of its high percentage of *Classopollis* (50 per cent) (Li Wenbeng 1983; Cao Zhengyao *et al.* 1982). This may be treated as a

[1] The Jehol fauna is characterized by *Lycoptera* (fish), *Eosestheria* (conchostracan), and *Euphemeropsis trisetalis* (insect). Important elements include *Diestheria* (conchostracan), *Ferganoconcha* (bivalve), *Sinamia, Peipiaosteus* (fish), and *Darwinula* (ostracod). Whether it belongs to the Late Jurrasic (Gu Zhiwei 1982; Wang Si'en 1983) or earliest Cretaceous (Hao Yichun *et al.* 1986) has been long under discussion. This book takes it as Late Jurassic.

Table 15.1 Typical biotas of Mesozoic and Cenozoic palaeoclimatic belts of China

	Wet and wet–dry tropical		Arid subtropical	Humid subtropical	Humid temperate
	Marine	Terrestrial	(red-bed belt)	(southern-type flora)	(northern-type flora)
Neogene	*Miogypsina*, *Globigerina*	Evergreen broadleaf forest		Deciduous broadleaf forest	Needleleaf and broad leaf mixed forest
Palaeogene	Nummulitids	Evergreen broadleaf plants (*Sabalites*, *Sinnamomum*)	*Palibinia*, *Ephedripites*, steppe Artiodactyla and Perisodactyla	Eocene NE China flora (Guo Shuangxing 1983)	N. Heilongjiang plants
Late Cretaceous late Early Cretaceous	Nummulitids Scleractinians		*Classopolis* and *Schizaeoispolites* abundant; *Tsindaosaurus*; *TPN* fauna and *Pseudohyria*	*Glyptostrobus*–*Metasequoia*	Beyond Chinese border
early Early Cretaceous	Rudistids		*Classopollis* > 50% *Weichselia*, *Peregrinoconcha* assemblage	Mixed flora: *Brachyphyllum* (dry–hot) *Acanthopteris* (humid–temperate)	*Acanthopteris*–*Nilssonia* assemblage (NE China)
Late Jurassic			*Classopollis* > 50%	*Ruffordia*–*Onychiopsis* flora	
				south-type (Zhejiang) Jehol fauna	north-type(Liaoning) Jehol fauna
Middle Jurassic	Megalodontids		*Ferganoconcha*–*Tutuella* (in north) *Eolamprotula*–*Psilunio*–*Cuneopsis*	*Ptilophyllum*–*Coniopteris*	*Coniopteris* (north NE China)
Early Jurassic	Heteromorphic ammonites		*Lufengosaurus*, *Mamenchisaurus*; unioniids in SW China	*Ptilophyllum*–*Coniopteris* assemblage	*Coniopteris*–*Phoenicopsis*
Late Triassic	Scleractinians			*Dictyophyllum*–*Clathropteris*	*Danaeopsis*–*Bernoullia* (*Symopteris*) assemblage
Early and Middle Triassic			German-type bivalves and ammonites	*Pleuromeia*–*Voltzia* assemblage	Korvonchang (Siberian) flora

peculiar subset of this type. The southern type is not necessarily located in the south. During Mesozoic arid periods, the humid subtropical zone may have been squeezed by the expanding red beds to the north of Beijing (J_2^2) or western Liaoning (K_2^2). However, data of these two periods are insufficient. It is also possible that this belt became incorporated with the temperate belt and lost its independence.

15.1.4 HUMID TEMPERATE

This is marked by northern-type floras in the Mesozoic periods. The Late Triassic *Danaeopsis–Bernoullia (Symopteris)* flora and the Early–Middle Jurassic *Coniopteris–Phoenicopsis* flora reflected less humid and cooler environments and have been attributed to the Siberian floral realm. Their southern subset yielded *D.–C.* and *P.–C.* elements and were taken as warm temperate. The late Middle Jurassic plant assemblages of north-eastern China, including Wandashan, Jixi, the Songliao Basin, and the Daxingan Mountains (J_2^2–J_3 according to other authors), were produced from what was generally regarded as a temperate coal-measure series, and they outcropped near the Wandashan Gulf, where Boreal *Arctocephalites* ammonites occur. These should belong to a temperate belt. During the time in interval when the Jehol fauna lived, the northern-type *Onychiopsis–Ruffordia flora* (Cao Zhengyao *et al.* 1982) belonged to the Siberian floral realm. Its distribution approximated that of the northern and transitional types of palynomorphs with few or no *Classopollis* (Li Wenben 1983) and adjoined the Wandashan Gulf which has yielded *Buchia*. It appears to represent the humid temperate belt. The south-western boundary of the *Onychiopsis–Ruffordia* fauna, including both southern and northern types, ran roughly from the southern margin of Junggar south-eastward to the outlet of the Zhujiang River. Meanwhile the south-western boundary of the Jehol fauna ran in a parallel direction a little distance back, from Hami of eastern Xinjiang to lower Yangtze. In areas south-west of these lines there were deposited red beds belonging to the arid subtropical zone. In the third arid period (K_1^2–K_2), the tropical–temperate boundary migrated northward beyond the Chinese border. Plants corresponding to the northern-type flora only developed in the Soviet Far East. Towards the Palaeogene, this boundary again moved back to the northernmost Heilongjiang province.

Taking the boundary of southern and northern type floras as the tropical (subtropical)–temperate boundary, we can see that during aridity expansion the humid subtropical zone, as represented by southern-type flora, generally shrank or became obscured (as in J_2^2 and K_2), and the southern margin of the temperate zone was close to the northern limit of arid-zone red beds.

15.2 THE MESOZOIC AND CENOZOIC PALAEOCLIMATE OF CHINA FROM THE VIEWPOINT OF PALAEOBIOGEOGRAPHY

China experienced three hot-and-arid–temperate-and-humid cycles in the Mesozoic and Cenozoic as viewed from palaeobiogeography.

15.2.1 FIRST CYCLE–TRIASSIC

The Early and Middle Triassic was a hot-arid time. South China was covered by an epicontinental sea in which flourished tropical–subtropical marine faunas of German type, a large part of which were of typical of restricted seas. Evaporites and red beds were widespread, and from the Early to the Middle Triassic they expanded northward, i.e. from Upper Yangtze (Feixianguan and Jialingjiang Formations) to Lower Yangtze (Huangmaqing and Upper Qinglong Formations); palaeomagnetic data shows that in the Triassic Yangtze was 90° anticlockwise to its modern situation, with Lower Yangtze heading northwards. Thus the tropical–subtropical arid climate of South China reached a climax in the Middle Triassic. Hainan Island was exceptional in that coal-measure facies developed (Lingwen and Jiuquejiang Groups). Qiangtang and Gangdise (Tibet) have tropical marine faunas, lack evaporites and red beds, but possess coal facies along the north-eastern margin (Caya, Gonjo), and therefore belonged to the wet tropical zone.

North and north-western China in the Early and Middle Triassic

There are two climatic interpretations concerning the widespread red-bed belt of China between palaeolatitudes 40°N and 70°N, including the Guodikeng-Shaofanggou Formations of Xinjiang, Xidagou-Dingjiayao Formations of Gansu, and Liujiagou–Ermaying Formations of North China. One treats them in the same way as the Jurassic–Cretaceous red-bed belts of China, i.e. all widespread Mesozoic red-bed belts with fluorishing dinosaurs and other tetrapods are regarded as dry tropical–subtropical. In the light of this interpretation, the Early and Middle Triassic would seem to be the climax of Mesozoic tropical expansion, extending to 40–70°N (and S too), and only Siberia, China's north-eastern corner, and Altai remained temperate. Advantages of this interpretation are two fold. In this way the red-bed belt of middle palaeolatitudes in China is put into the same hot-arid zone as other Eurasian red-bed belts of relatively lower latitudes, notably the European Buntsandstein and Verrucano (partly) and the Caspian Vetruge Formation. Thus the Triassic red beds of Eurasian acquire a reasonably continuous distribution. Secondly, it keeps consistent the interpretation of all Mesozoic red-bed belts of China. Nevertheless, such an interpretation contradicts palaeobiogeographical data. A contemporary reptile-bearing red-bed belt developed in corresponding palaeolatitudes of the southern hemisphere (Cox 1973), namely the Panchet and Yerrapalli Formations of India, the Middle and Upper Beaufort series of South Africa, the Manda Beds of East Africa, and the Santa Maria Formation of Brazil. This establishes a definite bipolarity favouring a non-tropical situation. The southern hemisphere belt was marked by a non-tropical *Dicroidium* flora and is close to the temperate marine fauna in the Himalayas and Madagascar. The northern hemisphere also adjoined the temperate marine fauna in South Qilian and the western Qinling Mountains. Both adjoining marine facies lacked evaporites.

These contradictions lead to the other interpretation accepted in this book. This is to regard these areas as middle-latitude arid temperate belts, similar to the situation in the Neogene and to modern analogues in Central Asia. They were located in the western portions of continents, with humid temperate areas to their east, (the terrestrial green clastic series of Korea and temperate marine facies of Primoria). This interpretation resolves the palaeobiogeographical contradiction and, besides, by restricting the arid subtropical within 30°S and N palaeolatitudes, the Triassic would not exceed the Jurassic and Cretaceous in warmth, which has been established by other evidence. A difficulty arises, however, in that the humid subtropical flora would be hard to find in this way, although the *Pleuromeia–Voltzia* assemblage of North China may represent a narrow zone of this type. We deem that due to the formation of Pangaea II and one of the greatest regressions in Earth's history, the Late Permian–Middle Triassic interval was one of the driest periods in the Phanerozoic, during which the arid belt extended into the temperate zone and greatly diminished the humid subtropical flora.

The Late Triassic was a humid time. Areas of 30–60°N, previously occupied by the Early–Middle Triassic arid belt, were replaced by the humid belt. The former was driven to the south (Khorat Group of Indochina) and far west (Morroco and North America). The tropical and temperate belts, represented respectively by the *Dictyophyllum–Clathropteris* and *Danaeopsis–Bernoullia (Symopteris)* floras, were delineated first along the Kunlun–Qinling Mountains, then turned abruptly north-eastward and extended to the vicinity of Vladivostok. This bend may have been effected by the Late Triassic west Pacific current and monsoon, or later strengthened by north-eastward movement along the Tanchen–Lujiang Fault.

Along the modern middle-latitude eastern coasts of the northern hemisphere continents, where northwards moving warm currents meet southwards moving cold currents, the former are offshore whereas the latter are nearshore. Thus the arrangement from west to east is: humid land margin-cold current-warm current. The Upper Triassic of Primoria and Japan showed such an arrangement. The terrestrial Inner

Belt of Japan has yielded the *D.–C.* flora (Mine, Nariwa), adjoined eastward by the temperate *Entomonotis–Tosapecten* fauna (Chugoku, Tambo, Joetsu) and further east, the Outer Belt was characterized by megalodontids, corals, and other Tethyan fauna intercalated with deep-sea deposits (south Kyushu). In southern Sikhote-Alin, the *D.–C.* flora prevailed west of Vladivostok, in south Primoria there occurred the temperate *Entomonotis–Tosapecten* fauna, and further eastward along the Pacific coast was a tropical fauna with megalodontids, corals, hydrozoans, sponges, etc. Such an arrangement calls attention to the possible existence of warm and cold palaeocurrents in the Mesozoic. It may even date back to the Early Triassic when a cool-water fauna spread along the Yangtze Valley–western Qinling seaway. A palaeobiogeographical interpretation is suggested here as a substitute to the terrain interpretation of the occurrence of tropical Tethyan fauna in middle latitudes along east Asian coast, suggested by many authors.

15.2.2 SECOND CYCLE—JURASSIC TO EARLY EARLY CRETACEOUS

The arid belt began to expand in the Jurassic. In the early Early Jurassic, red beds were distributed in eastern Yunnan (Lower Lufeng Formation), central Yunnan (Fengjiahe Formation), and western Guangxi (Baixing Formation), whereas a coal-measure series existed in central Guangdong (Jingji Formation), central Guangxi (Xiwang Formation), central Sichuan (Baitianba Formation), and western Hubei (Xiangxi Formation). In the late Early Jurassic the coal-measure series retreated to western Fujian (Upper Lishan Formation) and western Zhejiang (Majian Formation); to the west of them were red beds. The early Middle Jurassic coal-measure series were limited to areas north of the Yangtze River. Aridity reached a climax in the late Middle Jurassic (or late Middle Jurassic to early Late Jurassic according to other authors) when South and North China were covered by widespread red beds and the arid–humid boundary was located between North China and north-eastern China. In the Early–Middle Jurassic interval the tropical–temperate border moved northward by 15–20° latitude (Figs 10.3, 10.5, 10.10). This situation was supposedly caused by both southward plate migration and tropical expansion. In the southern hemisphere, southern Tibet suffered contemporaneous warming and was under tropical to warm temperate conditions.

The Late Jurassic–earliest Cretaceous in China was a more humid and temperate interval, during which the tropical–temperate boundary moved southward and the red-bed belt shrank south-westwards. Meanwhile, the temperate belt spread into southern Tibet where buchiids were abundant. This is in accordance with the widespread distribution of the Jehol fauna. An end-Jurassic–earliest Cretaceous temperature fall was echoed from other parts of the world (Fig. 11.1) and may have been on a global scale.

15.2.3 THIRD CYCLE—LATE EARLY CRETACEOUS TO RECENT

The Cretaceous was the most hot–arid period of China in the Mesozoic and Cenozoic. The tropical–subtropical belt involved the whole of China including southern Tibet. South and North China were covered by arid subtropical red beds. North Tibet belonged to a wet tropical and north-eastern China to a humid subtropical region.

The arid belt retreated in the Palaeogene, accompanied by expansion of the humid subtropical belt. The temperate belt returned first into northern Heilongjiang and possibly also Altai, then moved southwards, reaching the Yellow River region in the Miocene (Song *et al.* 1983). The Neogene was a global ice age. Meanwhile the uplift of the Tibet–Qinghai Plateau was another controlling factor. As a result, western China was under arid conditions. In eastern China, the subtropical–temperate boundary moved to the Yangtze drainage area, and humid tropical, subtropical, and temperate zones merged to cover the whole of eastern China as in the Late Triassic, but the average temperature was lower. This general tendency continues to the present day.

15.2.4 INFLUENCE OF PACIFIC COASTAL MIGRATION

Regularly, as a result of the monsoon influence, China's eastern coast should have experienced a humid climate, so it is interesting to know that such a situation did not exist for a large part of the Mesozoic, during which time the arid red beds extended to the east coast and may even have extended on a wider scale. This is interpreted here as a fluctuation of the Pacific influence, caused by coastal migration and the Yanshanian Orogeny (Fig. 15.3). According to Mesozoic data from China, Korea, and Japan, the east Asian coast was to the east of China's modern coast, swinging along the inner and outer zones of Japan, western Taiwan and Paracel Islands. During arid periods, judged from palaeogeographic maps, the distance between Liaoning–Jiling provinces and the palaeocoast is estimated to be 1000–2000 km; Guangdong and Guangxi *c.* 1000–1500 km; Fujian and Zhejiang more than 400 km. The last distance was underestimated because Taiwan was thrust westward during later orogenies, thus decreasing the distance in between. The modern width of the humid monsoon belt is 1000–2000 km, so the above distances restricted monsoon influence to beyond eastern China. The north-east striking coastal mountains formed during the Yanshanian Orogeny also barred the influence of humidity. An analogous situation can be observed on the modern east coasts of Australia and south-east Africa, where the humid belts are restricted to a narrow coastal zone and arid belts extend close to the east coast. It should also be noted that the periods when China possessed Pacific gulfs coincided with periods of humidity expansion. These periods were the Late Triassic to Early Jurassic (Ussuri–Nadanhada Gulf in north-eastern China and Guangdong–Hunan–Jiangxi Gulf in south-eastern China) and Late Jurassic to early Early Cretaceous (Wandashan Gulf in north-eastern China). The existence of gulfs indicates that the coastal line swung westwards, close to the modern coast, thus eastern China was relatively humid. The same situation applies to modern times, which are also humid.

15.3 THE MESOZOIC PALAEODRAINAGE OF CHINA FROM THE VIEWPOINT OF PALAEOBIOGEOGRAPHY

The non-marine Mesozoic aqueous faunas of China can be categorized into biotas of three drainage systems (Table 15.2).

15.3.1 TROPICAL EAST TETHYAN DRAINAGE SYSTEM

Faunas of this system originated from hot arid areas pertaining to the Tethyan drainage system, and existed from the Late Triassic to the Early Cretaceous. The brackish to freshwater Late Triassic *Yunnanophorus–Indosinion* bivalve assemblage is distributed south of the Kunlun–Qinling Mountains and west of the Xuefeng–Yunkai Mountains. During the Early and Middle Jurassic, there developed the freshwater *EPC* bivalves, *Paleolimnadia baitianbaensis* (J_1), *Euestheria ziliujingensis* (J_2) conchostracans, which extended beyond Kunlun–Qinling and spread northwards to Junggar, Turpan, western Gansu, Ordos and western Henan, thus partly overlapping with the *Ferganoconcha–Tutuella* fauna. This spread coincided with the Jurassic arid–tropical expansion and may also imply peneplainization of the above-mentioned watersheds. In the second humid–temperate cycle (J_3–K_1^1), this Tethyan fauna was again restricted to south-west of the Kunlun–Qinling and Xuefeng–Yunkai lines. Along the Tethyan margin there flourished the *Peregrinoconcha* bivalve assemblage, probably a mixture of marine and non-marine elements, while the freshwater *Eosestheriopsis* conchostracan assemblage extended into the Sichuan Basin. Accompanying the third arid–tropical expansion, the Early Cretaceous *TPN* fauna, evidently evolved from the *Peregrinoconcha* assemblage, broke through all watersheds and crossed the Jinshajiang suture, spreading from Indochina and North Tibet all the way to north-eastern China, the Soviet Far East, and Japan. However, the above lines still served as barriers to Early

Table 15.2 East Asian non-marine aqueous faunas of Mesozoic in China

	East Tethyan drainage system (tropical)	Pacific–Tethyan drainage system (subtropical)	Boreal drainage system (mainly temperate)
Late Cretaceous	*Pseudohyria* *TPN* unioniids	*Tenuestheria, Sinoestheria;* affined with Gondwanan elements	*Margaritifera, Plesielliptio, Cuneopsis*; *Estherites*; mainly in NE China
Early Cretaceous	*TPN* unioniids, *Peregrinoconcha*; (influence from Gondwanan and marine fauna); *Orthestheria, Anglestheria*; *Jingguella*	*Yanjiestheria, Migransia, Cratostracus*; SE China-type ostracods; Gondwanan affinity	*Yanjiestheria–Neodiestheria*; north-type Cyprididae
Late Jurassic	*Cuneopsis–Undulatula; Chuanestheria–Eosestheriopses; Damonella*	Jehol fauna (invaded from north)	Jehol fauna
Early and Middle Jurassic	*EPC* unioniids; *Palaeolimnadia, Euestheria*	*EPC* and *FT* elements(J_2) *Qiyangia, Palaeo-limnadiopsis* (J_1)	*Ferganoconcha–Tutuella*
Late Triassic	*Yunnanophorus–Indosinion*	*Jiangxiella, Guangdongella*; *Anyuanestheria*; *Oncocythere, Minicythere*	*Shaanxiconcha–Sibericoncha*; *Darwinula–Lutkevichinella–Tongchuania*

Cretaceous *Orthestheria* conchostracans and *Jingguella* ostracods, as well as Late Cretaceous *Aglestheria* conchostracans, restricting them to the southwestern side. The biotas of the tropical Tethyan drainage system diminished towards the Late Cretaceous. Decendants of the *TPN* fauna such as *Pseudohyria* remained in the territory of their antecedents, coexisting with elements from the north and east, and disappeared at the end of the Cretaceous.

15.3.2 SUBTROPICAL PACIFIC–TETHYAN DRAINAGE SYSTEM

Biotas of this system existed in south-eastern China during the Late Triassic–Early Jurassic and Cretaceous. The Late Triassic fauna was a brackish-estuarine one, represented by *Jiangxiella, Guangdongella* (bivalve), *Anyuanestheria* (conchostracan), *Oncocythere,* and *Minicythere* (ostracod). The Early Jurassic fauna was freshwater to brackish, typified by *Qiyangia* (bivalve) and *Palaeolimnadiopsis* (conchostracan) which crossed the Xuefeng line and spread to central Sichuan, southern Guangxi and north Vietnam. These faunas were soon replaced in the Middle Jurassic by northern invaders, which persisted throughout the rest of the Jurassic. The Early Cretaceous was marked by *Yanjiestheria–Migransia* (earlier) and *Cratostracus* (later) conchostracans, as well as ostracods of south-eastern China type, and the Late Cretaceous by *Tenuestheria–Sinoestheria* conchostracans. The Pacific–Tethyan system in the Cretaceous showed a more distinct affinity with Gondwana than with other drainage systems. Such affinity was shown in the conchostracans (Afrograptidae from Africa, *Migransia, Congestheriella)*, fish *(Paraclupea, Huashia gracilis)*, and ostracods (showing affinity with South Africa and Brazil). Even the *TPN* fauna was thought to be related to Gondwanan unioiniids.

15.3.3 BOREAL DRAINAGE SYSTEM

Faunas of this system were essentially temperate biotas. The earliest one, in the Late Triassic, seems to belong to Central Asian inland drainage. This includes the Late Triassic *Shaanxiconcha–Sibericoncha* bivalves distributed in North and north-western China, Central Asia, and western Siberia, and elements of *Darwinula–Lutkevichinella–Tongchuania* ostracods either in Central Asia, North and north-western China, or spreading further to western Europe. The Early–Middle Jurassic *Ferganoconcha–Tutuella* bivalves dispersed not only to northern parts of China but also to southern China, but *Ferganoconcha* was still unknown in south-western China. The Jehol fauna flourished in a vast area, including northern and south-eastern parts of China, but also failed to invade the east Tethyan system. Its south-west boundary, close to the northern-type floral line, marked the limit of the temperate belt. In the Early Cretaceous, following shrinkage of the humid temperate belt, faunas of the Boreal system retreated to north of the Yangtze. These were *Yanjiestheria-Neodiestheria* and northern-type cypridiid ostracods, mixed with *TPN* bivalve elements which had invaded from the south. In the Late Cretaceous, the elements of the Boreal system were mainly restricted to north-eastern China, characterized by unioniids analogous to those of the modern Palaeo-Arctic realm, i.e. *Plesielliptio* and *Magaritifera*, and *Estherites* conchostracans. During the Cenozoic, linked to southward migration of the humid temperate climate, these Palaeo-Arctic unioniids spread to North China and the Yangtze drainage area.

15.3.4 SUMMARY OF THE PALAEODRAINAGE

The development of the above three non-marine biotas has implications not only for paleoclimatic variations, but also for changes of relief, especially watersheds. The Kunlun–Qinling and Xuefeng–Yunkai water divides existed from the Late Triassic to the Early Jurassic, barring the tropical Tethyan drainage system from the others. Probably coexisting synchronously were the Jiangnan Massif, which subdivided the Pacific–Tethyan drainage from the lower Palaeo–Yangtze drainage (remnant of the Triassic Lower Yangtze seaway), and the Changbaishan Oldland, which separated the Central Asian inland freshwater fauna and *DB* flora from the Ussuri–Wandashan Gulf and its adjacent areas with Boreal fauna and *DC* flora. After the Middle Jurassic the latter two watersheds disappeared, so that faunas of the Boreal drainage system spread widely over northern and south-eastern parts of China. The first watershed may have been penetrated in the Early Jurassic but rebuilt again later. From the Middle Jurassic to the Palaeogene, faunas of the Boreal drainage system spread southwards to a belt roughly including Junggar, Turpan, western Gansu, North China, and north of the Lower Yangtze. This demarcation may be both climatic and a topographic water divide. Most of the Boreal drainage system seems to gather into the North Pacific, although a few rivers may have converged with rivers now belonging to the Arctic drainage system.

Putting the three drainage systems together, East Asia as a whole, including China, Central Asia, Mongolia, Korea, Japan, and adjacent areas, constituted a peculiar non-marine aqueous biogeographical unit in the Mesozoic and Cenozoic. The bivalves of East Asian type include *Yunnanoconcha–Indosinion* (T_3), *Shaanxiconcha–Sibericoncha* (T_3), Jehol* elements (J_3), and *TPN* fauna. Among conchostracans, Kobayashi (1975) established six Mesozoic groups in Asia, of which four are East Asian type: Daido group (T_3–J_1), Jehol* group (=*Eosestheria* assemblage), Kyongsang group (K_1=*Yanjiestheria* assemblage), and Sungari group (K_2=*Estherites* assemblage). Similar situations existed among ostracods, fish, and insects. A basic cause was the existence of the Pacific because it formed the gathering basin of two of the three drainage systems. From deposits of Late Cretaceous to Recent ages in various terrestrial basins of eastern China there have been discovered several horizons yielding brackish or even marine elements, such as forams, ostracods, polychaetes, fish, and seaweeds. These have been postulated to be due to marine incursions along the ancient water systems of Zhujiang, the Yangtze River and its tributaries, Songhuajiang, and rivers along the Chihli Gulf.

REFERENCES

Cao Zhengyao, Li Baoxian, and Guo Shuangxing (1982). The Cretaceous plant-bearing strata in China. In *Stratigraphic correlation chart in China* (ed. Nanjing Institute of Geology and Palaeontology, Academia Sinica), pp. 254–69. Science Press, Beijing (in Chinese).

Cox, C. B. (1973). Triassic tetrapods. In *Atlas of paleobiogeography,* (ed. A. Hallam), pp. 213–25. Elsevier, Amsterdam.

EBPGC (ed.) (1985). *Physical geography of China.* Science Press, Beijing (in Chinese).

Flakes, L. A. (1979). *Climates throughout geologic time.* Elsevier, Amsterdam.

Gu Zhiwei (1982). On the nonmarine Jurassic–Cretaceous boundary and mid-Cretaceous events in China. *Journal of Stratigraphy* **6**, (4), 247–57 (in Chinese).

Guo Shangxing (1983). Note on phytogeographic provinces and ecological environment of Late Cretaceous and Tertiary floras in China. In *Palaeobiogeographic provinces of China,* pp. 164–77. Science Press, Beijing (in Chinese).

Hao Yichun, *et al.* (1986). *The Cretaceous system of China—stratigraphy of China,* Vol. 12. Geological Publishing House, Beijing (in Chinese).

Kobayashi, T. (1975). Upper Triassic estheriids in Thailand and conchostracan development in Asia in the Mesozoic Era. *Geology and Paleontology of SE Asia* **16**, 57–90.

Li Wenben (1983). Microfloral areas of Early Cretaceous in China. In *Palaeobiogeographic provinces of China,* pp. 142–51, Science Press, Beijing (in Chinese).

Song Zhichen, Li Haomin, Zheng Yahui, and Liu Gengwu (1983). Miocene floristic regions of China. In *Palaeobiogeographic provinces of China,* pp. 178–84. Science Press, Beijing (in Chinese).

Strahler, A. N. (1954). *Introduction to physical geography* (3rd edn). Wiley, New York.

Wang Si'en (1983). Some Jurassic–Cretaceous conchostracans from Qinghai. *Acta Palaeontologica Sinica* **4**, 465–71 (in Chinese with English abstract).

Wu Shunqing (1983). On the Late Triassic, Lower and Middle Jurassic floras and phytogeography provinces of China. In *Palaeobiogeographic provinces of China*, pp. 121–30. Science Press, Beijing (in Chinese).

ADDENDUM

Chinese geographical names have been subjected to a variety of spelling systems, resulting in the appearance of many synonyms for the same localities which often cause confusion. From the many spelling systems, we have chosen the Pinyin system as the standard for spelling Chinese geographical and personal names, because it has been officially implemented by the Chinese government for decades in the Chinese mainland and is steadily gaining more recognition from the outside world, although there are many awkward x's and q's for Westerners to deal with. Thus, for example, Guangzhou is used instead of Canton. An index is attached; readers may also refer to *Gazetter of the People's Republic of China*, published by the Defense Mapping Agency of the USA, Washington DC (1979). There are a few exceptions, such as Tibet for Xizang, Inner Mongolia for Nei Mongol, Chilhi Bay for Bohai, Yangtze for Changjiang.

Names of Chinese mountains and lakes cause more trouble. Suffixes of mountains may appear as -ling, -shan, Mountains, and Ranges. To avoid confusion we have omitted the suffixes wherever possible and add 'Mountains' only if omitting it would lead to incomprehension. For example it would be Kunlun instead of Kunlunshan or Kunlun Mountains, but Tianshan, Qinling, not Tian, Qin, and Anyemaqen Mountains, not Anyemaqen.

Because Chinese stratigraphic names are unfamiliar to readers, a simplified correlation table is added to each chapter, which includes not all but many of the stratigraphic names used in the text.

INDEX OF CHINESE GEOGRAPHICAL LOCALITIES

	Northern latitude	Eastern longitude
Acheng	45°38′	126°58′
Aerjin Mountains	38°–40°	86°–93°
Aihui	50°12′	127°25′
Airao Mountains	23°–25°	101°–103°
Ajaguz	48°00′	80°30′
Alax Massif	38°–41°30′	100°–106°
Altay	47°15′	88°15′
Amdo	31°12′	91°36′
Anhui	29°–35°	115°–120°
Anyemaqen Mountains	33°30′–36°00′	98°–102°
Aohan Qi	42°03′	119°09′
Arkatag	36°18′	90°12′
Arxan	47°11′	119°58′
Badan Jaran	39°54′	102°20′
Bagong Hill	32°35′	116°50′
Baichen Bay	41°48′	81°52′
Baicheng	45°06′	122°08′
Baingoin	31°18′	90°00′
Baise	23°55′	106°33′
Baiyun′obo	41°48′	101°59′
Bange Lake (Cuo)	31°38′	90°28′
Bangong Lake (Cuo)	33°42′	78°48′
Baojing	28°40′	109°48′
Baokang	44°08′	123°18′
Baoshan	25°05′	99°10′
Baotou	40°42′	109°48′
Basu	30°02′	96°55′
Batang	30°00′	99°10′
Baxoi	30°00′	96°54′
Bayan Har Mountains	33°–35°	95°–102°
Bayan Obo	41°48′	110°00′
Beipiao	41°46′	121°45′
Beishan Mountains	41°30′	97°05′
Bogda Mountains	43°30′	90°00′
Borohoro Mountains	43°–45°	91°–94°
Bulong	31°30′	93°30′
Burhan Budai Mountains	35°–37°	95°–98°
Butha Qi	48°00′	122°08′

Ceheng	24°58′	105°50′
Celei	37°02′	80°32′
Chabu	27°03′	118°00′
Changbaishan	42°00′	128°06′
Changchun	43°58′	125°20′
Changjiang	19°10′	109°00′
Changning	24°45′	99°38′
Changxing (Jiangsu)	31°04′	119°55′
Changxing (Shilu Co., Hainan)	19°10′	109°00′
Chaoxian	31°06′	117°08′
Chaoyang	41°30′	120°25′
Chengxian	33°48′	105°40′
Chuxian	32°20′	118°14′
Cixian	36°14′	114°20′
Cuiluan	47°48′	128°40′
Cuxiong	25°00′	101°30′
Dabie Mountains	31°–32°	114°–116°
Dafang	27°04′	105°30′
Dajianggang	43°45′	125°40′
Danfeng	33°42′	110°18′
Dangyang	30°48′	111°42′
Daoxian	25°32′	111°35′
Daqing Mountains	41°–42°	111°–120°
Datong (of Qinghai)	37°04′	99°50′
Datong (of Shanxi)	40°05′	113°20′
Daxingan Mountains	44°–48°	121°–123°
Daye	30°05′	114°59′
Delingha	37°03′	97°03′
Dengqen	31°24′	95°30′
Dinggye	28°30′	87°45′
Dingri	28°40′	87°00′
Donganzhen	26°30′	111°13′
Dongning	44°02′	131°00′
Dongqian	31°09′	90°08′
Dong Ujimqin Qi	45°43′	116°56′
Dunhua	43°24′	128°24′
Duwa	37°18′	78°05′
Ebinur Lake	45°00′	82°40′
Ejin Qi	41°50′	100°50′
Emei	29°35′	103°25′
Erenhot	43°40′	112°00′
Ergun	50°–51°	120°–122°
Ertix River	47°52′	84°16′
Fangcheng	21°50′	108°20′
Fencheng	38°00′	101°04′

Fenghuang	27°55′	109°35′
Fujian	23°30′–28°30′	115°30′–121°
Funing	23°30′	105°35′
Fuqun	26°44′	107°32′
Fusong	42°16′	127°14′
Fuzhou	26°06′	119°18′
Gamba	28°18′	88°32′
Gangdise Mountains	29°30′–32°10′	80°–88°
Gansu	32°–43°	93°–109°
Gansu Corridor (Hexi Zhoulang)	37°–41°	96°–103°
Ganze	31°40′	110°00′
Ganzhou	25°48′	114°54′
Garze	31°30′	99°54′
Gaxunnur Lake	42°25′	101°00′
Geer	32°14′	80°00′
Gerze	32°30′	84°12′
Golmud	36°22′	94°55′
Gongbi	30°00′	83°30′
Gonggeer Mountains	38°07′	74°03′
Gonju	30°48′	98°10′
Guangde	30°09′	119°04′
Guangdong	20°–26°	109°–118°
Guangxi	21°–27°	104°–113°
Guangxing Commune	42°44′	117°42′
Guangyun	32°30′	105°58′
Guanniaohe	45°53′	127°30′
Guiyang	26°36′	106°42′
Guizhou	24°–30°	103°–110°
Guxian	28°57′	118°50′
Guyang	41°00′	110°36′
Gyangze	28°54′	89°36′
Gyirong	28°54′	85°12′
Hainan	19°–21°	108°–111°
Hami	40°25′	92°10′
Hancheng	35°30′	110°25′
Hangula	41°30′	103°40′
Hangzhou	30°03′	120°02′
Hebei	36°–42°30′	113°–120°
Hegenshan	42°–44°	110°–116°
Heilongjiang	43°–54°	115°–135°
Helan Mountains	39°00′	106°00′
Henan	31°–36°30′	110°–117°
Hengduan Mountains	24°–32°	96°–110°
Hexi	35°42′	112°55′
Hexian	31°07′	118°04′

Locality	Latitude	Longitude
Hinggan (see Xingan area)		
Hohxil Mountains	34°–36°	87°–93°
Hoit Taria	36°03′	96°07′
Honghe River	20°–26°	100°–107°
Horqin Youyi Qiangqi	45°55′	121°00′
Huadian	43°00′	126°42′
Huaihe River	32°–35°	112°–121°
Huaihua	27°32′	109°32′
Huainan	32°40′	117°05′
Huaiyang	33°43′	114°53′
Huanglin	35°00′	114°14′
Huangshi	30°12′	115°05′
Huatin	35°14′	106°38′
Hubei	29°–33°30′	108°30′–116°30′
Huicheng Basin	33°20′–34°	105°20′–106°40′
Huili	26°38′	102°14′
Hunan	24°30′–30°30′	108°30′–114°30′
Hunchun	42°48′	130°18′
Hunjiang	41°56′	126°24′
Huocheng	44°19′	80°59′
Huoxian	36°30′	111°50′
Huzhou	30°50′	120°08′
Ili River	43°48′	80°24′
Inner Mongolia	38°–53°	97°–125°
Jalaid Qi	46°44′	112°30′
Jiande	29°30′	119°18′
Jianghan Basin	30°–31°	113°–114°
Jiangnan (Oldland)	25°–30°	108°–119°
Jiangsu	30°30′–35°	116°–122°
Jiangxi	24°–30°30′	113°–118°30′
Jianshu	23°06′	102°08′
Jiantianba	30°18′	108°05′
Jianzha	35°55′	102°00′
Jiayin	48°48′	130°24′
Jilin	41°–47°	119°–131°
Jinggu	23°30′	100°40′
Jinghe	44°06′	82°09′
Jingmen	32°00′	112°06′
Jingyuan	36°04′	104°07′
Jinning	24°40′	102°35′
Jinping	22°46′	103°15′
Jinquan	38°50′	98°10′
Jinshajiang River	26°–33°	97°–103°
Jinxi	40°45′	120°50′
Jinxian	30°41′	118°23′

Jinzhai	31°07′	115°09′
Jiutai	44°06′	125°48′
Jixi	45°10′	130°55′
Jiyang	36°48′	117°10′
Jiyuan	35°00′	112°34′
Junggar	45°00′	88°00′
Junxian	32°30′	111°35′
Juyan Hai (see Gaxunnur Lake)		
Kaihua	29°02′	118°04′
Kaiyuan	23°40′	103°10′
Kalpin	40°30′	79°17′
Kam-dian (Oldland)	25°–29°	102°–104°
Kangma	28°32′	89°26′
Kashi	39°40′	76°12′
Kawabulake	41°35′	91°45′
Kelamaili Mountains	44°20′–44°40′	91°–92°
Keping (=Kalpin)	40°30′	79°17′
Kongur Mountains	38°42′	74°18′
Korgas	44°06′	80°24′
Kuqa	41°40′	83°10′
Kunlun Mountains	34°–38°	76°–96°
Kunming	25°05′	102°42′
Kunyang	24°40′	102°35′
Kuruktag	41°30′	89°00
Lancang	22°30′	99°54′
Lancanjiang River	21°30′	101°12′
Lantian	34°10′	109°20′
Laohugou	41°18′	119°21′
Laozhuanghu	40°50′	119°30′
Lazi	27°03′	87°25′
Leiqiong	20°–21°	110°
Leiwuqi	31°12′	96°32′
Lhasa	29°38′	91°05′
Lhaze	29°	87°30′
Lhunzhub	30°00′	91°12′
Liantuo	30°50′	111°10′
Liaoning	38°40′–45°20′	116°–126°
Lijiang	26°54′	100°20′
Liling	27°42′	113°30′
Lin′an	30°08′	119°42′
Linfen	36°06′	111°30′
Lingling	26°18′	111°18′
Lingxia	35°35′	103°20′
Linwu	38°05′	106°17′
Linyou	34°40′	107°50′

Ningbo	29°54′	121°30′
Ningqiang	32°34′	106°35′
Ningxia	35°–42°	102°–108°
Ningzhen Mountains	31°30′–32°30′	118°30′–119°30′
Nujiang River	24°–31°	91°–99°
Nyalam	28°12′	85°54′
Nyingchi	29°36′	94°24′
Ordos	34°–41°	106°–111°
Pinwu	32°13′	104°15′
Pulan	30°18′	81°10′
Puzhen	32°18′	118°42′
Qaidam	36°–40°	90°–98°
Qamdo	31°06′	97°06′
Qiangtang	32°–36°	79°–88°
Qianshan Mountains	39°–41°	121°–124°
Qiemo	38°10′	85°45′
Qilian Mountains	37°–40°	95°–103°
Qimantag Mountains	37°–38°	89°–92°
Qin′an	34°48′	105°36′
Qinghai	31°–40°	89°–102°
Qingyan	26°17′	106°51′
Qingyang	30°40′	117°50′
Qinhuangdao	39°54′	119°36′
Qinling Mountains	32°–36°	101°–113°
Qinzhou	21°57′	108°35′
Qishan	34°30′	107°40′
Qiubei	24°03′	104°13′
Qog Qi	41°30′	107°07′
Qomolangma (Peak)	27°09′	86°09′
Qujin	25°28′	103°45′
Raohe	46°48′	134°00′
Raoheishan	43°30′	30°26′
Ritu	33°25′	79°32′
Riwoqe	31°04′	96°05′
Rucheng	25°36′	113°42′
Ruqu	34°18′	102°12′
Rutog	33°05′	79°07′
Sandu	26°06′	113°18′
Sangzhi	29°23′	110°09′
Sanmenxia	34°50′	111°12′
Shaanxi	31°30′–40°	105°30′–111°30′
Shandong	34°–38°30′	115°–123°
Shangnan	33°30′	110°48′
Shangyi	41°05′	114°00′

Tuoba	31°16′	97°25′
Turpan	42°09′	89°02′
Ulanqab Meng	40°07′	113°02′
Ussuri River	45°–48°20′	132°40′–135°
Wandashan Mountains	46°18′	133°00′
Wangfenggang	30°50′	111°10′
Wangjiawan	24°40′	102°25′
Wangqing	43°20′	129°45′
Wanquan	40°50′	114°40′
Wenshan	23°20′	104°15′
Wuhan	30°30′	114°18′
Wulanchabu Meng	41°–43°	109°–114°
Wulin	23°10′	108°52′
Wuling Shan	25°30′	113°00′
Wunu′er	48°50′	121°10′
Wuqia (Ulugqat)	39°40′	75°20′
Wuwei	31°19′	117°54′
Wuzhou	23°30′	117°42′
Xainza	30°09′	88°06′
Xar Moron River	42°20′–43°40′	117°–123°
Xi Ujimqim Qi	44°06′	117°06′
Xialu	29°10′	88°45′
Xi′an	34°18′	108°54′
Xiangyun	25°30′	100°18′
Xiaoxingan Mountains (Xiao Hinggan Ling)	48°45′	127°00′
Xichang	27°54′	102°18′
Xichuan	32°55′	111°30′
Xigaze	29°15′	88°45′
Xilin Gorge	30°50′	111°10′
Xingan (county)	25°06′	10°06′
Xingan (area)	44°–53°	119°–130°
Xingkai Lake	45°00′	132°30′
Xining	36°30′	101°40′
Xinjiang	34°–50°	73°30′–96°30′
Xiugou	95°58′	35°36′
Xiuning	29°45′	118°10′
Xizang (Tibet)	26°30′–36°30′	78°–99°
Xuancheng	30°09′	118°07′
Xuanjiangping	30°45′	106°15′
Xuefeng Mountains	27°–28°40′	110°–111°20′
Xujiahe	32°46′	105°50′
Xuzhou	34°18′	117°12′
Yadon	27°29′	88°54′

Zhenfeng	25°22′	105°41′
Zhenkang	23°58′	98°59′
Zhenping	31°55′	109°30′
Zhongba	29°38′	84°04′
Zhongdian	27°50′	99°40′
Zhongning	37°05′	105°07′
Zhongwei	37°05′	105°02′
Zhongyicun	24°42′	102°12′
Zhongzan	29°10′	99°07′
Zhouqu	33°48′	104°15′
Zhujiang River	22°–26°	104°–115°
Zhusilenhairhan	41°50′	110°10′
Zixing	25°54′	113°24′
Ziyang	32°30′	108°30′
Zoige	31°30′	102°55′
Zouyun	40°00′	112°45′

INDEX